Degenerative Retinal Diseases

Degenerative Retinal Diseases

Edited by

Matthew M. LaVail
University of California, San Francisco
San Francisco, California

Joe G. Hollyfield
Cleveland Clinic Foundation
Cleveland, Ohio

and

Robert E. Anderson
University of Oklahoma Health Sciences Center
Oklahoma City, Oklahoma

Springer Science+Business Media, LLC

Degenerative retinal diseases / edited by Matthew M. LaVail, Joe G.
  Hollyfield, and Robert E. Anderson.
      p.   cm.
    "Proceedings of the VII International Symposium on Retinal
  Degenerations, held October 5-9, 1996, in Sendai, Japan"--T.p.
  verso.
    Includes bibliographical references and index.
    ISBN 978-1-4613-7718-4   ISBN 978-1-4615-5933-7 (eBook)

    DOI 10.1007/978-1-4615-5933-7
    1. Retina--Degenration--Congresses.   I. LaVail, Matthew M.
  II. Hollyfield, Joe G.   III. Anderson, Robert E. (Robert Eugene)
  IV. International Symposium on Retinal Degenerations (7th : 1996 :
  Sendai-shi, Miyagi-ken, Japan)
    [DNLM: 1. Retinal Degeneration--genetics--congresses. 2. Retinal
  Degeneration--physiopathology--congresses.   WW 270 D3173 1997]
  RE661.D3D45 1997
  617.7'35--dc21
  DNLM/DLC
  for Library of Congress   97-27725
                                                                  CIP

Proceedings of the VII International Symposium on Retinal Degeneration,
held October 5 – 9, 1996, in Sendai, Japan

ISBN 978-1-4613-7718-4

© 1997 Springer Science+Business Media New York
Originally published by Plenum Press, New York in 1997
Softcover reprint of the hardcover 1st edition 1997

10 9 8 7 6 5 4 3 2 1

Gordon and Lulie Gund

To Gordon and Lulie Gund for their untiring efforts over the past 26 years to seek the causes, treatments, and cures for retinal degenerative diseases, including retinitis pigmentosa, which blinded Gordon more than 20 years ago. One of the founders of The Foundation Fighting Blindness, formerly the National Retinitis Pigmentosa Foundation, Gordon serves as Chairman of its Board. Lulie also serves on the Board of Trustees and is President of the Foundation's New Jersey Affiliate. Gordon and Lulie Gund have led an international battle against blinding diseases, inspiring and motivating many others in the search for cures.

PREFACE

Since 1984, we have organized Satellite Symposia on Retinal Degenerations that are held in conjunction with the biennial International Congress of Eye Research. The timing and location of our Retinal Degeneration Symposia have allowed scientists and clinicians from around the world to convene and present their exciting new findings. The symposia have been arranged to allow ample time for discussions and one-on-one interactions in a relaxed atmosphere, where international friendships and collaborations could be established.

The II International Symposium (also known as the Sendai Symposium on Retinal Degeneration) was held in 1986 in Sendai, Japan, on the occasion of the retirement of Katsuyoshi Mizuno as Professor and Chairman of Ophthalmology at Tohoku Medical School. On October 5–9, 1996, we returned to Sendai, where the VII International Symposium was held at the Miyagi Zao resort hotel in the beautiful Mt. Zao region of northern Japan. This meeting was held on the occasion of the tenth anniversary of Makoto Tamai as Professor and Chairman of Ophthalmology at Tohoku Medical School. One afternoon of the meeting contained a special symposium in honor of Professor Tamai, who has significantly elevated the level and intensity of research on retinal degenerations at his university and in Japan during this past decade.

This book contains the proceedings of the VII International Symposium on Retinal Degeneration. A majority of those who spoke and presented posters at the meeting contributed to this volume, and we thank those authors for their efforts. The papers reflect a major emphasis on molecular genetics of retinal degenerations, particularly in the areas of candidate genes and in cloning and mapping. Significant new animal models are described, as are several studies on macular degeneration. Studies on animal models, both vertebrate and invertebrate, continue to provide important new findings on the mechanisms of photoreceptor degeneration and cell death, as well as the means to retard, prevent, or reverse the degeneration by potential new therapeutic methods, including transplantation, gene therapy, and survival factors. Diagnostic, clinical, cytopathologic, and physiologic aspects of retinal degenerations in human patients are also presented, as are several papers on structural, physiologic, and aging studies in normal tissues that are important correlates for the study of retinal degenerations.

The symposium received financial support from a number of organizations. We are particularly pleased to thank the Foundation Fighting Blindness, Baltimore, Maryland (Formerly the RP Foundation Fighting Blindness), for its continuing support of this and the previous biennial symposia, without which we could not have held these important meetings. Other international contributors we wish to thank include the German Retinitis Pigmentosa Foundation, the Swiss Retinitis Pigmentosa Foundation, the New Zealand Retinitis Pigmentosa Foundation, and the Louisiana State University Medical Center

Foundation (USA). For this particular meeting and the special symposium for Professor Tamai, we are particularly indebted to the support from the Miyagi Prefecture; the Alumni Association of the Department of Ophthalmology, Tohoku University School of Medicine; Tokyo Pharmaceutical Manufacturers Association; Osaka Pharmaceutical Manufacturers Association; Menicon; Chiba Vision Ricky Contact Lens; Pharmacia Upjohn (Japan); Carl Zeiss (Japan); and Alcon (Japan).

We want to thank Professor Makoto Tamai and his local organizing committee of Drs. Sei-ichi Ishiguro and Mitsuru Nakazawa and their colleagues for their untiring efforts. It was a pleasure working with these outstanding individuals who provided a most memorable, enjoyable, and scientifically productive experience in Japan.

Thanks also go to Ms. Gloria Riggs for her assistance with all of the correspondence and organization of the meeting. We also thank Plenum Press for publishing this volume.

Matthew M. LaVail
Joe G. Hollyfield
Robert E. Anderson

CONTENTS

Part I. Cytopathologic, Physiologic, Diagnostic, and Clinical Aspects of Retinal Degeneration

1. Histochemical Comparison of Ocular "Drusen" in Monkey and Human 1
   Robert F. Mullins and Gregory S. Hageman

2. TIMP-3 Accumulation in Bruch's Membrane and Drusen in Eyes from Normal
   and Age-Related Macular Degeneration Donors 11
   Motohiro Kamei, Suneel S. Apte, Mary E. Rayborn, Hilel Lewis, and
   Joe G. Hollyfield

3. Photoreceptor Rosettes in Age-Related Macular Degeneration Donor Tissues . . . 17
   Mary E. Rayborn, Kathy M. Myers, and Joe G. Hollyfield

4. Central Retinal Sensitivity Repeated Measurements as Long Term Follow-up in
   Retinitis Pigmentosa 23
   Enzo M. Vingolo, Andrea Perdicchi, Renato Forte, Patrizia Del Beato,
   Luigi Pannarale, and Roberto Grenga

5. Complete and Incomplete Type Congenital Stationary Night Blindness (CSNB)
   as a Model of "OFF-Retina" and "ON-Retina" 31
   Yozo Miyake, Masayuki Horiguchi, Satoshi Suzuki, Mineo Kondo, and
   Atsutoshi Tanikawa

6. A Family with X-Linked Cone Dystrophy Showing a Tapetal-like Reflex 43
   Mutsuko Hayakawa, Keiko Fujiki, Kenji Yanashima, Ikuko Kondo, and
   Atsushi Kanai

7. Autoimmune Retinopathy: Cystoid Macular Edema in Retinitis Pigmentosa
   Patients 51
   John R. Heckenlively, Nata Aptsiauri, and Paul A. Hargrave

8. Norrie Disease in Japan 57
   Norio Ohba, Yasushi Isashiki, Fumiyuki Uehara, and Kazuhiko Unoki

Part II. Animal Models of Retinal Degeneration

9. Development of a Model for Macular Degeneration 61
      P. E. Rakoczy, M. Lai, and I. J. Constable

10. Subretinal Iodoacetate: A Model of Retinal Degeneration in Cats 71
      Jun C. Huang, Masahiro Ishida, Sarah Goldfeder, Ilene K. Sugino, and
      Marco A. Zarbin

11. Hereditary Retinal Dystrophy of Swedish Briard Dogs: Exclusion of Six
      Candidate Genes by Molecular Genetic Analysis 81
      Andres Veske, Sven Erik G. Nilsson, Ulrich Finckh, Kristina Narfström,
      Simon Petersen-Jones, David Gould, David Sargan, and Andreas Gal

12. The VPP Mouse: A Transgenic Model of Autosomal Dominant Retinitis
      Pigmentosa 89
      Neal S. Peachey, Min Wang, and Muna I. Naash

13. Altered Regulation of Ion Channels in Cultured Retinal Pigment Epithelial Cells
      from RCS Rats 99
      O. Strauß, S. Mergler, M. Wienrich, and M. Wiederholt

14. Structures of the Oligosaccharides of Rhodopsin from Normal and RCS Rats 107
      Edward L. Kean, Tamao Endo, Naiqian Niu, Daniel T. Organisciak,
      Yuji Sato, and Akira Kobata

15. Indocyanine Green Videoangiography in the Royal College of Surgeons Rat ... 115
      Katsuhiro Yamaguchi, Keiko Yamaguchi, Takeo Satoh, and Shigeki Takahashi

16. Defective Choroidal Angiogenesis Precedes Retinal Pigment Epithelial
      Phagocytic Defect in Neonatal RCS Rats 121
      Margaret J. McLaren

17. Comparative Biology of Retinoid Deprivation and Replacement in Flies and
      Rodents 135
      William S. Stark

18. Rhodopsin-Dependent Models of Drosophila Photoreceptor Degeneration 145
      David R. Hyde, Scott Milligan, and Troy Zars

19. Recessive Degeneration of Photoreceptor Cells Caused by Point Mutations in
      the Cytoplasmic Domains of Drosophila Rhodopsin 159
      Joachim Bentrop, Karin Schwab, William L. Pak, and Reinhard Paulsen

Part III. Mechanisms of Retinal Degeneration and Cell Death

20. Nitric Oxide-Induced Increases in Retinal cGMP: A Role in Photoreceptor
      Degenerations 171
      I. G. Morgan and J. W. Wellard

21. Glycohistochemical Study of Light-Induced Retinal Degeneration: Removal
    System of Apoptotic Cells 181
    Fumiyuki Uehara, Norio Ohba, Toyoko Yanagita, Munefumi Sameshima,
    Naoto Iwakiri, Akiko Okubo, Yoshiko Maeda, Kazuhiko Unoki, and
    Taeko Miyagi

22. Light-Induced Retinal Degeneration Is Prevented in Mice Lacking *c-fos* 193
    Farhad Hafezi, Andreas Marti, Joachim P. Steinbach, Kurt Munz,
    Adriano Aguzzi, and Charlotte E. Remé

23. Ischemic Neuronal Death in the Fish Retina: In Respect of Oxygen Radicals and
    Glutathione 199
    S. Kato, Z.-Y., Zhou, K. Sugawara, Y. Yasui, N. Takizawa, K. Sugitani, and
    K. Mawatari

**Part IV. Candidate Genes, Cloning, Mapping, and Other
Molecular Genetics of Retinal Degeneration**

24. Isolation of Candidate Genes for Retinal Degenerations 205
    George Inana, Akira Murakami, Hitoshi Sakuma, Tomomi Higashide,
    Toshihiro Yajima, and Margaret J. McLaren

25. Studies on the Cone Cyclic GMP-Phosphodiesterase α′ Subunit Gene 227
    Debora B. Farber, Natik Piriev, Yong Qing Gao, Michael Danciger, and
    Andrea Viczian

26. Mutations in *PDE6A*, the Gene Encoding the α-Subunit of Rod Photoreceptor
    cGMP-Specific Phosphodiesterase, Are Rare in Autosomal Recessive
    Retinitis Pigmentosa 237
    M. Meins, A. Janecke, C. Marschke, M. J. Denton, G. Kumaramanickavel,
    S. Pittler, and A. Gal

27. Molecular Analysis of the Human PEDF Gene, a Candidate Gene for Retinal
    Degeneration: Localization to 17p13.3 245
    Joyce Tombran-Tink, Gerald Chader, and Robert Koenekoop

28. Screening of Candidate Genes on Japanese Retinal Dystrophies 255
    Yoshihiro Hotta, Keiko Fujiki, Mutsuko Hayakawa, Hitoshi Sakuma,
    Hiroyuki Kawano, Atsushi Kanai, Akira Murakami, Masaru Yoshii,
    Kiyoshi Akeo, Shigekuni Okisaka, Masayuki Matsumoto, Seiji Hayasaka,
    Yasushi Isashiki, Norio Ohba, Takashi Shiono, and Makoto Tamai

29. Strategies for the Genetic Analysis of Autosomal Recessive Retinitis
    Pigmentosa in Spanish Families 263
    Roser González-Duarte, Mónica Bayés, Amalia Martínez-Mir,
    Diana Valverde, Susana Balcells, Montserrat Baiget, Lluïsa Vilageliu, and
    Daniel Grinberg

30. Progress in Positional Cloning of RP10 (7q31.3), RP1 (8q11-q21), and VMD1
     (8q24)   277
     Stephen P. Daiger, Rachel E. McGuire, Lori S. Sullivan,
     Melanie M. Sohocki, Susan H. Blanton, Peter Humphries, Eric D. Green,
     Helen Mintz-Hittner, and John R. Heckenlively

31. Growth Factors in the Retina: Pigment Epithelium-Derived Factor (PEDF) Now
     Fine Mapped to 17p13.3 and Tightly Linked to the RP13 Locus   291
     Jacquie Greenberg, Rene Goliath, Joyce Tombran-Tink, Gerald Chader, and
     Rajkumar Ramesar

32. Genetic and Physical Localisation of the Gene Causing Cone-Rod Dystrophy
     (*CORD2*)   295
     James Bellingham, Sujeewa D. Wijesuriya, Kevin Evans, Alan Fryer,
     Greg Lennon, and Cheryl Y. Gregory

33. Usher Syndrome Type 1C: Localization to Chromosome 11p14 and
     Construction of a YAC Contig   303
     Radha Ayyagari, Anren Li, Ann Nestorowicz, Yan Li, Richard J. H. Smith,
     M. Alan Permutt, and J. Fielding Hejtmancik

34. Oguchi Disease, Retinitis Pigmentosa, and the Phototransduction Pathway   313
     Marion A. Maw and Michael J. Denton

35. A Patient with Progressive Retinal Degeneration Associated with Homozygous
     1147delA Mutation in the Arrestin Gene   319
     Y. Wada, M. Nakazawa, and M. Tamai

Part V. Transplantation, Gene Therapy, and Cell Rescue

36. bFGF Transfected Iris Pigment Epithelial Cells Rescue Photoreceptor Cell
     Degeneration in RCS Rats   323
     M. Tamai, K. Yamada, N. Takeda, H. Tomita, T. Abe, S. Kojima, and
     S.-I. Ishiguro

37. Transplantation of Neonatal Neural Retina in Photoreceptor Degeneration of
     Cats   329
     Kristina Narfström, Lena Ivert, Peter Naeser, and Peter Gouras

38. Mechanical Aspects of Retinal Pigment Epithelial Transplantation   339
     Devjani Lahiri-Munir, Lichun Lu, Charles A. Garcia, Antonios Mikos, and
     Emily Aguilar

39. Midkine from Various Sources in Constant Light-Induced Photoreceptor
     Degeneration of the Rat   347
     Kazuhiko Unoki, Hisako Muramatsu, Norio Kaneda, Shinya Ikematsu,
     Fumiyuki Uehara, Norio Ohba, and Takashi Muramatsu

40. The Influence of Oxygen on the Survival and Death of Photoreceptors:
     Evidence from Rat and Mouse 353
     Krisztina Valter, Kyle Mervin, Juliani Maslim, and Jonathan Stone

41. Histological Method to Assess Photoreceptor Light Damage and Protection by
     Survival Factors 369
     Matthew M. LaVail, Michael T. Matthes, Douglas Yasumura, Ella G.
     Faktorovich, and Roy H. Steinberg

Part VI. Structural, Physiologic, and Aging Correlates of Retinal Degeneration

42. Further Studies on the Phagocytosis of Photoreceptor Outer Segments by Rat
     Retinal Pigment Epithelial Cells 385
     Michael O. Hall, Toshka A. Abrams, Barry L. Burgess, and Alexey V. Ershov

43. Purification and Characterization of Matrix Metalloproteinase-3 (Stromelysin-1)
     from Bovine Interphotoreceptor Matrix 399
     Abdelkrim Smine and James J. Plantner

44. Senile Retinal Deficiencies in Astrocytes and Blood Vessels 409
     Elisabeth Rungger-Brändle and Peter M. Leuenberger

Corresponding Authors by Chapter Number 417

Index 423

HISTOCHEMICAL COMPARISON OF OCULAR "DRUSEN" IN MONKEY AND HUMAN[*]

Robert F. Mullins and Gregory S. Hageman[†]

Anheuser-Busch Eye Institute of Saint Louis University
Department of Ophthalmology
Saint Louis University School of Medicine
Saint Louis, Missouri 63104

INTRODUCTION

Age-related macular degeneration (AMD), the leading cause of irreversible blindness in the industrialized world, is characterized, in part, by the presence of drusen and basal laminar deposits (BLD) in the macula. Drusen are extracellular deposits which form between the retinal pigment epithelial (RPE) basal lamina and the inner collagenous zone of Bruch's membrane. Although the relationship between macular drusen and the development of AMD has been established in a number of studies (1–4), relatively little is known about their composition or origin. Thorough characterization of these age-related deposits has been hampered by a paucity of suitable human donor tissues and the lack of an appropriate animal model of the disease.

Various species have been examined morphologically and/or funduscopically for the presence of drusen that resemble those in humans. These include rabbits (5), mice (Hageman and Heckenlively, unpublished observations), Japanese quail (6), cats (7), horses (Kalsow and Hageman, unpublished observations), and monkeys (8–15). Of these species, only monkeys appear to have extracellular deposits that may resemble hard drusen. However, these deposits tend to be extremely rare and relatively small, compared to human drusen (9,13). To complicate the situation, many RPE abnormalities in monkeys mimic drusen funduscopically (11,16), making ophthalmoscopic examination alone insufficient for the detection or identification of drusen.

One focus of this laboratory has been directed toward the identification and characterization of drusen-associated molecules using histochemical and biochemical methods.

* Supported in part by NIH grants EY06463 and EY11515 (GSH) and an unrestricted grant to the Anheuser-Busch Eye Institute from Research to Prevent Blindness, Inc.

† Send correspondence to: Gregory S. Hageman, Ph.D., Anheuser-Busch Eye Institute of Saint Louis University, Saint Louis University School of Medicine, 1755 S. Grand Blvd., St. Louis, MO 63104. Telephone: (314) 577-8261; fax (314) 771-0596.

These studies have been based on the hypothesis that the identification of drusen-associated proteins may shed insight into the pathobiology of the aging eye and the development of AMD. These analyses have resulted in the identification of several major drusen-associated protein components, including vitronectin, amyloid P component, and complement C5, as well as a specific, restricted set of carbohydrate moieties including galactose, N-acetylglucosamine, glucose/mannose, and sialic acid. All of these constituents are present in all classes of hard and soft drusen (17–20). If monkey drusen are similar to those in human eyes, as has been suggested (8), one would assume that they should contain the same compliment of proteins and carbohydrate moieties.

In order to determine whether human and monkey drusen are related compositionally, lectin and antibody histochemical analyses of sections of human donor and monkey eyes were conducted. Of four lectins which bind intensely to both hard and soft drusen in humans, only ConA reacts intensely with monkey "drusen"; none of the antibodies shown to react with human drusen bind to drusen-like deposits found in monkey, suggesting that significant differences in sub-RPE, hard drusen-like deposits exist between these species.

MATERIALS AND METHODS

Tissues

Human donor eyes were obtained from Mid-America Transplant Services (St. Louis, MO), and were fixed within 3.5 hours of death. Of 108 cynomolgus monkeys examined, 4 possessed ocular deposits that resembled funduscopically hard, or nodular, drusen in humans. Eyes from these animals were enucleated within 15 minutes after euthanasia by barbiturate overdose. Animal treatment conformed to the guidelines of the Association for Research in Vision and Ophthalmology "Resolution on the Use of Animals in Research" (Invest. Ophthalmol. Vis. Sci. 30:2418, 1989), the NIH "Guide for the Care and Use of Laboratory Animals" (NIH publication 86-23), and guidelines established by the Saint Louis University Department of Comparative Medicine. Human and monkey eyecups were fixed in 100mM sodium cacodylate containing 4% paraformaldehyde, pH 7.4, for 2 hours. Contralateral eyes were fixed in one-half strength Karnovsky's fixative and processed for transmission electron microscopy as described previously (21). After rinsing 3x10 min in cacodylate buffer, pieces of tissue were infiltrated and embedded in acrylamide, as described previously (22).

Histochemistry

Cryostat-derived serial or stepped serial sections were mounted on Superfrost Plus slides (Fisher), and blocked for 15 min in 10mM sodium phosphate, pH 7.4, containing 0.85% NaCl, 1mM $CaCl_2$, 1mM $MgCl_2$ (PBS/M/C), and 1mg/ml globulin-free bovine serum albumin (BSA). Sections were then rinsed for 2x10 min in PBS/M/C, and incubated for 30 min in FITC-conjugated lectins or 60 min in unconjugated primary antibodies, both diluted in PBS/M/C (Table I). The antibodies used in this study included polyclonal antisera to complement C5, vitronectin, and amyloid P component. The lectins employed included *Canavalia ensiformis* agglutinin (Con A), *Triticum vulgaris* agglutinin (WGA), *Ricinus communis* agglutinin I (RCA-I) and *Limax flavus* agglutinin (LFA); these lectins bind human hard and soft drusen intensely. Lectin-stained slides were subsequently washed 2x10 min in PBS/M/C, and coverslipped using Immumount (Shandon; Pittsburgh, PA). After rinsing in PBS/M/C, antibody-labeled sections were incubated for 30 min with

FITC-conjugated goat anti-rabbit IgG (Cappel; Malvern, PA), prior to washing and mounting, as described above. Adjacent sections were stained with Oil Red O and Sudan Black B (Poly Scientific; Bay Shore, NY).

RESULTS

Histochemical Comparison of Human and Monkey Drusen

Of four monkeys with funduscopically-visible drusen, only one possessed a few rare, hard-drusen-like deposits. These drusen were comprised of highly heterogeneous extracellular material located between the RPE and the elastic layer of Bruch's membrane. The number of sub-RPE deposits observed histologically were inadequate to account for the number of "drusen" observed funduscopically. The RPE monolayer was intact but extremely attenuated over these deposits. The drusen present in the human donors were typical homogeneous, hyaline structures with essentially no empty space between the RPE and elastic lamina. In contrast, the monkey "drusen" consisted of large vacuoles containing discrete, clumped cellular components with some melanin and/or lipofuscin granules. The "drusen" observed in this monkey were slightly larger than those described in other studies (9,12,13), with heights of up to 40 μm and diameters up to 70 μm. Notably, photoreceptor inner and outer segments appeared to be missing or damaged over these deposits (Figure 1). No drusen were identified in monkeys that resembled hard clusters, soft clusters, or membranous soft drusen (23) observed in human eyes.

Sections of both human and monkey retina, RPE and choroid were stained with a number of lectin, antibody, and histochemical probes that have been shown to react human drusen (17–20); the results obtained are summarized in Table 1 and detailed below.

RCA-I

Human. Labeling of both hard and soft drusen was observed, although the intensity of the labeling was relatively weak. In some cases, spherical profiles in soft drusen showed stronger binding than the rest of the druse (data not shown). Strong labeling of the choriocapillaris and interphotoreceptor matrix was also observed.

Monkey. Intense labeling of the choriocapillaris was noted, as in human sections, but in contrast, only weak reactivity was observed around the periphery of the cellular clumps within these "drusen".

Table 1. Histochemical comparison of monkey and human hard drusen

Probe	Supplier	Dilution	Human	Monkey
WGA	Vector	1:50	++	+/–
RCA	Vector	1:50	+/–	+/–
ConA	EY Labs	1:50	+	+
LFA	EY Labs	1:5	++	+/–
anti Vn	Telios	1:50	+	–
anti SAP	Dako	1:50	+	–
anti C5	Dakopatts	1:50	+	–
Oil Red O	Poly Scientific	——	+	–
Sudan Black B	Poly Scientific	—	+	+

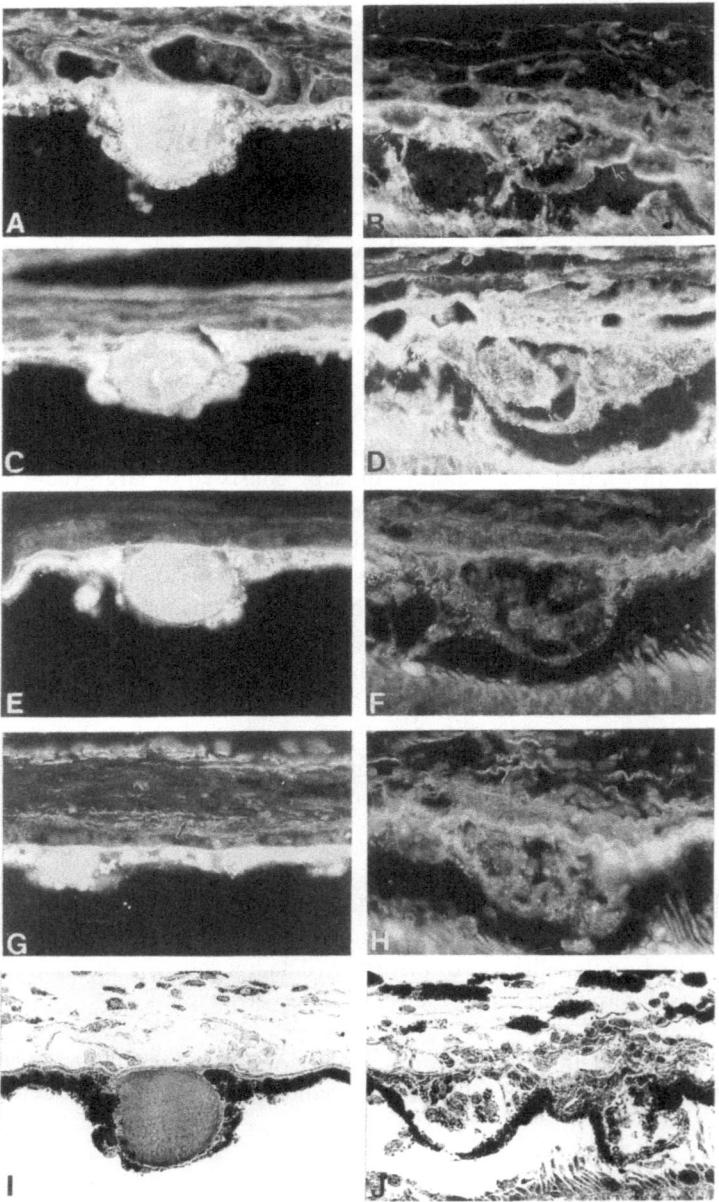

Figure 1. Fluorescence and bright field light micrographs depicting human (A,C,E,G,I) and monkey (B,D,F,H,J) drusen-like deposits. WGA (A,B) reacts intensely with human drusen (A), but only weakly in monkey drusen. Labeling is partially restricted to drusen peripheries (arrows). Con A (C,D) reacts with both human (C) and monkey (D) drusen. Anti-vitronectin antibodies react with human (E), but not monkey (F), drusen. Antisera to amyloid P component (G,H) react with drusen and to fibrillar components of the choroidal stroma in human eyes (G, arrows). Choroidal fibrils (arrows), but not drusen, are labeled by this antibody in monkey eyes (H). Sudan Black B (I,J) stains components of "drusen" in both humans (I) and monkeys (J).

WGA

Human. Intense labeling of all drusen was observed. This labeling tended to be homogeneous, particularly in hard drusen (Fig 1C). In addition, rod-associated glycoconjugates showed strong labeling, as did the choroidal stroma.

Monkey. Labeling of the peripheries ("rims") of drusen was noted; this labeling may be associated with BLD, or may be due to asymmetrical distribution of drusen glycoconjugates (Fig 1D). A similar pattern is observed occasionally in human drusen (20). In additon, rod-photoreceptor associated glycoconjugates showed strong labeling.

LFA

Human. Labeling of hard and soft drusen was observed. Labeling was also seen in the interphotoreceptor matrix as well as in choroidal blood vessel walls and the choroidal stroma.

Monkey. Labeling was observed along Bruch's membrane and in interdigitating strands of fibrillar material between clumps of drusen-like material. Strong labeling of the interphotoreceptor matrix was also noted.

Con A

Human. Moderate labeling of all drusen and of blood vessel walls was observed (Fig 1A). Intense labeling of rod photoreceptors was also noted.

Monkey. Moderate labeling of clumped material within drusen was noted (Fig 1B). In addition, in areas which more closely resembled human drusen (large, flat mounds of sub-RPE material), labeling was seen throughout the deposit. Rod photoreceptors and the choroidal stroma also exhibited binding.

Anti-Vitronectin Antibody

Human. Intense labeling of all drusen was observed. Moderate labeling of fibrils in the choroidal stroma was also noted (Fig 1E).

Monkey. No labeling of drusenoid material was detected. Faint labeling of some strands in the choroid was detected (Fig 1F).

Anti-Amyloid P Component Antibody

Human. Intense labeling of all drusen was observed. Intense labeling of some choroidal fibrillar profiles was also noted (Fig 1G).

Monkey. No labeling of drusen was noted; positive labeling of choroidal extracellular matrix fibrils was exhibited (Fig. 1H).

Anti-Complement C5 Antibody

Human. Intense labeling of all drusen in the section was observed.

Monkey. No labeling of drusen was observed. Weak labeling of blood vessel lumena was detected.

Oil Red O

Human. Modest labeling of all drusen in the tissue section was observed; some reaction with photoreceptor inner segments was noted.

Monkey. No labeling of drusen was detected; some reaction with photoreceptor inner segments was noted.

Sudan Black B

Human. Strong labeling of drusen was observed (Fig 1I).

Monkey. Strong labeling of globular components within the deposit was observed (Fig 1J).

Morphological Manifestations of Funduscopic "Drusen"

Three monkey eyes containing numerous, funduscopically-visible drusen did not possess any hard drusen, as assessed by sectioning. In one animal, a number of areas resembled drusen in 1 μm sections, as evidenced by the elevation of melanosomes away from Bruch's membrane. When viewed by transmission electron microscopy, however, these "drusen" were actually abnormal intracellular accumulations of membranous profiles resembling smooth endoplasmic reticulum (Fig. 2A). The mitochondria in these cells were pushed basally and were massed along the basal aspect of the cells, while the melanosomes were concentrated along the apical margin of the cells, adjacent to the interphotoreceptor space. These regions of hypertrophied smooth endoplasmic reticulum hypertrophy resemble those observed in liver cells following exposure to toxins (24), and most likely were responsible for the funduscopic manifestation of "drusen".

Another feature observed in one of the monkey eyes with funduscopic drusen was the appearance of distinctive folds in the neural retina, in which the photoreceptor outer segments formed extracellular clumps (Fig. 2B). Since no typical drusen were present in this eye, it seems that these areas of outer segment accumulation might have presented as drusen funduscopically.

Finally, cysts in the inner retina were seen in one animal (Fig. 2C); it is possible that these areas of degeneration might resemble drusen when viewed funduscopically. It should also be noted that areas of basal laminar deposits, characterized by the accumulation of "long-spacing collagen", were also observed in some monkey eyes (Fig. 2D). These deposits were frequently observed in association with vacuolated RPE cells.

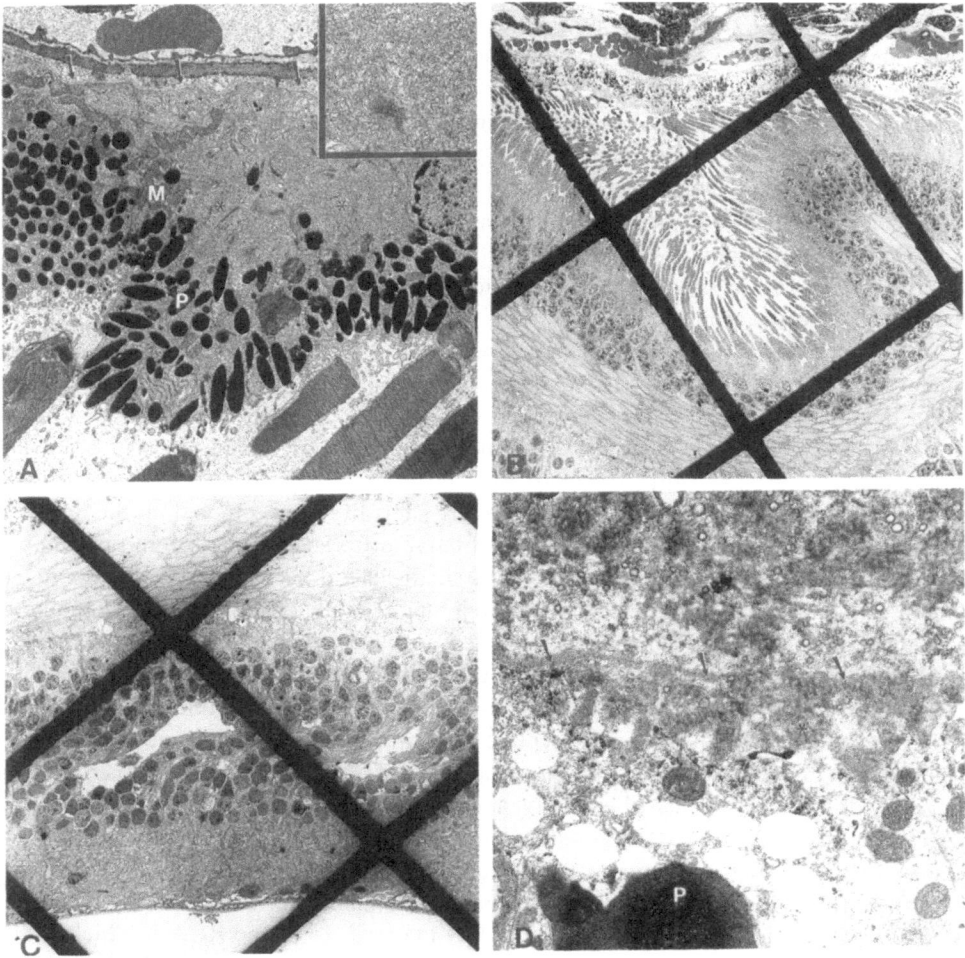

Figure 2. Potential morphological manifestations of "drusen" in monkey eyes with funduscopicically-detectable drusen. Hypertrophy of smooth endoplasmic reticulum (A, asterisks), apparent with transmission electron micros- copy, causes the elevation of melanosomes and the distortion of RPE cell shape, resembling typical drusen at low magnificantions (mitochondria, M; melanosomes, P; RPE basal lamina, arrows: inset: higher magnification of membranous profiles). Folds in the neural retina (B) and cysts in the inner retina (C) are present in some monkey eyes with funduscopically-detectable drusen; these eyes showed no evidence of typical, sub-RPE drusen. Basal laminar deposits (D, asterisks) were also seen in association with vacuolated RPE cells in some monkeys (basal lamina, arrows; melanosome, P).

DISCUSSION

Funduscopically-visible drusen have been described in primates by a number of in- vestigators. In several of these studies, only ophthalmoscopic evidence for identifying these as typical drusen was presented (8,10,14). The pitfalls in assessing morphology from funduscopic appearance in primates have been discussed (11,16). A number of morpho- logical features which appear to give the appearance of drusen on funduscopic examina- tion can be manifested by other structures. Hirata and Feeney-Burns (16) observed yellow-white spots in the macula which, on histologic examination, were determined to be vacuolated RPE cells, rather than sub-RPE deposits. Ishibashi (11) proposed that monkeys possess three types of histologically distinct abnormalities which resemble human drusen ophthalmoscopically; these included 1) vacuolated RPE cells, 2) mounds of PAS-positive material which appear similar to human drusen, and 3) discrete focal accumulations of cellular debris. We have identified other cellular abnormalities which are present in eyes with funduscopically-detectable drusen. These consisted of RPE cells severely distended by abnormal accumulations of smooth endoplasmic reticulum and, in other cases, of accumulations of photoreceptor outer segment material in areas of retinal folds. Similar retinal folds have also been noted in mice with funduscopic white spots (25) and in humans (Hageman, unpublished observation).

It is clear that some drusen identified histologically in monkeys resemble those in humans to a certain extent (15). However, these deposits generally tend to be much smaller than human drusen, rarely deforming the RPE monolayer to any significant extent (9,12,13). Engel et al. (9) described drusen-like deposits in Rhesus monkey maculae which were no more than 4μm in height and 20μm in diameter. This distinction in size may be a consequence of the monkey's relatively short life span, precluding the development of larger drusen, or it may reflect a qualitative difference in the origin or composition of these deposits. In addition, only rare, small hard drusen have been observed in monkeys. Because large, soft drusen represent a much greater risk for the development of AMD than small, hard drusen (26), it is unclear to what extent observations in monkey maculae might be applicable to understanding maculopathy in humans.

Ishibashi (11) proposed that discrete focal accumulations of cellular debris seen in monkey eyes lead to the formation of typical drusen. In order to understand better drusen formation and to assess the monkey as a model for AMD, we characterized this third type of monkey drusen deposit histochemically, and compared it with typical human drusen. Of the lectins which consistently bind to drusen in human eyes, only ConA shows a strong reaction with globular components of these monkey drusen. Interestingly, WGA showed a similar pattern of labeling to that seen previously in human donor tissues, with a strong reaction along the apical margin of the drusen-like material. LFA and RCA-I, which exhibited weak heterogeneous labeling of some drusen components, did not show a similar reaction to that seen in human eyes. Immunoreactivity of polyclonal antisera raised against vitronectin, amyloid P component, and complement C5 all showed a strong reaction with human, but not monkey, drusen; labeling of fibrils in the choroidal stroma with antisera to amyloid P component and, to a lesser extent, with antisera to vitronectin, suggests that these antibodies do react with monkey proteins. While it is possible that molecules identified in human drusen might accumulate at later stages of monkey drusen development — assuming that the discrete cellular debris represents early drusen — it is significant that lectins and antibodies which react with human drusen do so at all sizes, phenotypes and presumed stages of development. The major implication of these results is that human and monkey hard drusen are compositionally dissimilar.

In summary, we believe that the following observations are pertinent to the consideration of the monkey as a potential animal model for age-related macular degeneration. First, typical hard drusen are found in monkey eyes only very rarely, and the number of these drusen is not sufficient to account for the density of drusen-like deposits seen funduscopically. Second, a number of ultrastructural cellular and extracellular abnormalities in monkey eyes may result in the appearance of funduscopic drusen. Third, only hard drusen are seen in these animals — large and/or soft drusen are not observed. Fourth, drusen-like deposits in monkey are compositionally different from those in humans, both in their immunohistochemical properties and glycoconjugate profiles. Taken together, the results presented herein suggest that, while human and monkey drusen share some funduscopic, morphological and compositional features, it is not clear that they represent the same morphological or biochemical entity. In light of this dissimilarity, application of studies on age-related funduscopic changes in monkeys to understanding human disease should be treated with caution.

ACKNOWLEDGMENT

The authors wish to acknowledge Mrs. Bobbie Schneider for expert technical assistance in transmission electron microscopy.

REFERENCES

1. Pauleikhoff, D., Barondes, M.J., Minassian, D., Chisholm, I.H., and Bird, A.C., 1990, Drusen as risk factors in age-related macular disease, *Am. J. Ophthalmol.* **109**:38–43.
2. Bressler, S.B., Maguire, G., Bressler, N.M., and Fine, S.L., 1990, Relationship of drusen and abnormalities of the retinal pigment epithelium to the prognosis of neovascular macular degeneration, *Arch. Ophthalmol.* **108**:1442–1447.
3. Vinding, T., 1990, Occurrence of drusen, pigmentary changes and exudative changes in the macula with reference to age-related macular degeneration: An epidemiological study of 1000 aged individuals. *Acta Opthalmol.* **68**:410–414.
4. Bressler, N.M., Silva, J.C., Bressler, S.B., Fine, S.L., and Green, W.R., 1994, Clinicopathologic correlation of drusen and retinal pigment epithelial abnormalities in age-related macular degeneration. *Retina* **14**:130–142.
5. Tabatay, C.A., D'Amico, D.J., Hanninen, L.A., Kenyon, K.R., 1987, Experimental drusen formation induced by aminoglycoside injection, *Arch. Ophthalmol.* **105**: 826–830.
6. Fite, K.V., Bengston, C.L., and Cousins, F., 1994, Drusen-like deposits in the outer retina of Japanese quail, *Exp. Eye. Res.* **59**:417–424.
7. Collier, L.L., King, E.J., and Prieur, D.J., 1986, Age-related changes of the retinal pigment epithelium of cats with Chediak-Higashi syndrome, *Invest. Ophthalmol. Vis. Sci.* **27**: 702–707.
8. Hope, G.M., Dawson, W.W., Engel, H.M., Ulshafer, R.J., Kessler, M.J., and Sherwood, M.B., 1992, A primate model for age related macular drusen. *British J. Ophthalmol.* **76**:11–16.
9. Engel, H.M., Dawson, W.W., Ulshafer, R.J., Hines, M.W., and Kessler, M.J., 1988, Degenerative changes in maculas of rhesus monkeys. *Ophthalmologica* **196**:143–150.
10. Monaco, W.A. and Wormington, C.M., 1990, The rhesus monkey as an animal model for age-related macular degeneration, *Optometry and Vision Science.*, **67**:532–537.
11. Ishibashi, T., Sargente, N., Patterson, R., and Ryan, S.J., 1986, Pathogenesis of drusen in the primate. *Invest. Ophthalmol. Vis. Sci.* **27**:184–193.
12. Olin, K.L., Morse, L.S., Murphy, C., Paul-Murphy, J., Line, S., Bellhorn, R.W., Hjelmeland, L.M., and Keen, C., 1995, Trace element status and free radical defense in elderly rhesus macaques *(Macaca mulatta)* with macular drusen,*Proc. Soc. Exp. Biol. Med.* **208**:370–377.
13. Ulshafer, R.J., Engel, H.M., Dawson, W.W., Allen, C.B., and Kessler, M.J., 1987, Macular degeneration in a community of rhesus monkeys: ultrastructural observations, *Retina* **7**:198–203.
14. Duvall, J. and Tso, M.O., 1985, Cellular mechanisms of resolution of drusen after laser coagulation: an experimental study, *Arch. Ophthalmol.* **103**:694–703.
15. Stafford, T.J., Anness, S.H., and Fine, B.S., 1984, Spontaneous degenerative maculopathy in the monkey, *Ophthalmology* **91**:513–521.
16. Hirata, A. and Feeney-Burns L., 1992, Autoradiographic studies of aged primate macular retinal pigment epithelium, *Invest. Ophthalmol. Vis. Sci.* **33**:2079–2090.
17. Hageman, G.S., Mullins, R.F., Clark, W.G., Johnson, L.V. and Anderson, D.H., 1995, Drusen share molecular constituents common to atherosclerotic, elastotic and amyloid deposits, *Invest. Ophthalmol. Vis. Sci. (Suppl.)* **36**: S432.

18. Mullins, R.F., Johnson, L.V., Anderson, D.H. and Hageman, G.S., 1995, Molecular composition of drusen: histochemical and enzymatic characterization of glycoconjugates, *IInvest. Ophthalmol. Vis. Sci. (Suppl.)* **36**: S432.

19. Hageman, G.S., Mullins, R.F., Johnson, L.V. and Anderson, D.H., 1996, Cell biology of AMD: Analyses of age-related extracellular deposits, *Invest. Ophthalmol. Vis. Sci. (Suppl.)* **37**: S450.

20. Mullins, R.F., Johnson, L.V., Anderson, D.H., and Hageman, G.S., Characterization of drusen-associated glycoconjugates, *Ophthalmology* **In press**.

21. Lazarus, H.S., Sly, W.S., Kyle, J.W., and Hageman, G.S., 1993, Photoreceptor degeneration and altered distribution of interphotoreceptor matrix proteoglycans in the mucopolysaccharidosis VII mouse, *Exp. Eye Res.* **56**: 531–541.

22. Johnson, L.V. and Blanks, J.C., 1984, Application of acrylamide as an embedding medium in studies in lectin and antibody binding in the vertebrate retina, *Curr .Eye Res .***3**:969–974.

23. Sarks, J.P., Sarks, S.H., and Killingsworth, M.C., 1994, Evolution of soft drusen in age-related macular degeneration, *Eye* **8**:269–83.

24. Ghadially, F.N., 1988, Ultrastructural pathology of the cell and matrix, 3rd ed., Butterworth, New York.

25. Chang, B., Hageman, G.S., Heckenlively, J.R., Hawes, N.L., Peng, C., Reoderick, T.H., Davisson, M.T., 1996, A mouse model for retinitis punctata albescens; a new pathologic finding for white dots, *Invest. Ophthalmol. Vis. Sci. (Suppl.)* **37**: S505.

26. Green, W.R. and Enger, C., 1993, Age-related macular degeneration histopathologic studies. *Ophthalmology* **100**:1519–1535.

TIMP-3 ACCUMULATION IN BRUCH'S MEMBRANE AND DRUSEN IN EYES FROM NORMAL AND AGE-RELATED MACULAR DEGENERATION DONORS

Motohiro Kamei,[1] Suneel S. Apte,[2] Mary E. Rayborn,[1] Hilel Lewis,[1] and Joe G. Hollyfield[1]

[1]Division of Ophthalmology
[2]Department of Biomedical Engineering
Cleveland Clinic Foundation
Cleveland, Ohio 44195

INTRODUCTION

Age-related macular degeneration (ARMD) is the leading cause of central vision loss in individuals over the age of 50 in the Western hemisphere.[1] ARMD is characterized by the accumulation of debris (drusen) within Bruch's membrane during the early stages of the disease, with the subsequent development of choroidal neovascularization or atrophy of the choriocapillaris and retinal pigment epithelium (RPE) at later stages. The latter events are thought to be causally involved in the death of macular photoreceptors. While the ophthalmological changes during the progression of the various forms of ARMD have been described, and end-stage histopathological descriptions of ARMD are available, little is known about the cellular mechanisms associated with the normal maintenance and turnover of Bruch's membrane, the events that lead to drusen formation, or the cause of subretinal neovascularization and atrophy of the RPE and choriocapillaris.

Bruch's membrane is a complex extracellular matrix which is a product of both the RPE and the endothelial cells of the choriocapillaris. Matrix metalloproteinases (MMPs) constitute a family of secreted enzymes which are involved in degrading components of the extracellular matrix (ECM) in the normal course of matrix turnover and renewal.[2] MMPs are also implicated during the initial stages of neovascularization, where they are required, along with other proteases, for degradation of components of the vessel basement membrane as a prerequisite for new vessel outgrowth.[3] Regulation of the activity of these extracellular enzymes is accomplished through the activity of another group of secreted proteins referred to as TIMPs (tissue inhibitors of metalloproteinases).[4] TIMPs are thought to constitutively suppress excessive degradation of ECM and inhibit the development of neovascularization.[5] TIMPs non-covalently bind to specific MMPs with high affinity which results in the inhibition of their enzymatic activity.[4,6]

Mutations in the gene coding for TIMP-3 were recently found in some families with Sorsby's fundus dystrophy, which is an inherited disease, characterized by thickening of Bruch's membrane and submacular neovascularization.[7] *In situ* hybridization studies in mice indicate that the RPE is a major site of TIMP-3 expression[8] and immunocytochemical studies indicate that TIMP-3 immunoreactivity is present in normal Bruch's membrane.[9]

The expression of TIMP-3 in the RPE and its accumulation in Bruch's membrane, coupled with the demonstration that a mutation in TIMP-3 is causally involved in Sorsby's fundus dystrophy, an inherited disease which affects the interface between the RPE and choriocapillaris, suggests that the analysis of MMPs and TIMPs may be important in understanding other pathologies which affect the RPE-choroidal interface. This is a preliminary report of the observation of high levels of TIMP-3 immunoreactivity in the debris (drusen) which accumulates between the RPE and Bruch's membrane in donor tissues with ARMD.

TISSUES AND METHODS

The tissues used in this analysis are as follows: normal eyes were from two male and two female donors, ages 28 to 73 years, enucleated and placed on ice between 1 to 18 hours postmortem; ARMD eyes were from four male donors between 72 and 96 years old, recovered between 3 and 11 hours postmortem. Normal donor tissue was obtained from the Lions Eye Bank, Houston, Texas and the Cleveland Eye Bank, Cleveland Ohio. Tissues from ARMD donors were obtained through the Eye Donor Program of The Foundation Fighting Blindness, Hunt Valley, Maryland. Normal histology was performed using mixed aldehyde fixation, and light microscope analysis of 0.5–1.0 μm thick sections stained with toluidine blue and analyzed with light microscopy. Immunohistochemistry was performed on unfixed cryosections through the macula using a mouse monoclonal anti-human TIMP-3 antibody (clone 136–13144, Fuji Chemicals, Ltd., Japan). The avidin-biotin method was used and immunoreactivity was resolved with horseradish peroxidase-amino ethyl carbazole (AEC) reaction product visualized with light microscopy. In the color images presented, the reaction product produces a deep red to magenta color. For controls, non-immune mouse IgG was applied instead of the primary antibody.

RESULTS

A comparison of the RPE-choroidal interface in macular samples from normal and ARMD donor eyes is presented in Fig. 1. In the region below the macula in the normal donor tissue without drusen, Bruch's membrane is recognized as a continuous lamina on the basal side of the RPE with an approximate thickness of 3–4μm (Fig. 1A). In tissues from donors of ARMD, the region below the macula can contain either single or continuous drusen which separates the RPE from the densely staining middle zone of Bruch's membrane (Fig. 1B). In some instances, the drusen accumulation appears to be 3 to 5 times the thickness of the RPE cells.

Bruch's membrane in both normal and ARMD tissues showed strong immunoreactivity with the TIMP-3 antibody as described below. In the normal tissue (Figs. 2A and B) the RPE is closely opposed to Bruch's membrane which appears as a bright red line below

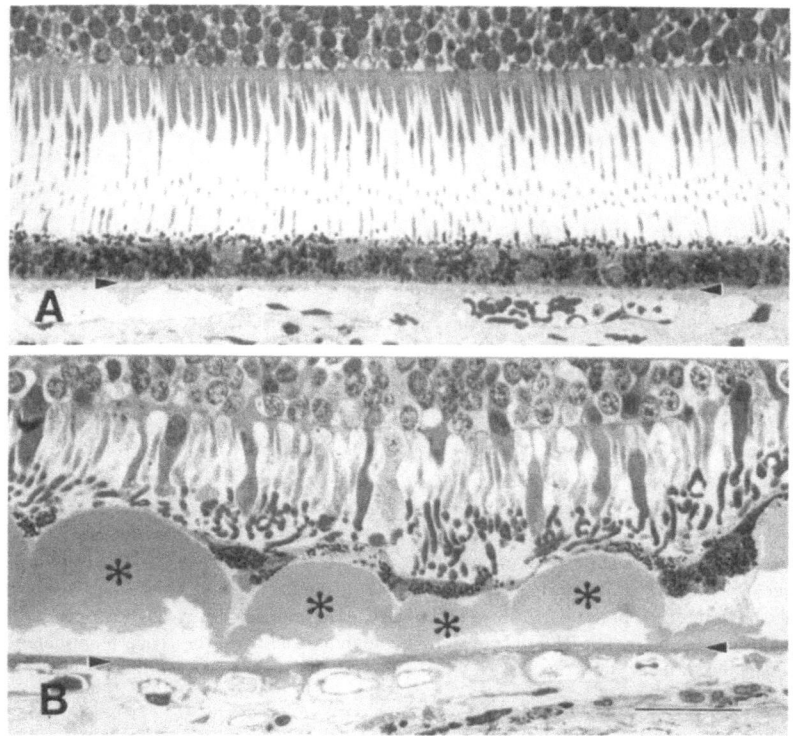

Figure 1. Light photomicrograph of the macula in a normal (A) and an age-related macular degeneration eye (B). Continuous drusen accumulation (asterisks) and thickened Bruch's membrane are present beneath the retinal pigment epithelium (RPE) of the macular degeneration eye (B). Horizontal arrowheads designate the middle of Bruch's membrane in both micrographs.

the RPE (Fig. 2A), indicating strong TIMP-3 immunoreactivity within this extracellular matrix compartment. We also detected an occasional druse below the RPE in the normal tissue which was highly immunoreactive with the TIMP-3 antibody (not shown). In normal control tissues incubated with non-immune mouse IgG instead of the TIMP-3 monoclonal antibody, Bruch's membrane shows no immunoreactivity and appears only as a brightly birefringent line below the RPE (Fig. 2B). In the ARMD tissue (Figs. 2C and D), an extensive accumulation of continuous debris elevates the RPE from Bruch's membrane (designated with asterisks). In addition to TIMP-3 immunoreactivity in a band that represents Bruch's membrane (arrows), the accumulated debris below the RPE is also strongly immunoreactive (Fig. 2C). No immunolabeling was observed in the negative controls in either Bruch's membrane or in the debris in the ARMD tissues (Fig. 2D). In several locations, TIMP-3 immunoreactivity was located beyond Bruch's membrane in finger like extensions that surrounded choroidal vessels (not shown). No TIMP-3 immunoreactivity was detected in the neurosensory retina, at more sclerad regions of the choroid, or in the sclera. The RPE did not show immunoreactivity, although low levels of signal, if present, would have been difficult to detect because of the presence of the brown to orange colored melanin granules within the RPE cytoplasm (Figs. 2A-D). We were unable to demonstrate any immunoreactivity in normal or ARMD tissues which were fixed with aldehydes.

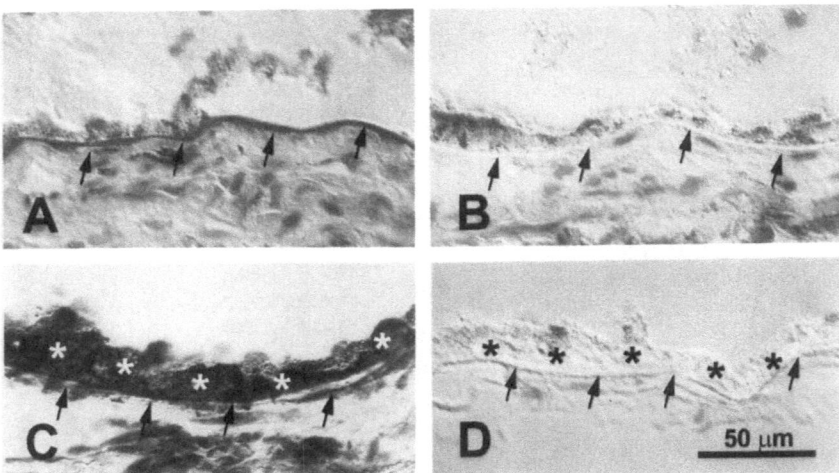

Figure 2. Immunohistochemistry demonstrating the localization of tissue inhibitor of metalloproteinases-3 (TIMP-3) in a normal (A,B) and an age-related macular degeneration eye (C,D). Red immunoreactive products are recognized in Bruch's membrane (arrows) of both the normal (A) and ARMD eye (C). The negative control (B,D) shows distinctly thickened Bruch's membrane (arrows) and no immunolabeling. Continuous drusen accumulation is observed beneath the RPE in the ARMD eye (C,D) designated with asterisks, which is intensely immunoreactive with the TIMP-3 antibody. The RPE did not show any immunoreactivity. For a color representation of this figure, please see the insert facing this page.

DISCUSSION

TIMP-3 immunoreactivity was present in Bruch's membrane and drusen in both the ARMD and normal human eyes. Studies in the mouse retina indicate that TIMP-3 mRNA expression can be detected at high levels in the RPE of normal mouse eyes.[8] While TIMP-3 is known to be expressed in a variety of tissues,[10] the observation of elevated expression in the RPE suggests that much of the TIMP-3 present in Bruch's membrane may be synthesized and delivered to this extracellular site by the RPE.

What is the significance of TIMP-3 localization in Bruch's membrane? Since TIMP-3 has been shown to be anti-angiogenic both *in vitro* and *in vivo*,[11,12] it is likely that one of the roles of TIMP-3 in Bruch's membrane is to provide a molecular barrier to angiogenesis and neovascularization, preventing vascular invasion of the RPE and interphotoreceptor matrix by the choroidal circulation. The presence of TIMP-3 within Bruch's membrane places this MMP inhibitor at a critical location between the choriocapillaris and the RPE. At this site, it could inactivate MMPs released from vascular endothelial cells in response to angiogenic signals and thus prevent new vessel ingrowth from the choroidal circulation.

TIMP-3 may also be involved in regulating the activity of MMPs involved in the degradation of matrix components during normal turnover and renewal of Bruch's membrane. The accumulation of debris (drusen) in Bruch's membrane in normal and ARMD tissues could in part be due to an imbalance in the TIMP-MMP ratio leading to a reduction in MMP activity and an accompanying thickening of Bruch's membrane.

Because of the significance of Bruch's membrane permeability in the trafficking of metabolites between the choroid and RPE, along with the known early alterations in this lamina in ARMD, with the subsequent loss of macular photoreceptors, a full understanding of the regulation of turnover of the components which comprise this complex

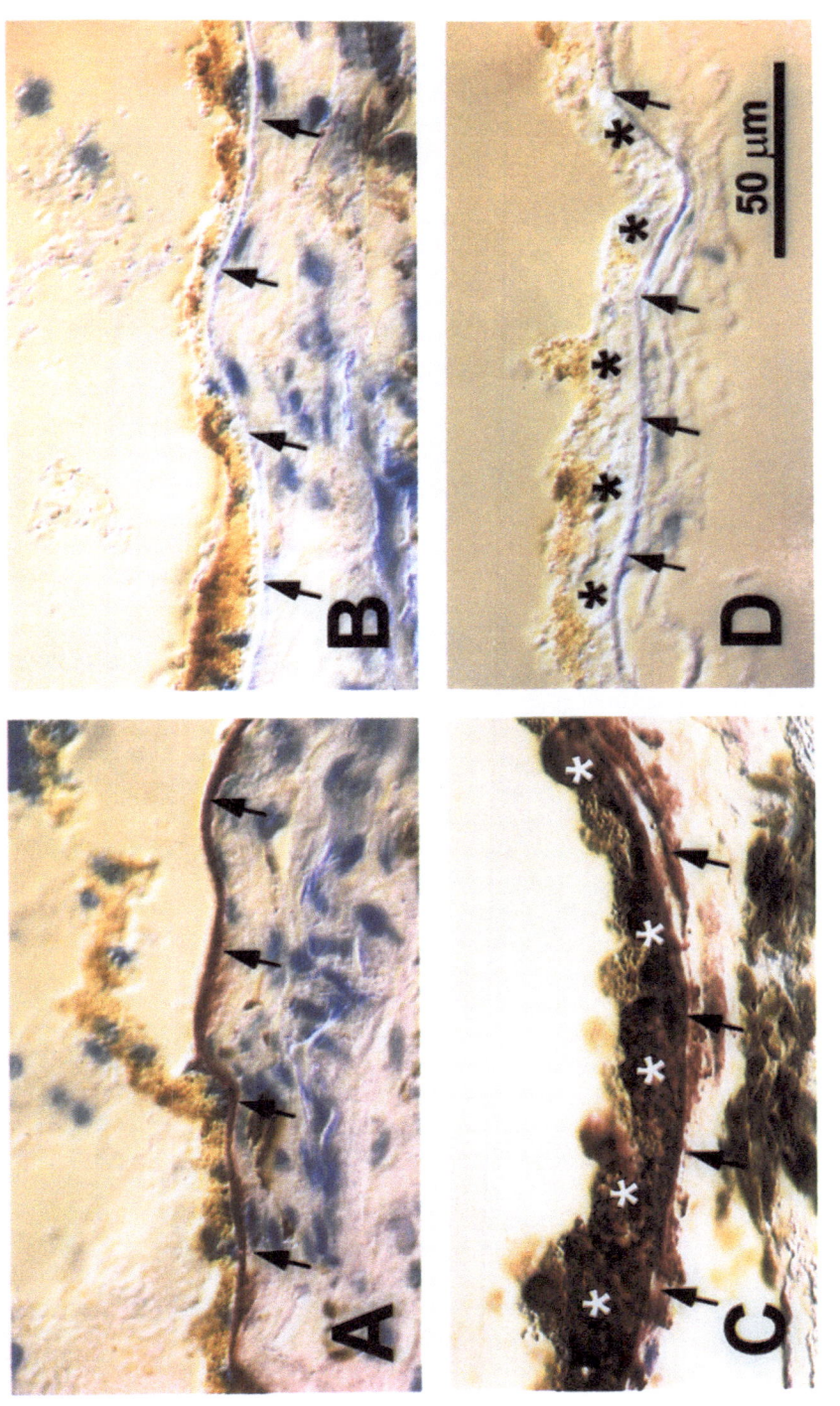

Figure 2. Immunohistochemistry demonstrating the localization of tissue inhibitor of metalloproteinase-3 (TIMP-3) in a normal (A, B) and an age-related macular degeneration (ARMD) eye (C, D). Red immunoreactive products are recognized in Bruch's membrane (arrows) of both the normal (A) and ARMD eye (C). The negative control (B, D) shows distinctly thickened Bruch's membrane (arrows) and no immunolabeling. Continuous drusen accumulation is observed beneath the RPE in the ARMD eye (C, D, designated with asterisks), which is intensely immunoreactive with the TIMP-3 antibody. The RPE did not show any immunoreactivity.

layer is needed. The highly specific localization of TIMP-3 in Bruch's membrane of normal tissues and the apparent increased concentration of this MMP inhibitor in drusen in ARMD tissues, suggests that a complete understanding of the MMPs, TIMPs, as well as other matrix degrading enzymes which function below the RPE, are an important area for further study and will be needed for an understanding of several pathologies which occur at this important ocular interface.

ACKNOWLEDGMENTS

This study was supported by grants from The Foundation Fighting Blindness, Hunt Valley, MD, the Retina Research Foundation, Houston, TX and NIH AR 44436. We thank Dr. K. Iwata of Fuji Chemicals, Ltd. for providing the TIMP-3 antibody. We also thank Ms. Ruth Anna Carlson and Ms. Adeline Leonetti for their generous gift to the macular degeneration research program at The Cleveland Clinic Foundation.

REFERENCES

1. Bressler NM, Bressler SB, Fine SL, 1988, Age-related macular degeneration. *Surv Ophthalmol* **32**:375–397.
2. Birkedal-Hansen H, Moore WGI, Bodden MK, 1993, Matrix metalloproteinases : a review. *Oral Biol Med* **4**:197–250.
3. Klagsbrun M and Folkman J, 1990, *Handbook of Experimental Pharmacology* **95**:549–586.
4. Woessner JF Jr, 1991, Matrix metalloproteinases and their inhibitors in connective tissue remodeling. *FASEB Letters*, **5**:2145–2154.
5. Moses MA and Langer R, 1991, A metalloproteinase inhibitor as an inhibitor of neovascularization. *J Cell Biochem*. **47**:230–235.
6. Hayakawa T, 1994, Tissue inhibitor of metalloproteinases and their cell growth-promoting activity. *Cell Struct Funct* **19**:109–114.
7. Weber BHF, Vogt G, Pruett RC, Stohr H and Felbor U, 1994, Mutations in the tissue inhibitor of metalloproteinases-3 (TIMP-3) in patients with Sorsby's fundus dystrophy. *Nature Genetics* **8**:352–356.
8. Della NG, Campochiaro PA and Zack DJ, 1996, Localization of TIMP-3 mRNA expression to the retinal pigment epithelium. *Invest Ophthalmol Vis Sci*, **37**:1921–1924.
9. Fariss RN, Apte SS, Olsen BR, Iwata K and Milam AH, 1997, Tissue inhibitor of metalloproteinases-3 is a component of Bruch's membrane of the eye. *Am J Path* **150**:323–328.
10. Leco KJ, Khokha R, Pavloff N, Hawkes SP and Edwards DR, 1994, Tissue inhibitor of metalloproteinases-3 (TIMP-3) is an extracellular matrix-associated protein with a distinctive pattern of expression in mouse cells and tissues. *J Biol Chem*. **269**:9352–9360.
11. Apte SS, Olsen BR and Murphy G, 1995 , The gene structure of tissue inhibitor of metalloproteinases (TIMP-3) and its inhibitory activities define the distinct TIMP gene family. *J Biol Chem*. **270**:14313–14318.
12. Anand-Apte B, Pepper MS, Voest E, Montesano R, Olsen BR, Murphy G, Apte SS and Zetter B, 1997, Inhibition of angiogenesis by tissue inhibitor of metalloproteinases-3. *Invest Ophthalmol Vis Sci*. **38**: in press.

3

PHOTORECEPTOR ROSETTES IN AGE-RELATED MACULAR DEGENERATION DONOR TISSUES

Mary E. Rayborn,[1] Kathy M. Myers,[2] and Joe G. Hollyfield[1]

[1]Department of Ophthalmic Research
Cleveland Clinic Foundation
Cleveland, Ohio 44195
[2]Oklahoma Medical Research Foundation
Oklahoma City, Oklahoma 73130

INTRODUCTION

Age-related macular degeneration (ARMD) is the leading cause of blindness in individuals over 50 years of age in industrialized countries.[1] In the USA, ARMD is estimated to be responsible for vision loss in 6 to 10 million individuals. Even though the occurrence is vast, little is known about the etiology of ARMD. ARMD may represent several different diseases but is generally classified as either a "wet" or "dry" form. The dry form, which is most common and represents 90% of the macular degeneration population, is characterized by pigmentary changes, drusen and atrophy of the macular photoreceptors. Vision loss occurs gradually over the course of many years. The wet form is less common but results in a more acute visual loss, which involves newly formed blood vessels below the macula that leak fluid and blood into the macular area.

We have initiated an analysis using end-stage ARMD donor eyes in order to begin a classification, using a variety of histochemical and immunocytochemical probes, of the components which accumulate within Bruch's membrane. In the course of this analysis, we observed an unusual feature in this material which, to our knowledge, has not been reported in ARMD tissue: the presence of photoreceptor rosettes.

MATERIALS AND METHODS

ARMD eyes used in this study were obtained through the donor program of The Foundation Fighting Blindness. Ages of the donors ranged from 72 to 92. Of the eleven eyes examined, the presence of rosettes was found in only one donor. This donor was a 92 year old male and the globes were enucleated at 1.5 hours postmortem. The eyes were held on ice for 5.25 hours, then they were slit at the pars plana and placed in fixative (0.5% glutaraldehyde, 2% paraformaldehyde in 0.1M phosphate buffer, pH 7.2). The tissue was stored in this fixative in the cold for approximately 2 months. At the time of sampling, 5 mm trephined areas of the macular region were removed. The tissue was rinsed in 0.1M phosphate buffer, postfixed in 1% OsO_4 in 0.1M phosphate buffer, dehydrated in a graded series of ethyl alcohols, and embedded in epon/araldite. For light microscopy, 1 micron thick sections were cut and stained with toluidine blue O. Selected areas were then thin sectioned (100 nm thickness) and stained with uranyl acetate and lead citrate and examined with a Zeiss 902 electron microscope.

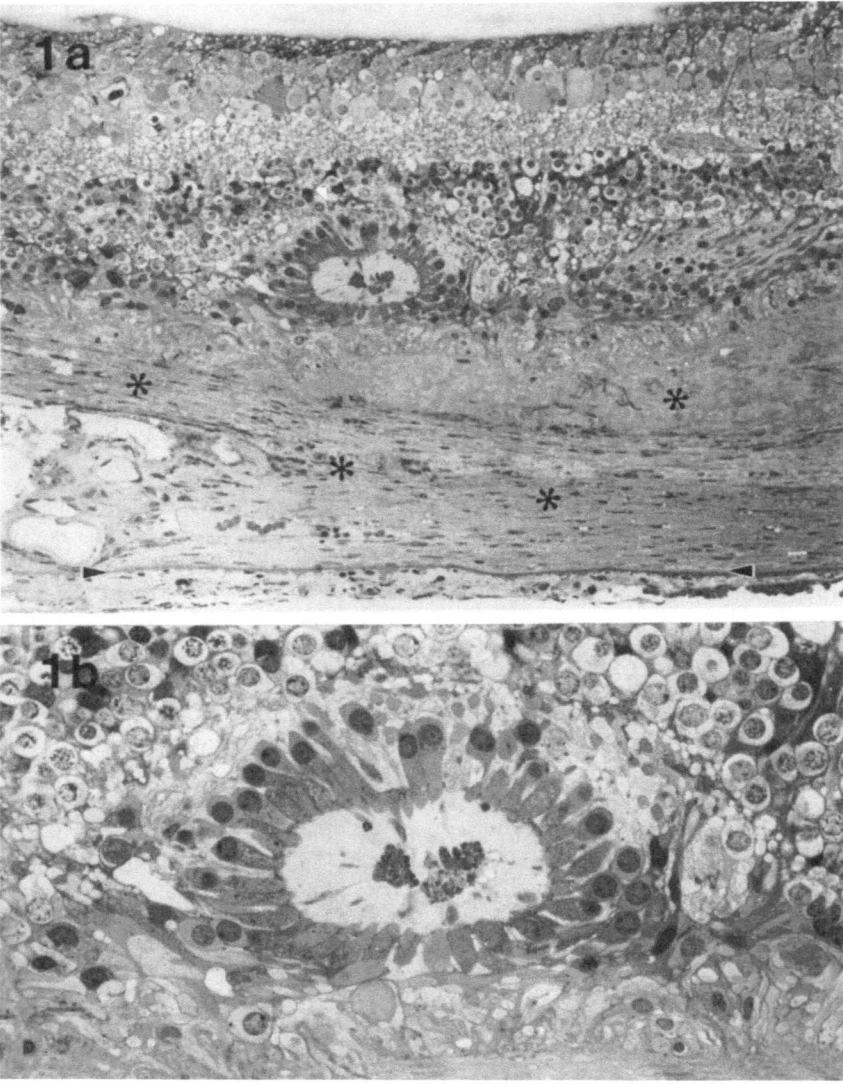

Figure 1. Light microscope images of the retina of a 92 year old male with age-related macular degeneration. Fig. 1a. Section through the macular region from the vitreal surface (above) through the choriocapillaris (below). Arrows in lower border of micrograph designate the remnants of Bruch's membrane. A large fibrotic scar (asterisks), containing blood vessels, lies between Bruch's membrane and the remaining retina. A large rosette, composed of approximately 30 degenerate photoreceptor cells, surrounding an oval lumen, with inner segments of some of the photoreceptors protruding into this compartment, is present and shown at higher magnification in Fig. 1b. Magnification 1a X 400, 1b X 1000.

RESULTS AND DISCUSSION

A number of features of end-stage ARMD tissues have been described. These include the presence of drusen, basal laminar deposits, a general thickening of Bruch's membrane, fibrotic scar tissue below the retina, geographic loss of the RPE and extensive loss of macular photoreceptors. In the eleven eyes we studied, all of the above pathological changes were noted. In addition, in one of the donors we observed arrays of photoreceptor rosettes. To the best of our knowledge, this is the first report of rosette formation in eyes from ARMD affected tissue.

With light microscopy the rosettes were spherical or oval in shape (Figs.1a and 1b). Surrounding the rosettes, the surviving retina was unstructured with no discernible retinal lamina apparent. In the macular region, the pigment epithelium was missing. A wide band of fibrotic tissue containing blood vessels was located between the retina and the choroid. In most areas a choriocapillaris was not evident, but large vessels were present in the outer choroid.

The walls of the rosettes were comprised of elongate, highly degenerate photoreceptors cells surrounding a luminal compartment (Fig. 1b). The photoreceptors were linked to Müller cell processes by *zonula adherentes* arranged in a single plane (Fig. 2). Villous processes of Muller cells were also found projecting into the lumen of the rosettes. The lu-

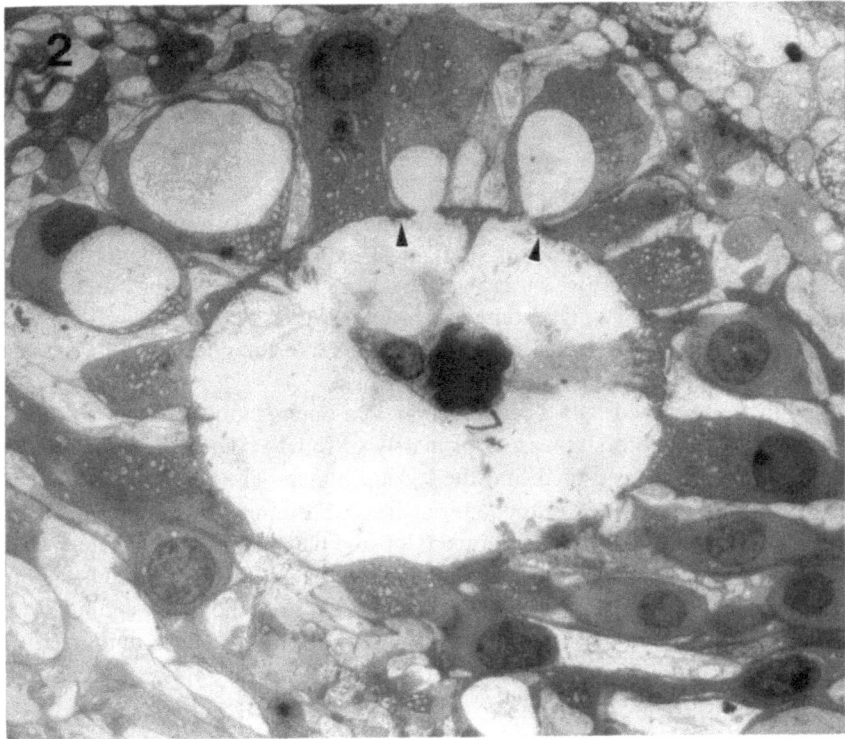

Figure 2. Electron micrograph of a rosette from the same donor described in Fig. 1, showing the highly degenerate photoreceptors surrounding the lumen. Arrows point to a zone of junctional complexes. Some of the cells bordering the lumen contain large vacuoles which appear to be continuous with the rosette lumen. Magnification X 2500.

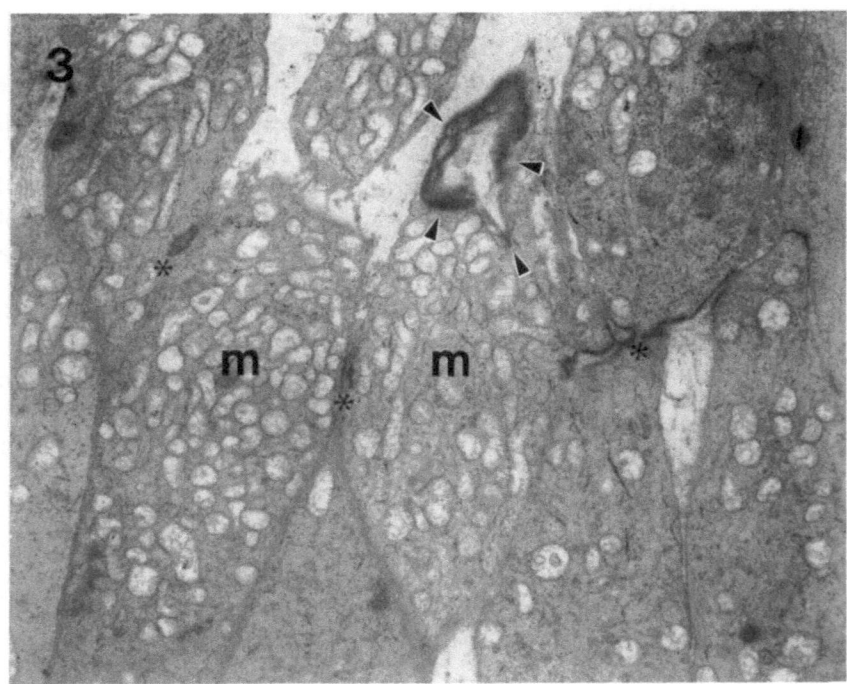

Figure 3. Electron micrograph of photoreceptors at their interface with the rosette lumen. Large clusters of mito-chondria are evident (m) and the apical borders of the cells are associated with each other by junctional complexes (asterisks). A cilium and disorganized outer segment membranes are denoted with the arrowheads. Magnification X10,000.

men appeared electron lucent and contained fine reticular material. The photoreceptor cells making up the rosette contained short, disorganized outer segments that projected into the lumen of the rosette (Fig. 3). In many instances, these photoreceptors also contrib-uted to horizontal arrays of long axons which appeared to be remnants of Henle's fiber layer. This latter observation suggests that at least some of the photoreceptors surviving in these rosettes are cones.

Although we observed only a few disorganized outer segments, as mentioned above, the degenerate photoreceptors consisted primarily of inner segment components, some of which were bulbous and protruded into the rosette lumen. The inner segments contained rounded to elongate clusters of mitochondria, free ribosomes, a fine granular cytoplasm and few other cellular organelles. Nuclei were located basally in the cell. There were some electron opaque residual bodies present in the inner segment. Occasionally basal bodies of cilia, either in cross or longitudunal section, were observed in the inner segments.

Rosettes are a common feature of retinoblastoma tumors. These tumors can present varying degrees of differentation with well-formed rosettes standing out against a back-ground of undifferentiated neuroblastic cells.[2,3] In well differentiated tumors, the cells comprising the rosettes show some photoreceptor features. Rosettes have also been de-scribed in primate retinas that have been exposed to ionizing radiation during fetal life.[4] Recent retinal transplantation studies also report that rosettes frequently form when fetal retinal tissue is transplanted to the adult retina.[5] Rosettes can be generated experimentally when embryonic neural tissue is dissociated to single cells and allowed to reaggregate in suspension culture. While the causes of rosette formation are not fully understood, these structures are an apparent attempt of polarized cells to form or remain associated with an apical surface.

The only other degenerative retinal disease in which rosettes have been observed is from a donor with autosomal dominant retinitis pigmentosa.[6] We now add ARMD to the list of retinal pathologies in which photoreceptor rosettes have been identified.

ACKNOWLEDGMENTS

We thank The Foundation Fighting Blindness, Hunt Valley, MD for their support of these studies.

REFERENCES

1. Bressler, N.M., Bressler, S.B., and Fine, S.L., 1988, Age-related macular degeneration, *Surv. Ophthalmol.* 32:375–397.
2. Reese. A.B., 1951, *Tumors of the Eye,* Paul B. Hoeber, Inc., (medical book dept. of Harper & Brothers) New York.
3. McLean, I.W., 1996, Retinoblastomas, retinocytomas, and pseudoretinoblastomas, in: *Ophthalmic Pathology,* Volume 2 (W. Spencer, ed.), pp.1332–1438, W.B.Saunders Co., Philadelphia.
4. Rugh, R., Skaredoff, L., 1969, X-rays and the monkey fetal retina. *Invest. Ophthalmol.Vis. Sci.* 8:31–40.
5. Bergström, A., Ehinger, B., Wilke. K., Zucker, C.L., Adolph, A.R., Aramant, R., and Seiler, M., 1992, Transplantation of embryonic retina to the subretinal space in rabbits. *Exp. Eye Res* . 55:29–37.
6. Milam, A.H. and Jacobson, S.G., 1990, Photoreceptor rosettes with blue cone opsin immunoreactivity in retinitis pigmentosa, *Ophthalmology* 97:1620–1631.

4

CENTRAL RETINAL SENSITIVITY REPEATED MEASUREMENTS AS LONG TERM FOLLOW-UP IN RETINITIS PIGMENTOSA

Enzo M. Vingolo, Andrea Perdicchi, Renato Forte, Patrizia Del Beato, Luigi Pannarale, and Roberto Grenga

Inherited Retinal Disorders Center
Chair of Ophthalmology
University "La Sapienza" of Rome

INTRODUCTION

In Retinitis Pigmentosa the progressive photoreceptor degeneration determines severe functional loss and visual impairment and consequently retinal sensitivity damage.

The main complications involved in central vision disturbances are caused by macular atrophy or edema, vitreous changes and cataract. As a result, patients show progressive loss of discrimination and depressed foveal sensitivity acuity.

Many histological and physiological studies show that even macular photoreceptors are damaged early in RP and degenerate progressively. Such foveal cone function abnormalities are supported by histopathological studies that demonstrated reduced packing density of foveal cone photoreceptors in RP patients with normal visual acuity.

Moreover, foveal cone function may also be abnormal in early stages of the disease, despite normal visual acuity.

A complete ERG test may be more sensitive than Snellen acuity in detecting photoreceptor function abnormalities, but ERG usually is severely depressed also in early stages of the disease, resulting in a not very effective way to determine progression of the disease.

In this view several studies underline the importance of measuring the visual field in RP patients as previously reported by Sung and Sloan (1967), Tanino (1977), Pearlman (1979), Berson (1985), Massof (1990).

Measurement of the visual field in the natural history of RP, is dramatically important. In this view Sung and Sloan (1967) on 25 RP patients noticed that the visual function loss rate varies both among and within patients, with periods of rapid decay with intervals of time with no significance changes. Pearlman in 1979 performed linear regressions on measured visual field radial amplitude related to time, then the reported age of night blindness onset, concluding that there were no statistically significant differences in the rate of visual field loss among the genetic subtypes of RP. Tanino (1977) adfirmed that

Degenerative Retinal Diseases, edited by LaVail *et al.*
Plenum Press, New York, 1997

dominant RP patients, as a group, progressed the slowest, X-linked RP patients as a group progressed the most rapidly, and recessive RP patients, as a group progressed at an intermediate rate. Berson (1985) concluded that statistically significant losses in visual area could be resolved over a period of 2 yrs, but no less. The average rate of loss in visual field area for all RP patients was 4.6% per year. For the initial test, 23% of the RP patients had normal visual field areas. No attempt was made to compare visual field loss rates across genetic subtypes.

Massof (1990) computed an analysis of the dynamics of prospectively-measured visual field loss in 172 repeatedly tested typical RP patients, concluding that the first order dynamics of the progressive retinal degeneration is the same in all types of RP and that the visual loss in RP, regardless of subtype, progresses at a very predictable rate, following an exponential decay function. Moreover in our previous study, correlating visual field area to the ERG amplitude (Vingolo 1992, Iannaccone 1995), we noted a direct correlation between visual field area and ERG amplitude, demonstrating the possibility to evaluate the evolution of retinal damage with visual field aarea measurement.

The introduction of computerized perimetry in clinical practice and its diffusion seems to offer more opportunities in order to study visual field loss. Moreover Software analysis programs for transmission, storage and processing of computerized visual field.

Recently these programs performing Linear regression analysis of MD were utilized to detect and analyze the visual field decay in different forms of open angle chronic glaucoma (Weber 1993) and retrospective analysis of visual field (EXPERIMETRY, PERIDATA and DELTA programs) has been applied on different group of patients affected by retinal involvement in ocular and sistemic diseases.

In a preevious study we reported data on a group of patients with clear signs of retinitis pigmentosa evaluating retinal sensibility of the central 30° and correlated its statistical significance to clinical picture.

METHODS

The study included 14 patients (28 eyes) with RP (mean age 32.33±11.49 yr), and 14 (28 eyes) normal subjects (mean age 31.61±9.06 yr) with age and sex matched to RP patients.

Exclusion criteria were Presence of posterior subcapsular cataract with a grading > 2 (FIARP classification) presence of Vitreoretinal membrane, presence of a cystoid macular oedema evident ophthalmoscopically.

In 53.6% of the patients tested, a genetic transmission of the disease had been discovered. In the other 46.4% a typical genetic transmission could not be proved.

The small differing rates, of the genetic subtypes did not offer the possibility of studying the relationship between visual field decay and genetic transmission. It is however known that in the RP clear stage the visual field has a common exponential decay for all the types of RP. Instead, the genetic transmission seems to have a role in the different distribution of the "critical age" (i.e. onset age) (Massof 1990).

All participants received a clinical evaluation including visual acuity, direct and indirect ophthalmoscopy and slit-lamp biomicroscopy. Best-corrected visual acuity was 20/30 or better.

The patients were submitted to an initial training visual field test (screening strategy), in order to accustom them to this tecnique and therefore reduce the learning effect.

All subjects had been tested for three years once every six months with computerized Humphrey Visual Field Analyzer, program 10–2. Resulting data were processed with

Experimetry Visual Field serial analysis considering main retinal function parameters: Mean Sensitivity, Mean Loss, Short-term Fluctuation and Patterned Standard Deviation. Reliable fields remaining in the study had < 15% false positive answers and < 20% false negative answers and < 30% fixation losses. Visual field data was stored electronically and analyzed using the program Experimetry.

Refractive errors were corrected for near vision during retinal sensitivity testing, and each eye was tested individually, the right eye was always tested first. Serial analysis and evolution series were obtained by Experimetry software, Main parameters are indicated as means ± SD. Statistical significance was assessed with Student's T test.

MS = Mean Sensitivity: Evalues the arithmetic mean of the difference between measured values and normal values at each single different tested location. Is directly correlated to the retinal function.

SF = Short Term Fluctuation: Represents the average of local scatter observed when the same threshold is measured indipendently during a single examination of the visual field. Increase of SF is considered as early sign of change (Flammer 1985 , Jenni 1983)

PSD = Petterned Standard Deviation: a local non-uniformity of visual field defect. An increase of PSD could be due to a real deviation of sensibility values or to a scatter

CPSD = Corrected Petterned Standard Deviation: Is PSD corrected for SF (Flammer 1985). CPSD helps to separate real deviations from deviations to scatter (Augustiny and Flammer 1985).

DATA TREND option offers the linear regression analysis of all local values and global indices values vs time (MS, PSD, CPSD, SF).

RESULTS

There was a substantial difference between performances of retinitis pigmentosa patients when compared with normal subjects: Regression Analysis of Mean Sensitivity during the follow-up showed a moderate, but not significative, tendency to worsening.

Highly statistically significative data were noted for central retinal sensitivity regarding Mean sensitivity value, between two groups ($P < 0.0001$).

Moreover in the RP group we noted a significantly higher Short-term fluctuation ($p > 0.053$) even in presence of high Best Corrected Visual Acuity.

Finally there was no significant difference between the results for right and left eyes of the 14 subjects ($p > 0.05$).

In RP patients, the visual field was relatively stable or deteriorated slowly. Indeed in three years of the study, MS deteriorated statistically only in 26% of the eyes studied. This with a percentage of decay of MS/Time was of 14.8% in three years, resulting in a percentage of 4.9%/year. (Berson 1985: 4.6%/year).

This decay of the MS in the analysis of the visual field is does not depend on best corrected visual acuity that in our sample remained the same (decay value not significative). This in our opinion seems as a confirmation of the unpredictibility declining of visual acuity in RP as stated by Marmor 1980.

The relatively low values of PSD and CPSD can be considered signes of diffuse true visual field damage. In this view progressive statistical decrease of PSD and CPSD shows that a major uniformity of defect is reached. In fact a strong relationship exists between statistical PSD and CPSD that evidentiates a true deterioration of the visual field. Moreover CPSD helps to separate real deviations from deviations to scatter (Augustiny and Flammer 1985).

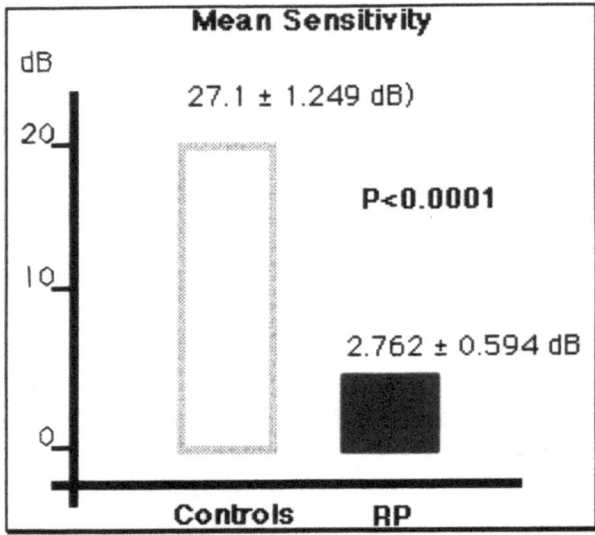

The presence of PSD and CPSD significant damage in all 14 patients associated with a statistically significant MS damage, leads us to believe that in this group of RP patients the deterioration of the visual field in RP is a result of the deepening of the defect at the first examination and not due to any deterioration in the relatively healty visual field area.

CONCLUSION

In our study the Visual Field analysis in Retinitis Pigmentosa patients demonstrated that retinal sensitivity remains moderately stable during the follow up time.

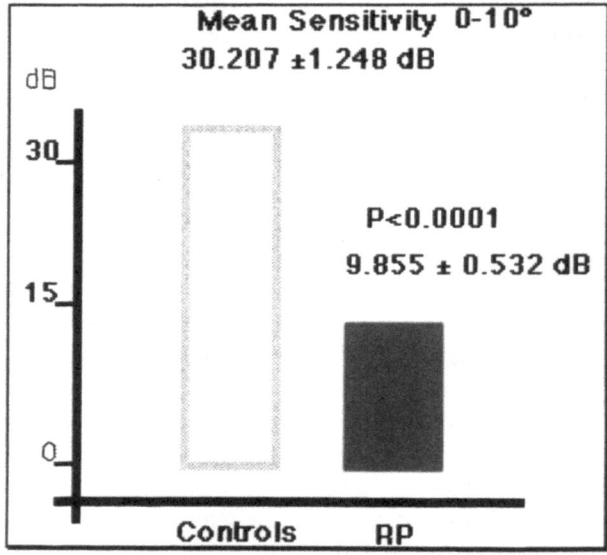

Figure 2. Central retina (0°–10°) mean sensitivity in RP and control group.

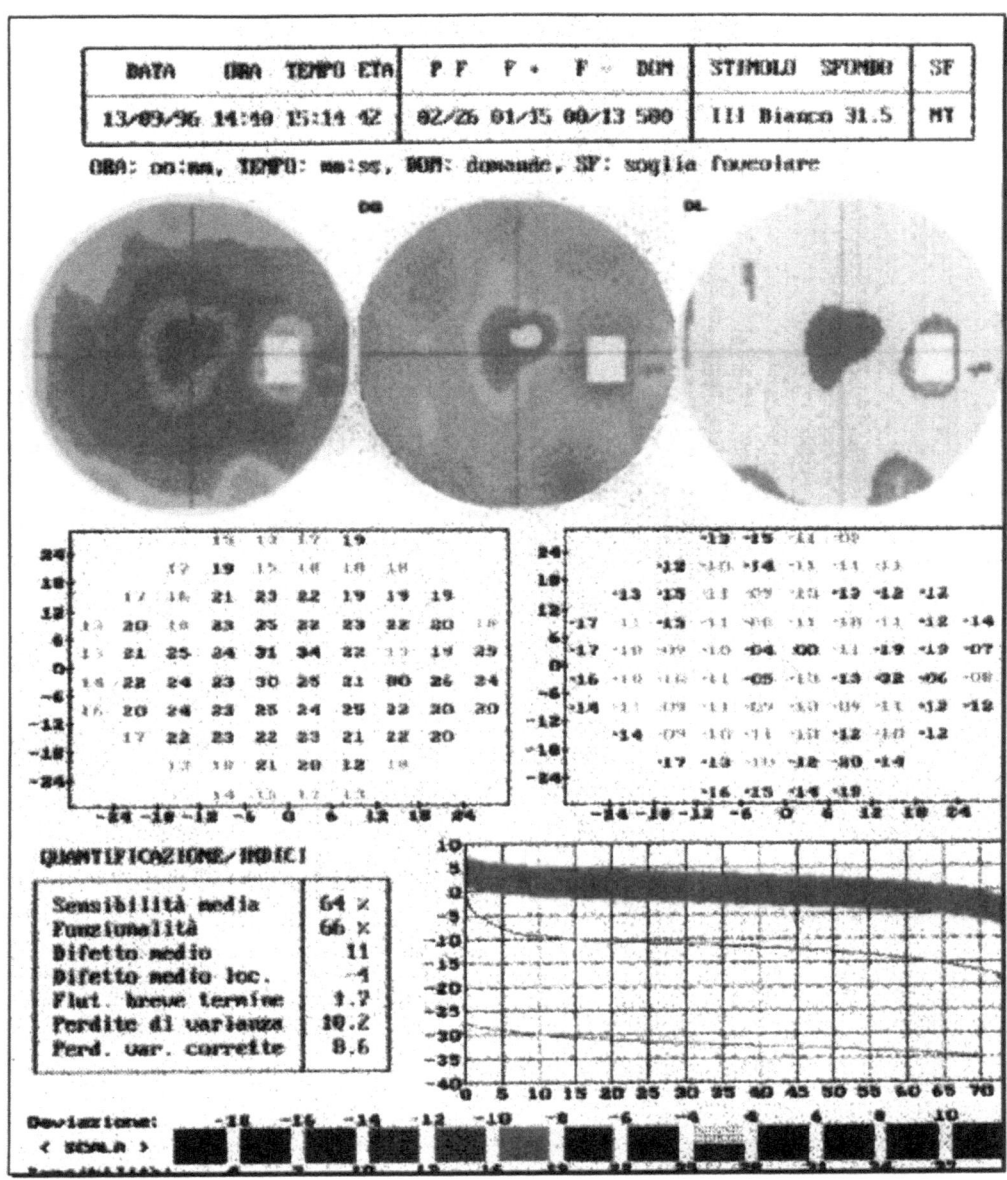

Figure 3. Visual field analysis.

There was a significant decrease of retinal sensitivity in less than 30% of cases respect normal values. Thus even if the central retinal area is apparently spared, as indicated from an ophthalmological and FA retinal examination the function might be clearly reduced.

Moreover, by the means of computerized perimetry and regression analysis interpretation consent us to evaluate retinal function; and as indicated by not significative changes during the follow up, we can be able to monitor accurately the evolution of degenerative process.

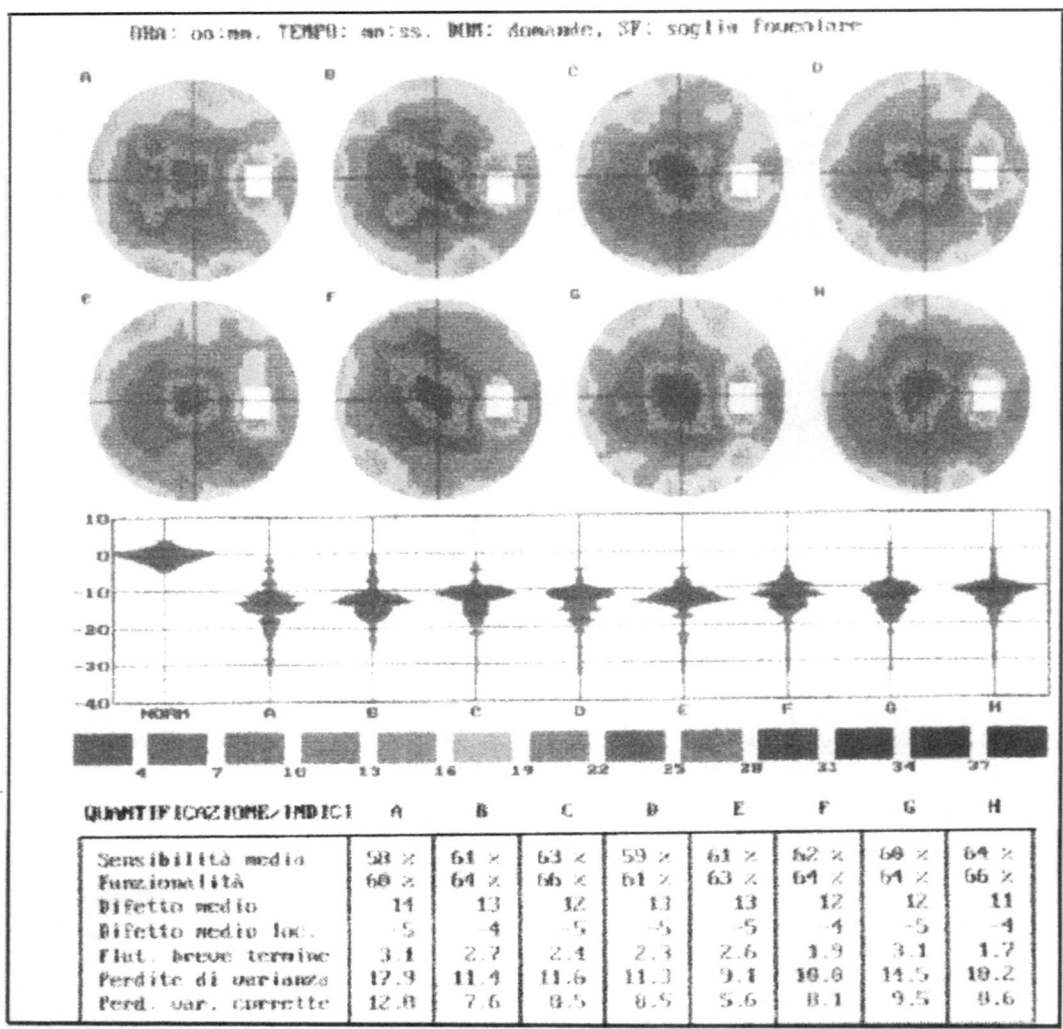

Figure 4. Visual field evolution.

However evaluating RP patients with serial analysis of computerized visual field pointed out that Mean Sensitivity damage with concomitant presence of Controlled Pattern Standard Deviation and Pattern Standard Deviation.

In our interpretation this could be due to a deepening of the defect presented by the patient at first examination, more than to a deterioration of others relatively functioning areas.

REFERENCES

Fishman GA: Retinitis Pigmentosa. Visual loss. Arch Ophthalmon 95: 1185–1188 (1978).

Iannaccone A., Rispoli E., Vingolo E.M., Onori P., Steindl K., Rispoli D., Pannarale M.R: Correlations between Goldmann perimetry and maximal electroretinogram response in retinitis pigmentosa. Doc. Ophth. 90 (2): 129–142 (1995).

Madreperla SA, Palmer RW, Massof RW, Flikelstein D: Visual acuity loss in Retinitis Pigmentosa. Arch Ophthalmol 108: 358–361 (1990).

Massof RW, Dagnelie G, Benzschawel T, Palmer RW, Flikelstein D: First order of visual field loss in Retinitis Pigmentosa. Clin Vis Sci 5: 1–26 (1990).

Marmor MF: Visual acuity and field loss in retinitis pigmentosa Arch Ophthalmon 109: 13 (1991).

Massof RW, Flikelstein D, Starr SJ, Kenyon KR, Fleischman JA, Maumenee IR: Bilateral symmetry of vision disorders in typical Retinitis Pigmentosa. British Journal of Ophthalmology 63: 90–96 (1979).

Marmor MF: Visual loss in Retinitis Pigmentosa: American Jornal of Ophthalmology 89: 692–698 (1980).

Pannarale M.R., Vingolo E.M., Rispoli E., Forte R., Iannaccone A., Pannarale L., Del Porto G., Steindl K.: Cataract Complicated by Retinitis Pigmentosa. in Degenerative Retinopathies: Advances in clinical and genetic research. CRC Press Ed. Boca Raton Ann Arbor Boston pp.145–149 (1993).

Pearlman JT, Axelrod RN: frequency of central visual impairment in Retinitis Pigmentosa. Arch Ophthalmol. 95: 894 (1977).

Perdicchi A., Pannarale L., Amodeo S., Rispoli M., Vingolo E.M.: Computerized perimetry in retinitis pigmentosa. ARVO Meeting 1994. Invest. Ophthalmol. Vis. Scie. 35 (4): 4185. (1994).

Vingolo E.M., Iannaccone A., Onori P., Steindl K., Rispoli E.: Correlations between perimetric and electroretinographic measurements in retinitis pigmentosa. ARVO Meeting 1992. Invest. Ophthalmol. Vis. Scie. 33 (4): 745 (1992).

5

COMPLETE AND INCOMPLETE TYPE CONGENITAL STATIONARY NIGHT BLINDNESS (CSNB) AS A MODEL OF "OFF-RETINA" AND "ON-RETINA"

Yozo Miyake, Masayuki Horiguchi, Satoshi Suzuki, Mineo Kondo, and Atsutoshi Tanikawa

Department of Ophthalmology
Nagoya University School of Medicine
Nagoya, Japan

INTRODUCTION

The Schubert-Bornschein type congenital stationary night blindness (CSNB) shows an essentially normal fundus and a negatively shaped, mixed rod-cone electroretinogram (ERG), in which the amplitude of a-wave is normal and larger than that of b-wave (1). We previously reported that CSNB cases with negative ERG can be divided into two types, complete or incomplete, based on the difference in rod visual functions, cone mediated ERGs, ERG oscillatory potentials, degree of refractive errors, and family survey (2). It has been concluded that complete and incomplete types are different clinical entities, having demonstrated several explicit differences between the two CSNB types (3,4).

Although we were partially uncertain as to what these differences indicate in response to the pathogenesis, recent elucidation of the retinal neuro-synapses and their correlation to ERG components has opened a new door for interpreting the pathogenesis of CSNB. In this article, we discuss the rod and cone visual pathways in complete and incomplete CSNBs with a special emphasis on the function of ON- and OFF-bipolar cells.

REVIEW OF ERGs

Full-Field Rod and Cone ERGs

Conventional ERGs were recorded with full-field stimuli after 30 minutes of dark adaptation. The rod (scotopic) ERG was recorded with a blue light at an intensity of $5.2 \times 10^{-3} \text{cd/m}^2 \cdot \text{sec}$. The rod-cone mixed single flash (bright white) ERG was recorded with a white stimulus at an intensity of $44.2 \text{ cd/m}^2 \cdot \text{sec}$. The cone ERG and the 30-Hz

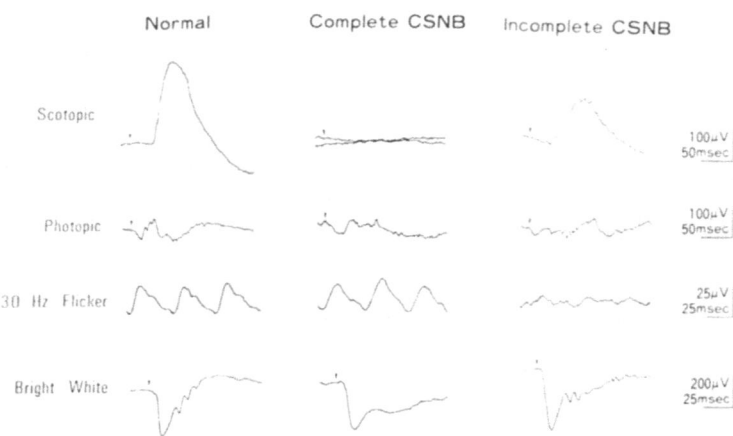

Figure 1. Conventional full-field ERGs in a normal subject (left), a complete (middle) and an incomplete (right) CSNB patient.

flicker ERG were recorded with a white stimulus intensity of 4 cd/m^2 · sec and 0.9 cd/m^2 · sec, respectively.

Figure 1 shows typical examples of full-field ERGs in a normal subject, a complete and incomplete CSNB patient. A complete CSNB patient disclosed nonrecordable rod (scotopic) ERG with relatively restored cone (photopic and 30-Hz flicker) ERGs, whereas an incomplete CSNB patient demonstrated subnormal rod ERG with extremely deteriorated cone ERGs. The rod-cone mixed single flash (bright white) ERG showed a negative configuration in both types; however the oscillatory potentials (OPs) were clearly recordable only in the incomplete CSNB patient. These differences were also observed in our larger series containing both types of patients (2).

Rod Visual Pathway

In order to evaluate the ERG function of the rod visual pathway, the ERG intensity series were recorded after one hour of dark adaptation (5). The light source used was a 10-µsec xenon strobe, mounted on top of a Ganzfeld integrating sphere. The stimulus intensity was controlled with neutral density filters, and the maximum intensity of the series was 44.2 cd/m^2 · sec; 16 to 32 responses were averaged. Figures 2 and 3 show ERG intensity series, elicited by relatively dim (Fig. 2) and intense (Fig. 3) stimuli, of a normal subject, a complete and an incomplete CSNB patient. The calibration of amplitude and time scale in the two figures differ.

In the normal subject, the cornea-negative scotopic threshold response (STR) was recordable at −8.2 log units, and the maximum STR amplitude was 24µV (Fig. 2). The STR peak time near the threshold was approximately 162 msec, which shortened as the stimulus intensity increased. At an intensity of −5.8 log units, the b-wave became clearly visible for the first time. Figure 3 shows that the b-wave had saturated essentially at −3.4 log units and at greater intensities of −0.8 log units, the OPs became clearly visible on the ascending slope of the b-wave. The a-wave was observed at −1.7 log units and progressively increased in amplitude. In the complete CSNB patient, neither STR nor the b-wave was recorded when the stimulus intensity was low (Fig. 2). At a moderate stimulus inten-

Dim Stimulus Range

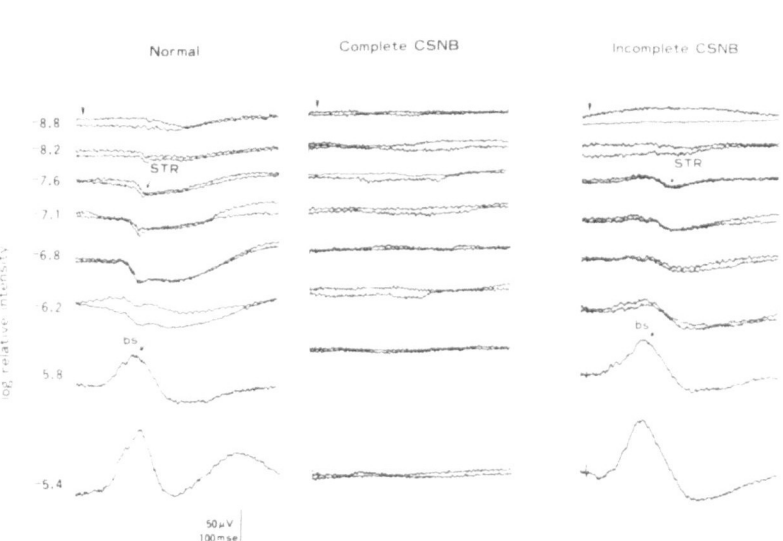

Figure 2. ERG intensity series elicited by relatively dim stimuli in a normal subject (left), a complete (middle) and an incomplete (right) CSNB patient.

Relatively Intense Stimulus Range

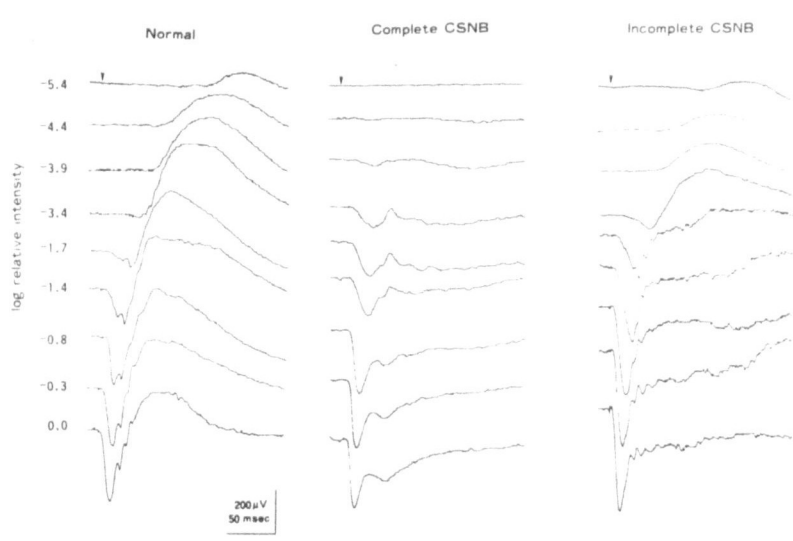

Figure 3. ERG intensity series elicited by relatively intense stimuli in a normal subject (left), a complete (middle) and an incomplete (right) CSNB patient. The calibration of amplitude and time scale in Figures 2 and 3 differ.

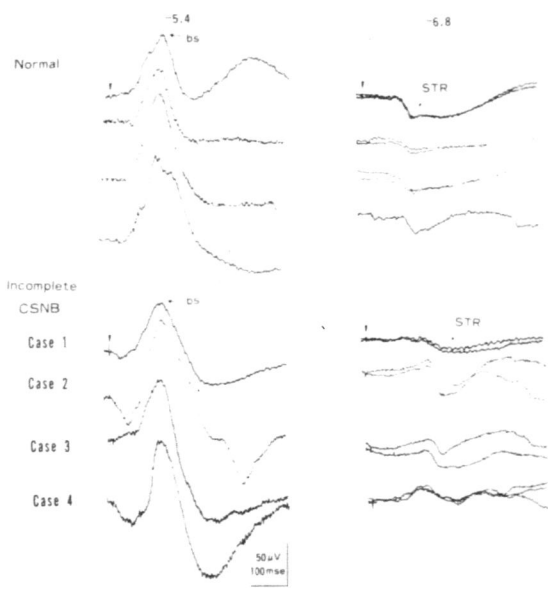

Figure 4. Near-threshold b-wave (left) and STR (right) of four normal subjects and four incomplete CSNB patients.

sity of −4.4 log units (Fig. 3), both a- and b-waves began to appear, with the former presenting normal and increasing amplitude. However, the b-wave had saturated quickly, resulting in a negative configuration when the stimulus intensity was relatively strong. OPs were undetectable. In the incomplete CSNB patient, the STR started to appear at −7.6 log units, showing a slightly higher threshold than that of normal subject (Fig. 2). The STR amplitude at −7.6 log units was nearly the same as that of the normal subject; however, the peak time was approximately 80 msec longer than normal. The b-wave began to appear at −5.8 log units as in the normal subject, with normal amplitude and peak time. At greater intensities, the b-wave amplitude became lower than normal, saturating at −3.4 log units, whereas the a-wave amplitude continued to increase progressively, resulting in a negative configuration. The OPs were clearly recordable.

Figure 4 shows the STR (−6.8 log units) and near-threshold b-wave (bs, −5.4 log units) of four normal subjects and four incomplete CSNB patients. In the latter patients, the STR amplitude was normal but the negative STR peak time showed enormous delay, the amplitude and peak time of bs were within the normal range (5).

Cone Visual Pathway

As mentioned above (2,4) although the cone (photopic and 30-Hz flicker) ERG is relatively good in complete CSNB, it is very deteriorated in incomplete CSNB (Fig. 1). Instead of a brief flash stimulus (photopic ERG) or 30-Hz flicker stimuli, which were used in conventional full-field ERG, we attempted to use rectangular monochromatic repetitive stimuli under photopic condition (photopic long-flash ERG) to isolate the ON- and OFF-responses in the cone visual pathway (6). The colored-light stimuli, produced by a 560 nm monochromatic filter using a 500 w xenon arc, was passed in front of the eye and the white background illumination was led into the fiber optics mounted in a Ganzfeld system. The intensities of colored stimulus and white background illumination were 870 mW/m^2

Photopic Long-Flash LM-Cone ERG

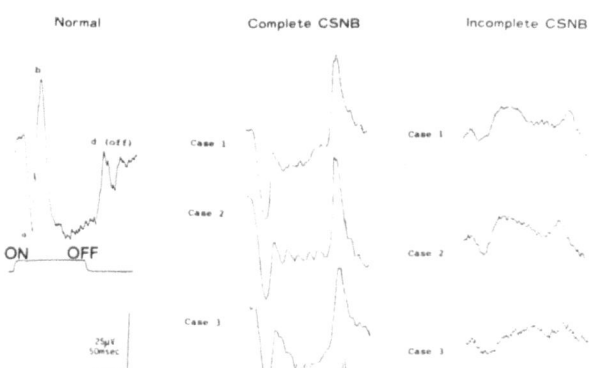

Figure 5. Photopic long-flash ERGs in a normal subject (left), four complete CSNB patients (middle), and four incomplete CSNB patients (right).

and 130 mW/m², respectively. The stimulus frequency of the rectangular stimuli was 4-Hz with 125 msec of light on and 125 msec light off; 32 responses were averaged by a signal processor.

Figure 5 shows photopic long-flash ERGs in a normal subject, four complete CSNB patients, and four incomplete CSNB patients. The complete CSNB patients showed normal a-wave, small b-wave, and large OFF-(d) wave. On the contrary, the incomplete CSNB patients showed small a-wave, relatively large b-wave, and small OFF-(d) wave.

FURTHER STUDIES OF CONE VISUAL PATHWAY

Foveal Cone Densitometry

In order to study the kinetics of foveal cone pigments, we performed foveal cone densitometry as follows (7). The light from a 500 w xenon arc was divided into reference (803 nm, 10 td), measuring (562 nm, 950 td) bleaching (white, 10^6 td) beams, and were projected through an optical fiber leading to a modified fundus camera. After the bleaching light was projected onto the ocular fundus (central 3.5°) for 5 minutes, the reflected light (central 1°) was measured by a photomultiplier, and the regeneration time and the two-way density (8) were abstracted from the records. The two-way density, which is the decadic logarithm of the measuring-reference beam ratio under bleached conditions, eliminated the artifacts caused by blinking.

Figure 6 shows the foveal cone densitometry in a normal subject, a complete CSNB patient, and an incomplete CSNB patient. The mean±SD of the two way density (log) and the time constant (sec) in 45 normal subjects were 0.34±0.09 and 139±79.3 respec-

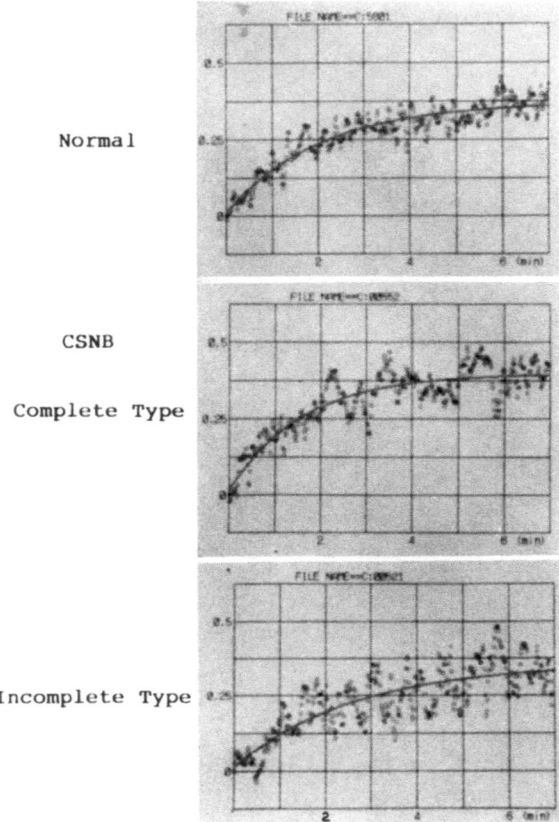

Normal

CSNB

Complete Type

Incomplete Type

Figure 6. Foveal cone densitometry in a normal subject (top), a complete (middle) and an incomplete (bottom) CSNB patient.

tively (7). The foveal cone densitometry of a complete and incomplete CSNB patient was within the normal range in terms of two-way density and time constant.

CSNB and ON-, OFF-Bipolar Cells

Figure 7 is a simplified schema showing the rod and cone visual pathways in mammalian retina (9). The photoreceptors transmit visual information to the second-order neurons, the retinal bipolar cells. The rods contact only the depolarizing (ON-rod) bipolar cells through a sign-inverting (−) synapse (ON-synapse). The cones contact both depolarizing (ON-cone) and hyperpolarizing (OFF-cone) bipolar cells through a sign-inverting (−) (ON-synapse) and sign-preserving (+) synapse (OFF-synapse), respectively. The rod bipolar cells do not synapse directly onto the ganglion cells, but a distinctive type of narrow-field amacrine cell or the so-called AII-amacrine cells are interposed. AII-amacrine cells produce glycerinergic chemical synapses with OFF-cone bipolar cells and OFF-ganglion cells and contact the ON-cone bipolar cells via large gap junction (9).

Each of the two different synapses to bipolar cells is selectively sensitive to different glutamate analogs (10,11). APB (2-amino-4-phosphonobutyric acid) can block the sign-inverting synapse (ON-synapse), and KYN (kynurenic acid) the sign-preserving synapse (ON-synapse), selectively. Sieving (12) studied the changes of photopic long-flash ERG by treating monkey eyes with APB and KYN. Figures 8 and 9 show Sieving's findings

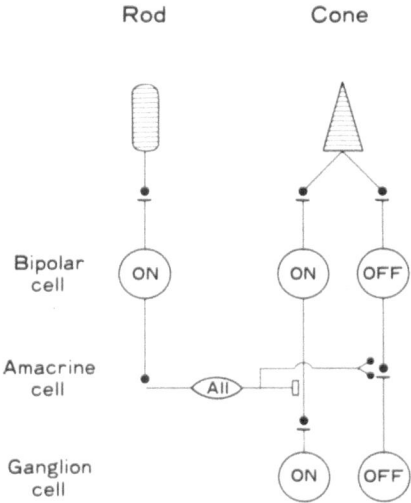

Figure 7. Schema showing the rod and cone visual pathways in mammalian retina. See details in text.

and the ERG results of our complete and incomplete CSNB patients. The ERG of both monkey and human controls was quite similar in shape, showing a-wave, b-wave and OFF-(d) wave. The ERG of monkey eye treated by APB to block the ON-synapse was similar to that of complete CSNB patient (Fig. 8), and the ERG of monkey eye treated by KYN to block the OFF-synapse similar to that of incomplete CSNB patient (Fig. 9). Namely, the APB-treated eye and the complete CSNB showed large a-wave, small b-

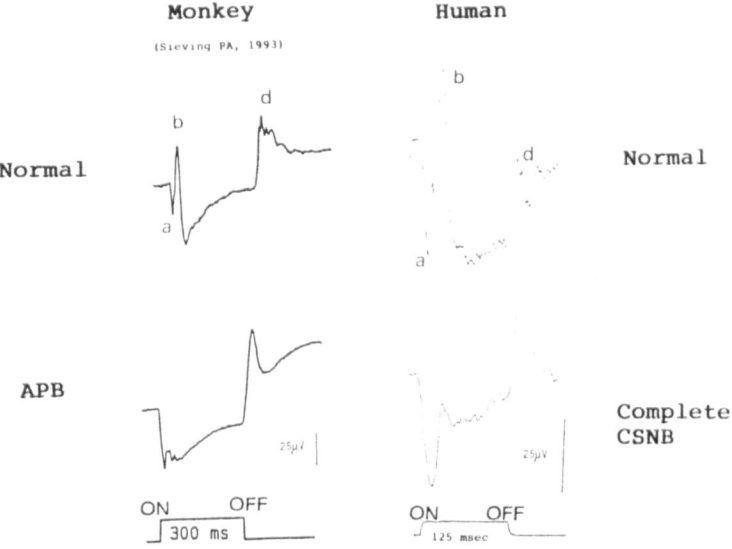

Figure 8. Photopic long-flash ERG in a control of monkey eye (left, upper), APB treated eye (left, lower), a control of normal human subject (right, upper), and a complete CSNB patient (right, lower). The ERGs of monkey are the copy of reference 12 (Sieving, 1993).

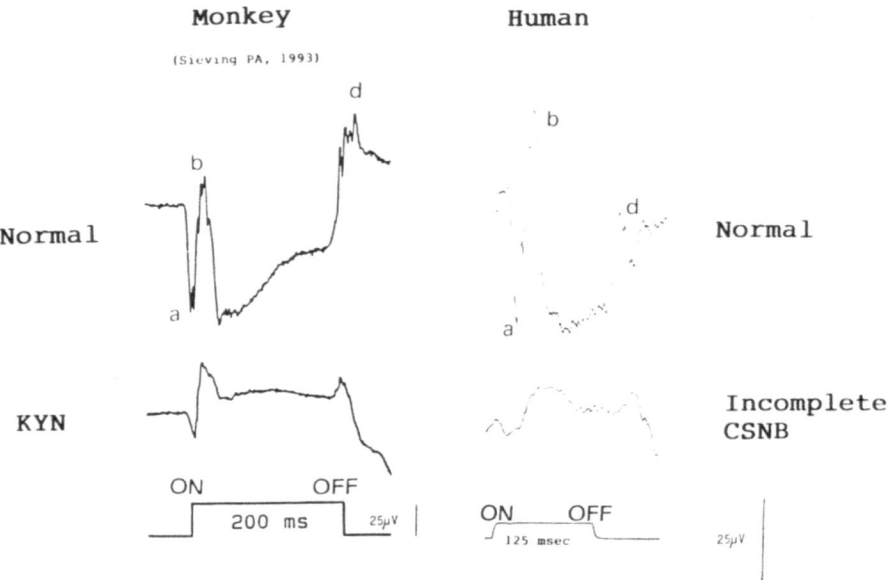

Figure 9. Photopic long-flash ERG in a control of monkey eye (left, upper), KYN treated eye (left, lower), a control of normal human subject (right, upper), and an incomplete CSNB patient (right, lower). The ERGs of monkey are the copy of reference 12 (Sieving, 1993).

wave, and large OFF-(d) wave, while KYN treated eye and the incomplete CSNB disclosed small a-wave, relatively large b-wave with an elevating plateau, and small OFF-(d) wave with an unusually negative-forming waveform swing at stimulus termination.

Specialty of Blue-Cone (S-Cone) ERG

Having recorded S-cone ERG using a contact lens electrode with a built-in blue-emitting (450 nm) diode (LED) (intensity: $426\mu W/cm^2$) under the intense yellow background illumination (intensity: $7.1\mu W/cm^2$) with 3-Hz rectangular stimuli (on:off=1:1) and averaged 32 responses, we have proven that the ERG elicited under this condition is S-cone mediated (13). Figure 10 shows the S-cone ERG from five normal subjects, five complete CSNB patients, and three incomplete CSNB patients. The S-cone ERG was absent in all complete CSNB patients while it was well preserved in incomplete CSNB patients.

DISCUSSION

Rod Visual Pathway

In both complete and incomplete CSNB, the amplitude of a-wave in mixed rod-cone ERG is normal, indicating that the rod a-wave is essentially normal. It is thus conceivable that the disturbed rod vision is likely to be caused by the post-receptoral processing of the rod-mediated signals. The STR, named by Sieving et al (14,15), originates in an area postsynaptic to the rod bipolar cells, around the inner plexiform layer or between the inner plexiform layer and the ganglion cell layers. In complete CSNB, neither the rod b-wave

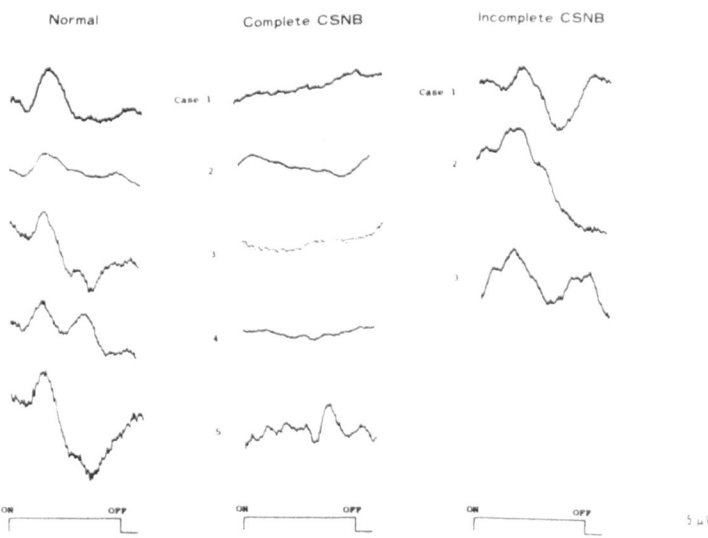

Figure 10. S-cone ERGs from five normal subjects (left), five complete CSNB patients (middle), and three incomplete CSNB patients (right).

nor the STR was recordable, suggesting that the rod signal cannot reach the rod bipolar cells because of the defect of ON-synapse to the rod bipolar cells (5). In incomplete CSNB, it was surprising that a near-threshold b-wave was normal in amplitude and in peak time, which may result from the extreme delay of the negative peak of STR as we discussed previously (5) (although the rod ERG of conventional full-field ERG was subnormal, probably due to a moderately high stimulus intensity). Namely, the negative peak of STR and the positive peak of near-threshold b-wave presented nearly the same implicit time in normal subjects (Fig. 2, 4), indicating that the b-wave is a summation of the positive PII component and the negative STR. When the negative peak of STR was delayed as seen in incomplete CSNB, the amplitude of the positive b-wave may consist only of the positive PII component (5). Even so, however, the amplitude of near-threshold b-wave still remained within normal range, since the amplitude of STR is much smaller than that of the b-wave (Fig. 2, 4). The almost normal near-threshold b-wave with abnormal STR (delayed peak time) suggests that the defect of the rod visual pathway in incomplete CSNB is proximal to the rod bipolar cells.

Cone Visual Pathway

In spite of the moderately low visual acuity and abnormal cone (photopic and 30-Hz flicker) ERG, the foveal cone densitometry was normal in both types of CSNB. This result indicates that the kinetics of the foveal cone pigments is essentially normal. We also studied the suppressive rod-cone interaction and found it to be normal as well. These findings all suggest that as with the rod visual pathway, the defect of the cone visual pathway is in the post-receptoral processing.

Instead of using a brief flash stimulus or 30-Hz flicker stimuli in conventional full-field ERG, we recorded the photopic long-flash ERG to separate the ON- and OFF-components, and discovered that the wave shape was quite different between complete and incomplete CSNBs (6) (Fig. 5). The wave shape of the complete CSNB resembled that of monkey ERG treated by APB to block the ON-synapse (Fig. 8), while the wave shape of incomplete CSNB was similar to that of monkey ERG treated by KYN to block the OFF-synapse to the cone bipolar cells (Fig. 9). These results suggest that complete CSNB and incomplete CSNB have a defect of ON-synapse and OFF-synapse, respectively, to the bipolar cells. This hypothesis has been supported by the specialty of the S-cone ERG (Fig. 10). Although red and green cones contact both ON- and OFF-synapses, the blue cone only contacts the ON-synapse (16). When the complete CSNB has a defect of ON-synapse to the cone bipolar cells, the S-cone ERG may not be recordable; when the incomplete CSNB has a defect of OFF-synapse to the cone bipolar cells, the S-cone ERG may be well preserved.

ON- and OFF-Retina

Because it is conceivable that complete CSNB has a defect of ON-synapse to rod and cone bipolar cells, only the OFF-visual pathway is adequately functioning (OFF-retina). The incomplete CSNB, on the other hand, appears to be more complicated in pathophysiology. Selective defect of the OFF-synapse to cone bipolar cells and possibly normal rod bipolar cells (normal near-threshold b-wave) suggest that the ON-synapse to rod and cone bipolar cells may function well (ON-retina). Since STR and OFF-cone bipolar cells are abnormal in incomplete CSNB, however, the glycerinergic chemical synapse from AII-amacrine cells with OFF-cone bipolar cells (9) (Fig. 7) could play a key role to elucidate the pathophysiology of incomplete CSNB.

SUMMARY

Congenital stationary night blindness (CSNB) of complete and incomplete types are different clinical entities. In complete CSNB, ERG rod b-wave and scotopic threshold response (STR) were absent while the rod a-wave was normal, indicating dysfunction of rod ON-bipolar cells. Cone driven ERG recorded with long-flash stimuli (photopic long-flash ERG) showed extremely reduced b-wave (ON) with large d-wave (OFF), resembling monkey ERG blocked by APB, and the S-cone ERG was absent, indicating dysfunctions of cone ON-bipolar cells. Above results suggest that complete CSNB has "OFF-retina". Incomplete CSNB showed normal near-threshold b-wave with abnormal STR, suggesting that rod ON-bipolar cell can function adequately. The photopic long-flash ERG presented a shape similar to that of monkey eye blocked by KYN, and the S-cone ERG was well preserved. Namely, incomplete CSNB has normal rod and cone ON-bipolar cells with abnormal cone OFF-bipolar cells (ON-retina). Above results suggest that complete CSNB and incomplete CSNB can serve as a model disease of "OFF-retina" and "ON-retina", respectively.

REFERENCES

1. Schubert, G., Bornschein, H., 1952, Beitrag zur Analyse des menschlichen Electroretinogram, Ophthalmologica 123:396–413.
2. Miyake, Y., Yagasaki, K., Horiguchi, M., Kawase, Y., Kanda, T., 1986, Congenital stationary night blindness with negative electroretinogram, Arch. Ophthalmol. 104:1013–1020.

3. Miyake, Y., Horiguchi, M., Ota, I., Shiroyama, N., 1987, Characteristic ERG flicker anomaly in incomplete congenital stationary night blindness, Invest. Ophthalmol. Vis. Sci. 28:1816–1823.

4. Tremblay, F., LaRoche, R.G., De Becker, I., 1995, The electroretinographic diagnosis of the incomplete form of congenital stationary night blindness, Vision Res. 35:2383–2393.

5. Miyake, Y., Horiguchi, M., Terasaki, H., Kondo, M., 1994, Scotopic threshold response in complete and incomplete types of congenital stationary night blindness, Invest. Ophthalmol. Vis. Sci. 35:3770–3775.

6. Miyake, Y., Yagasaki, K., Horiguchi, M., Kawase, Y., 1987, On- and Off-responses in photopic electroretinogram in complete and incomplete types of congenital stationary night blindness, Jpn. J. Ophthalmol. 31:81–87.

7. Saito, A., Miyake, Y., Wang, J.X., Yagasaki, K., Matsumoto, Y., Horio, N., Horiguchi, M., 1995, Foveal cone densitometer and changes in foveal cone pigments with aging, J. Jpn. Ophthalmol. Soc. 99:212–219.

8. van Norren, D., van de Kraats, J., 1989, Continuously recording retinal densitometry, Vision Res. 29:369–374.

9. Mueller, F., Wassele, H., Voigt, T., 1988, Pharmacological modulation of the rod pathway in the cat retina, J. Neurophysiol. 59:1657–1672.

10. Slaughter, M.M., Miller, R.F., 1981, 2-amino-4-phosphonobutyric acid: A new pharmacological tool for retina research, Science 211:182–185.

11. Slaughter, M.M., Miller, R.F., 1983, An excitatory amino acid antagonist blocks cone input to sign-conserving second-order retinal neurons, Science 219:1230–1232.

12. Sieving, P.A., 1993, Photopic ON- and OFF-pathway abnormalities in retinal dystrophies, Trans. Am. Ophthalmol. Soc. 91:701–773.

13. Horiguchi, M., Miyake, Y., Kondo, M., Suzuki, S., Tanikawa, A., Koo, H.M., 1995, Blue light emitting diode built-in contact lens electrode can record human S-cone electroretinogram, Invest. Ophthalmol. Vis. Sci. 36:1730–1732.

14. Sieving, P.A., Frishman, L.J., Steinberg, R.H., 1986, Scotopic threshold response of proximal retina in cat, J. Neurophysiol. 56:1049–1061.

15. Sieving, P.A., Nino, C., 1988, Scotopic threshold response (STR) of the human electroretinogram, Invest. Ophthalmol. Vis. Sci. 29:1608–1614.

16. Evers, H.U., Gouras, P., 1985, Three cone mechanisms in the primate electroretinogram: two with, one without off-center bipolar responses, Vision Res. 26:245–254.

6

A FAMILY WITH X-LINKED CONE DYSTROPHY SHOWING A TAPETAL-LIKE REFLEX

Mutsuko Hayakawa,[1] Keiko Fujiki,[1] Kenji Yanashima,[1] Ikuko Kondo,[2] and Atsushi Kanai[1]

[1]Department of Ophthalmology
Juntendo University School of Medicine
3-1-3, Hongo, Bunkyo-ku, Tokyo 113, Japan
[2]Department of Hygiene
Ehime University School of Medicine
Shizugawa, Shigenobu, Onsen-gun, Ehime 791-02, Japan

INTRODUCTION

X-linked cone dystrophy (XLCD) is an infrequent cause of progressive central visual loss in males. Other clinical features include central scotoma, color vision disturbance, macular degeneration and abnormalities on cone mediated electroretinogram (ERG) (1–4). This eye disease is distinguishable from the autosomal form by a family pedigree in which affected males appear on the maternal side without male to male transmission. Two XLCD families, with a greenish-golden tapetal-like reflex, were reported by Heckenlively et al. (3) in 1986. The cases described were considered to represent a newly recognized entity of XLCD associated with the Mizuo-Nakamura phenomenon in which the tapetal-like reflex disappeared after long dark adaptation (3). DNA analysis of XLCD families has also been reported, and the causative gene (COD1) has been localized to the short arm of the X-chromosome (5–7).

No XLCD cases have been reported, to date, in Japan. Herein, we present clinical features of a family with XLCD associated with a tapetal-like reflex, and the results of linkage analysis conducted using MAOA (8), a polymorphic marker on the short arm of the X chromosome, are discussed.

CASE REPORT

Case 1, I.H. (Fig 1, III-9, proband) a 40-year-old Japanese male, who is now deceased, presented to our clinic in 1984 with a chief complaint of visual loss in both eyes. The family history included two brothers and a maternal uncle with central visual loss (Fig 1). The pa-

Degenerative Retinal Diseases, edited by LaVail et al.
Plenum Press, New York, 1997

43

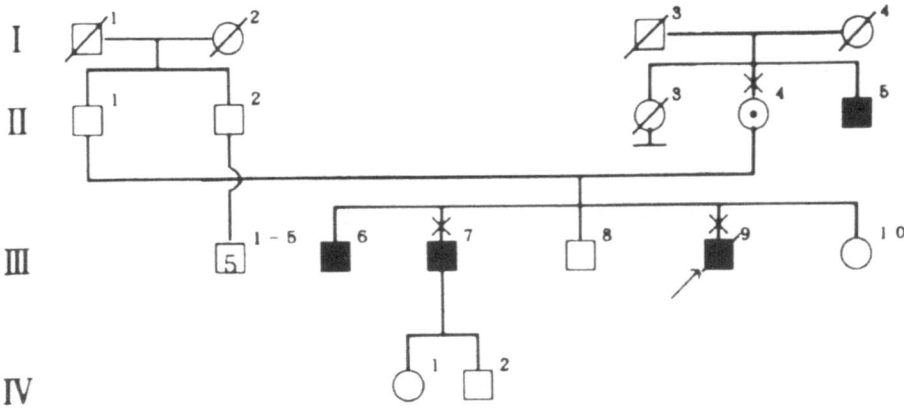

Figure 1. Family pedigree. Open squares indicate normal males, solid squares: affected males, open circles: normal females, circle with dot: obligate carrier female, arrow: proband, X mark: examined, and slashes: deceased.

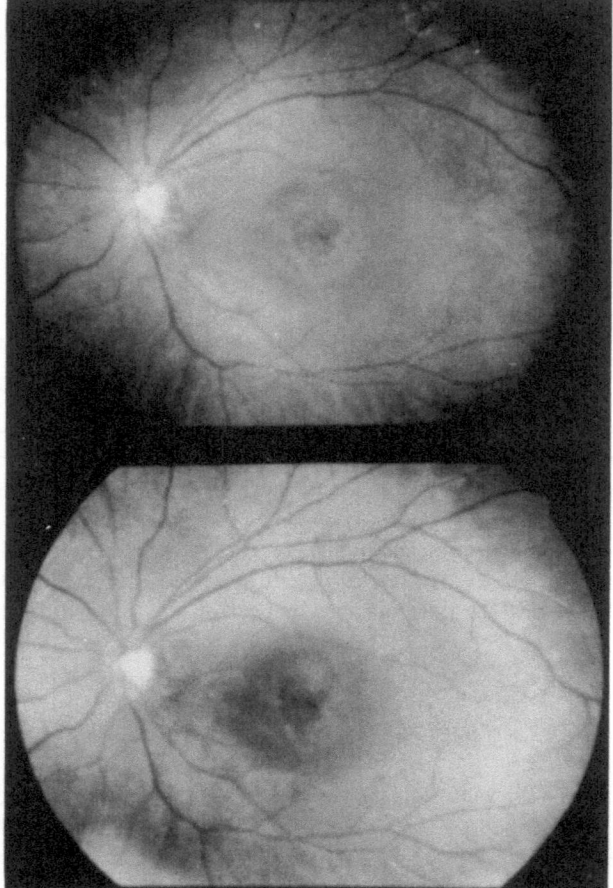

Figure 2. Fundus photographs of the left eye (case 1) show a round macular atrophic lesion associated with the tapetal-like reflex in the posterior pole area (upper photograph), in which the reflex increased after long dark adaptation (lower photograph).

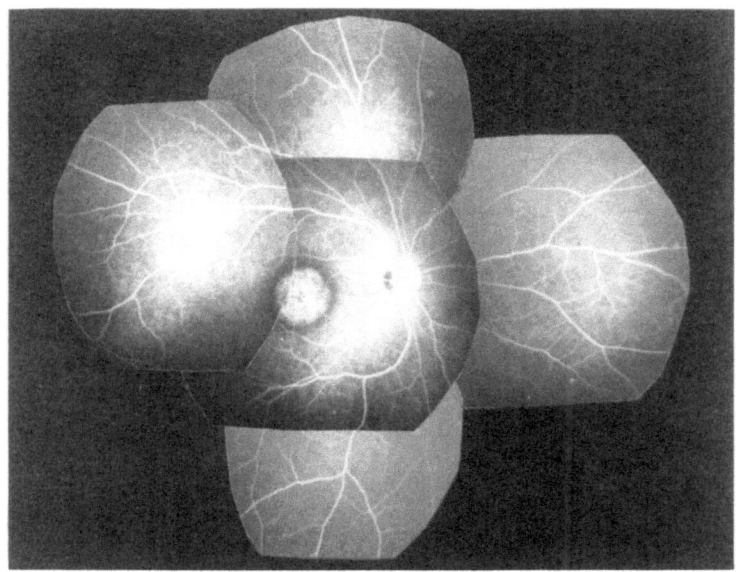

Figure 3. Fluorescein angiogram of the right eye (case 1) shows a round macular hyperfluorescent lesion associated with mildly mottled hyperfluorescence of the surrounding area.

tient's visual loss had first been noted at approximately age 35, and subsequently progressed. At the first visit to our clinic, his corrected visual acuity was 0.1 bilaterally; refraction RE sph-2.00 cyl-0.75 axis 180°, LE sph-2.00 cyl-0.50 axis 180°. Funduscopic examination revealed round macular atrophic lesions associated with a tapetal-like reflex in both eyes (Fig 2). These macular lesions were hyperfluorescent on fluorescein angiography (Fig 3). Goldmann perimetry demonstrated bilateral central scotomas with normal peripheral fields (Fig 4). The panel D-15 test showed a deutan color vision defect. The first phase of Goldmann-Weekers dark adaptometry was absent, while the final threshold was slightly elevated. After pupillary dilatation with 1% tropicamide and 20 minutes of dark adaptation, a scotopic ERG was

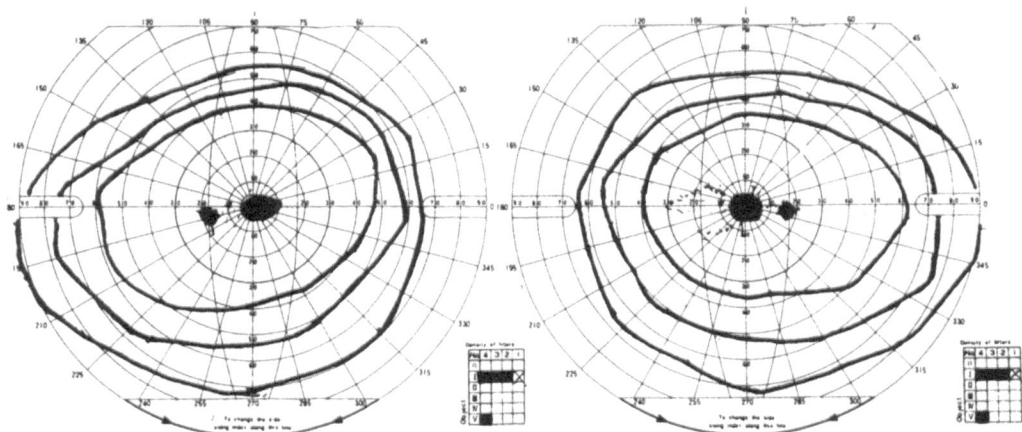

Figure 4. Goldmann perimetry of both eyes (case 1) demonstrates central scotoma.

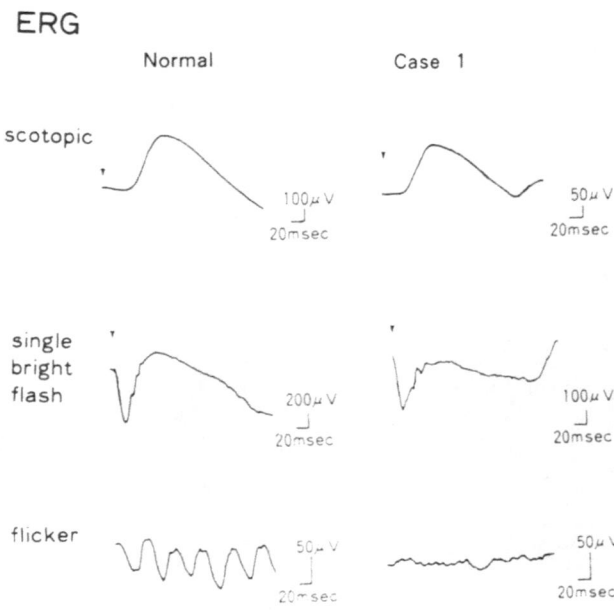

ERG

Figure 5. In case 1, single flash and scotopic ERGs show moderately diminished amplitudes. Flicker ERG shows severely diminished amplitude.

recorded with a 0.3-J blue dim light (full field stimulus) and 50 responses were averaged. The scotopic b-wave amplitude was moderately reduced. A single-flash ERG was recorded with a 40-J, full field stimulus. The amplitude of the b-wave was also moderately diminished. A flicker ERG was recorded with a 30-Hz white-light stimulus and 100 responses were averaged. The cone response was severely reduced (Fig 5). Electro-oculography was not done because of severe photophobia during preexamination bleaching. Twelve hours of dark adaptation produced an increase in the tapetal-like reflex (Fig 2).

Case 2, I.Y. (Fig 1, III-7), the 53-year-old elder brother of the proband first visited our clinic in 1992 with a chief complaint of bilateral central visual loss. He had first noticed a gradual deterioration in his visual acuity at age 51. His corrected visual acuity RE 0.7, LE 0.6; refraction RE sph -8.00 cyl -1.75 axis 65°, LE sph -12.00 cyl -1.00 axis 40°. Goldmann perimetry demonstrated a pericentral scotoma in his right eye and relative central scotoma in his left eye. Goldmann-Weekers dark adaptometry revealed a slightly elevated final threshold. Color vision testing (100 hue test) showed a tritanopia-like disorder in his right eye and moderate color defect with a nonspecific axis in his left eye. Funduscopic examination revealed a tigroid fundus, absence of the normal foveal reflex and a mildly tapetal-like reflex bilaterally (Fig 6). Fluorescein angiography showed bilateral macular bull's eye lesions (Fig 7). The tapetal-like reflex was slightly increased after 12 hours of dark adaptation (Fig 6). The amplitude of the flicker ERG was moderately reduced and those of scotopic and single flash ERG were mildly reduced. The Arden index of the electro-oculogram was 1.3 in the right eye and 1.4 in the left eye.

Case 3, I.M. (Fig 1, II-4), the 80-year-old mother of cases 1 and 2 had first been diagnosed as having high myopia at age 20 years. She had first noticed a gradual visual loss in her forties. Her corrected visual acuities were RE 0.02, LE finger counting at 5 cm. She used lenses with a prescription of -14.00 diopters sphere. Slit lamp examination revealed severe brown cataracts bilaterally. Chorioretinal atrophy, due to very high myopia was apparent on funduscopic examination (Fig 8). The axial length of the right eyeball was

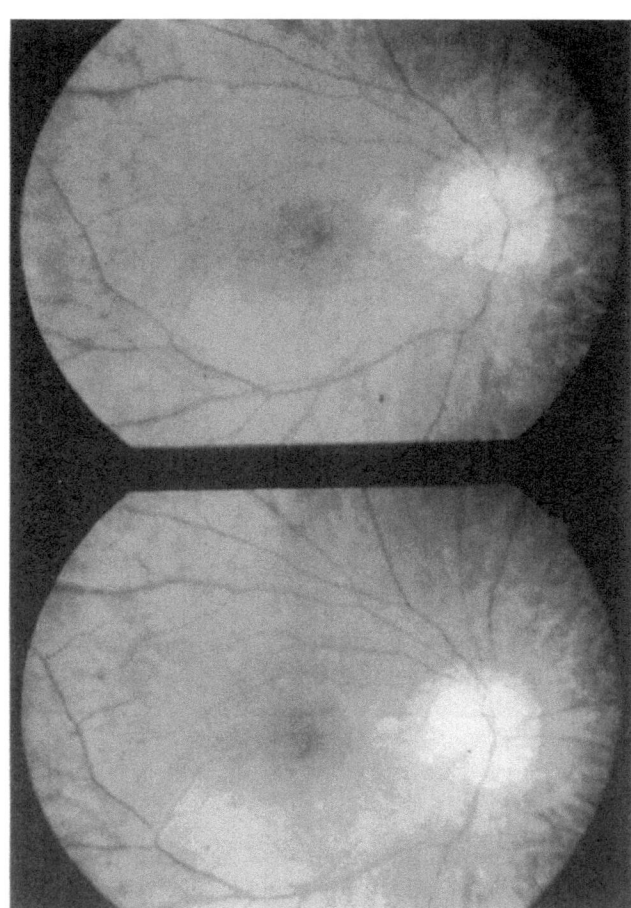

Figure 6. Fundus photographs of the right eye (case 2) show minor macular atrophy with tapetal-like reflex in the posterior pole area (upper photograph), in which the reflex slightly increased after long dark adaptation (lower photograph).

Figure 7. Fluorescein angiogram of the right eye (case 2) shows ring shaped macular hyperfluorescent lesion associated with mildly mottled hyperfluorescence of the surrounding area.

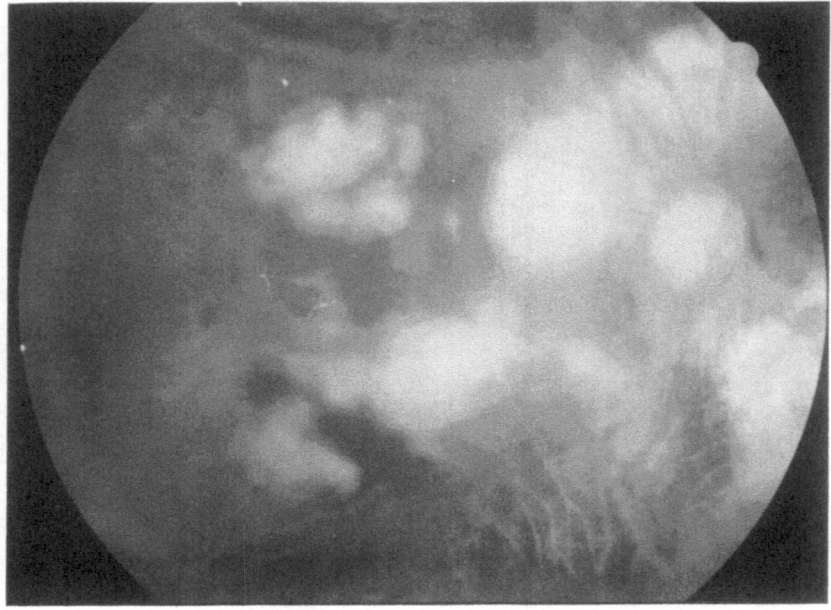

Figure 8. Fundus photograph of the right eye (case 3) shows advanced chorioretinal atrophy caused by high myopia.

30.06mm and that of the left was 29.37mm. Visual field testing revealed a defect in the nasal portion of the left field, and nasoperipheral narrowing and a central scotoma on the right. The single flash ERG was subnormal bilaterally. Color vision testing was not done due to brown cataract and severe visual loss.

LINKAGE ANALYSIS

The MAOA gene containing a CA dinucleotide repeat has been mapped to Xp11.2-p11.4 (7) and tightly linked to the COD1 locus (a maximum lod score of 3.04 at Q=0.0), as reported by Hong et al. (7). We performed linkage analysis in our family using MAOA as a polymorphic marker. Genomic DNA was extracted from blood samples obtained from cases 1 and 2 and their mother (case 3), and polymerase chain reaction (PCR) amplification was performed in an automated thermocycler (Model PJ 1000, Perkin Elmer Cetus) using the method reported by Hins et al. (7).

The PCR products were separated on 3.5% Metaphore agarose gels and stained with ethidium bromide. According to the classification of Hins et al. (7), genotypes of MAOA were B2 (310bp) in the father and B4 (312bp) and B5 (314bp) in the mother, and the affected sons had the same allelle, B5 (314bp), as the mother (data not shown). No recombination was observed between the COD1 gene in this family and the genetic marker locus MAOA, giving a maximum lod score of 0.301 at Q=0.0.

DISCUSSION

Heckenlively et al. (3) described the characterstic clinical features of XLCD as 1) a tapetal-like reflex, 2) progressive loss of central visual acuity, 3) macular degenerative lesions, 4) an increased cone threshold on dark adaptation and 5) decreased amplitudes on flicker and photopic ERG. Their case had a greenish-golden tapetal-like reflex and abnormal amplitudes on flicker and scotopic ERG. These clinical findings resemble those of our two male cases. Therefore, our male cases were diagnosed as XLCD. However, there was a difference between their cases and ours, in that both of ours had a tapetal-like reflex which increased rather than disappearing after long dark adaptation. This reversed Mizuo phenomenon has been described as one of the characteristics of the X-linked retinitis pigmentosa carrier state (9), while the Mizuo-Nakamura phenomenon in which the tapetal-like reflex disappears after dark adaptation has been reported to be a characteristic of Oguchi disease, one of the forms of congenital stationary night blindness which has an autosomal recessive inheritance pattern (10,11). The tapetal-like reflex of XLCD patients was reported by Jacobson et al. (4), but there was no description of the relation between this reflex and dark adaptation. It is difficult to determine whether the difference between the two depends on either the different subgroups of XLCD or other mechanisms underlying the tapetal-like reflex. The clinical features of male cases in our family appear to overlap those of previously reported cases of COD1 (7).

As to the XLCD carrier state, the following findings have been reported to be characteristic; 1) mild macular and fundus abnormalities(4), 2) myopia (1), 3) mild to moderate central visual loss (1,4), 4) decreased cone ERG amplitude(1,4) and 5) abnormal color vision (1,4). The mother of our male cases had myopia, visual loss and decreased amplitude on single flash ERG. Her severe visual loss was considered to be attributable to her age (80 years), i.e., advanced chorioretinal atrophy induced by high myopia as well as poor translucency due to severe brown cataracts, in addition to the XLCD carrier state. In middle age, she had noticed neither photophobia, severe central visual loss nor severe color vision disturbance, which constitute the typical features of cone dystrophy. Therefore, the clinical data suggest her ocular findings to reflect the XLCD carrier state rather than autosomal dominant cone dystrophy. Judging from the family history, the mode of disease inheritance was likely to be X-linked recessive in this family (Fig 1).

In addition, our linkage study using MAOA polymorphism, revealed the clinical manifestations to have been transmitted from the mother to two of her sons via one of her X chromosomes, suggesting that the gene causing this disease is linked to the MAOA gene locus at Xp11.2-p11.4 (8). Although further study involving additional family members is necessary for precise identification of the gene locus for COD1 in this family, the results of the linkage study appear to support the clinical diagnosis of XLCD. Furthermore, combining our results with those of a previous study (7), MAOA yielded a maximum lod score of 3.34, at $Q_{max}=0.0$.

ACKNOWLEDGMENTS

This work was supported by a Grant-in Aid for Scientific Research (B08457468) from the Ministry of Education, Science and Culture of Japan (Dr. Hayakawa).

REFERENCES

1. Pinkers, A., Groothuizen, G.G.C., Timmerman, G.J.M.E.N., and Deutman, A.F., 1981, Sex-difference in progressive cone dystrophy, II. Ophthalmic Paediatr. and Genet. 1:25–36.
2. Pinkers, A., and Deutman, A.F., 1987, X-linked cone dystrophy: An overlooked diagnosis?, Int. Ophthalmol. 10:241–243.

3. Heckenlively, J.R., and Weleber, R.G., 1986, X-linked recessive cone dystrophy with tapetal-like sheen: A newly recognized entity with Mizuo-Nakamura phenomenon, Arch. Ophthalmol. 104:1322–1328.

4. Jacobson, D.M., Thompson, H.S., and Bartley, J.A., 1989, X-linked progressive cone dystrophy: Clinical characteristics of affected males and female carrier, Ophthalmology 96:885–895.

5. Bartly, J., Gies, C., and Jacobson, D., 1989, Cone dystrophy (X-linked) (COD 1) maps between DXS 7 (L1.28) and DXS 206 (Xj 1.1) and is linked to DXS 84(754), Cytogenet. Cell Genet. 51:959.

6. Bergen, A.A.B., Meire, F., ten Brink, J., Schuurman, E.J.M., van Ommen ,G.J.B., and Dellewon, J.W. 1993, Additional evidence for a gene locus for progressive cone dystrophy with late rod involvement in Xp21.1-p11.3, Genomics 18:463–464.

7. Hong, H.K., Forrel, R.E., Paul, T.O., and Gorin, M.B., 1994, Clinical diversity and chromosomal localization of X-linked cone dystrophy (COD1), Am. J. Hum. Genet. 55:1173–1181.

8. Hinds, H.L., Hendriks, R.W., Craig, I.W., and Chen, Z.Y. 1992, Characterization of a highly polymorphic region near the first exon of the human MAOA gene containing a GT dinucleotide and a novel VNTR motif, Genomics 13:896–897.

9. Duke-Elder, S., and Dobree, J.H., 1967, Pigmentary retinal dystrophy, in:System of Ophthalmology, Volume 10 (Duke-Elder, S., ed.), p577–622, Henry Kimpton, London.

10. Oguchi, C., 1907, Ueber einen Fall von eigenartiger Hemeralopie, Nippon Ganka Gakkai Zasshi, 10:123–134.

11. Mizuo, G., and Nakamura, B., 1914, Pathogenesis and a new phenomenon in connection with the adaptation to darkness of Oguchi disease, Nippon Ganka Gakkai Zasshi 18:73–126.

7

AUTOIMMUNE RETINOPATHY

Cystoid Macular Edema in Retinitis Pigmentosa Patients

John R. Heckenlively,[1] Nata Aptsiauri,[2] and Paul A. Hargrave[2]

[1]The Jules Stein Eye Institute
UCLA Medical Center
Los Angeles, California 90024
[2]Department of Ophthalmology
University of Florida
Gainesville, Florida 32610

ABSTRACT

Patients with panretinal degeneration who also had bilateral foveal cysts or cystoid edema were compared to normal controls and to patients with retinitis pigmentosa (RP) who did not have foveal cysts. Their serum was examined for the presence of antibodies that immunoreacted with retinal proteins. Twenty-seven of 30 patients with bilateral cystoid edema and retinal degeneration had serum anti-retinal protein antibody activity, compared to 3/50 healthy donors and only 4 of 30 patients with RP lacking foveal cysts. The high correlation observed suggests that there is an autoimmune component of the cystoid edema present in RP patients that is absent in controls and in patients with RP without cystoid edema. The significance of the antibodies to retinal proteins in the pathophysiology of the disease process is currently under study.

INTRODUCTION

Cystoid macular edema (CME) is a relatively common condition, but the causes of this disorder are not clearly understood. Cystoid edema is seen on occasion in the post operative period in extracapsular cataract extraction patients. CME was more common when intracapsular techniques were the preferred cataract surgical method, with incidences reported in the range of 2% to 10% (1). The Irvine-Gass syndrome, which originally emphasized the occurrence of CME in patients who underwent intracapsular cataract extractions, was one of the first defined syndromes which described the typical finding of multilobular foveal cysts seen best in later phases of the fluorescein angiogram (2, 3).

CME has been reported in a variety of associated conditions including chronic uveitis and pars planitis, branch vein occulsions, perifoveal telangiectasia, post retinal detachment repair, diabetic retinopathy, entrapment of vitreous in the cataract or other anterior segment incisions (4–6). A more common associated condition, however, is retinitis pigmentosa (RP). The reported prevalence of CME in RP has ranged from 13% to 25% (7, 8). It is not clear why CME is more commonly found in RP patients.

Recently we have been investigating and characterizing the presence of anti-retinal protein antibodies (ARPAs) in a number of conditions including RP. Initially our studies were focused on patients presenting with cancer-associated retinopathy (CAR) (9, 10). Studies were extended to include patients with findings similar to the CAR syndrome, including presence of ARPAs, but who have not been found to have cancer. Some of these latter patients had retinal edema, thus we hypothesized that ARPAs might be playing a role in cystoid edema in patients with panretinal degeneration. We initiated a prospective study to investigate ARPAs in all patients who presented in clinic to the first author with bilateral cystoid edema or cysts and panretinal degeneration.

EXPERIMENTAL

Patients

The patients in this study were all examined at the Jules Stein Eye Institute at the University of California at Los Angeles between 1993–1995. After informed consent, blood samples were donated by patients with bilateral macular cysts and RP. Serum was separated and analyzed as coded samples. Of the 30 consecutive patients with panretinal degeneration and foveal cysts, 6 patients had a history of chronic intermittent low-grade uveitis (mild iritis) as well as panretinal degeneration with or without obvious pigment deposits. Most patients had simplex retinitis pigmentosa.

All patients and controls had a standard ophthalmologic examination including visual acuity, biomicroscopy, intraocular pressures, ophthalmoscopy, fundus photographs and fluorescein angiography. All patients had Goldmann visual field testing, and all but one had electroretinographic testing.

Two groups were used for comparison. A control group consisted of normal individuals with no history of visual problems. A second group for comparison consisted of RP patients with retinal degeneration similar to that of the test group, but in whom there was no evidence of foveal cysts.

Laboratory Methods

Serum samples, masked from the lab personnel, were used for immunostaining of blots containing human retinal proteins separated by molecular weight. Antigen-antibody complexes on the blots were visualized by a colorimetric reaction. This enabled determination of the presence of circulating antibodies that could react with retinal antigens.

In order to do this, human retinas were solubilized in octyl glucoside (10, 11). The mixture of retinal proteins was separated on a molecular weight basis by SDS-polyacrylamide gel electrophoresis on slab gels. A portion of each gel was stained to visualize all proteins present. The remainder was used to electrophoretically transfer the proteins to a membrane polymer support. This membrane was pretreated to minimize nonspecific binding, then incubated with diluted (1:100) patient's serum. Antigen-antibody complexes on the immunoblot were visualized using a phosphatase-conjugated secondary antibody and a phosphatase substrate that developed a colorimetric reaction.

The identity of several stained protein bands was determined using authentic puri-fied retinal proteins and antibodies specific to these proteins. Two-dimensional gels were subjected to immunoblot analysis as described above, where necessary. Proteins investi-gated were interphotoreceptor retinoid binding protein, IRBP (141 kDa), arrestin/S-anti-gen (48 kDa), enolase (46 kDa), carbonic anhydrase II (30 kDa), and recoverin (23kDa).

RESULTS

Twenty-seven of 30 patients with RP and foveal cysts had antibodies to retinal pro-teins compared to 4 out of 30 of the RP patients without foveal cysts. This represents an extraordinarily high correlation between the presence of bilateral foveal cysts in retinal degeneration and the presence of serum antiretinal protein antibodies (ARPAs), compared to RP patients without foveal cysts ($P<.00001$, two-tail Chi-square test).

In the foveal cyst group, 18 of the 27 had two or more immunolabelled bands. The most common retinal proteins showing antigenicity were carbonic anhydrase II (30 kDa) and enolase (46 kDa), both occurring in 17/27 (63%) of patients. Of the 4 of 30 RP pa-tients without foveal cysts who showed ARPAs, only one had immunoreactivity against both 30 kDa and 46 kDa, a pattern typical for RP patients with cysts.

There was no significant difference in other measured parameters, e.g., intraocular pressure, between the two patient groups. The severity of the patients' foveal cysts and retinal degeneration did not always correlate with the intensity of the ARPA activity. Some patients showed weak staining of a single band, while others showed very intense staining Patients with multiple immunoreactivities might have several weak and one strong reaction. Although the most commonly stained proteins were enolase (46 kDa) and carbonic anhydrase II (30 kDa), one patient had antibody against IRBP (141 kDa) and 2 patients had antibodies against arrestin/S-antigen (48 kDa). Several patients' sera had an immunoreactive band near molecular weight 23 kDa, but the use of authentic standards demonstrated that their antibody activity was not anti-recoverin. Among the RP patients without foveal cysts, 3 of the 4/30 who showed ARPA had immunolabeling against un-known proteins of molecular weights not typically seen in the foveal cyst group.

Twenty-three percent of patients had a history of uveitis which was chronic and low grade in nature, with intermittent cell and flare, and occasionally mild vitritis. None of these patients showed serum immunoreactivity to arrestin/S-antigen, which was only found in two cases who did not have a history of uveitis. All the patients with intermittent uveitis had diffuse retinal atrophy with (2/7) or without (5/7) bone-spicule-like pigmentary deposits, abnormal ERGs and visual field losses. In many of these cases it is likely that the chronic uveitis contributed to the panretinal degeneration.

DISCUSSION

The primary finding of this study is the strong correlation between the presence of anti-bodies against retinal proteins in the serum of patients with panretinal degeneration who also have foveal cysts compared to those who have panretinal degeneration in the absence of foveal cysts. This raises questions about the origin of these antibodies and their significance for the disease process. It is possible that a breakdown of the blood-retinal barrier accompany-ing retinal degeneration allowed retinal proteins to enter the circulation with a resultant im-mune response. Whether such an immune response then led, in some cases, to cystoid edema and exacerbated retinal degeneration remains to be determined. Further examination of some of the above patients, however, documents that RP patients with ARPAs had on average a rate of visual field loss about three times greater than patients without ARPAs (11).

The presence of circulating antibodies might simply be a marker for the disease but in itself have no deleterious consequences. However, when anti-recoverin antibodies were incubated with cells from a retinal cell line, the antibodies were taken up and produced destruction of the cells in a dose- and time-dependent manner (12). It remains to be seen if a similar mechanism might operate to produce cystoid macular edema and enhanced retinal degeneration in patients already experiencing retinal degeneration.

The presence of anti-retinal antibodies in RP cases has been known to be fairly common, with 37% of patients testing positive by indirect immunofluorescence (13, 14). Autoantibodies to proteins have been detected in a number of autoimmune diseases: e.g., anti-C1-inhibitor antibodies in angioedema (15).

Antibodies to carbonic anhydrase II were found in 63% of our patient sample. Carbonic anhydrase II has a wide tissue distribution and is critical to acid-base balance and ion and fluid movement, regulating osmotic pressure and fluid secretion (16). Carbonic anhydrase II is widely distributed in the eye, and is abundant in epithelial cells of the ciliary body, the Müller cells of the retina, and in a subset of cone photoreceptor cells (17–19). Recently, autoantibodies to carbonic anhydrase I and II have been detected in sera of patients with a variety of autoimmune diseases. Thirty percent of patients with systemic lupus erythematosus, polymyositis, and systemic sclerosis, and 21% of patients with Sjögren's syndrome, 69% of patients with endometriosis, but only 11.8% of control individuals had antibodies to CAII (20–22).

In addition to antibodies to carbonic anhydrase, antibodies to enolase were found in 63% of our patient population. The glycolytic enzyme enolase is present in the cytoplasm of mammalian cells as a dimeric protein composed of three possible subunits, designated alpha, beta and gamma. β_2-enolase is predominant in muscle. In the nervous system, α_2-enolase characterizes glial cells and developing neurons. γ_2-enolase, also called neuron-specific enolase (NSE), as well as the hybrid form $\alpha\gamma$-enolase, are normally found in neurons and neuroendocrine cells (23, 24). NSE has been considered to be a reliable marker for these cell types and their malignant counterparts (25, 26). Serum antibodies to enolase may be found in a number of diseases, including cancer-associated retinopathy (10). Both enzymes, carbonic anhydrase II and enolase, are abundant cytoplasmic proteins. They can be released into the blood circulation following cell death and the immune system can produce antibodies against them.

The finding of ARPAs in cases of retinal degeneration with foveal cysts may have treatment implications. Frequently, steroids such as prednisolone drops, subtenons depo-steroid, or prednisone are used depending on the severity of the disease. Clearly, the latter two approaches will be more effective if the ARPAs, which are systemically derived, are playing a major causative role. Topical prednisolone does not penetrate the retina to any significant extent, but is useful in cases that also have low grade uveitis. Until more is known about the mechanism and the best immunosuppressive medications to use, traditional approaches of titrating against the clinical effect is a reasonable approach.

ACKNOWLEDGMENTS

This work was supported in part by the RP Foundation Fighting Blindness, Baltimore, Maryland, grants from the National Eye Institute of the National Institutes of Health

EY06225, EY06226, EY08571, and an unrestricted departmental award from Research to Prevent Blindness, Inc.

REFERENCES

1. Taylor, D.M., Sachs, S.W. and Stern, A.L., 1984, Aphakic cystoid macular edema. Longterm clinical observations., *Surv Ophthalmol. Suppl.* **28**: 437–441.
2. Gass, J.D.M. and Norton, E.W.D., 1966, Cystoid macular edema and papilledema following cataract extraction, *Arch Ophthalmol* **36**: 646–661.
3. Irvine, S.R., 1953, A newly defined vitreous syndrome following cataract surgery, *Am J. Ophthalmol* **36**: 599–619.
4. Gutman, F.A. and Zegarra, H., 1984, Macular edema secondary to occlusion of the retinal veins, *Surv. Ophthalmol. Suppl.* **28**: 462–470.
5. Coscas, G. and Gaudric, A., 1984, Natural course of nonaphakic cystoid macular edema, *Surv. Ophthalmol. Suppl.* **28**: 471–484.
6. Schepens, C.L., Avila, M.P., Jalkh, A.E. and Trempe, C.L., 1984, Role of the vitreous in cystoid macular edema, *Surv. Ophthalmol. Suppl.* **28**: 499–504.
7. Heckenlively, J.R., 1988, Clinical Findings in Retinitis Pigmentosa, in: *Retinitis Pigmentosa,* (J.R. Heckenlively, eds.), pp. 80–85, J. B. Lippincott Co., Philadephia.
8. Fetkenhour, C.L., Choromokos, E., Weinstein, J. and al., e., 1977, Cystoid macular edema in retinitis pigmentosa, *Trans Am Acad Ophthalmol Otolargyngol* **83**: 515–521.
9. Adamus, G., Guy, J., Schmied, J.L., Arendt, A. and Hargrave, P.A., 1993, Role of Anti-Recoverin Autoantibodies in Cancer-Associated Retinopathy, *Invest Ophthalmol Visual Sci* **34**: 2626–2633.
10. Adamus, G., Aptsiauri, N., Guy, J., Heckenlively, J., Flannery, J. and Hargrave, P.A., 1996, The Occurrence of Serum Autoantibodies against Enolase in Cancer-Associated Retinopathy, *Clinical Immunology and Immunopathology* **78**: 120–129.
11. Heckenlively, J.R., Aptsiauri, N., Nusinowitz, S., Peng, C. and Hargrave, P.A., 1996, Investigations of Anti-retinal Antibodies in Pigmentary Retinopathy and other Retinal Degenerations, *Transactions American Ophthalmological Society* **94**: in press.
12. Machnicki, M., Siegell, G.M. and Adamus, G., 1996, Evidence for Pathogenic Effect of Anti-recoverin Antibodies in Cancer Associated Retinopathy (CAR), *Invest. Ophthalmol. & Vis. Sci.* **37**: S535.
13. Chant, S.M., Heckenlively, J.R. and Meyers-Elliott, R.H., 1985, Autoimmunity in hereditary retinal degeneration, *Brit. J. Ophthalmol.* **69**: 19–24.
14. Heckenlively, J.R., Solish, A.M., Chant, S.M. and Meyers-Elliott, R., 1985, Autoimmunity in hereditary retinal degenerations. II. Clinical Studies: Antiretinal antibodies and fluorescein angiogram findings, *Brit. J. Ophthalmol.* **69**: 758–764.
15. Mandle, R., Baron, C., Roux, E., Sundel, R., Gelfand, J., Aulak, K., Davis, A.E., Rosen, F.S. and Bung, D.H., 1994, Acquired C1 inhibitor deficiency as a result of an autoantibody to the reactive center region of C1 inhibitor, *J. Immunol* **152**: 4680–4685.
16. Dogson, S.J., 1991, The carbonic anhydrases: overview of their importance, in: *The carbonic anhydrases: Cellular and molecular genetics,* (S.J. Dogson, *et al.,* eds.), pp. Plenum Press, New York.
17. Friedland, P.J. and Maren, T.H., 1984, The role of carbonic anhydrase in lens ion transport and metabolism, *Ann NY Acad Sci.* **429**: 582–586.
18. Wistrand, P.J., Schenholm, M. and Lonnerholm, G., 1986, Carbonic anhydrase isoenzymes CAI and CAII in the human eye, *Invest. Ophthalmol. Vis. Sci.* **27**: 419–428.
19. Nork, T.M., McCormick, S.A., Chao, G.-M. and Odom, V., 1990, Distribution of carbonic anhydrase among human photoreceptors, *Invest. Ophthalmol. Vis. Sci.* **31**: 1451–1458.
20. Inagaki, Y., Jinno-Yoshida, Y., Hamasaki, Y. and Ueki, H., 1991, A novel autoantibody reactive with carbonic anhydrase in sera from patients with systemic lupus erythematosus and Sjorgren syndrome, *J. Dermatol. Sci.* **2**: 147–154.
21. Itoh, Y. and Reichlin, M., 1992, Antibodies to carbonic anhydrase in systemic lupus erythematosus and other rheumatic diseases, *Arthritis and Rheumatism* **35**: 73–82.
22. Kiechle, F.L., Quattrociocchi-Longe, T.M. and Brinton, D.A., 1994, Carbonic anhydrase antibody in sera from patients with endometriosis, *Immunopathology* **101**: 611–615.
23. Peshavaria, M., Quinn, G.B., Reeves, I., Hinks, L.J. and Day, I.N.M., 1990, Molecular biology of the human gene family: nerve (γ), muscle (β) and general (α) isoforms, *Biochemical Society Transactions* **18**: 254–255.

24. McAleese, S.M., Dunbar, B., Fothergill, J.E., Hinks, L.J. and Day, N.M., 1988, Complete amino acid sequence of the neurone-specific I isoenzyme of enolase (NSE) from human brain and comparison with the non-neuronal A form (NNE), *Eur. J. Biochem.* **178**: 413–417.

25. Nou, E., Stenholtz, L., Bergh, J., Nilsson, K. and Pahlman, S., 1990, Neuron-specific enolase as a follow up marker in small cell bronchial carcinoma, *Cancer* **65**: 1380–1385.

26. Matfield, R.H. and McKeran, R.M., 1992, Serum neuron specific enolase as a quantitative marker of neuronal damage in rat stroke model, *Brain Research* **577**: 249–252.

8

NORRIE DISEASE IN JAPAN

Norio Ohba, Yasushi Isashiki, Fumiyuki Uehara, and Kazuhiko Unoki

Department of Ophthalmology
Kagoshima University Faculty of Medicine
Sakuragaoka 8-35-1, Kagoshima-shi 890, Japan

INTRODUCTION

Norrie disease (ND) is a genetic disease characterized by bilateral congenital blindness. The salient clinical features early in life consist of retrolental vascularized mass due to maldeveloped retina, followed by complications of cataracts and corneal opacities leading to bulbar atrophies in adult. Some patients have psychomotor retardation and progressive hearing loss as part of a multisystem syndromic disorder. This vitreoretinal dysplastic disease is transmitted as a simple X-linked pattern of inheritance, so that hemizygous males are affected whereas heterozygous females remain asymptomatic.[1-2]

The ND gene, assigned to the short arm of the X-chromosome, has recently been isolated and characterized by positional cloning, which consists of three exons encoding a protein of 133 amino acids.[3-4] ND patients show mutations of the ND gene. It is remarkable that there is a wide variation of the gene mutation among families.[5-12]

ND is rare but spreads worldwide. Patients have been reported in Europe, North America and Asia. In Japan, the disease was not recognized until our report in 1980 of two families.[13] Since then, two Japanese families have been reported.[11,14] In addition, we reported a rare female patient with rare chromosomal disorder.[15] Hercin, these Japanese ND families are reviewed for clinical and molecular genetic assessments.

NORRIE DISEASE IN JAPAN

ND Family 1

This family has lived in Kagoshima, located in the southern Japan. The proband was a male infant, born blind in both eyes in 1978. Both pupils were unresponsive to light with aberrant fibrous tissues on the surface of ectopic irides, and the retinas were maldeveloped with dense, whitish-yellow retrolental masses containing proliferated vessels. Electroretinograms were nonrecordable. His maternal grandfather had been blind with atrophic cyc-

Degenerative Retinal Diseases, edited by LaVail *et al.*
Plenum Press, New York, 1997

balls. After a report of this family in 1980,[13] another boy was born in 1989 with features of ND. The patients of this family were unremarkable in mental or hearing ability.

Molecular genetic analyses according to the standard methods in reference to the candidate ND gene revealed in affected males a mutation at the initiation codon of the exon 2 of the ND gene (ATG to GTG) and in the obligate mother of the proband a mixture of normal and mutant type of ND gene as expected for heterozygote.[10]

ND Family 2

This family has also lived in Kagoshima, but allegedly unrelated with the above ND family 1. The proband was a male infant, born blind in bilateral eyes because of severe vitreoretinal dysplasia. His maternal cousin and maternal uncle were also affected with ND and had been virtually blind since birth. This family was found in 1978, and reported in 1980.[13] In the subsequent follow-up, the three patients in this family remained otherwise unremarkable in neurological and otological examinations.

Assessment of the ND gene revealed, to our surprise, a mutation identical to that in the ND family 1.[10] These two families had lived in the same area of the country for at least several centuries and they claimed no relationship. It is, however, possible in view of the rarity of the disease that the identical gene mutation represents a founder effect.

ND Family 3

This family has lived in the central Japan. The proband was a three-month-old male, born with shallow anterior chambers, ectopia irides and vascular proliferated yellow masses in the retrolental space. A maternal uncle and a maternal cousin had ocular disorders compatible with ND.

Direct sequencing and PCR-restriction detection of the ND gene revealed a mutation at the codon 95 of exon 3 (TGC to CGC) in the proband and a mixture of this mutant and normal gene in the proband's mother.[11]

ND Family 4

This family was from Chiba prefecture, located in the central Japan near Tokyo. The male proband was born blind in 1991 and diagnosed as ND at the age three months. His maternal uncle was congenitally blind with bilateral ocular shrinkage.

The proband showed a mixture of normal and shifted fragment in isodensity on SSCP of the exon 2 of the ND gene, and he had a wild-type clone and a mutant-type clone (Ser57Pro). The other patient's clones had only normal nucleotide sequences, and the mutation at the codon 57 was not found in any other family members or controls. In the two patients, 15 or 9 kb region overlapping the exon 2 of ND gene was not amplified by long PCR methods, whereas other family members, including the obligate carriers, showed target fragments of the regions. Southern blots of *EcoRI* or *Hind III* digested DNA from the family members showed each single signal at 9.5 kb or at 4.0 kb without any other abnormal signals; densities of hybridized signals were intense in two female carriers, intermediate in two ND patients, and low in one normal male in the family. DXS1003, a DNA polymorphic marker located at Xp 11.4-11.23, showed a common repeat in the two patients and two female carriers, compatible with X-linked transmission of a near ND allele. The findings suggest a tandem duplication within the ND gene in two patients. The size of duplication may be too long to access the full length, and thus we could not determine the duplication junction.

A Female with ND Associated with X Autosome Translocation

A female was born blind with typical clinical and histopathologic features of ND, with negative family history. This patient had a peculiar chromosomal aberration, 46XX, t(X, 10), a balanced reciprocal chromosomal translocation. Since the karyotype of his parents normal, it was suggested that a de novo chromosomal translocation was disrupted the vitreoretinal dysplasia gene itself.[15] This interesting patient was seen before the era of molecular genetics, and materials were not available for further work-up.

COMMENT

These families are all that have been established for ND to date in Japan, consistent with a view that ND is a rare disease. It is, however, emphasized that the clinical diagnosis of ND is frequently difficult, in particular nonfamilial cases, to be differentiated from many distinct diseases with similar clinical manifestations. The diagnostic difficulties may be overcome with the advancement of molecular diagnosis. There is a wide variability of ND mutations among families, and phenotype-genotype correlation remains to be elucidated, as summarized in Table 1. The pathogenesis of ND is to be clarified with further

Table 1. Norrie disease gene mutations and clinical manifestations

Authors Family I.D.	Mutation codon	Type	Effect	Mental retardation	Deafness	Family history
Meindl et al.[5]						
167	60	missense	Val to Glu	+	−	−
204	29	nonsense	truncated protein	+	?	+
605	96	missense	Cys to Tyr	+	−	+
1650	44	missense	Tyr to Cys	+	−	+
Berger et al.[6]						
1127	90	missense	Arg to Pro	−	−	+
1245	75	missense	Ser to Cys	+	+	−
1443	9	insertion	truncated protein			
2248	intron 2		splicing disorder	+/−	−	+
2843	57	nonsense	truncated protein	+/−	+	+
2942	61	missense	Leu to Phe	−	+	+
3710	74	missense	Arg 74 Cys	−	+	+
9500	133	one base deletion	adding 127 AA	+	?	+
A. La.	110	nonsense	truncated protein	−	−	−
Cyp.	96	missense	Cys to Tyr	+	−	+
G.	110	nonsense	truncated protein		+	+
P.Ru.	97	one base deletion	truncated protein	+	?	+
Fuentes et al.[7]						
89ND1	121	missense	Arg to Gln	−	−	+
90ND2	58	missense	Lys to Asn	−	+	+
Wong et al.[8]						
ND	128	nonsense	truncated protein	+	−	−
Fuchs et al.[9]	13	missense	Leu to Arg	?	+	+
Isashiki et al.[10]						
Family 1	1	initiation	truncated protein	−	−	+
Family 2	1	initiation	truncated protein	−	−	+
Isashiki et al.[11]	95	missense	Cys to Arg	−	−	+
Joos et al.[12]	39	missense	Cys to Arg	+	−	+

[a]The number and case identification corresponds to the reference cited.

understanding of the functional role of the ND gene. It is of interest in this connection that X-linked familial exudative vitreoretinopathy, clinically distinct from ND, is suggested to be caused by the same gene.[16]

REFERENCES

1. Warburg, M., 1966, A congenital progressive oculo-acoustico-cerebral degeneration, *Acta Ophthalmol . Suppl.* **89**: 1-147.
2. Ohba, N., and Isashiki Y., 1996, A literature review of Norrie disease, *J. Jpn. Ophthalmol. Soc.* **100**: 101-110.
3. Berger, W., Meindl, A., van de Pol, T.J.R., Cremers, F.P.M., Ropers H.H., Derner, C., Monaco, A., Bergen A.A.B., Lebo, R., Warburg, M., Zergollern, L., Lorenz, B., Gal, A., Bleeker-Wagemakers, E.M., and Meitinger, T, 1992, Isolation of a candidate gene for Norrie disease by positional cloning, *Nature Genet.* **1**: 199-203.
4. Chen, Z-Y., Hendricks R.W., Jobling M.A., Powell, J.F., Breakefield, X.O., Sims, K.B., and Craig, I.W., 1992, Isolation and characterization of a candidate gene for Norrie disease, *Nature Genet.* **1**: 204-208.
5. Meindl, A., Berger, W., Meitinger, T., van de Pol, D., Achatz, H., Drner, C., Haasemann, M., Hellebrand. H., Gal, A., Cremers, F., and Ropers, H-H, 1992, Norrie disease is caused by mutations in an extracellular protein resembling C-terminal globular domain of mucins, *Nature Genet.* **2**: 139-143.
6. Berger, W., van de Pol, D., Warburg, M., Gal, A., Bleeker-Wagemakers, L., de Silva, H., Meindl, A., Meitinger, T., Cremers, F., and Ropers, H-H., 1992, Mutations in the candidate gene for Norrie disease, *Hum. Mol. Genet.* 1: 461-465.
7. Fuentes, J.J., Volpini, V., Fernndez-Toral, F., Coto, E., and Estivill, X., 1993, Identification of two new missense mutations (K58N and R121Q) in the Norrie disease (ND) gene in two Spanish families, *Hum. Mol. Genet.* **2**: 1953-1955.
8. Wong, F., Goldberg, M.F., Hao, Y., 1993, Identification of a nonsense mutation at codon 128 of the Norrie's disease gene in a male infant, *Arch. Ophthalmol.* **111**: 1553-1557.
9. Fuchs, S., Xu, S.Y., Caballero, M., Salcedo, M., La O, A., Wedemann, H., and Gal, A., 1994, A missense point mutation (Leu13Arg) of the Norrie disease gene in a large Cuban kindred with Norrie disease, *Hum. Mol. Genet.* **3**: 655-656.
10. Isashiki, Y., Ohba, N., Yanagita, T., Hokita, N., Doi, N., Nakagawa, M., Ozawa, M., and Kuroda, N., 1995, Novel mutation at the initial codon in the Norrie disease gene in two Japanese families, *Hum. Genet.* **95**: 105-108.
11. Isashiki, Y., Ohba, N., Yanagita, T., Hokita, N., Hotta, Y., Hayakawa, M., Fujiki, K., and Tanabe, U., 1995, Mutations in the Norrie disease gene: a new mutation in a Japanese family, *Br. J. Ophthalmol.* **79**: 703-708.
12. Joos, K.M., Kimura, A.E., Vandenburgh, K., Bartley, J.A., Stone, E.M., Ocular findings associated with a Cys39Arg mutation in the Norrie disease gene, *Arch. Ophthalmol.* **112**: 1547-1579.
13. Fujita, S., Fujiwara, N., and Ohba, N, 1980, Norrie's disease: report of cases in two Japanese families, 1980, *Jpn. J. Ophthalmol.* **24**: 22-28.
14. Kuroda, N., Isobe, M., Watanabe, Y., Ishikiriyama, S., Horie, H., and Kimura, T., A case of Norrie's disease, 1993, *Jpn. J. Clin. Ophthalmol.* **47**: 1110-1111.
15. Ohba, N., and Yamashita, T., 1986, Primary vitreoretinal dysplasia resembling Norrie's disease in a female: association with X autosome chromosomal translocation, *Br. J. Ophthalmol.* **70**: 64-71.
16. Chen, Z-Y., Battinelli, E.M., Fielder, A., Bundey, S., Sims, K., Breakefield, X.O., and Craig, I.W., 1993, A mutation in the Norrie disease gene (NDP) associated with X-linked familial exudative vitreoretinopathy, *Nature Genet.* **5**: 180-183.

9

DEVELOPMENT OF A MODEL FOR MACULAR DEGENERATION

P. E. Rakoczy,* M. Lai, and I. J. Constable

Molecular Biology
Lions Eye Institute
University of Western Australia
2 Verdun St, Nedlands, 6009, Australia

INTRODUCTION

Age Related Macular Degeneration (AMD) is the leading cause of blindness in the developed world. With increasing life expectancy the prevalence of AMD (15–30%) in the age group of >75 years (Bressler[1], Klein[2], Vingerling[3], Mitchell[4]) will significantly increase, causing enormous social and financial problems for the community. In spite of the significance of the problem, to date the pathogenesis of AMD remains unknown and the disease is essentially untreatable.

Traditional investigative cell biological techniques have demonstrated the accumulation of several types of debris within the retinal pigment epithelium and Bruch's membrane with the advancement of age. These debris are lipofuscin (Weiter et al.,[5]), basal laminar deposit (Sarks et al.,[6]) and drusen (Gass et al.,[7]; Pauleikhoff et al.,[8]). One of the most ubiquitous cellular signs of aging is the accumulation of lipofuscin (Katz et al.,[9]; Weiter et al.,[5]). Although, the exact structure of lipofuscin is unknown there is now substantial evidence that in the retinal system it represents the residues of incompletely processed photoreceptor outer segments (POS) (Katz and Eldred,[10]). Lipofuscin accumulation has been shown to correlate remarkably well with the annular pattern of AMD. It was recently demonstrated that increased phagocytic and metabolic load on the retinal pigment epithelial RPE cells in the macula causes preferential age related accumulation of lipofuscin in the RPE, which ultimately leads to adjacent photoreceptor death (Dorey et al.,[11]; Delori et al., [12]). The structure of basal laminar deposit remains unknown. However a strong correlation between the presence of the basal laminar deposit and AMD has been demonstrated (Sarks et al.,[6]). The progressive accumulation of lipofuscin within the RPE cells is thought to be implicated in a second manifestation of the aging retina, that of drusen formation (Pauleikhoff et al.,[13]). Although, the origin and structure of drusen are

* Ph: (619) 346 2802; fax: (619) 346 1545; E-mail: rakoczy@uniwa.uwa.edu.au.

Degenerative Retinal Diseases, edited by LaVail et al.
Plenum Press, New York, 1997

widely debated, the consensus is that the development of drusen correlates closely with abnormal pigment epithelial function. Recent electron microscopic pictures demonstrated the presence of membranous structures in drusen similar to those found in photoreceptor outer segments. Also the presence of photoreceptor outer segment specific lipids (Pauleik-hoff *et al.*,[8,13]) strongly suggests the outer segment origin of drusen. Drusen formation cannot be considered to be a benign sign of aging as drusen aggregation in the posterior pole is recognized clinically as the first sign of AMD (Gass[7]; Bressler *et al.*,[14]).

Although these three types of debris appear at different locations in the retina and they have different compositions, their common features are that they all have been implicated in the development of AMD, they are most likely to be photoreceptor-derived and they accumulate within or in the proximity of the RPE cells. It has been generally accepted that the presence of these debris will impair RPE function. However, it remains debated if this effect is due to the sheer volume of the debris and/or the aggregation of biologically active components within them (Boulton,[15] Dorey,[11] Eldred,[16]). To be able to understand the effect of cellular debris accumulation on the normal function of RPE cells it is fundamental to develop a model which is easy to use both *in vitro* and *in vivo*, and in which debris accumulation can be achieved in a relatively short period of time and maintained.

The aim of this study was to evaluate the suitability of antisense DNA technology to induce photoreceptor-derived debris accumulation in RPE cells.

METHODS

RPE Cell Culture

Eyes of 7 and 18 year old donors were obtained from the Lions Eye Bank of Western Australia. Preparation of primary RPE cultures and subculturing of RPE cells was described earlier (Rakoczy *et al.*,[17]). The culture medium used for subsequent passage of RPE cell cultures and all experiments (unless otherwise stated) consisted of 45% Ham's F12, 45% DMEM, 10% FCS, 0.4% glucose, 10 mM HEPES and 50 mg/ml gentamicin. The epithelial origin of all RPE cell cultures was confirmed by positive cytokeratin staining using AE 1 and 3 monoclonal antibodies to middle and high molecular weight keratins.

Transfection of RPE Cells

RPE cells were incubated in a humidified atmosphere of 5% CO_2 till they were 60 to 80% confluent. The RPE cells were then washed twice in serum- and antibiotic-free OPTI-MEM (GIBCO-BRL, Gaithersburg, MD, USA). Lipofectin reagents (GIBCO-BRL) diluted 1 in 10 in OPTI-MEM to a final volume of 100 μl, were gently mixed with 5 μg antisense cathepsin D and pHβApr-1-neo DNA (vector) (Gunning *et al.*,[18]) diluted in OPTI-MEM to a final volume of 100 μl and incubated at room temperature for 15 minutes (mins). Following this, an additional 800 μl of OPTI-MEM was added to the mixture and gently overlaid on to the washed RPE cells. The cells were incubated for 20 hours (hrs) in a humidified atmosphere of 5% CO_2 at 37 °C before the transfection media was removed and replaced with growth medium. After a further 48-hr incubation, the cells were trypsinised and subcultured at 1:5 in growth medium and Geneticin (GIBCO-BRL) at a final concentration of 600 μg/ml. Successfully transfected cells selected with Geneticin were maintained in growth medium and Geneticin at a final concentration of 300 μg/ml. Confluent transformed cultures were frozen for storage and subcultured for further analysis.

Challenge of RPE Cultures with Rod Outer Segments (ROS) with or without the Presence of Inhibitors

From bovine eyes, ROS were prepared using the modified method adapted from O'Brien et al.[19,20] and stored in 2.5% sucrose in incomplete PBS in liquid nitrogen. The confluent RPE cultures were challenged with ROS (10^7/ml) with or without the presence of leupeptin, pepstatin, antisense (AS1) (5'CAAACCAGCCGTTTCATCT) and sense (AS2) (5'AGATGAAACGGCTGGTTTG) oligonucleotides complementary to cathepsin S and an antisense 2 oligonucleotide (AS3) (5'CGCCGCCATGCAGCCC) complementary to cathepsin D at a concentration of 10 µM. The cultures were incubated for 3 days. Following incubation the cells were harvested in dim light and immediately measured for the presence of autofluorescence with fluorocytometry.

Measurement of Autofluorescent Debris Accumulation by Fluorocytometry

RPE cultures were harvested by trypsinisation at day three. Three parallel cultures were measured at each concentration. Trypsinised cells were pelleted by centrifugation and taken up in 250 µl Isoton (balanced salt solution) per sample, and filtered through 44 µm nylon mesh filters. The autofluorescence of 10,000 RPE cells harvested from each well was then measured at 530 nm (with 30 nm band width) on a fluorescence activated cell sorter (FACS) using light excitation at 488 nm.

Injection of Oligonucleotides into Rat Vitreous

Seven-week old RCS-rdy+ non-pigmented rats from our own colony were used for these experiments. The animals were anaesthetized by intraperitoneal injection of sodium pentobarbital (50 mg/kg body weight). The right eyes were injected with 2 µl of saline or 2 µl of saline containing 66 µg (10 nmol) of oligonucleotides AS1 and AS2 respectively into the vitreous. Injected animals were allowed to recover from anaesthesia and at the required post-injection time points, they were given an overdose of sodium pentobarbital. The injected eyes were enucleated and processed for further analysis. Two animals were used for each experimental set up.

Confocal Microscopy

At 7 days after intravitreal injection of a FITC-labeled AS1, the animals were deeply anaesthetised with halothane. The eyes were enucleated and snap frozen in OCT compound by placing in isopentane equilibrated with liquid nitrogen. The frozen blocks were sectioned ona cryostat and 7–10 µm thick sections were immediately analyzed by confocal laser scanning microscopy CLSM, using a MRC-1000 confocal microscope BioRad, Hercules, Ca. USA equipped with a Krypton/Argon-laser. A 10x Plan Apo lens NA 0.45, air, Nikon, Tokyo, Japan as well as a 20x Plan Apo lens NA 0.75, air, Nikon were used.

Light Microscopy

Following enucleation, whole eyes were immersed in 2.5% glutaraldehyde and 1% paraformaldehyde in 0.125 M sodium cacodylate buffer, pH 7.35 for 2 hours. The cornea and lens were then dissected free and the eye cup trimmed for orientation purposes. The tissue was fixed overnight at 4°C and then post fixed for 1 hour in 1% osmium tetroxide at room temperature. After ethanol dehydration, the tissue was embedded in epoxy resin Two μm sections were cut using a LKB 2088 Ultratome (LKB-Produkter, Brommer, Sweden) and stained with toluidine blue. The number of phagosomes that accumulated in the RPE cells of each specimen was determined by counting. Ten sets of counts were made at 100-fold magnification and the mean and standard error (SE) were calculated. Each set consisted of the total number of phagosomes in 100 μm length of RPE.

Immunohistochemistry and Cytochemistry

Immunohistochemical staining of cryostat sections and cultured RPE cells were processed and photographed as described earlier. Briefly, transfected RPE cells were seeded onto chamber slides and challenged with ROS for 12 hrs as described above. Following challenge the cells were washed with PBS and further incubated in growth medium At 5 days post wash the transfected RPE cells were washed again in PBS. RPE cells and cryostat sections were fixed in methanol at -20°C for 30 s and permeabilized by incubating with 47.5% ethanol in water for 15 min at room temperature. Cells were rehydrated in PBS / 0.1% BSA at room temperature for 30 min and then incubated at room temperature for 30 min with 10% normal goat serum in PBS containing 0.1% BSA. The goat serum was then drained off and replaced with the primary antibody overnight at 4°C. The cells were then washed three times in PBS containing 0.1% BSA followed by incubation with secondary antibodies (anti-rabbit IgG TRITC conjugate, Sigma Chemical Company, St. Louis, Mo., USA) at room temperature for 1 hr. Slides were examined at 570 nm excitation wavelength and photographed.

RESULTS

Accumulation of ROS-Derived Debris in ROS Challenged Cultured RPE Cells

RPE monolayers were challenged with ROS for 3 days and the accumulation autofluorescent debris was measured by FACS. It was found that the presence of leupeptin and antisense oligonucleotide AS1 resulted in a significant increase in the autofluorescent debris content of the cells (Table 1). However, AS2 sense oligonucleotide, pepstatin and a

Table 1. Accumulation of rod outer segment (ROS) derived debris in ROS challenged retinal pigment epithelial cells in the presence of lysosomal enzyme inhibitors

Target	Inhibitor	FACS[a]	IHC[b]
Cysteine proteases	leupeptin	23.5%[c]	+[d]
	AS1	39.2%	+
Aspartic proteases	pepstatin	3.8%	background
Antisense sequences	AS3	2.1%	N/D
	transfected cells	−25.1%	+++

[a] Fluorescence Activated Cell Sorter.
[b] Immunohisto/cytochemistry using a polyclonal antibody raised against bovine rod outer segments.
[c] Autofluorescent signal increase measured above background.
[d] The number of crosses represents an increase in the intensity of the signal.

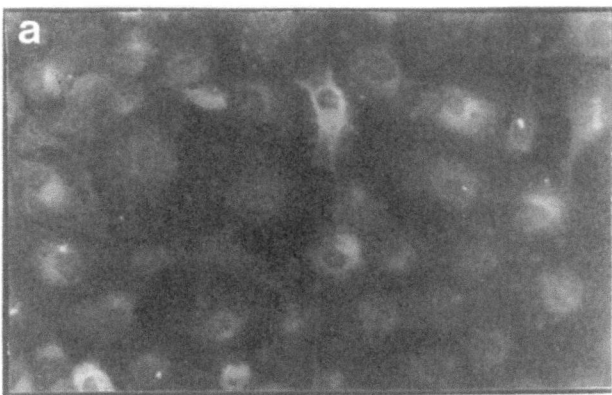

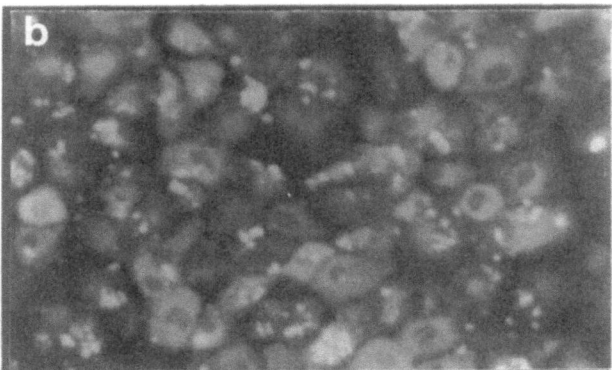

Figure 1. Immunocytochemical staining of RPE cells using a polyclonal antibody raised to bovine rod outer segments (ROS) 5 days post challenge. (a) Immunofluorescent staining of RPE cells incubated in the presence of 10 μM AS1. (b) Immunofluorescent staining of RPE cells incubated in the presence of 10 μM AS1 antisense oligonucleotide and challenged with ROS.

cathepsin D complementary antisense oligonucleotide (AS3) failed to induce an autofluorescent debris accumulation.

RPE monolayers were challenged with ROS for 12 hrs, rinsed and continued to be incubated in the medium. Five days post challenge immunocytochemistry was used to demostrate the presence of immunologically active ROS derived debris in the cells. It was found that RPE cells which were challenged and incubated in the presence of leupeptin or AS1 had significant amounts of ROS-derived debris present (Fig.1), which were easy to visualise. In contrast in RPE cells which were incubated in the absence of inhibitors, or in the presence of pepstatin, AS2 or AS3 only, small, residual amounts of immunologically active ROS-derived debris were present (Table 1).

Accumulation of ROS-Derived Debris *in Vivo*

It was found that the intravitreal injection of a fluoresceine-labeled AS1 oligonucleotide resulted in the accumulation of the oligonucleotide in the retinal pigment epithelium (Fig.2a) where it remained visible for an extended period of time (up to 28 days). The presence of the oligonucleotide was accompanied by the accumulation of phagosome-like structures (Fig.2b).

The number of phagosome-like structures in AS1-injected animals was significantly higher than in non-injected, saline or sense control oligonucleotide (AS2)-injected animals at 28 days post injection (Fig.3). Immunohistochemical staining of frozen retinas demon-

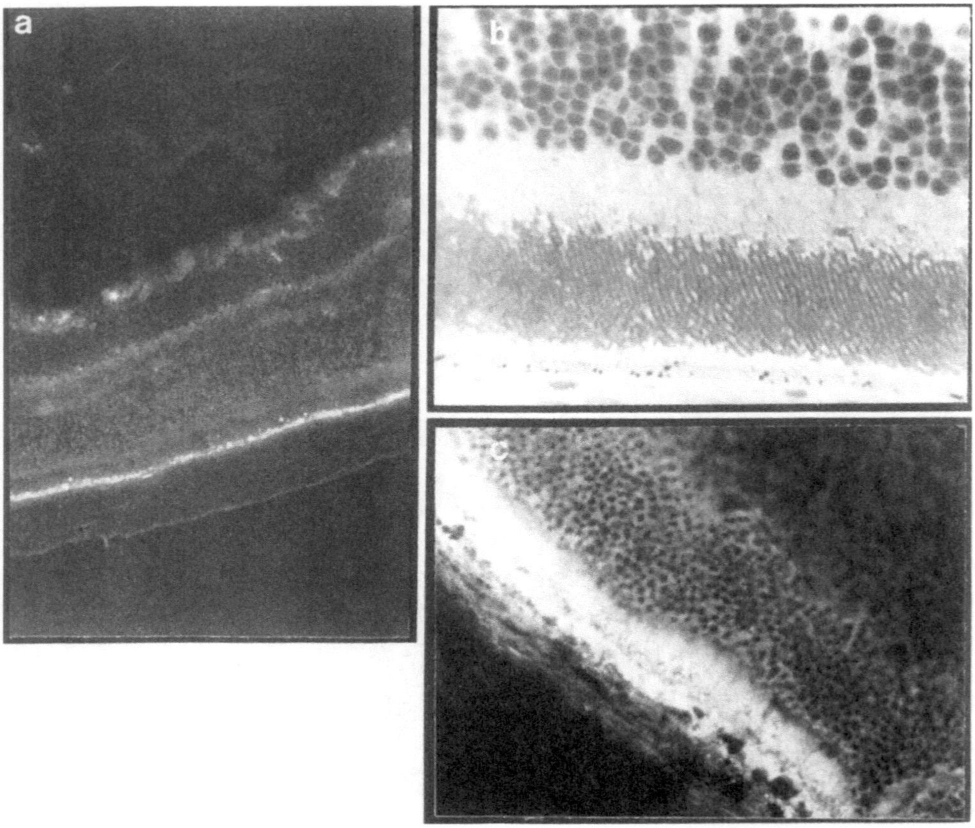

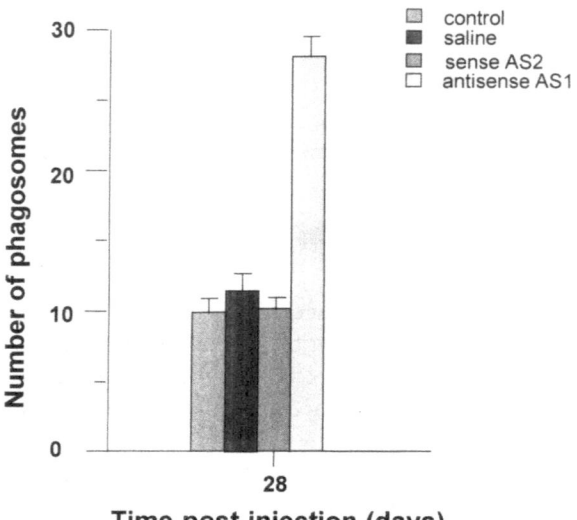

Figure 2. Microscopical examination of the retina of RCS-rdy+rat injected with 10 nmol AS1 antisense oligonucleotide at 28 days post injection. a: Confocal microscopic demonstration of the specific accumulation of Fluoresceine-labeled AS1 in the RPE layer (7 days post injection). b: Light microscopic demonstration of debris
accumulation in the RPE layer of an AS1-injected animal (note the large number of dark phagosomes). c: Imuno-
histochemical demonstration of photoreceptor derived particles in the RPE layer of an AS1-injected animal.

Figure 3. Graphical presentation of the accumulation of phagosome-like structures in
the RPE layer of uninjected control, saline
antisense (AS1) and sense (AS2) oligonu-
cleotide-injected RCS-rdy+ rats at 28 days
post injection.

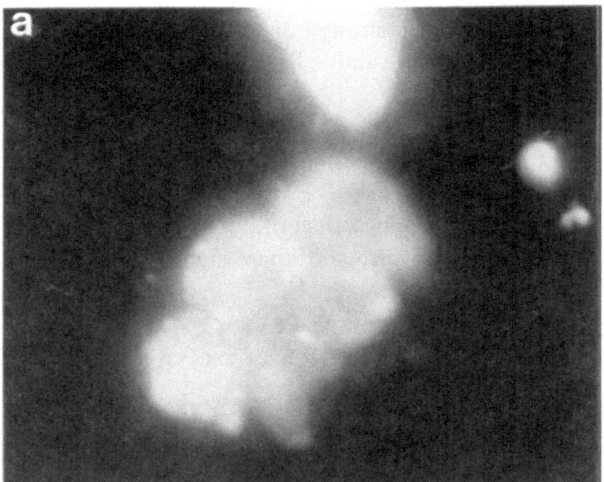

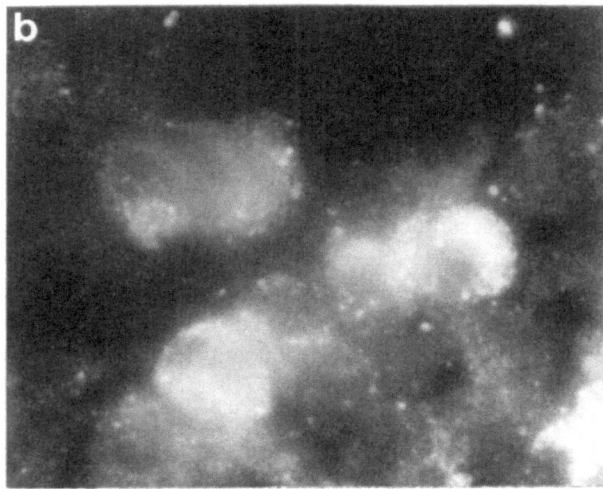

Figure 4. Immunocytochemical staining of antisense cathepsin D transformed RPE cells using a polyclonal antibody raised to bovine rod outer segments (ROS). (a) Demonstration of the presence of ROS derived particles in transformed RPE cells following 6 hours of challenge with ROS. (b) Demonstration of the presence of undigested ROS derived particles in transformed RPE cells at 5 days post challenge.

strated the presence of an increased amount of undigested, immunologically active outer segment particles in the RPE layer of AS1-injected animals at 28 days post injection (Fig.2c) when compared to the controls (data not shown).

Accumulation of Debris in Antisense Cathepsin D Transformed RPE Cells

RPE cells were transformed using a pHβApr-1-neo vector construct carrying cathepsin D in antisense direction. The presence, activity and effect of antisense cathepsin D in the transfected cells was demonstrated by molecular biological techniques including PCR, RTPCR and Western blot analysis, respectively (data not shown). Transfected cells retained their phagocytosing ability. However phagocytosis did not lead to the accumulation of autofluorescent debris (Table 1). Immunocytochemical analysis of transfected cells revealed that the cells phagocytosed ROS vigorously (Fig 4a). Following 12 hrs of challenge and 5 day post challenge incubation, there was no visible immunocytochemical signal present in vector transfected cells (data not shown). In contrast, in antisense cathepsin D transfected cells, a strong immunocytochemical signal (Fig.4b) demonstrated the presence of immunologically active ROS-derived debris.

DISCUSSION

The devastating effect of debris accumulation in the retina has been the focus of several investigations. Most of these studies were performed on RCS rats which are an excellent model to investigate changes occurring in the neural retina as a result of an impaired phagocytosis (Mullen and LaVail,[21]). However, neither this nor any other animal model enable us to study the effects of excessive debris accumulation which might occur within the RPE cells.

In an effort to induce photoreceptor-derived debris accumulation in the RPE layer, peptide-based lysosomal enzyme inhibitors have been used (Katz and Shanker,[22]). These studies demonstrated that the presence of leupeptin, a cysteine protease inhibitor, can induce the accumulation of autofluorescent phagosomes-like structures in the RPE cells both *in vitro* (Rakoczy *et al.*,[23]) and *in vivo* (Katz and Shanker,[22]) and provided circumstantial evidence about the outer segment origin of the debris. In this work, using immunocytochemistry we demonstrated the photoreceptor origin of the autofluorescent debris accumulating in leupeptin-treated RPE cells. These results confirm the difference between leupeptin-induced autofluorescent debris and the immunologically inactive lipofuscin. Inhibition of cathepsin D, the major lysosomal enzyme responsible for opsin proteolysis, was unsuccessful either with pepstatin or an antisense oligonucleotide (AS3).

To be able to develop an animal model in order to study POS debris accumulation, it is necessary to ensure the long term, steady inhibition of the target enzyme. One of the major shortcomings of peptide-based inhibitors is their limited stability in living cells thus preventing the performance of long term studies. In this work the efficacy of antisense DNA technology was tested *in vivo*. The ability of RPE cells to take up large molecules (Feeney and Mixon,[24], Rakoczy *et al.*,[17], Kennedy *et al.*,[25]), including oligonucleotides (Rakoczy *et al.*,[23]) *in vitro* suggests that RPE cells might be a suitable target for antisense therapy also *in vivo*. We have demonstrated that intravitreal injection of antisense oligonucleotides will result in the site-specific accumulation of the oligonucleotides in the retinal pigment epithelium, the longevity of an antisense oligonucleotide is superior to peptide-based inhibitors and that they can remain biologically active for an extended period of time (Figs.2 and 3). It was also demonstrated both *in vitro* and *in vivo* that the accumulating debris was of outer segment origin (Figs.1 and 4). However the efficacy of AS1 oligonucleotide *in vivo* was inferior to that previously detected in cultured RPE cells (Rakoczy *et al.*,[23]). While the use of AS1 *in vitro* caused the cytoplasm of cultured RPE cells to be clearly packed with ROS-derived phagosomes, the *in vivo* experiments could only demonstrate a significant increase (Fig.3). These differences, most probably, were due to a lower than optimal AS1 concentration in the *in vivo* system.

To overcome difficulties related to maintaining a biologically active concentration of the inhibitor and to provide inhibition for an extended period of time (months or years rather than days) modification of the genetic make up of RPE cells might offer an alternative. Transfection of cells with antisense molecules provides stable expression of the antisense sequence (Gunning *et al.*,[18], Saleh *et al.*,[26]). The expression: (i) is regulated by a specific promoter, (ii) is independent of the genomic sequence, (iii) eliminates the use of complicated protective groups, and (iv) usually provides higher specificity. We applied this technology to inhibit the expression of cathepsin D in RPE cells. Transfected RPE cells demonstrated that the antisense cathepsin D successfully inhibited cathepsin D production in these cells. There was no cathepsin D signal detected in the Western blot analysis (data not shown). In a functional analysis of transfected RPE cells, we found no increase in the autofluorescent debris content of antisense cathepsin D-transfected cells following ROS challenge. However immunocytochemistry demonstrated that the cells phagocytosed ROS and that the digestive process of phagocytosed outer segments was slowed down. These results suggest that, although ROS-derived debris accumulation can be induced by cathepsin D inhibition, the resulting debris is not autofluorescent.

Although none of the above described attempts provided the ideal technique to induce outer segment-derived debris accumulation in the RPE layer, these results demonstrated that molecular biology offers an effective tool for the production of precise *in vitro* and *in vivo* models which are possibly relevant to macular degeneration.

ACKNOWLEDGMENTS

The authors are grateful to Prof Michael Hall for the bovine outer segment polyclonal antibody. This work was supported by the National Health and Medical Research Council of Australia and by the Ulverscroft Foundation of the United Kingdom.

REFERENCES

1. Bressler, N.M., S.B. Bressler, S.K. West, S.L. Fine, and H.R. Taylor. 1989, The grading and prevalence of macular degeneration in Chesapeake Bay watermen. *Arch. Opthalmol.* **107:** 847–852.
2. Klein, R., B.E.K. Klein, and K.L.P. Linton. 1992. Prevalence of age-related maculopathy. The Beaver Dam Eye Study. *Opthalmology* **99:** 933–943.
3. Vingerling, J.R., I. Dielemans, A. Hofman, D.E. Grobbee, M. Hijmering, C.F.L. Kramer, and P.T.V.M. de Jong. 1995, The prevalence of age-related maculopathy in the Rotterdam study. *Ophthalmology* **102:** 205–210.
4. Mitchell, P., W. Smith, K. Attebo, and J.J. Wang. 1995, Prevalence of age-related maculopathy in Australia. *Ophthalmology* **102:** 1450–1460.
5. Weiter, J.J., F.C. Delori, G.L. Wing, and K.A. Fitch. 1986, Retinal pigment epithelial lipofuscin and melanin and choroidal melanin in human eyes, *Invest Ophthalmol. Vis. Sci.* **27:** 145–152.
6. Sarks, S.H. 1976, Ageing and degeneration in the macular region: a clinico-pathological study, *Br. J. Ophthalmol.* **60:** 324–341.
7. Gass, J.D.M. 1973, Drusen and disciform macular detachment and degeneration, *Arch Ophthalmol.* **90:** 206–217.
8. Pauleikhoff, D., M.J. Barondes, D. Minassian, I.H. Chisholm, and A.C. Bird. 1990, Drusen as risk factors in age-related macular disease, *Am. J. Opthalmol.* **109:** 38–43.
9. Katz, M.L. and W.G. Robison. 1984, Age-related changes in the retinal pigment epithelium of pigmented rats, *Exp. Eye Res.* **38:** 137–151.
10. Katz, M.L. and G.E. Eldred. 1989, Retinal light damage reduces autofluorescent pigment deposition in the retinal pigment epithelium, *Invest Ophthalmol. Vis. Sci.* **30:** 37–43.
11. Dorey, C.K., G. Wu, D. Ebenstein, A. Garsd, and J.J. Weiter. 1989, Cell loss in the aging retina, *Invest. Ophthalmol. Vis. Sci.* **30:** 1691–1699.
12. Delori, F.C., C.K. Dorey, G. Staurenghi, O. Arend, D.G. Goger, and J.J. Weiter. 1995, *In vivo* fluorescence of the ocular fundus exhibits retinal pigment epithelium lipofuscin characteristics, *Invest. Ophthalmol. Vis. Sci.* **36:** 718–729.
13. Pauleikhoff, D., A. Harper, J. Marshall, and A.C. Bird. 1990, Aging changes in Bruch's membrane, *Ophthalmology* **97:** 171–178.
14. Bressler, N.M., B. Munoz, M.G. Maguire, S.E. Vitale, O.D. Schein, H.R. Taylor, and S.K. West. 1995, Five-year incidence and disappearance of Drusen and retinal pigment epithelial abnormalities, *Arch. Ophthalmol.* **113:** 301–308.

15. Boulton, M., N.M. McKechnie, J. Breda, M. Bayly, and J. Marshall. 1989, The formation of autofluorescent granules in cultured human RPE, *Invest Ophthalmol. Vis. Sci.* **30**: 82–89.

16. Eldred, G.E. and M.R. Lasky. 1993, Retinal age pigments generated by self-assembling lysosomotropic detergents, *Nature* **361**: 724–726.

17. Rakoczy, P., C. Kennedy, D. Thompson-Wallis, K. Mann, and I. Constable. 1992, Changes in retinal pigment epithelial cell autofluorescence and protein expression associated with phagocytosis of rod outer segments *in vitro*, *Biol. Cell* **76**: 49–54.

18. Gunning, P., J. Leavitt, G. Muscat, S. Ng, and L. Kedes. 1987, A human b-actin expression vector system directs high-level accumulation of antisense transcripts, *Proc. Natl. Acad. Sci. USA* **84**: 4831–4835.

19. O'Brien, P.J. and C.G. Muellenberg. 1973, Incorporation of glucosamine into rhodopsin in isolated bovine retina, *Arch. Biochem. Biophys.* **158**: 36–42.

20. O'Brien, P.J., C.G. Muellenberg, and J.J. Bungenberg de Jong. 1972, Incorporation of leucine into rhodopsin in isolated bovine retina, *Biochemistry* **11**: 64–70.

21. Mullen, R.J. and M.M. LaVail. 1976, Inherited Retinal Dystrophy: Primary Defect in Pigment Epithelium Determined with Experimental Rat Chimeras, *Science* **192**: 799–801.

22. Katz, M.L. and M.J. Shanker. 1989, Development of lipofuscin-like fluorescence in the retinal pigment epithelium in response to protease inhibitor treatment, *Mech. Ageing Dev.* **49**: 23–40.

23. Rakoczy, P.E., K. Mann, D.M. Cavaney, T. Robertson, J. Papadimitreou, and I.J. Constable. 1994, Detection and possible functions of a cysteine protease involved in digestion of rod outer segments by retinal pigment epithelial cells, *Invest Ophthalmol. Vis. Sci.* **35**: 4100–4108.

24. Feeney, L. and R.N. Mixon. 1976, An in vitro model of phagocytosis in bovine and human retinal pigment epithelium, *Exp. Eye Res.* **22**: 533–548.

25. Kennedy, C.J., P.E. Rakoczy, T.A. Robertson, J.M. Papadimitriou, and I.J. Constable. 1994, Kinetic studies on phagocytosis and lysosomal digestion of rod outer segments by human retinal pigment epithelial cells *in vitro*, *Exp. Cell Res.* **210**: 209–214.

26. Saleh, M., S.A. Stacker, and A.F. Wilks. 1996, Inhibition of growth of C6 glioma cells *in vivo* by expression of antisense vascular endothelial growth factor sequence, *Cancer Res.* **56**: 393–401.

10

SUBRETINAL IODOACETATE

A Model of Retinal Degeneration in Cats

Jun C. Huang, Masahiro Ishida, Sarah Goldfeder, Ilene K. Sugino, and
Marco A. Zarbin

Department of Ophthalmology
New Jersey Medical School
University of Medicine and Dentistry of New Jersey
Newark, New Jersey

ABSTRACT

To create a model of retinal degeneration, sodium iodoacetate was used to selectively abolish the photoreceptors (PR) of the cat retina in the area centralis. Iodoacetate (IA) was administered by subretinal injection in the area centralis. Subretinal delivery involved injection of IA into the subretinal space of a 4 mm diameter retinal bleb, using a dose of 0.1, 1, 1.6, 2 or 5 μg IA/bleb. IA was dissolved in 50 μl of BSS or a viscous hyaluronate solution (Healon®, 0.5%). Survival time was 2 to 6 weeks, followed by histologic study of the retina. Subretinal IA injection induced well localized and selective PR destruction with preservation of RPE and other retinal cells. Total PR degeneration within the retinal bleb was present 6 weeks after injection of a mixture of IA (1.6, 2, or 5 μg) with Healon®. The viscosity of the solution may have prolonged the confinement of IA to the subretinal space and promoted a localized IA effect. Subretinal IA dissolved in BSS caused less severe and less consistent PR degeneration. Healon® or BSS alone did not induce PR damage. Thus, a model with PR-lacking area centralis has been created by subretinal injection of IA with a viscous solution. This model may subserve investigations of retinal degeneration as well as retinal transplantation.

INTRODUCTION

An appropriate animal model can facilitate the development of human photoreceptor (PR) transplantation techniques. The model should be reproducible and relatively inexpensive. The recipient retina should be PR deficient and holangiotic. Ideally, the remainder of the retina and the retinal pigment epithelium (RPE) is normal in such animals to mimic the condition found in patients with retinitis pigmentosa, the likely first human recipients of such transplants.[1] In addition, the size of the eye should approximate that of the human to permit the development of clinically useful instruments. The animal's visual acuity should be readily and accurately assessable. Ideally the animal should also exhibit different sorts of photoreceptor degenerations to foster the development of a generally applicable technology.

The anatomy and electrophysiology of the cat retina has been well studied[2-6], and feline visual acuity has been quantitated in a number of behavioral studies.[7-9] Some cats exhibit hereditary retinal degeneration, such as Abyssinian *Rdy* and *Rd* cats.[10-12] It is, however, very expensive and time consuming to breed and maintain a colony of such animals. Furthermore, the precise mechanism that causes PR degeneration (i.e., abnormalities in the PRs vs. RPE) in these animals is unknown. As an alternative, an animal model may be produced using pharmacologic agents to selectively induce PR degeneration.

Iodoacetic acid, often as sodium iodoacetate (IA), is a metabolic inhibitor which was first shown to produce selective effects on the function of the mammalian retina by intravenous administration.[13] Depending on the dose, the toxic effect of intravenous IA is either reversible or is associated with PR cell death. Appropriate doses will destroy PRs with minimal damage to RPE and other retinal cells. IA has been shown to cause impairment of PR structure and function in several species including cat, rabbit, and monkey, with the cat PRs being most sensitive to the drug.[13-18] It appears, however, that the dose required for PR degeneration by intravenous injection can cause severe systemic side effects which may hinder its application as a useful model for retinal degeneration.

Attempts have been made to reduce the dose of IA by intracarotid arterial[19,20] or intravitreous IA injection.[21,22] With intracarotid injection in the rabbit, the dose used was still about 50% of that for the intravenous approach.[19,20] With intravitreous injection in the rabbit, the IA dose required was greatly reduced. The corresponding electroretinogram was diminished, and the retina was degenerate.[21,22] Nonetheless, intravitreous iodoacetate in the rabbit also damaged inner retinal neurons, such as bipolar and ganglion cells, as well as PR cells.[22]

We carried out studies to induce PR degeneration using IA, comparing intravenous, intravitreal, and subretinal routes of IA administration. This chapter focuses on the results of subretinal IA injection. Subretinal IA injection is a new experimental paradigm as far as we know. We found that subretinal IA injection is an appropriate and perhaps ideal route of administration by which to create a PR deficient area centralis in cats.

METHODS

Normal domestic cats (2.5–4.5 kg) were purchased from Liberty Lab, NY. The animals were housed under cyclic light conditions: 12 hours on and 12 hours off.

Subretinal IA (Sodium Iodoacetate, Sigma I7768) Injection

A balanced salt solution (BSS) or 0.5% Healon® (sodium hyaluronate, a viscoelastic liquid for intraocular use; Pharmacia Ophthalmics Inc., Ca) was used to dissolve IA. BSS (Alcon Surgical, Fort Worth, Texas) has been widely used during intraocular surgery in humans and appears to be well tolerated clinically. It is well tolerated in cats.[23,24] Each milliliter contains sodium chloride (0.64%), potassium chloride (0.075%), calcium chloride dihydrate (0.048%), magnesium chloride hexahydrate (0.03%), sodium acetate trihydrate (0.39%), and sodium citrate dihydrate (0.17%). The solution of IA/Healon® was prepared as follows: 1 part IA in BSS was mixed with 1 part 1% Healon® (10 mg/ml), followed by vortex and brief centrifugation; the final solution contained the desired concentration of IA in 0.5% Healon®. All solutions were stored in the dark at 4°C and used within 3 hours of preparation.

Animals were sedated with IM ketamine (20 mg/kg) and xylazine (1 mg/kg), followed by intubation. Anesthesia was maintained with isoflurane (1–1.5%) in oxygen (1.5–2 liter/min). The animals breathed spontaneously and were monitored with pulse oximetry. Under aseptic conditions, animals underwent pars plana vitrectomy. A 30 gauge cannula was used to incise the retina over the tapetum and BSS was injected to create a 4 mm diameter retinal bleb. We utilized a slow injection technique to minimize hydraulic displacement of RPE from Bruch's membrane. Two such blebs were created over the tapetum separated by 1–3 mm. IA, dissolved in 50 µl of either BSS or 0.5% Healon®, was then injected through the retinotomy into one of the blebs. The IA dose was: 0.1, 1.0, 1.6, 2.0, or 5.0 µg. When injecting a solution of BSS, a 30 g cannula was used. When injecting a solution of Healon®, a 27 g cannula was used due to the viscosity of Healon®. The other bleb received either BSS or Healon® as a control. The animals were followed clinically daily for 3 days. Animals were sedated with IM ketamine (20 mg/kg) and xylazine (1 mg/kg) for fundus photography and FA at 2 or 6 weeks prior to sacrifice. Animals were sacrificed with an overdose of sodium pentobarbital (250mg/kg, i.v.).The experiments conformed to the ARVO Statement for the Use of Animals in Ophthalmic and Vision Research.

Morphologic Analysis

After enucleation, the globes were fixed with 2.5% glutaraldehyde and 2% formaldehyde, pH 7.4. Selected retinal areas were embedded in JB4 and sectioned at 2 µm. The sections were stained with toluidine blue. Sections from all specimens were examined with a Zeiss Axioskop light microscope. We also studied photographs obtained with a Hamamatsu video camera connected to a Seikosha VP-4500 printer. The geographic location of each section was recorded. The severity of retinal damage after IA treatment was graded by comparison among all areas and recorded in the photographically. In this way, degree of retinal and RPE damage in each geographic location was compared with that in other areas and mapped out.

RESULTS

Subretinal IA injection resulted in selective PR degeneration that was well localized within the retinal bleb. The doses used included 0.1, 1.0, 1.6, 2.0 and 5.0 µg, with survival times of 2 or 6 weeks. Initially, we dissolved the drug in BSS for injection. We observed that, after injection, some retinal blebs partially collapsed even before the end of surgery. Presumably, some solution with the drug leaked into the vitreous cavity. It was not evident that this intravitreal IA damaged retinal cells in the other areas of the eye, probably because the amount was too small to be effective and because intraocular irrigation was still occurring during the surgery, thus flushing the drug out of the vitreous cavity. However, this situation could result in an inconsistent IA concentration inside the bleb. Experiments in which the same amount of IA (e.g., 5 µg) in BSS was injected into similarly sized blebs yielded quite different pathologic results. For example, one bleb exhibited near total PR cell degeneration, while another exhibited only vacuolation of OSs, ISs, and the OPL and no apparent ONL changes (not shown). This variability was likely due to the diffusion of IA out of the bleb. This hypothesis led us to use a solution of IA dissolved in a viscous agent, 0.5% Healon® for subsequent subretinal injections.

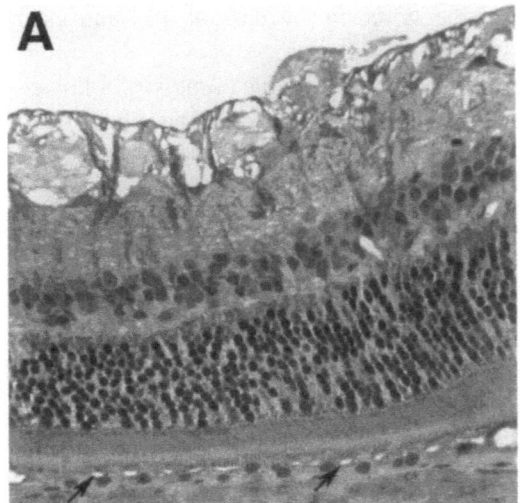

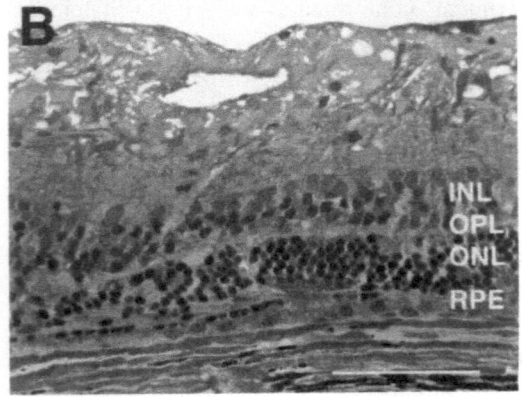

Figure 1. Histology of retinal blebs 6 weeks after a subretinal injection of 1.6 µg iodoacetate in BSS or BSS alone. (A) BSS only: All retinal structures appear normal, except for some microdetachments (arrows). Some vacuoles in the NFL are artifacts. (B) IA with BSS: PR cells are degenerate. There is loss of the OSs, ISs, and most of the OPL, and thinning and disorganization of the ONL with pyknosis. Other retinal cells and the RPE appear normal. Bar = 50 µm.

Histological study confirmed that localized retinal detachments created by Healon® could last for 6 weeks. Furthermore, the IA effect on the PR cells was consistent among different animals in different experiments. Figures 1 and 2 show the histology of retinal blebs that were treated with 1.6 µg IA, BSS, or Healon® for 6 weeks. Retinal histology in blebs injected with BSS or Healon® was essentially normal (Figs. 1A and 2A), although some microdetachments with BSS (Fig. 1A) and a shallow total detachment with Healon® (Fig. 2A) were present. PR cells were severely damaged in the bleb treated with iodoacetate in BSS; OSs, ISs, and most of OPL were degenerate or absent, and ONL was thinned and disorganized (Fig. 1B). In contrast to the partial PR destruction observed after IA/BSS injection, complete PR degeneration was induced by subretinal injection of IA with Healon® (Fig. 2B). Almost all PR cells were destroyed with only a few pyknotic ONL nuclei remaining. The outer retina was gliotic. Other retinal cells appeared mildly affected with some vacuolation present in the inner plexiform (IPL) and nerve fiber layers (NFL), but the vacuolation was sometimes minimal (see Fig. 4). Blebs created with Healon®, with or without IA, were still detached 6 weeks after injection (Figs. 2A, B), whereas the blebs created with BSS, with or without IA, exhibited essentially complete retinal reattachment (Figs. 1A, B).

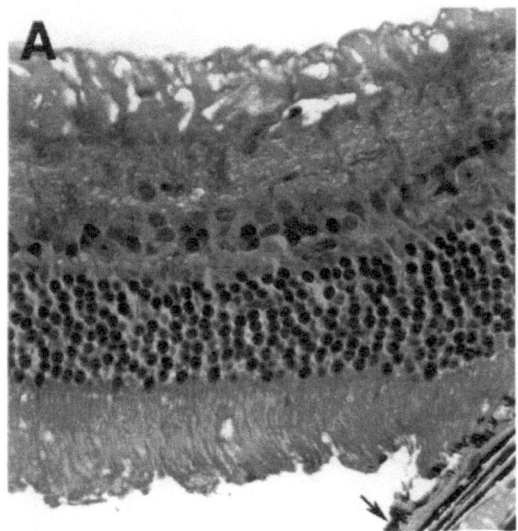

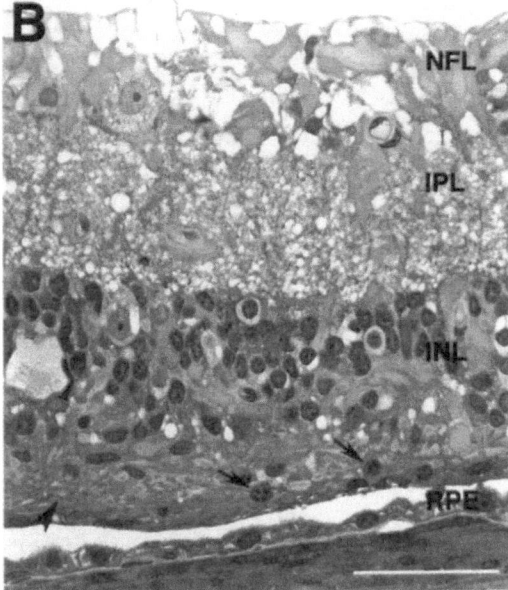

Figure 2. Histology of retinal blebs 6 weeks after a subretinal injection of 1.6 µg iodoacetate in Healon® or Healon® alone. (A) Healon® alone: The bleb is still detached from the RPE (arrow). PRs and other retinal cells appear normal. (B) IA with Healon®: Almost all PR cells are absent with a few pyknotic PR nuclei remaining (arrows). There is some vacuolation in the IPL and NFL. The INL is slightly thickened. Some gliosis is present (arrowhead) in place of the degenerated PR cells which are separated from the RPE monolayer by a space. Note that retinal blebs with Healon® are still detached from the RPE throughout the bleb (i.e., about 4 mm in diameter) while blebs with BSS (Fig. 1) are reattached to the RPE. Bar = 50 µm.

Figure 3 shows the fundus photographs of the blebs receiving 2.0 µg IA with either BSS or Healon®. The bleb treated with IA in BSS showed RPE mottling in the area of previous retinal detachment (Fig. 3A). The bleb treated with IA in Healon® showed similar RPE mottling except in the center where persistent low lying retinal detachment obscured the subjacent RPE and choroid (Fig. 3B). This was confirmed histologically (Fig. 4). Fluorescein angiography revealed no significant dye leakage from the choroidal or retinal circulation in both blebs, nor in other blebs similarly treated (not shown). This was consistent with the histologic findings that no neovascularization in the retina or choroid, nor RPE defects, were present. It also indicates that retinal degeneration occurs in this model with preservation of the inner and outer blood-retina barriers.

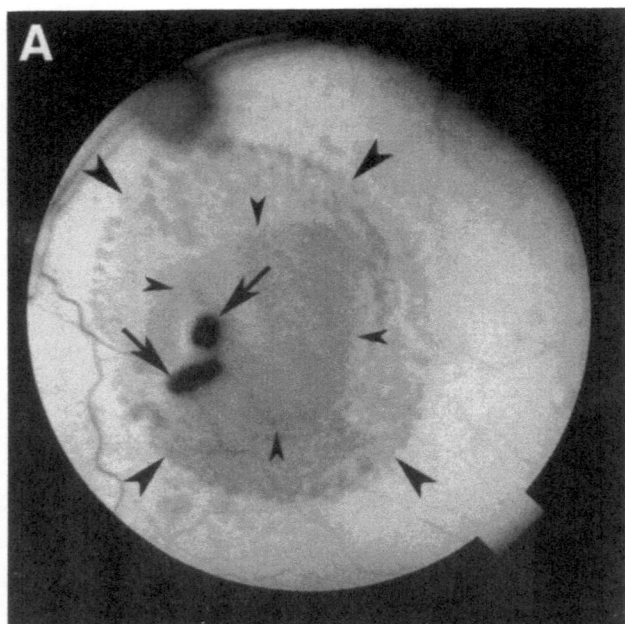

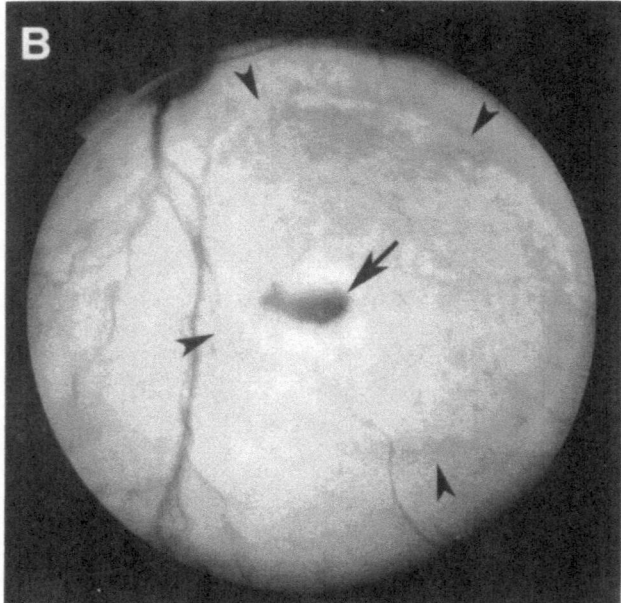

Figure 3. Fundus photographs of retinal blebs treated with 2.0 µg subretinal iodoacetate (IA) for 6 weeks. The bleb margins are evident (arrowheads), and the size of both blebs is about 4 mm in diameter. The large pigmented dots (arrows) inside the blebs are old breaks in the RPE and tapetum created by the cannula while making the bleb. The pigmentation arises from the choroid as verified histologically. (A) IA with BSS: The bleb is outlined by arrowheads. The color of the retinal bleb is close to that of the surrounding normal areas. Stippled RPE mottling is evident in the area of the previous retinal detachment. (B) IA with Healon[a]: Stippled RPE mottling is evident in the area of the bleb except centrally where persistent low lying retinal detachment obscures a view of the underlying RPE-choroid (small arrowheads). Note that histology revealed a persistent separation of retina and RPE throughout the bleb (Fig. 4).

Histology of the retinal bleb treated with 2.0 µg IA and Healon[®] and shown clinically in Fig. 3B is presented in Figure 4. PR cells were absent throughout the entire 4 mm diameter bleb (Fig. 4A). The PR cells appeared to be replaced by some degree of gliosis forming a membranous structure. The INL and RPE cells survived well with minimal vacuolation (Fig. 4B). There was a mild increase in the thickness (i.e., cellularity) of the INL. The bleb containing IA and BSS had partial PR degeneration similar to that shown in Figure 1B (not shown).

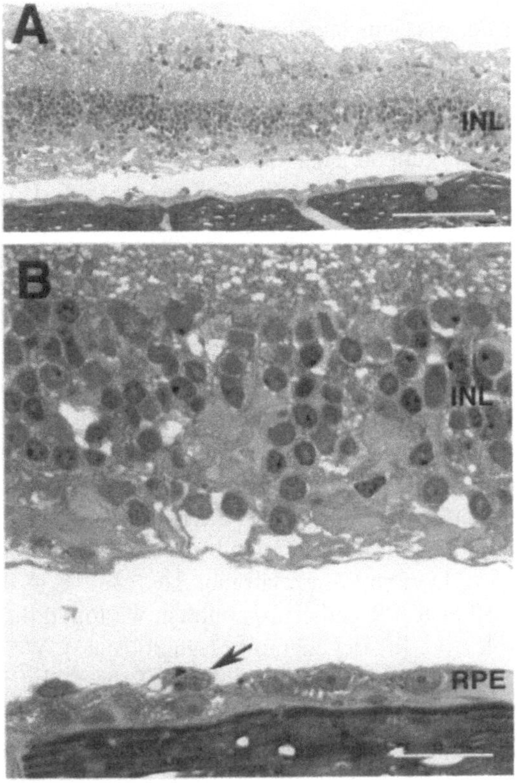

Figure 4. Histology of retinal bleb shown in Fig. 3B treated with 2.0 µg subretinal iodoacetate and Healon® for 6 weeks. The bleb was 4 mm in diameter and separated by a space from the RPE throughout. (A) Low magnification. PR cells are completely degenerate while other retinal cells including RPE appear normal. There is mild vacuolation in the INL. Bar = 100 µm. (B) High magnification. No PR cells are seen, and glial tissue (arrowhead) forms a membrane outlining the margin of the remaining outer retina. The INL is slightly thickened, but cell nuclei and cytoplasm appear essentially normal with minimal vacuoles. RPE cells appear normal without degenerative or hypertrophic changes. A macrophage with possible phagocytized cellular nucleus is present (arrow). Bar = 20 µm.

Thus, using IA in Healon®, we obtained consistent outer retinal degeneration in 4 cats with IA doses of 1.6 µg (n=2 cats), 2.0 µg, and 5.0 µg. All doses caused PR degeneration as seen in Figures 1–4. In the animal receiving 5 µg IA, the RPE cells were hypertrophic (cuboidal in shape) and exhibited occasional acinus formation. This morphology was seen in both IA/BSS and IA/Healon blebs. This RPE hypertrophy was not seen in any cat receiving a lower IA dose. In contrast to IA/Healon, IA/BSS injection often resulted in inconsistent and incomplete PR degeneration in a total of 8 cats, with doses of 0.1 µg, 1.0 µg, 1.6 µg (n=2), 2.0 µg, and 5.0 µg (n=3).

DISCUSSION

Systemically administered IA has been reported to induce selective PR destruction.[13–18] Damage to the PR cells includes suppression of PR responses in the electroretinogram at low doses [13,14] and permanent PR destruction at higher doses. Early changes, 3 to 6 hours after intravenous injection, involve disorganization and lysis of the OS discs, swelling and vacuolation of the matrix, endoplasmic reticulum and Golgi complex of the ISs, mitochondrial disintegration, and lysis of the synaptic vesicles.[25,26] The rods are more susceptible than the cones, but both are destroyed completely at proper doses.[13–15,25]

The mechanism of IA action is thought to be inhibition of glycolysis, especially anaerobic glycolysis.[13,17,18] Retinal cells, particularly PR cells, have a very high rate of glycolytic metabolism,[17,18,27] which may account for their susceptibility to IA toxicity. IA uptake in the retina, probably mainly in the PR cells, is greater than that in the RPE, choroid, iris, lens, and brain tissue.[26] The pattern of degeneration, i.e., selectivity for PR cells and higher sensitivity for rods than cones, led Noell to suggest that retinal degeneration induced by IA simulates retinitis pigmentosa.[15]

We have reproduced similar pathologic changes in the PR cells by intravenous IA in the cat (unpublished observations). One week after IA injection, PR cells were degenerating with destruction of OSs, ISs, and the OPL. However, systemic toxicity (severe neurological symptoms and hematolgical disorders) with intravenous IA rendered this approach unsatisfactory, although intravenous IA has been employed in most prior studies. The systemic toxicity was mentioned briefly by Noell[13], including severe central nervous, intestinal, and renal symptoms lasting for several days. Nonetheless, our experience using even lower doses than those employed by Noell leads us to believe that the intravenous injection of IA is not an ideal way to obtain retinal degeneration. We contemplated unilateral carotid artery injection that, according to the literature, could reduce the drug dose by about 50%.[19,20] Because of possible systemic side effects, we did not pursue this approach. We studied other drug delivery routes to greatly reduce the needed doses and systemic toxicity of IA. Intraocular injection (i.e., intravitreous [unpublished] and subretinal) reduced the dose needed to induce PR degeneration by 1,000 fold or more with no systemic side effects.

In rabbits intravitreous IA resulted in damage to the ganglion and bipolar cells as well as to PR cells.[22] In contrast, we found that intravitreal IA (5–20 μg) caused relatively selective PR degeneration (unpublished). We think this difference may be due to the unusual anatomy and physiology of the rabbit retina. (For example, under anoxia or postmortem, rabbit retina undergoes deterioration from the outer towards the inner retina,[28] while animals with a dual retinal blood supply, including cats[29] and humans,[30] exhibit changes from the inner towards the outer retina.) Nonetheless, the severity of intravitreal IA-induced PR degeneration exhibited unpredictable topographical variation.

Subretinal IA injection yielded a reproducible model of PR degeneration in the area centralis (Figs. 1–4). PR cells were selectively and predictably damaged (by morphologic criteria) with 1.6, 2.0, or 5.0 μg of IA dissolved in Healon[®]. The ideal IA dose may depend on several factors, especially (1) the size of retinal bleb, which may vary somewhat with different experiments and different surgeons, and (2) the retention of the drug inside the bleb after injection, which may be best accomplished by using a viscous liquid as a carrier. We made the blebs about 4 mm in diameter and used 0.5% Healon[®] solution to retain the drug inside the bleb. This delivery approach resulted in a consistent effect of IA on PR cells. We think that a dose of 1.6 or 2.0 μg per bleb may be optimal for selective PR destruction. In blebs exhibiting PR degeneration, other retinal cells appeared largely normal morphologically, except for some vacuolation in the IPL and possibly in the NFL (Figs. 1, 2 and 4). RPE remained a monolayer of cells without hypertrophy in most cases. Some gliosis was present with PR degeneration. This situation may actually be more relevant to transplantation in humans, as gliosis also occurs after PR degeneration in some hereditary retinal degeneration such as retinitis pigmentosa.[31] We emphasize that our conclusions regarding the selectivity of this model for PR damage are based solely on histopathological data. Electrophysiological experiments to assess the function of the surviving retinal and RPE cells are in progress.

IA injected in a salt solution (BSS) vehicle leaked from the bleb into the vitreous cavity, as the bleb often collapsed partially soon after subretinal IA injection. The IA doses applied to the subretinal space were probably inconsistent, thus exerting a variable and weaker effect on PR cells, compared with that seen in experiments using a Healon[®] vehicle (Fig. 2B and Fig. 4). In control groups, injection of BSS alone did not cause damage to the retina, as all retinal blebs reattached with only some microdetachments. Retinal morphology was normal (Fig. 1A). Injection of Healon® kept the retinal bleb elevated for a long period of time, e.g., 6 weeks, at which point the animals were sacrificed. It is interesting that this prolonged low lying retinal detachment did not cause significant changes in retinal morphology. No shortening or swelling of the OSs or ISs, nor thinning of the ONL and OPL were evident. Thus, the PR degeneration described above was clearly a consequence of the action of IA, and Healon® per se did not cause retinal damage.

ACKNOWLEDGMENTS

The authors are grateful to Dr. Eva Ryden and Dr. Matthew Panarella for their helpful discussions and suggestions on the animal experiments with iodoacetate, and to Drs. Ruihong Yao and Xue guang Zhang for their assistance in the surgical procedures.

Supported by Research to Prevent Blindness, Inc. and the Lions Eye Research Foundation.

REFERENCES

1. Das, T.P., del Cerro, M., Lazar, E.S., Jalali, S., DeLoreto, D.A., Little, C.W., Sreedharan, A., del Cerro, C. and Rao, G.N., 1996, Transplantation of neural retina in patients with retinitis pigmentosa, *Invest. Ophthalmol. Vis. Sci.*, **37**: S96.
2. Braekevelt, C.R., 1990, Fine structure of the retinal photoreceptors of the domestic cat (felis catus), *Anat. Histol. Embryol.*, **19**: 67–76.
3. Braekevelt, C.R., 1990, Retinal epithelial fine structure in the domestic cat (felis catus), *Anat. Histol. Embryol.*, **19**: 58–66.
4. Siliprandi, R., Bucci, M.G., Canella, R. and Carmignoto, G., 1988, Flash and pattern electroretinograms during and after acute intraocular pressure elevation in cats, *Invest. Ophthalmol. Vis. Sci.*, **29**: 558–565.
5. Arakawa, K., Peachey, N.S. and Celesia, G.G., 1993, Spatial frequency response functions obtained from cat visual evoked potentials, *Electroencephalography Clin. Neurophysiol.*, **88**: 143–150.
6. Vaegan and Millar, T.J., 1994, Effect of kainic acid and NMDA on the pattern electroretinogram, the scotopic threshold response, the oscillatory potentials and the electroretinogram in the urethane anaesthetized cat, *Vis. Res.*, **34**: 1111–1125.
7. Blake, R., Cool, S.J. and Crawford, M.L.J., 1974, Visual resolution in the cat, *Vis. Res.*, **14**: 1211–1217.
8. Mitchell, D.E., Giffin, F., Wilkinson, F., Anderson, P. and Smith, M.L., 1976, Visual resolution in young kittens, *Vis. Res.*, **16**: 363–366.
9. Pasternak, T. and Horn, K., 1991, Spatial vision of the cat: Variation with eccentricity. *Vis. Neurosci.*, **6**: 151–158.
10. Barnett, K.C. and Curtis, R., 1985, Autosomal dominant progressive retinal atrophy in Abyssinian cats, *J. Heredity*, **76**: 166–170.
11. Narfstrom, K., 1983, Hereditary progressive retinal atrophy in the Abyssinian cat. *J. Hered.*, **74**: 273–276.
12. Narfstrom, K., 1985, Progressive retinal atrophy in the Abyssinian cat. Clinical characteristics, *Invest. Ophthalmolo. Vis. Sci.*, **26**: 193–200.
13. Noell, W.K., 1951, The effect of iodoacetate on the vertebrate retina, *J. Cell. Comp. Physiol.*, **37**: 283–297.
14. Noell, W.K., 1952, The impairment of visual cell structure by iodoacetate, *J. Cell Comp. Physiol.* **40**: 25–45.
15. Noell, W.K., 1953, Experimentally induced toxic effects on structure and function of visual cells and pigment epithelium, *A. J. Ophthalmol.*, **36**: 103–116.
16. Graymore, C. and Tansley, K., 1959, Iodoacetate poisoning of the retina, I. Production of retinal degeneration, *Brit. J. Ophthalmol.*, **43**: 177–185.
17. Webb, J.L., 1966, Iodoacetate and Iodoacetamide, in: Webb, J.L. (ed), *Enzymes and metabolic inhibitors*, **3**: 1–283, Academic Press, New York, .
18. Yamamoto, F. and Honda, Y., 1993, Effects of intravenous iodoacetate and iodate on pH outside rod photoreceptors in the cat retina, *Invest. Ophthalmol. Vis. Sci.*, **34**: 2009–2017.

19. Tieri, O., de Berardinis. E., Vecchione, L. and Polzella, A., 1962, Eludes sur le metabolisme et la fonction des cellules visuelles-I. Recherches sur le mecanisme d'action de l'acide iodacetique sur la retine de lapin, *Vision Res.*, **2**: 373–382.

20. Gaipa, M., Tieri, O., Vecchione, L. and Polzella, A., 1963, Etudes sur le metabolisme et la fonction des cellules, *Vision Res.*, **3**: 285–288.

21. Bock, J., Bornschein,H. and Hommer, K., 1962, Experimentalle untersuchungen uber eine transvitreale beeinflussung des elektroretinogramms, *Doc. Ophthal.*, **16**: 35–52.

22. Ponte, F. and Lauricella, M., 1968, Intravitreous administration in the study of experimental physiopathology of the retina, II. Effect of iodoacetic acid. *Annali. di. Ottalmologia.*, **94**: 477–489.

23. Panozzo, G., Sugino, I. and Zarbin, M.A., 1992, Effect of mechanical RPE debridement on photoreceptor survival, *Invest. Ophthalmol. Vis. Sci.*, **33**: 1128.

24. Zhang, X.G., Sugino, I.K. and Zarbin, M.A., 1994, A clinicopathological correlation of localized mechanical and hydraulic retinal pigment epithelium (RPE) debridement in cats. *Invest. Ophthalmol. Vis. Sci.*, **35**: 1335.

25. Lasansky, A. and DeRobertis, E., 1958, Submicroscopic changes in visual cells of the rabbit induced by iodoacetate, *J. Biophysic. Biochem.*, **5**: 245–261.

26. Orzalesi, N., Calabria, G.A. and Grignolo, A., 1970, Experimental degeneration of the rabbit retina induced by iodoacetatic acid, *Exp. Eye Res.*, **9**: 246–253.

27. Voaden, M.J., 1979, Vision: the biochemistry of the retina, in: Bull. E.T., Laguado. J.R,, Thomas, J.O. and Tipton, K.F. (eds), *Companion to biochemistry, II.*, pp. 451–473, Longman Group Ltd., London, New York.

28. Johnson, N.F. and Grierson, I., 1976, Postmortem changes in the rabbit retina, *Acta Ophthalmol.*, **54**: 529–541.

29. Reinecke, R.D., Kuwabara, T., Cogan, D.G. and Weis, D.R., 1962, Retinal vascular patterns. V. Experimental ischaemia of the cat eye, *Arch. Ophthalmol.*, **67**: 70–475.

30. Huang, J.C., 1990, A study of structure and function in postmortem human retina, *Ph.D. Thesis*, University of London, U.K.

31. Marshall, J. and Heckenlively, J.R., 1988, Pathologic findings and putative mechanisms in retinitis pigmentosa, in: *Retinitis Pigmentosa*, (J.R. Heckenlively, ed.), pp. 37- 67, J.B. Lippincott Co., New York.

11

HEREDITARY RETINAL DYSTROPHY OF SWEDISH BRIARD DOGS

Exclusion of Six Candidate Genes by Molecular Genetic Analysis

Andres Veske,[1*] Sven Erik G. Nilsson,[2] Ulrich Finckh,[1] Kristina Narfström,[3] Simon Petersen-Jones,[4] David Gould,[4] David Sargan,[4] and Andreas Gal[1]

[1]Institut für Humangenetik
Universitäts-Krankenhaus Eppendorf
Butenfeld 42, D-22529, Hamburg, Germany
[2]Department of Ophthalmology
University of Linköping
S-581 85 Linköping, Sweden
[3]Department of Surgery and Medicine
Faculty of Veterinary Medicine
Swedish University of Agricultural Sciences
S-750 07 Uppsala, Sweden
[4]Department of Clinical Veterinary Medicine
University of Cambridge
Cambridge CB3 0ES, United Kingdom

SUMMARY

During the past years, molecular genetic studies have revealed disease causing mutations in several genes encoding various retina-specific proteins in inherited retinal dystrophies both in different animal species and in human. In dogs, different types of retinal dystrophies have been described affecting primarily the photoreceptors and/or pigment epithelium. A recessively inherited retinal dystrophy has been reported in a strain of Swedish Briards. The aim of our study was to determine, whether or not defects in the genes encoding canine rhodopsin, peripherin/RDS, arrestin, ROM1, the β-subunit of rod photoreceptor cGMP-specific phosphodiesterase (βPDE), and the α-subunit of rod cGMP-gated cation channel are cause of the inherited retinal dystrophy in this strain of dogs.

* On leave of absence from the Institute of Molecular and Cell Biology, University of Tartu, Tartu, Estonia.

Degenerative Retinal Diseases, edited by LaVail *et al.*
Plenum Press, New York, 1997

We analysed cDNA or genomic DNA of the above mentioned genes by single strand conformation polymorphism (SSCP) analysis and direct sequencing of PCR products to screen the coding region as well as most splice donor and acceptor sites for mutations in a small kinship of dogs consisting of six offspring of an affected Swedish Briard dog. Our analysis revealed various sequence polymorphisms in all genes except arrestin. Our results show that the disease phenotype does not cosegregate with any of the intragenic sequence variants identified by us in the above-mentioned genes. In a view of these data, it seems unlikely that mutations of any of these genes are causative for the retinal degeneration observed in the Swedish Briard kindred studied.

INTRODUCTION

The hypothesis that specialization of photoreceptor rod and cone cells makes them more vulnerable to inherited and acquired diseases has been widely proclaimed (1), and evidence to support this assumption is now emerging. Mutations of genes encoding various photoreceptor-specific proteins have been identified in various types of retinal degenerations, dystrophies, and dysplasias in different species. In dogs, inherited retinal dystrophies comprise a heterogeneous group of diseases with the terminal phenotype of blindness (2). Most of these conditions follow an autosomal recessive inheritance, although X-linked and autosomal dominant forms have also been described (3). However, up to now, disease related mutations have been identified in only one canine gene. Suber et al. (1) reported a nonsense mutation in the gene encoding the β-subunit of rod photoreceptor cGMP-specific phosphodiesterase (βPDE) in the early onset rod/cone dysplasia of Irish setter.

The slowly progressive retinal dystrophy found in Swedish Briard dogs is an autosomal recessive disorder with complete penetrance (4–6). Although night blindness is a constant finding in affected animals there is a largely variable expression as to the other clinical findings. First generation affected dogs displayed night blindness but had normal or only mildly reduced day-light vision. Mating of an affected dog from this generation with his offspring led to complete blindness or severely reduced daylight vision, in addition to night blindness in litter-mates. Analysis by d.c. electroretinography showed signs suggestive of severely delayed phototransduction (7). The disorder seems to be different from other canine hereditary retinal degenerations in that it affects the retinal pigment epithelium (RPE) in addition to changes in the photoreceptor layer (4, 5).

It is very likely that canine retinal dystrophies are caused by mutations in the canine equivalents of genes known to cause retinal disease in other animal species and human. To test this hypothesis, we performed molecular genetic analysis of six retina-specific dog genes, i.e. those encoding rhodopsin, peripherin/RDS, ROM-1, arrestin, the α-subunit of cGMP-gated cation channel, and the β- subunit of PDE, the homologues of which have been shown to be mutated in different retinal dysplasias in other animals and/or in human. We performed segregation analysis on six offspring of an affected Swedish Briard dog to obtain genetic evidence for or against the involvement of these genes in the pathogenesis of this type of retinal dystrophy.

METHODS

Dog genomic DNA was extracted from peripheral blood by conventional methods. Retinal poly(A)$^+$ mRNA from affected and control animals were extracted by using a magnetic separation method (Dynabeads Oligo[dT$_{25}$]; Dynal) and converted to cDNA using standard reverse transcription protocols. Specific cDNA fragments were amplified by the method of rapid amplification of cDNA ends (RACE). The oligonucleotide primers used were designed based on sequence homology in evolutionarily conserved regions of the corresponding gene in other species together with cDNA-end ligated adaptor-specific primers (8). Nested or seminested polymerase chain reaction (PCR), high fidelity Pfu DNA polymerase (Promega), and touch-down PCR procedure (9) were used to increase both specificity and yield of PCR reactions. Amplified sense and antisense cDNA strands from two different non-affected animals were sequenced directly on Applied Biosystems 377 and 373A Automated DNA Sequencers. The cDNA sequence information obtained was used to create primers for nonradioactive single strand conformation polymorphism (SSCP) analysis, and in some cases, for sequencing introns from genomic DNA by primer walking. The entire coding region together with intron-exon boundaries were amplified by PCR for the rhodopsin, ROM1, peripherin/RDS, and βPDE genes. In the case of arrestin, only the coding region was examined, whereas coding region and one intron were analyzed for the cGMP-gated channel α-subunit gene. Amplified fragments were examined by SSCP using two different conditions (6% polyacrylamide, 5% glycerol, room temperature or 8% polyacrylamide, 5% glycerol, +4°C) to improve mutation detection efficiency (10). Gels were silver stained. The size of PCR fragments used for SSCP were 120–380 bp. All fragments with mobility shift were sequenced and the changes found were, if possible, confirmed by restriction enzyme digestion. cDNA or genomic sequences of dog arrestin (EMBL X98460), rhodopsin (EMBL X71380), peripherin/RDS (EMBL U36577); βPDE (EMBL L13262), and cGMP-gated channel α-subunit (EMBL X99913; EMBL X99914) are available from the databank.

RESULTS AND DISCUSSION

A simplified pedigree of the Swedish Briard dogs analyzed here is shown in Figure 1. Due to the high degree of inbreeding, affected animals are expected to be homozygous not only for the disease causing gene mutation but also for nonpathogenic sequence variants in the disease gene. Therefore a heterozygous pattern for any sequence change excludes the gene in question as the one implicated in the retinal disorder in this strain.

Mutations in the arrestin genes (*arr1* and *arr2*) of Drosophila lead to light-dependent retinal degeneration (11). More recently a mutation in the human arrestin gene has

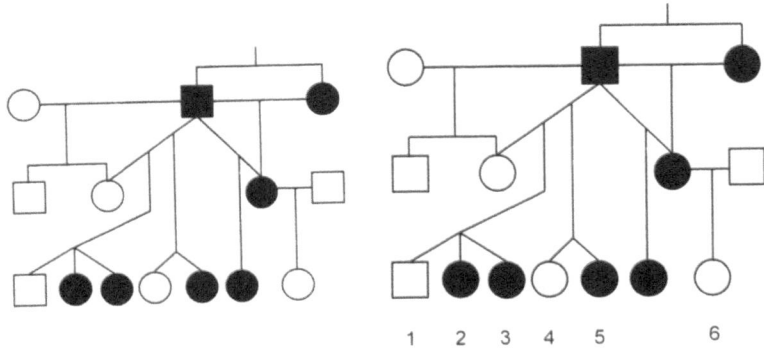

Figure 1. Simplified pedigree of Swedish Briard dogs examined in the present study.

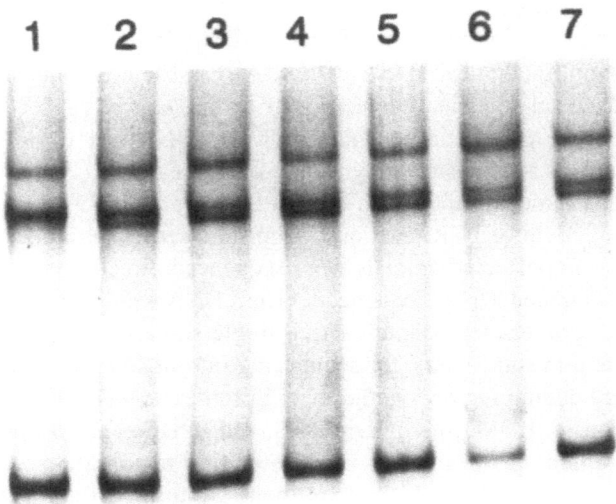

Figure 2. SSCP analysis of intron 2 of the canine peripherin/RDS gene. DNA samples of affected animals nos. 2 and 3 show a different mobility pattern of single-stranded fragments than that obtained for the affected dog no. 5 suggesting a DNA sequence alteration in this latter. Lanes 1–6: animals nos. 5, 6, 1, 2, 4, and 3, respectively.

been described as cause of a rare autosomal recessive form of congenital stationary blindness, Oguchi disease (12). Therefore the gene encoding retinal arrestin was a good candidate for the retinal dystrophy in Swedish Briards. cDNA analysis by SSCP and direct sequencing of overlapping fragments amplified by PCR did not reveal any difference between affected and unaffected animals.

Mutations in the peripherin/RDS gene have been implicated in different forms of human and murine retinal disorders (13–15). In addition, noncomplementation of heterozygous mutations of the peripherin/RDS and ROM1 (rod outer segment membrane protein 1) genes have been described in some cases of retinitis pigmentosa (RP) with digenic inheritance (16, 17). SSCP analysis of the peripherin/RDS gene revealed two different mobility patterns of single-stranded DNA fragments amplified from intron 2. As shown in Fig. 2, pattern 1 was seen in the sample of dog no. 5, while all other animals, including the unrelated and unaffected control presented with pattern 2. As dog no. 5 is affected by retinal dystrophy but carries a different allele from the other affected animals, a mutation of the peripherin/RDS gene as cause of the disease seems to be unlikely in this strain of dogs.

In the canine ROM1 gene, we have identified several polymorphic changes in the 5'-untranslated region (5'-UTR) and in exons 1 and 2. None of these variants cosegregate with the disease phenotype (Fig. 3). Recently, Moghrabi et al. have excluded the involvement of the peripherin/RDS gene in the early retinal degeneration (erd) found in Norwegian elkhounds (18).

Nonsense and missense mutations in the gene encoding βPDE have been identified as cause of autosomal recessive retinitis pigmentosa in some families in human as well as in the rod/cone dysplasia of Irish setter and the retinal dystrophy of the rd mice (3, 19, 20). In the gene for dog βPDE, we found a G/C transversion at nucleotide -45 in the 5'-UTR. As affected dog no. 2 carried this polymorphism in heterozygous form, segregation analysis excluded βPDE as candidate for the disease (Fig. 3).

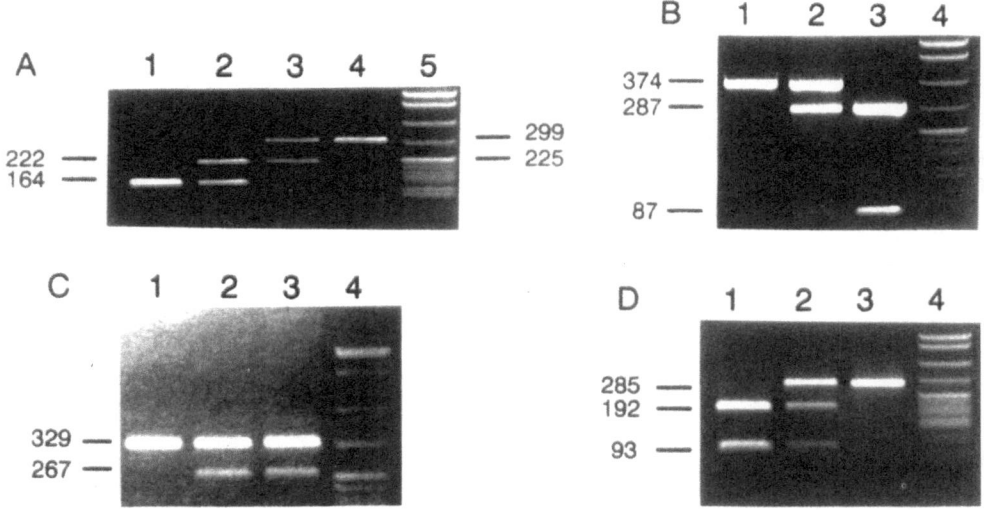

Figure 3. Detection of sequence polymorphisms in four canine genes by restriction enzyme analysis. (A) Detection of two point mutations in the ROM1 gene. A 272 bp fragment from the 5'-UTR was digested by Alu I and separated by agarose gel electrophoresis. As the -30A allele defines an additional restriction site in the amplicon, allelic fragments of 24, 26 (not visible on the picture), 58, and 222 bp are generated. The 299 bp fragment from exon 2 is cleaved by Cla I resulting in fragments of 225 and 74 bp. The 926C-allele lacks the Cla I restriction site. Lane 1: Homozygote -30A; lane 2: heterozygote -30A/G; lane 3: heterozygote 926T/C; lane 4: homozygote 926T; lane 5: molecular weight marker (pBR322, MspI digested). (B) Detection of a same-sense (Ala168) point mutation (600A/G) in the canine rhodopsin gene. A 374 bp amplicon was digested by Ban I and the fragments were separated by agarose gel electrophoresis. Due to the recognition site present at this position in the 600G- allele, allelic fragments of 287 and 87 bp are generated, while the 600A allele is not cut by the enzyme. Lane 1: Homozygote 600A-pattern; lane 2: heterozygote sample for 600A/G; lane 3: homozygote for 600G; lane 4: molecular weight marker (pBR322, MspI digested). (C) Detection of a point mutation in the 5'-UTR of the canine gene for ßPDE. A 329 bp fragment was digested by AlwN I and separated on agarose gel. Due to the recognition site present at this position on the -45C allele, allelic fragments of 267 and 62 bp (not visible on the picture) are generated. Allele -45G is not cut by the enzyme. Lane 1: homozygote -45G pattern; lanes 2 and 3: animals heterozygous for -45G/C; lane 4: molecular weight marker (pBR322, MspI digested). (D) Confirmation of the point mutation 294T/C in intron 4 of the canine cGMP-gated channel α-subunit gene. Amplified 285-bp fragment was digested with Hinf I and separated on 1,5% agarose gel. Nucleotide T at position 294 defines a recognition site for Hinf I and generates restriction fragments of 192 and 93 bp. Lane 1: homozygote for 294T; lane 2: heterozygous sample for 294T/C; lane 3: homozygote for 294C; lane 4: molecular weight marker (pBR322, MspI digested).

About 100 mutant alleles of the rhodopsin gene have been identified as cause of retinitis pigmentosa in human. Almost all of them are dominant, although a few recessive mutations have also been reported (for a recent review see ref. 21). In the canine rhodopsin gene (22), we found a silent third position G/A change at nucleotide 600 (Ala168) in exon 1. Dogs nos. 1, 3, 4, and 5 are homozygous 600A, dogs nos. 2 and 6 are heterozygous 600A/G, while the unaffected and unrelated control was homozygous 600G/G (Fig. 3). Again, segregation data excludes rhodopsin as disease gene for the retinal dystrophy in Swedish Briards.

Recently it has been shown by Dryja et al. (23) that mutations in the gene encoding the α-subunit of rod cGMP-gated cation channel are causative in some forms of autosomal recessive retinitis pigmentosa in human. SSCP analysis and direct sequencing of intron 4

Table 1. Polymorphic nucleotide changes detected in five canine retina-specific genes. Nucleotide numbers correspond to that given in the EMBL database. + and − indicate creation and loss of a restriction site, respectively

Gene	Exon/intron	Nucleotide change	Restriction enzyme	Finding
Peripherin/RDS	exon 1	37 A/C	—	Heterozygous: normal control
Phosphodiesterase, β-subunit	5′-UTR	−45 G/C	—	G/C: nos. 1 and 2
	exon 1	267 T/C	+AlwN I	Heterozygous: normal control
Rhodopsin	exon 2	600 A/G	+Bbv I, −Ban I	A/A: nos. 1, 3, 4, 5
				A/G: nos. 2, 6
				G/G: control
ROM1	5′-UTR	−77 G/A	+Bsr I, −BstN I	Heterozygous for all changes:
		−30 A/G	+Alu I	nos. 2, 3, and unaffected
	exon 1	160 G/T	+Alw26 I, −Nla IV	and unrelated control
		444 A/G	+Ban I	
	exon 2	914 G/A	—	
		926 T/C	−Cla I	
		1019 T/C	—	
		1058 C/T	+Nla III	
cGMP-gated channel, α-subunit	intron 4	270 C/T	+Xmn I	Heterozygous for all changes:
		294 T/C	+Hinf I	nos. 1, 2, 3, 5
		374 T/C	—	

of the canine cGMP-gated cation channel α-subunit gene showed polymorphic changes, which, again, did not cosegregate with the disease phenotype (Table 1 and Fig. 1).

Animals with inherited photoreceptor degeneration and blindness have been looked for and bred systematically because of their potential usefulness in understanding the molecular pathology of RP, the most common group of inherited blinding diseases in human. Among these animals, dogs have an important place representing a model in intermediate-size animals quite suitable for potential therapeutic approaches. In the current study, we evaluated six genes whose mutations have been implicated in retinal degeneration in different species. Based upon the data reported here, our results exclude six genes expressed specifically in the retina as disease causing genes in the progressive retinal dystrophy of Swedish Briard dogs. Nevertheless it is possible that mutations in abovementioned genes or in the regulatory regions thereof are responsible for other forms of canine retinal degenerations or dystrophies. We feel that the candidate gene/cDNA approach remains an appropriate strategy for identifying causative genes, especially in view of the relative paucity of genetic markers available in dog for linkage analysis and positional cloning. Additional candidate genes to be tested in the future include the genes for other components of the visual transduction pathway and photoreceptor-specific structural proteins.

ACKNOWLEDGMENTS

This study was financially supported by grants from the Deutsche Forschungsgemeinschaft (Ga 210/5-4) and the Swedish Medical Research Council (projects 12x-734 and 19X-09938). This article is based in part on a doctoral study by A.V. in the Faculty of Biology, University of Hamburg (Germany).

REFERENCES

1. Suber, M.L., Pittler, S.J., Qin, N., Wright, G.C., Holcombe, V., Lee, R.H., Craft, C.M., Lolley, R.N., Baehr, W., and Hurwitz, R.L., 1993, Irish setter dogs affected with rod/cone dysplasia contain a nonsense mutation in the rod cGMP phosphodiesterase β-subunit gene, *Proc. Natl. Acad. Sci. USA* **90**: 9968–9972.
2. Aguirre, G.D., 1976, Inherited retinal degenerations in the dog, *Trans. Am. Acad. Ophthalmol. Otolaryngol.* **81:** 667–676.
3. Clements, P.J.M., Sargan, D.R., Gould, D.J., and Petersen-Jones, S.M., 1996, Recent advances in understanding the spectrum of canine generalized progressive retinal atrophy, *J. Small Anim. Pract.* 37:155–162.
4. Wrigstad, A., Nilsson, S.E.G., and Narfström, K., 1992, Ultrastructural changes of the retina and the retinal pigment epithelium in Briard dogs with hereditary congenital night blindness and partial day blindness, *Exp. Eye Res.* **55:** 805–818.
5. Wrigstad, A., Narfström, K., and Nilsson, S.E.G., 1994, Slowly progressive changes of the retina and retinal pigment epithelium in Briard dogs with hereditary retinal dystrophy, *Doc. Ophthalmol.* **87:** 337–354.
6. Narfström, K., Wrigstad, A., Ekesten, B., and Nilsson S.E.G., 1994, Hereditary retinal dystrophy in the Briard dog: clinical and hereditary characteristics, *Vet. Comp. Ophthalmol.* **4:** 85–92.
7. Nilsson, S.E.G., Wrigstad, A., and Narfström, K., 1992, Changes in the d.c. electroretinogram in Briard dogs with hereditary congenital night blindness and partial day blindness, *Exp. Eye Res.* **54:** 291–296.
8. Frohman, M.A., Dush, M.K., and Martin, G.R., 1988, Rapid production of full-length cDNA from rare transcripts: Amplification using a single gene-specific oligonucleotide primer, *Proc. Natl. Acad. Sci. USA* **85:** 8998–9002.
9. Don, R.H., Cox, B.J., Wainwright, B.J., Baker, K., and Mattick, J.S. 1991, „Touchdown" PCR to circumvent spurious priming during gene amplification. *Nucl. Acids Res.* **19**, 4008.
10. Bunge, S., Fuchs, S., and Gal, A., 1996, Simple and nonisotopic methods to detect unknown gene mutations in nucleic acids, in: *Methods in molecular genetics* (K.W. Adolph ed.), pp. 26–39, Academic Press, Orlando.
11. Dolph, P.J., Ranganathan, R., Colley, N.J., Hardy, R.W., Socolich, M., and Zuker, C.S., 1993, Arrestin function in inactivation of G protein-coupled receptor rhodopsin in vivo. *Science* **260**: 1910–1916.
12. Fuchs, S., Nakazawa, M., Maw, M., Tamai, M., Oguchi, Y., and Gal, A., 1995, A homozygous 1-base pair deletion in the arrestin gene is a frequent cause of Oguchi disease in Japanese, *Nature Genet.* **10:** 360–362.
13. Farrar, G.J., Kenna, P., Jordan, S.A., Kumar-Singh, R., Humphries, M.M., Sharp, E.M., Sheils, D.M., and Humphries, P., 1991, A three-base-pair deletion in the peripherin-RDS gene in one form of retinitis pigmentosa, *Nature* 354: 478–480.
14. Kajiwara, K., Sandberg, M.A., Berson, E.L., and Dryja, T.P., 1993, A null mutation in the human peripherin/RDS gene in a family with autosomal dominant retinitis punctata albescens, *Nature Genet.* **3**: 208–212.
15. Wells, J., Wroblewski, J., Keen, J., Inglehearn, C., Jubb, C., Eckstein, A., Jay, M., Arden, G., Bhattacharya, S., Fitzke, F., and Bird, A., 1993, Mutations in the human retinal degeneration slow (RDS) gene can cause either retinitis pigmentosa or macular dystrophy, *Nature Genet.* **3**: 213–218.
16. Bascom, R.A., Liu, L., Heckenlively, J.R., Stone, E.M., and McInnes, R.R., 1995, Mutation analysis of the ROM1 gene in retinitis pigmentosa, *Hum. Mol. Genet.* **4:** 1895–1902.
17. Kajiwara, K., Berson, E.L., and Dryja, T.P., 1994, Digenic retinitis pigmentosa due to mutations at the unlinked peripherin/RDS and ROM1 loci, *Science* **264:** 1604–1608.
18. Moghrabi, W.N., Kedzierski, W., and Travis, G.H., 1995, Canine homolog and exclusion of retinal degeneration slow (*rds*) as the gene for early retinal degeneration (*erd*) in the dog, *Exp. Eye Res.* **61:** 641–643.
19. McLaughlin, M.E., Ehrhart, T.L., Berson, E.L., and Dryja, T.P., 1995, Mutation spectrum of the gene encoding the β subunit of rod phosphodiesterase among patients with autosomal recessive retinitis pigmentosa, *Proc. Natl. Acad. Sci. USA* **92**: 3249–3253.
20. Pittler, S.J., and Baehr, W., 1991, Identification of a nonsense mutation in the rod photoreceptor cGMP phosphodiesterase β-subunit gene of the rd mouse, *Proc. Natl. Acad. Sci. USA* **88:** 8322–8326.
21. Gal, A., Apfelstedt-Sylla, E., Janecke, A.R., and Zrenner, E., 1997, Rhodopsin mutations in inherited retinal dystrophies and dysfunctions, *Prog. Retin. Eye Res.* **16:** 51–79.
22. Petersen-Jones, S.M., Sohal, A.K., and Sargan, D.R., 1994, Nucleotide sequence of the canine rod-opsin-encoding gene, *Gene* 143: 281–284.
23. Dryja, T.P., Finn, J.T., Peng, Y.-W., McGee, T.L., Berson, E.L., and Yau, K.-W., 1995, Mutations in the gene encoding the α subunit of the rod cGMP-gated channel in autosomal recessive retinitis pigmentosa, *Proc. Natl. Acad., Sci. USA* **92**: 10177–10181.

12

THE VPP MOUSE

A Transgenic Model of Autosomal Dominant Retinitis Pigmentosa

Neal S. Peachey,[1] Min Wang,[2] and Muna I. Naash[2]

[1]Research Service (151)
Hines VA Hospital
Hines, Illinois 60141
Department of Neurology
Loyola University of Chicago
Maywood, Illinois 60153
[2]Department of Ophthalmology and Visual Sciences
College of Medicine
University of Illinois at Chicago
Chicago, Illinois 60612

INTRODUCTION

Retinitis pigmentosa (RP) refers to a class of blinding retinal disorders which lead to the same clinical phenotype (1). RP may be inherited as a genetic trait in autosomal recessive (ar), autosomal dominant (ad) and X-linked fashions. While it had long been suspected that RP initially involves the photoreceptor cells, it was not until defects in photoreceptor-specific gene products were identified in patients with adRP (2) or arRP (3,4) that this suspicion was confirmed. This new information has reoriented the approach taken to clinical studies of RP patients since the disease may now be identified by the specific point mutation involved (5). In addition, for rhodopsin mutations which act as dominantly-inherited traits, this new information has led to the development of strains of transgenic mice expressing a mutant form of the rhodopsin gene and which develop a progressive photoreceptor degeneration (6–9). The purpose of this chapter is to summarize a series of studies that have been carried out on one of these transgenic lines, the so-called 'VPP' mice (7). The available evidence indicates that the VPP mice may be a particularly useful animal model of adRP because they bear a close phenotypic similarity to adRP patients carrying the P23H point mutation and because a large body of information has now been produced concerning the characteristics of the degenerative process and how the time course may be influenced by environmental and genetic factors.

Degenerative Retinal Diseases, edited by LaVail *et al.*
Plenum Press, New York, 1997

VPP TRANSGENIC MICE

The VPP Transgene

A 15 kb mouse opsin genomic fragment was used to generate the VPP line. This fragment contains 6.0 kb of the promoter region, all the introns and exons, and 3.5 kb of the 3′ untranslated region containing all of the multiple polyadenylation signals (10). Mutations were introduced by site directed mutagenesis in PCR reactions. As diagrammed in Figure 1A, three mutations resulted in amino acid substitutions, valine at position 20 by

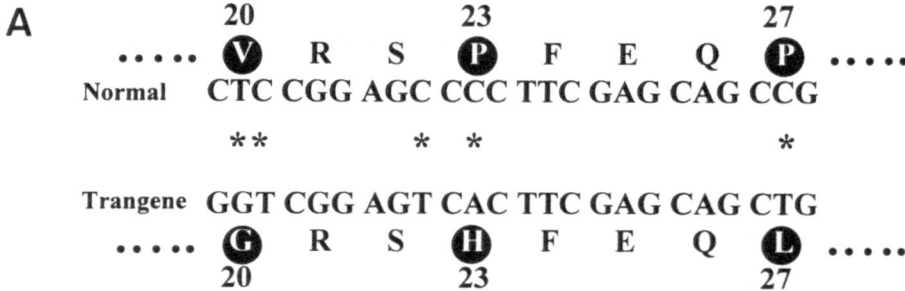

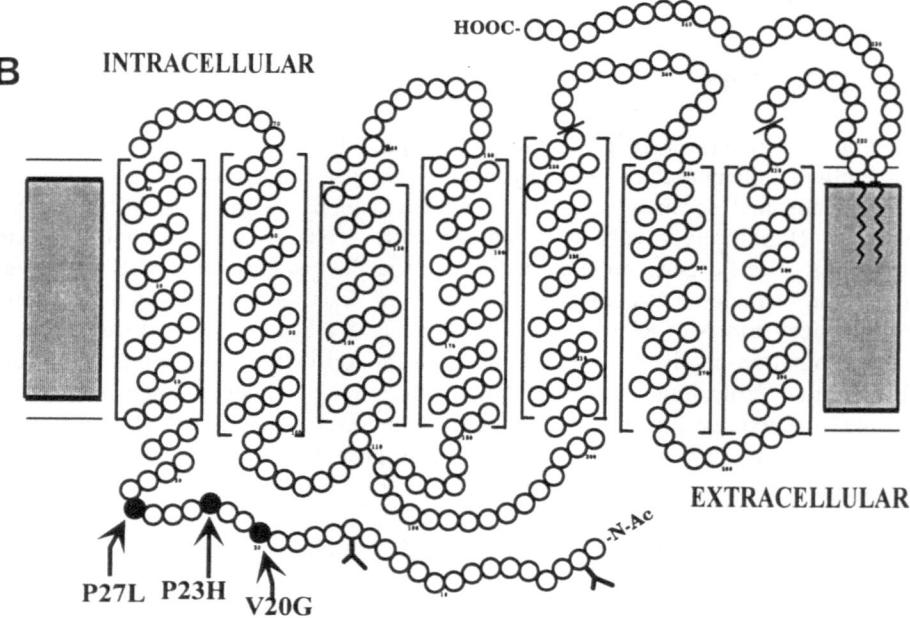

Figure 1. A. Nucleotide and amino acid sequences of normal mouse opsin and VPP mutant opsin at the N-terminus region where the mutations were introduced. B. Schematic diagram of the rhodopsin molecule. Filled circles indicate the amino acids that were changed in the VPP mutant.

glycine (V20G), proline at position 23 by histidine (P23H), and proline at position 27 by leucine (P27L). P23H was included because that point mutation is commonly encountered in adRP patients (11, 12). Moreover, P23H is undoubtedly responsible for the VPP phenotype, since neither the V20G nor the P27L mutation has ever been associated with any retinal abnormality (5) and since different strains of mice with glycine at position 20 (i.e., V20G) show no signs of retinal degeneration (10, 13).

An additional mutation did not alter serine at position 22 and was included to differentiate the transgene and the endogenous gene at the mRNA level (7, 14). The final mutation did not alter histidine at position 100, but deleted the second NcoI site in exon 1 to create a restriction length polymorphism (RFLP. Digestion of the PCR fragment with NcoI results in the formation of a band of higher molecular weight on an agarose gel [689-, 431-, and 197-bp bands are derived from normal mice, while VPP mice have an additional 886-bp band (7, 14)]. The transgene is inherited as a dominant trait with approximately 50% transmittance. For example, in a sample of 20 litters derived from mating normal mice and mice heterozygous for the VPP transgene, 168 progeny were produced with 87 (51.8%) being normal and 81 (48.2%) expressing the VPP trait. The availability of normal littermates is a convenient feature of this line since non-transgenic littermates from the same genetic background and age provide readily available control subjects.

Biochemical Studies of VPP Mice

DNA blot analysis was used to determine the number of copies of the transgene that were integrated into the mouse genome. As shown in Naash et al. (14), between 2 and 5 copies of the transgene appear to be integrated into a single site.

At the mRNA level, the total amount of transgene transcription is the same as that for the two native alleles for normal rhodopsin (15). Moreover, the total amount of rhodopsin synthesized is not different from that encountered in normal mice, with approximately half normal and half transgenic (15), replicating what is thought to occur in the human condition. These observations suggest that the mouse opsin promoter confers normal regulation of transgene expression, and indicate that the VPP phenotype does not result simply from rhodopsin overexpression (6).

Programmed cell death, or apoptosis, has been shown to occur in several mouse models of photoreceptor degeneration (16,17). An apoptotic pathway is also activated in the VPP mice; Naash et al. (18) reported that the number of cells undergoing apoptotic cell death is greatly increased in VPP mice. In animals reared under normal laboratory conditions, this cell death is more severe in the inferior retina (18).

Histological Studies of VPP Mice

Histologically, the VPP mice are defined by the development of a normal number of photoreceptors, which are maintained to ~P20 (postnatal day 20). Thereafter, the number of cell bodies in the outer nuclear layer (ONL) declines slowly with age (7). As noted above, the degenerative process does not appear to affect the retina uniformly. For example, Figure 2 presents a vertical retinal cross-section centered on the optic nerve for a 3 month old VPP mouse. At this age, the ONL of the central retina is reduced to 3–4 rows of photoreceptor nuclei in the inferior portion and 4–5 rows in the superior region. In comparison, Figure 3 presents peripheral sections taken from the same retina. The peripheral retina (both inferior and superior) retains an approximately two-fold greater number of photoreceptor nuclei, indicat-

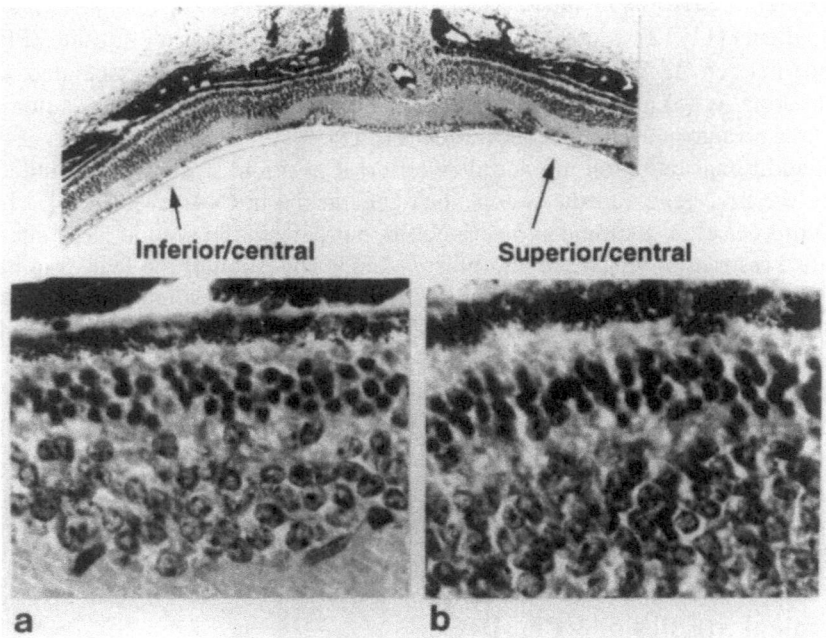

Figure 2. Cross section of central retina of 3 month old VPP mouse. The photoreceptor layer includes only 3–4 rows of nuclei.

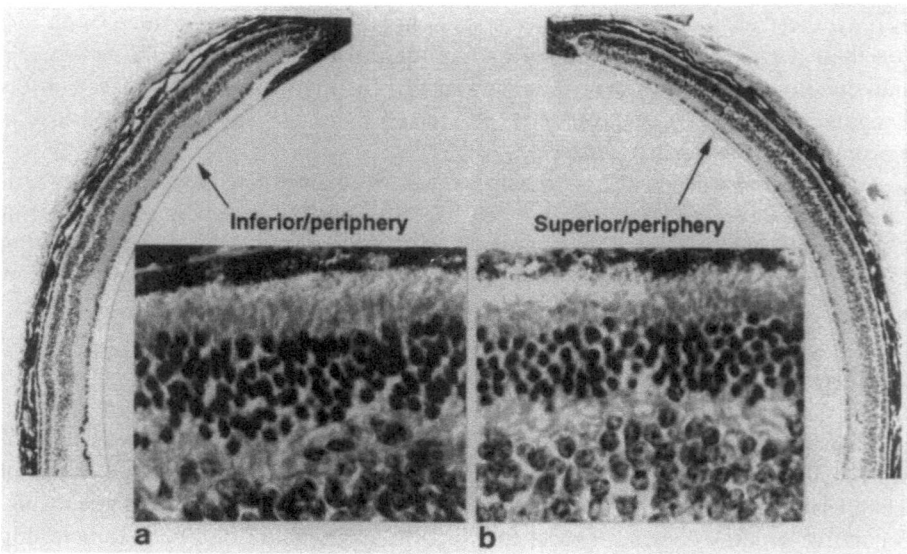

Figure 3. Cross section of peripheral retina of the same VPP mouse shown in Figure 2. A greater number of photoreceptor cells are retained in the peripheral retina, indicating that the degenerative process is slower in the peripheral retina.

ing that the peripheral retina undergoes a slower rate of degeneration. It is noteworthy that P23H patients typically have defects that are more severe in the inferior central retina (19).

Functional Studies of VPP Mice

The electroretinogram (ERG) has been used to assess noninvasively the time course of photoreceptor degeneration in the VPP mice. In VPP mice, the amplitude of the rod-mediated ERG a-wave was reduced by ~50% at P30 and declined slowly thereafter (7, 20). In addition, the use of standard clinical conditions allows the mouse cone ERG to be recorded in isolation (21), and to thus compare the extent of rod and cone involvement. In VPP mice, the cone ERG remained within the normal range at ages where rod ERG measures were clearly abnormal (Fig. 4, 20). Both the slow course of rod degeneration and the relative sparing of cone function are features that are typical of P23H patients (2, 22).

In RP patients, the reduction in rhodopsin content of the retina acts to reduce visual sensitivity due to the decreased likelihood of light capture ('quantum catch') in the outer segment. However, an important question concerns whether sensitivity is elevated to a greater extent than would be predicted by 'quantum catch' considerations alone. It has been suggested (23) that a constitutively active form of rhodopsin will induce photoreceptor degeneration, a possibility that can be tested by examining the relationship between rhodopsin density and visual sensitivity. The basic question is whether or not threshold elevations are related to the decreased rhodopsin content of the retina by the probability function. In this situation, a 50% reduction in rhodopsin density corresponds to a two-fold (0.3 log unit) reduction in visual sensitivity, a 90% reduction corresponds to a 10-fold (1.0

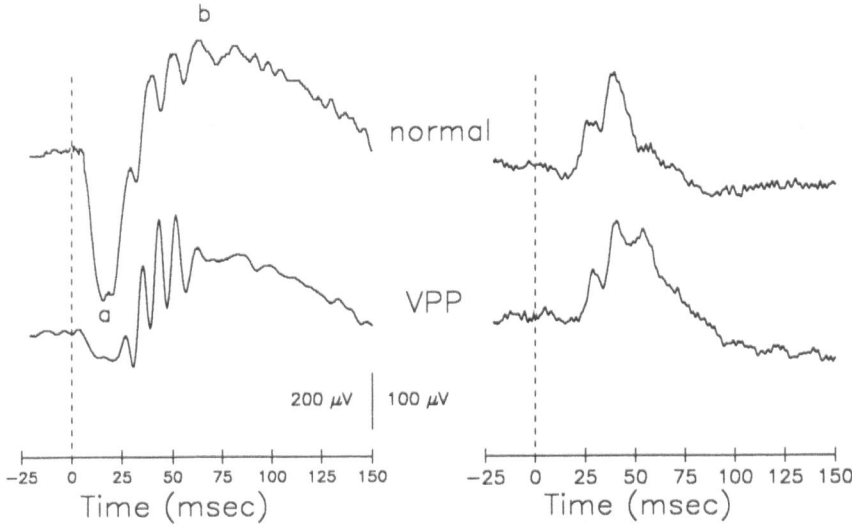

Figure 4. ERGs obtained from a normal and from a VPP mouse at 3 months of age. Lefthand waveforms were recorded under dark-adapted conditions and thus reflect primarily rod-mediated activity. The letters 'a' and 'b' identify the major a-wave and b-wave components of the ERG waveform. At this age, there is a pronounced decline in rod-mediated ERG components. Righthand responses were obtained using stimulus conditions that isolate a cone-mediated ERG (21). At this age, cone ERGs obtained from VPP mice were normal in amplitude, indicating a relative preservation of cone photoreceptors. Note that different amplitude scales were used to plot dark- and light-adapted responses.

Table 1. Summary of comparisons between VPP mice and P23H patients

Feature	VPP mice	P23H patients
Inheritance pattern	autosomal dominant	autosomal dominant
Ratio of expression of mutant to normal	1:1	1:1
Slow photoreceptor degeneration	✓	✓
Initial sparing of cones threshold elevations follow	✓	✓
Quantal relationship	✓	✓
Normal phototransduction gain	✓	—
Slow dark adaptation	✓	✓
Slow a-wave recovery	✓	✓

log unit) reduction, and so on. VPP data fell along the probability function (20), as do data for P23H patients (24). These results provide another point of similarity between the VPP mice and the human condition and indicate that constitutive activation does not play a role in the degenerative process.

The ERG has recently been used to assess the characteristics of rod phototransduction and the kinetics of response recovery. For example, the leading edges of a-waves recorded to high intensity stimuli have been analyzed in terms of a computational model (25, 26) derived from properties of the rising phase of the photocurrent response of isolated rods (27). Two parameters of this model correspond to maximum response amplitude and gain. Consistent with other ERG measures and retinal histology, the maximum response amplitude parameter was decreased below normal in VPP mice (28). However, the gain of phototransduction was not different from normal littermates, indicating a normal amplification of the visual signal in VPP rods (28). This result appears to distinguish the VPP mice from P23H patients, whose responses have lower than normal gain (29). The characteristics of response recovery have been analyzed using a double-flash paradigm in P23H patients (29) and in VPP mice (28). In both cases, a-wave amplitude recovery is slower than normal. Evidence that these changes do not reflect degeneration *per se* is provided by the observation that mice expressing a mutant rds/peripherin gene develop a progressive photoreceptor degeneration yet retain normal recovery kinetics (30). Finally, the ERG was used to follow the dynamics of rod dark adaptation in VPP mice by recording the ERG b-wave to a dim flash. The results were similar to psychophysical human data (24) in that the recovery of rod ERG amplitude after a partial bleach was much slower in VPP mice than in normal littermates (20).

As summarized in Table 1, studies to date have identified a number of features which are common between the VPP mice and P23H patients. The available evidence indicates that the fundamental abnormality imposed by the P23H mutation is to change the balance between disk synthesis and disk shedding (20), most likely due to a slowed process of disk formation. In this scenario, the primary limitation on rod-mediated vision is the decreased number of outer segment disks available to capture light incident on the retina. However, although the VPP rod outer segment contains fewer disks, for each disk the activation stages of phototransduction remains normal. The slowed recovery kinetics measured in the VPP mice and in P23H patients indicates that the P23H mutation interferes with the processes involved with the deactivation of photoexcited rhodopsin. Although the precise mechanism has not been identified, the delays could reflect, for example, delays in the kinetics of rhodopsin phosphorylation, and/or in the binding and action of arrestin. Identification of the precise abnormality will require additional work, but may provide important information regarding how the P23H point mutation leads to a panretinal photoreceptor degeneration.

ALTERING THE DISEASE TIME COURSE

Retinal Light Exposure

A key reason for developing an animal model that resembles the homologous human condition is that strategies for altering the disease time course can be readily tested. When a positive result is found, this may have a direct application to human patients. It has long been suspected that excessive light exposure may speed the course of retinal degeneration (31). Clinical observations of P23H patients (32) indicated that light exposure may exacerbate this form of RP. To examine this issue in a controlled fashion, two studies were conducted on the VPP mice. In the first, mice were reared in constant darkness from birth (18). At specific time points (2, 4 and 6 months of age), a battery of experimental procedures was used to compare animals reared in darkness with those reared in typical laboratory conditions (12L:12D) which have no deleterious effects on normal mice. At each age examined, the results showed that dark-reared animals retained a higher degree of photoreceptor function, whether assessed by histology, rhodopsin densitometry or ERG techniques (18), although the degenerative process continued to progress even in total darkness. In comparison, light deprivation did not alter the disease time course in several other mouse models of photoreceptor degeneration including the *rds* mouse (33), the nervous mouse (34), the vitiligo mouse (35), and two lines of transgenic mice expressing a mutant opsin gene (9, 36).

The second study examined the question of whether increased light exposure will lead to a faster degeneration than is seen under typical light conditions (37). VPP mice were maintained in constant light for a 24-hour period, and their retinas examined at later time points. The results showed that constant light exposure significantly accelerated the degenerative process in VPP mice (37). In comparison, light exposure had no effect on normal mice nor on another line of transgenic mice which experience a similar rate of rod photoreceptor degeneration that is caused by a different mechanism (38). Taken together, these results demonstrate that some feature of the P23H mutation renders the photoreceptors more sensitive to light, and indicate that P23H patients may retain vision for a longer period of time by minimizing the amount of light to which the retina is exposed.

Genetic Factors

An important question raised by the identification of gene defects that underlie RP concerns the intra-familial variability often seen in patients with the same gene defect (22, 32). As noted above, environmental factors such as light exposure clearly contribute to this variability. In addition, it is possible that gene products that interact with the expression of the mutant opsin may also play a role (39). In support of this, a more rapid photoreceptor degeneration was observed when the VPP transgene was co-expressed with the genetic defect causing albinism (40). Compared with pigmented transgenic animals, albino VPP mice reared on the same C57BL/6 background demonstrated a more severe degeneration as measured by the survival of photoreceptor cells, the retention of ERG amplitude, and the rhodopsin content of the retina (40). This difference is light-independent since it was observed when albino and pigmented VPP mice were reared in darkness as well as under normal cyclic light conditions (40).

CONCLUSIONS

The availability of a transgenic model of a human disorder provides a powerful tool with which to understand the human condition and to evaluate possible therapeutic approaches. As reviewed above, the available evidence indicates that the photoreceptor degeneration of VPP mice has many features in common with the human form of adRP due to a P23H mutation. The observation that the rate of degeneration is linked to the degree of retinal light exposure suggests that human P23H patients will benefit from reduced light exposure. In the future, other potential therapies may be evaluated using the VPP mice to determine the extent to which they may be applied to the human condition induced by the P23H mutation.

REFERENCES

1. Heckenlively, J.R., 1988, Retinitis Pigmentosa, Lippincott, Philadelphia.
2. Dryja, T.P., McGee, T.L., Reichel, E., Hahn, L.B., Cowley, G.W., Yandell, D.W., Sandberg, M.A. and Berson, E.L., 1990, A point mutation of the rhodopsin gene in one form of retinitis pigmentosa. Nature 343: 364–366.
3. McLaughlin, M.E., Sandberg, M.A., Berson, E.L., and Dryja, T.P., 1993, Recessive mutations in the gene encoding the b-subunit of rod phosphodiesterase in patients with retinitis pigmentosa. Nature Genet. 4: 130–134.
4. Dryja, T.P., Finn, J.T., Peng, Y.-W., McGee, T.L., Berson, E.L. and Yau, K.-W., 1995, Mutations in the gene encoding the a subunit of the rod cGMP-gated channel in autosomal recessive retinitis pigmentosa. Proc. Natl. Acad. Sci. U.S.A. 92: 10177–10181.
5. Berson, E.L., 1993, Retinitis pigmentosa. Invest. Ophth. Vis. Sci. 34: 1659–1676.
6. Olsson, J.E., Gordon, J.W., Pawlyk, B.S., Rood, D., Hayes, A., Molday, R.S., Mukai, S., Cowley, G.S., Berson, E.L., and Dryja, T.P., 1992, Transgenic mice with a rhodopsin mutation (Pro23His): A mouse model of autosomal dominant retinitis pigmentosa. Neuron 9: 815–830.
7. Naash, M.I., Hollyfield, J.G., Al-Ubaidi, M.R., and Baehr, W., 1993, Simulation of human autosomal dominant retinitis pigmentosa in transgenic mice expressing a mutated murine opsin gene. Proc. Natl. Acad. Sci. U.S.A. 90: 5499–5503.
8. Sung, C.-H., Makino, C.L., Baylor, D. and Nathans, J., 1994, A rhodopsin gene mutation responsible for autosomal dominant retinitis pigmentosa results in a protein that is defective in localization to the photoreceptor outer segment. J. Neurosci. 14: 5818–5833.
9. Li, T., Franson, W.K., Gordon, J.W., Berson, E.L., and Dryja, T.P., 1995, Constitutive activation of phototransduction by K296E opsin is not a cause of photoreceptor degeneration. Proc. Natl. Acad. Sci. U.S.A. 92: 3551–3555.
10. Al-Ubaidi, M.R., Pittler, S.J., Champagne, M.S., Triantafyllos, J.T., McGinnis, J.F, and Baehr, W., 1990, Mouse opsin: gene structure and molecular basis of multiple transcripts. J. Biol. Chem. 265: 20563–20569.
11. Dryja, R.P., Hahn, L.B., Cowley, G.S., McGee, T.L., and Berson. E.L., 1991, Mutation spectrum of the rhodopsin gene among patients with autosomal dominant retinitis pigmentosa. Proc. Natl. Acad. Sci. U.S.A. 88: 9370–9374.
12. Sung, C.-H., Davenport, C.M., Hennessey. J.C., Maumenee, I.H., Jacobson, S.G., Heckenlively, J.R., Nowakowski, R., Fishman, G., Gouras, P. and Nathans, J., 1991, Rhodopsin mutations in autosomal dominant retinitis pigmentosa. Proc. Natl. Acad. Sci. U.S.A. 88: 6481–6485.
13. Baehr, W., Falk, J.D., Bugra, K., Triantafyllow, J.T., and McGinnis, J.R., 1988 Isolation and analysis of the mouse opsin gene. FEBS Lett. 238: 253–256.
14. Naash, M.I., Al-Ubaidi, M.R., Hollyfield, J.G., and Baehr, W., 1993, Simulation of autosomal dominant retinitis pigmentosa in transgenic mice, in: Retinal Degeneration, (J.G. Hollyfield, R.E. Anderson, and M.M. LaVail, eds.), pp. 201–210, Plenum Press, New York.
15. Cheng, T. and Naash, M.I., 1995, Quantitative analysis of mRNA levels of the transgenic and endogenous opsin genes in retinas of transgenic mice. Invest. Ophth. Vis. Sci. 36: S273. (ARVO abstracts).
16. Chang, G.-Q., Hao, Y., and Wong, F., 1993, Apoptosis: Final common pathway of photoreceptor death in rd, rds, and rhodopsin mutant mice. Neuron 11: 595–605.

17. Portera-Cailliau, C., Sung, C.-H., Nathans, J., and Adler, R., 1994, Apoptotic photoreceptor cell death in mouse models of retinitis pigmentosa. Proc. Natl. Acad. Sci. U.S.A. 91: 974–978.

18. Naash, M.I., Peachey, N.S., Li, Z.-Y., Gryczan, C.C., Goto, Y., Blanks, J., Milam, A.H. and Ripps, H., 1996, Light-induced acceleration of photoreceptor degeneration in transgenic mice expressing mutant rhodopsin. Invest. Ophth. Vis. Sci. 37: 775–782.

19. Stone, E.M., Kimura, A.E., Nichols, B.E., Khadivi, P., Fishman, G.A., and Sheffield, V.C., 1991, Regional distribution of retinal degeneration in patients with the proline to histidine mutation in codon 23 of the rhodopsin gene. Ophthalmology 98: 1806–1813.

20. Goto, Y., Peachey, N.S., Ripps, H., and Naash, M.I., 1995, Functional abnormalities in transgenic mice expressing a mutant rhodopsin gene. Invest. Ophth. Vis. Sci. 36: 62–71.

21. Peachey, N.S., Goto, Y., Al-Ubaidi, M.R., and Naash, M.I., 1993, Properties of the mouse cone-mediated electroretinogram during light adaptation. Neurosci. Lett. 162: 9–11.

22. Berson, E.L., Rosner, B., Sandberg, M.A., and Dryja, T.P., 1991, Ocular findings in patients with autosomal dominant retinitis pigmentosa and a rhodopsin gene defect (pro-23-his). Arch. Ophth. 109: 92–101.

23. Fain, G.L. and Lisman, J.E., 1993, Photoreceptor degeneration in vitamin A deprivation and retinitis pigmentosa: the equivalent light hypothesis. Exp. Eye Res. 57: 335–340.

24. Kemp, C.M., Jacobson, S.G., Roman, A.J., Sung, C.-H., and Nathans, J., 1992, Abnormal rod dark adaptation in autosomal dominant retinitis pigmentosa with proline-23-histidine rhodopsin mutation. Amer. J. Ophth. 113: 165–174.

25. Breton, M.E., Schueller, A.W., Lamb, T.D., and Pugh, E.N. Jr., 1994, Analysis of ERG a-wave amplification and kinetics in terms of the G-protein cascade of phototransduction. Invest. Ophth. Vis. Sci. 35: 295–309.

26. Hood, D.C. and Birch, D.G., 1994, Rod phototransduction in retinitis pigmentosa: estimation and interpretation of parameters derived from the rod a-wave. Invest. Ophth. Vis. Sci. 35: 2948–2961.

27. Pugh, E.N., Jr., and Lamb, T.D., 1993, Amplification and kinetics of the activation steps in phototransduction. Biochim. Biophys. Acta 1141: 111–149.

28. Goto, Y., Peachey, N.S., Ziroli, N.E., Seiple, W.H., Gryczan, C., Pepperberg, D.R. and Naash, M.I., 1996, Rod phototransduction in transgenic mice expressing a mutant opsin gene. J. Opt. Soc. Amer. A 13: 577–585.

29. Birch, D.G., Hood, D.C., Nusinowitz, S., and Pepperberg, D.R., 1995, Abnormal activation and inactivation mechanisms of rod transduction in patients with autosomal dominant retinitis pigmentosa and the pro-23-his mutation. Invest. Ophth. Vis. Sci. 36: 1603–1614.

30. Birch, D.G., Kedzierski, W., Nusinowitz, S., Anderson, J.L., and Travis, G.H., 1995, Rod photoreceptor function in transgenic mice expressing mutant rds/peripherin. Invest. Ophth. Vis. Sci. 36: S641 (ARVO abstracts).

31. Berson, E.L., 1980, Light deprivation and retinitis pigmentosa. Vision Res. 20: 1179–1184.

32. Heckenlively, J.R., Rodriguez, J.A., and Daiger, S.P., 1991, Autosomal dominant sectoral retinitis pigmentosa: two families with transversion mutation in codon 23 of rhodopsin. Arch. Ophth. 109: 84–91

33. Sanyal, S. and Hawkins, R.K., 1986, Development and degeneration of retina in rds mutant mice: Effects of light on the rate of degeneration in albino and pigmented homozygous and heterozygous mutant and normal mice. Vision Res. 26: 1177–1185.

34. LaVail, M.M., White, M.P., Gorrin, G.M., Yasumura, D., Porrello, K.V., and Mullen, R.J., 1993, Retinal degeneration in the nervous mutant mouse: I: Light microscopic cytopathology and changes in the interphotoreceptor matrix. J. Comp. Neurol. 333: 168–181.

35. Smith, S.B., Cope, B.K., and McCoy, J.R., 1994, Effects of dark-rearing on the retinal degeneration of the C57BL/6-mi^{vit}/mi^{vit} mouse. Exp. Eye Res. 58: 77–84.

36. Gryczan, C., Kuszak, J.R., Novak, I., Peachey, N.S., Goto, Y., and Naash, M.I., 1995, A transgenic mouse model for autosomal dominant retinitis pigmentosa caused by a three-base pair deletion in codon 255/256 of the opsin gene. Invest. Ophth. Vis. Sci. 36: S243 (ARVO abstracts).

37. Wang, M., Lam, T.T., Tso, M.O.M., and Naash, M.I., 1997, Expression of a mutant opsin gene increases the susceptibility of the retina to light damage. Visual Neurosci. 14: 55–62.

38. Zhang, C., Al-Ubaidi, M.R., Quiambao, A.B., Lam, T.T., Zhang, S.R., and Tso, M.O.M., 1994, The effects of light exposure on photoreceptor degeneration in transgenic mice with SV40 T antigen. Invest. Ophth. Vis. Sci. 35:, 1517 (ARVO abstracts).

39. Applebury, M.L., 1992, Variations in retinal degeneration. Curr. Biol. 2: 113–115.

40. Naash, M.I., Ripps, H., Li, S., Goto, Y., and Peachey, N.S., 1996, Polygenic disease and retinitis pigmentosa: albinism exacerbates photoreceptor degeneration induced by the expression of a mutant opsin in transgenic mice. J. Neurosci. 16: 7853–7858.

13

ALTERED REGULATION OF ION CHANNELS IN CULTURED RETINAL PIGMENT EPITHELIAL CELLS FROM RCS RATS

O. Strauß,[1] S. Mergler,[1] M. Wienrich,[2] and M. Wiederholt[1]

[1]Institut für Klinische Physiologie
Universitätsklinikum Benjamin-Franklin
Freie Universität Berlin
Hindenburgdamm 30, 12200 Berlin, Germany
[2]ZNS Pharmacologie
Boehringer Ingelheim
Binger Straße, 55216 Ingelheim am Rhein, Germany

INTRODUCTION

The Royal College of Surgeon (RCS) rat suffers from an inherited retinal degeneration because their retinal pigment epithelium (RPE) is unable to phagocytose shed photoreceptor outer membranes (1,2,3,4). The genetical cause leading to this functional defect is unknown. The phagocytosis of photoreceptor outer membranes is regulated by the calcium/inositol-phosphate second messenger system (5,6,7). Generation of inositolphosphates represents the "on" signal (6) whereas an increase of the cytosolic free calcium represents the "off" signal for phagocytosis (5). The RPE cells from RCS rats are able to bind shed photoreceptor membranes but are unable to ingest these membranes (1,2,3,4). Thus, RPE cells express receptors for photoreceptor membranes but the second messenger system seems to be unable to initialize the ingestion of bound membranes (7). Several lines of evidence point to a altered second messenger metabolism in RPE cells of RCS rats. Abnormal inositol phosphate generation (8), a reduced cAMP generation (9), changed protein phosphorylation (10) and, in consequence, a changed growth factor responsiveness have been reported (11,12).

Many functions of RPE cells are regulated by ion conductances of the cell membrane. Therefore, patch-clamp studies (13) have been performed to characterize ion channels in RPE cell membranes. In many species potassium (14,15,16,17,18), sodium (19,20), calcium (21,22) and chloride (23,24,25,26) channels have been described. Preliminary data concerning the regulation of chloride channels showed that the inositolphosphate/calcium second messenger system activates calcium-dependent chloride channels (25). Comparing the ion conductance of RPE cells from non-dystrophic and RCS rats, we found that RPE of RCS rats express an increased membrane conductance for calcium (27).

Since it has been shown that RCS RPE cells express an altered second messenger system, we investigated the interaction of ion conductances and the second messenger system in RPE cells from RCS rats. Since the same second messenger system which regulates phagocytosis also activates calcium-dependent chloride channels we compared the chloride current activation by inositol-1,4,5-triphosphate (IP3) in RCS and non-dystrophic control rats. In addition, we tried to find out wether the increased calcium conductance is due to the modified ion channel regulation by an altered second messenger system. It has been shown that proteinphosphorylation is changed in RPE cells from RCS rats. Since calcium channels are mainly regulated by protein kinase C (PKC) (28) we compared the effects of PKC inhibition or PKC activation on barium currents from L-type calcium channels in both rat strains. A detailed knowledge of the changes in second messenger system of RPE cells from RCS rat will help to formulate a candidate gene causing the retinal degeneration in this rat strain.

METHODS

Patch-clamp recordings were performed in 3–9 days old primary cultures of RPE cells from RCS and non-dystrophic BDE ("hooded") rats. Primary cultures were established according to the method of Edwards (29) as previously described (25). For patch-clamp recordings glass cover-slips with cultured RPE cells were placed into a perfusion chamber on stage of an inverted microscope. The cells were superfused by a potassium-free solution to eliminate superimposed potassium currents. The bath solution was similar to Ham's culture medium (in mM): 135 NaCl, 0.6 $MgCl_2$, 33 HEPES, 10 TEACl, 6.1 glucose; adjusted to pH = 7.2 with Tris. The final $CaCl_2$ concentrations were 0.3 (for calcium conductance measurements), 1 or 10 mM. Calcium-free solutions contained 1 mM EGTA. For measurements of calcium conductance 10 mM $BaCl_2$ was added as charge carrier. Protein kinase modulators and calcium channel modulators were added freshly to the bath solution from stock solutions. The potassium-free pipette solution contained (in mM): 100 Cs-methansulfonate, 20 NaCl, 2 $MgCl_2$, 10 HEPES, 5.5 EGTA, 0.5 $CaCl_2$; adjusted to pH = 7.2 with Tris. IP3 was added freshly from stock solutions. The perforated-patch configuration was established using 150μg/ml nystatin in the pipette solution.

RESULTS

In the first set of experiments were compared activation of chloride currents by intracellular application of IP3 in RPE cells from non-dystrophic and RCS rats. During the whole-cell configuration IP3 was applied via the patch pipette. For this purpose, the whole-cell configuration was established with 10μM IP3 in the patch-pipette. Directly after establishing the whole-cell configuration no voltage-dependent currents were observed (Figure 1A). After 60 sec the membrane conductance started continously to increase. After 10 min the current activation was maximal (Figure 1A). These currents could be indentified as calcium-dependent chloride currents (25). The currents were activated by an IP3-induced calcium influx into the cell. Under extracellular calcium-free conditions IP3 was unable to induce chloride currents (25). Thus, the activation of chloride currents was used as tool to study the reaction of the IP3/calcium second messenger system. No difference could be observed when the maximal amplitude of IP3-induced chloride current was compared in RPE cells from RCS and non-dystrophic rats. Therefore, we compared the speed of chloride current activation in both rat strains (Figure 1B). This was performed using extracellular calcium concentrations of 1 mM and 10 mM. The speed of chloride current ac-

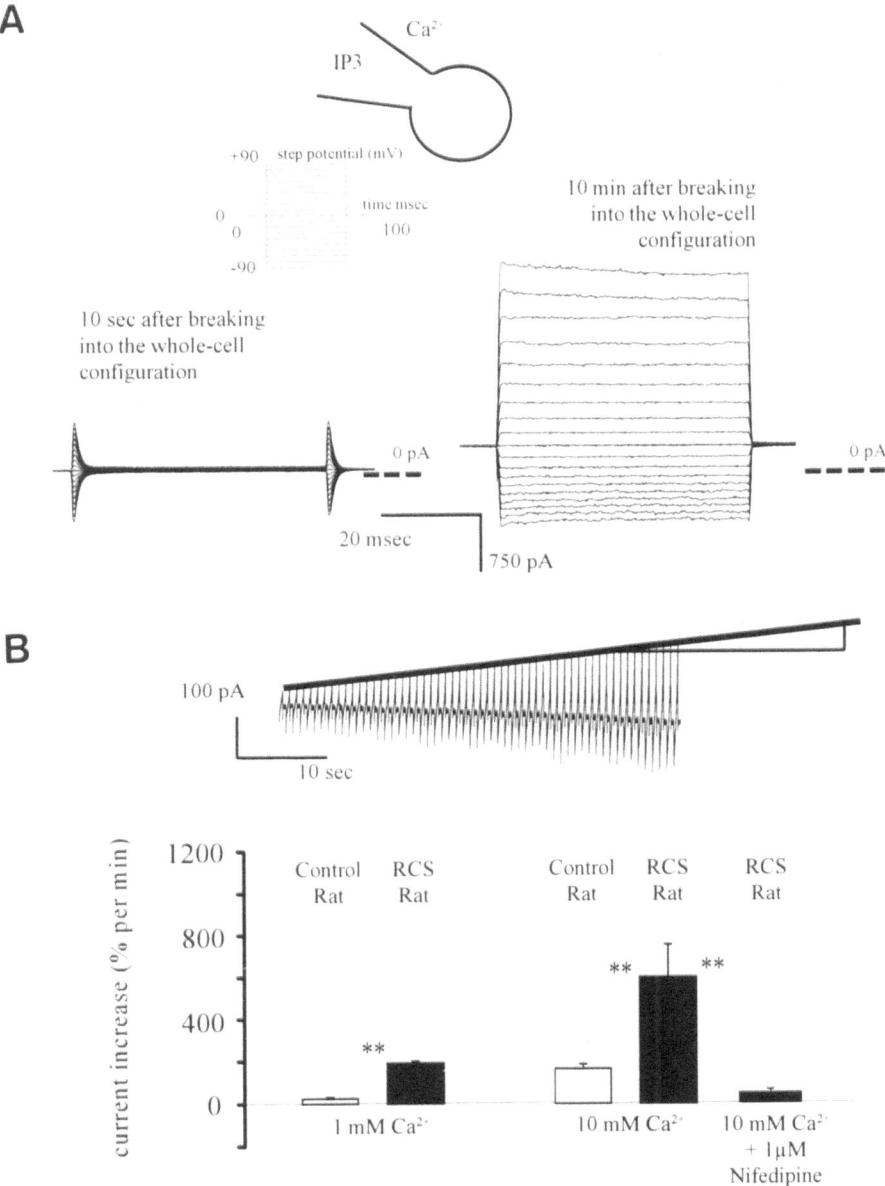

Figure 1. Comparison of the speed of chloride current activation in RPE cells from RCS and non-dystrophic rats: (A) Activation of chloride currents by intracellular application of IP3. The whole-cell configuration was established with 10μM IP3 in the patch-pipette. Directly after breaking into the whole-cell configuration no voltage-dependent currents could be observed (left panel) when the cell was electrically stimulated using a voltage-step protocol (middle panel). From a holding potential of 0 mV the cell was depolarized with nine voltage-steps of 10mV increasing amplitude and 50 msec duration. This was followed by nine voltage-steps of 10 mV increasing amplitude and 50 msec duration to hyperpolarize the cell. After 7–10 min in the whole-cell configuration a the sustained current acivation reached maximal amplitudes (right panel). (B) Comparison of the speed of chloride current activation in RPE cells from RCS and non-dystrophic rats.Upper panel: Estimation of the speed of chloride current activation: For continous recording of the membrane conductance the cell was electrically stimulated every 0.5 sec using a ramp-like stimulation between the potentials −145 mV and +55 mV. The speed was calculated as per cent outward current increase/ min to the control current (directly after breaking into the whole-cell configuration). Lower panel: Comparison of the speed of chloride current activation in RCS and non-dystrophic rats using two different extracellular calcium concentrations or nifedipine (1μM) with 10 mM Ca^{2+} in the bath solution.

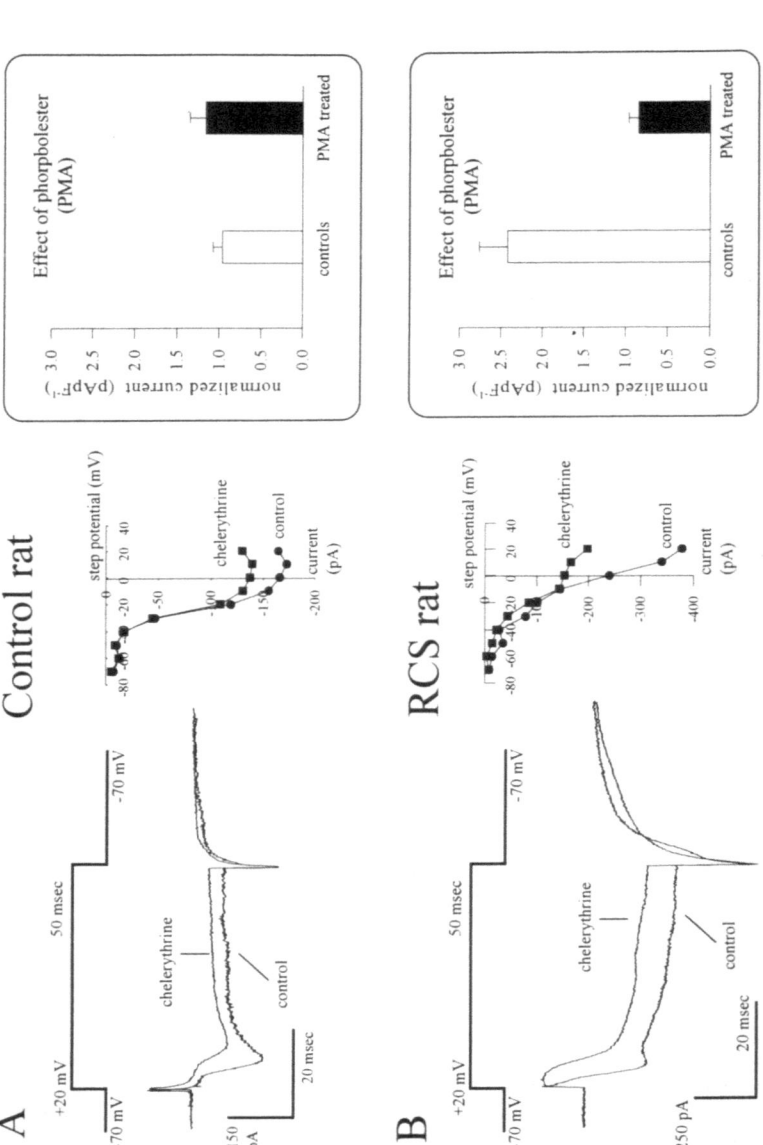

Figure 2. Comparison of the influence of protein kinase C on L-type calcium channels in RPE cells from non-dystrophic and RCS rats. (A) Effect of PKC inhibition and upregulation on L-type calcium channel currents in RPE cells from non-dystrophic rats. L-type calcium conductance was measured in the perforated patch configuration with 10 mM Ba²⁺ as charge carrier. The barium currents were identified as L-type currents using the specific L-type calcium channel activator BayK 8644 (1μM). Currents were evoked by a voltage-step from a holding potential of −70 mV to +20 mV (upper panel). Application of chelerythrine (10μM) to the bath solution led to a decrease of the current amplitude (left panel). Current/voltage plot (maximal current amplitudes are plotted against ten step potenials to evoke barium currents) summarizes the effect of chelerythrine (middle panel). Comparison of normalized maximal current amplitude (measured by an voltage-step from −70 to +10 mV) in untreated cells and in cells preincubated in PMA (1μM) for 30 min for activation of PKC. (B) Effect of PKC inhibition and upregulation on L-type calcium channel currents in RPE cells from RCS rats. Application of chelerythrine to the bath solution led to a decrease of the current amplitude (left panel). Current/voltage plot (maximal current amplitudes are plotted against ten step potenials to evoke barium currents) summarizes the effect of chelerythrine (middle panel). Comparison of normalized maximal current amplitude (measured by an voltage-step from −70 to +10 mV) in untreated cells and in cells preincubated in PMA for 30 min for activation of PKC.

tivation was expressed as per cent increase of the control outward current amplitude per minute. In both rat strains, the speed of chloride current activation was higher with 10 mM Ca^{2+} than with 1 mM Ca^{2+} in the bath solution. With a higher gradient of calcium from extra- to intracellular space calcium enters the cell faster and accelerates activation of calcium-dependent chloride currents. However, in RPE cells from RCS rats the speed of chloride current activation was higher than in RPE cells from non-dystrophic rats. Thus, the reaction of the IP3/calcium second messenger system led to changed reaction of the RPE cell fom RCS rats. Since we reported an increased calcium conductance in RCS RPE cells, we tested the possibility that calcium conductance may have changed the reaction of the IP3/calcium second messenger system. We performed the same experiments as shown above with the L-type calcium channel blocker nifedipine (1μM) in the bath solution. Nifedipine reduced the speed of chloride current activation. Thus, increased calcium conductance is involved in a modulation of the second messenger system.

For further investigation of the interaction of the second messenger system with ion channels we analysed the mechanisms leading to increased membrane conductance for calcium. Since L-type calcium channels are mainly regulated by protein kinases we compared the effects of inhibitors and activators of protein kinases on the L-type calcium conductance in both rat strains. L-type calcium channel conductance was measured using Ba^{2+} as charge carrier in the perforated-patch configuration, which maintains integrity of the cell. Barium currents through L-type channels were identified by using the L-type channel activator Bay K 8644 (1 μM). Then we investigated the effects of protein kinase C on L-type currents (Figure 2). Application of the 10μM PKC-specific blocker chelerythrine (30) decreased the L-type barium current amplitude in RPE cells from both rat strains. The decrease was not significantly different in cells from both rat strains. In addition, we tested the effect of upregulation of PKC. For this purpose the cells were preincubated for 30 min in 1μM phorpbol-12-myristate-13-acetate (PMA) before performing patch-clamp experiments. In RPE cells from non-dystrophic control rats the preincubation had no effect on the L-type current amplitude. In contrast, in cells from RCS rats preincubation in PMA led to a significant decrease of the L-type current amplitude. Thus, the increased calcium conductance in RPE cells from RCS rats is due an altered regulation of channel activity by PKC.

DISCUSSION

We demonstrated differences in properties of the second messenger system in RPE cells from RCS and non-dystrophic rats. An altered reaction of the IP3/calcium second messenger system was due to an increased L-type calcium conductance in RPE cells from RCS rats. The increased calcium conductance in RPE cells from RCS rats seems to be due to a modified calcium channel regulation by PKC.

We have previously demonstrated that RPE cells from RCS rats express an increased membrane conductance for calcium (27). This conductance could be identified as L-type calcium conductance. In the present paper we provide evidence that the calcium conductance changes the reaction of the IP3/calcium second messenger system using the activation of calcium-dependent chloride channels in RPE cells from both rat strains (25). In a previous work we demonstrated that intracellularly applied IP3 led to release of calcium from cytosolic calcium stores which in turn led to an influx of extracellular calcium into the cell for chloride channel activation (25). Using the speed of chloride channel activation we could show that the reaction of the IP3/calcium second messenger system was dependent on the entry pathway into the cell. A higher gradient of calcium led to a faster activation of chloride channels. In RPE cells from RCS rats we observed a faster chloride channel activation due to increased membrane conductance for calcium.

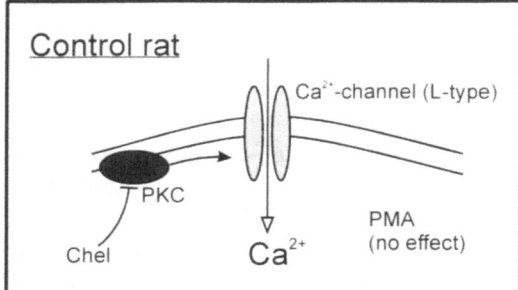

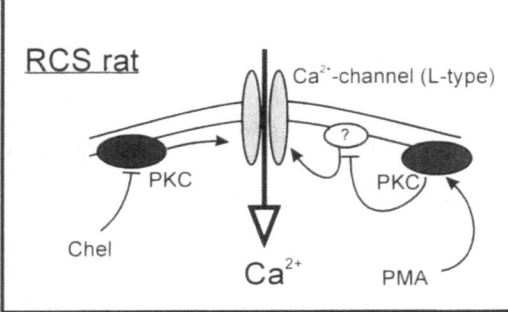

Figure 3. Scheme of hypothetical L-type calcium channel regulation in RPE cells from RCS and non-dystrophic rats. Upper panel: Regulation of calcium channel activity by PKC in RPE cells from non-dystrophic rats. In the resting cell PKC phosphorylates the calcium channel and thus activate calcium channels. Chelerythrine ("Chel") led to a decrease of calcium currents by inhibition of PKC. PMA had no effect, because channels are sufficiently phosphorylated in the resting cell. Lower panel: Regulation of calcium channel activity by PKC in RPE cells from RCS rats. In the resting cell, calcium channels are activated by PKC-dependent phosphorylation. An increased calcium conductance can be observed because a second factor ("?") led to an additional activation of the phosphorylated calcium channel. Chelerythrine ("Chel") led to a decrease of the calcium channel currents by PKC inhibition. The factor has no effect anymore because calcium channels are not sufficiently phosphorylated. PMA led to a decrease of calcium channel current, because the factor is inhibited by upregulation of PKC. The factor can be a cAMP-dependent, cGMP-dependent, Ca^{2+}/calmodulin-dependent or tyrosin protein kinase.

Increased membrane conductance in RPE cells from RCS rats seems to be due to a changed calcium channel regulation by PKC (Figure 3). It has been shown that PKC is the major regulator of L-type calcium channel activity (28). In both rat strains we found that inhibition of PKC led to a decrease of L-type calcium channel currents. However, using phorbolesters (PMA) for upregulation of PKC we could detect a different effect of PMA on L-type channel currents in RPE cells from RCS rats and non-dystrophic rats. In RPE cells from RCS rats upregulation of PKC led to a decrease of calcium channel currents. In RPE cells from non-dystrophic rats no difference could be observed between PMA treated and non-treated cells. Probably PKC has a different effect on calcium currents in RPE cells from RCS rats and non-dystrophic rats. In non-dystrophic rats calcium channels are sufficiently phosphorylated so that only inhibition (but not activation) of PKC has an effect on calcium conductance. In RPE cells from RCS rats PKC acts directly on the channel. In addition, it seems that a second factor (regulated by PKC) led to increased membrane conductance for calcium in RPE cells from RCS rats. The factor can only activate phosphorylated calcium channels. In the resting cell, the channels are sufficiently phosphorylated, the factor can enhance the calcium current in RPE cells from RCS rats. Activation of PKC led to a decrease of the calcium currents by PKC-dependent inhibition of the factor. Inhibition of PKC led to a decrease of the calcium currents by dephosphorylation of the calcium channel similar to the effect on RPE cells from control rats. Such a factor may be a different type of protein kinase (cAMP-dependent PK, cGMP-dependent PK, Ca^{2+}/calmodulin-dependent PK or protein tyrosine kinase) which acts cooperatively on the calcium channel.

It has been shown that phagocytosis is regulated by the IP3/calcium second messenger system (5,6,7). An increase of the cytosolic inositol-phosphate concentration represents the "on"-signal (6) whereas an increase of the cytosolic free calcium and subsequent activation of PKC represents the "off"-signal for phagocytosis (5). We previously discussed that increased calcium conductance in RPE cells from RCS rats could generate an "off"-signal for phagocytosis by changing the reaction of the IP3/calcium second messenger system (27). Using the activation of calcium-dependent chloride channels we demonstrated that increased calcium conductance could change the reaction of this second messenger system (25). However, we could detect a changed effect of PKC on calcium channel regulation. Thus, the change of the IP3/second messenger system by the increased calcium conductance may be a secondary effect. It has been shown that activation of PKC represents the "off"-signal for phagocytosis (6). In addition, Heth and Schmidt (10) could demonstrate that protein phosphorylation is changed in RPE cells from RCS rat. We could show that PKC activation has an effect on calcium channel regulation in RPE cells from RCS rats and not in RPE cells from non-dystrophic rats. Thus, it seems that a changed regulation of PKC or different effects of PKC activation may be the cause for the unability for phagocytosis of photoreceptor outer membranes by the RPE in RCS rats.

ACKNOWLEDGMENTS

This work was supported by the DFG grant Wi/992 and the German Retinitis pigmentosa Society (DRPV). The authors thank Prof. H. Kettenmann and Dr. T. Weiser for helpful discussions. The expert technical assistance of H. Morhardt, A. Krolik and M. Boxberger is greatfully acknowledged.

REFERENCES

1. Edwards, R.B. and Szamier, R.B., 1977. Defective phagocytosis of isolated rod outer segments by RCS rat retinal pigment epithelium in culture. *Science* **197**: 1001-1003.
2. Goldman, A.I. and O'Brien, P.J., 1978. Phagocytosis in the retinal pigment epithelium of the RCS rat. *Science* **201**: 1023-1025.
3. Chaitin, M.H. and Hall. M.O., 1983. Defective ingestion of rod outer segments by cultured dystrophic rat pigment epithelial cells. *Invest. Ophthalmol. Vis. Sci.* **24**: 812-820.
4. Hall, M.O. and Abrams, T.A., 1991. RPE cells from normal rats do not secrete a factor which enhances the phagocytosis of ROS by dystrophic rat RPE cells. *Exp. Eye Res.* **52**: 461-464.
5. Hall, M.O., Abrams, T.A. and Mittag, T.W., 1991. ROS ingestion by RPE cells is turned off by increased protein kinase C activity and by increased calcium. *Exp. Eye Res.* **52**: 591-598.
6. Rodriguez de Turco, E.B., Gordon, W.C. and Bazan, N.G., 1992. Light stimulates in vivo inositol lipid turnover in frog retinal pigment epithelial cells at the onset of shedding and phagocytosis of photoreceptor membranes. *Exp. Eye Res.* **55**: 719-725.
7. Heth, C.A. and Marescalchi, P.A., 1992. Generation of inositol triphosphate in cultured RCS retinal pigment epithelium. *Invest. Ophthalmol. Vis. Sci.* **33**: 1203.
8. Heth C.A. and Marescalchi P.A., 1994. Inositol triphosphate generation in cultured rat retinal pigment epithelium. *Invest.Ophthalmol.Vis. Sci.* **35**: 409-416.
9. Gregory, C.Y., Abrams, T.A. and Hall, M.O., 1992. cAMP production via the adenyl cyclase pathway is reduced in RCS rat RPE. *Invest.Ophthalmol. Vis. Sci.* **33**: 3121-3124.
10. Heth, C.A. and Schmidt, S.Y., 1992. Protein phosphorylation in retinal pigment epithelium of Long-Evans and Royal College of Surgeon rats. *Invest. Ophthalmol. Vis. Sci.* **33**: 2839-2847.
11. Leschey, K.H., Hackett, S.F., Singer, J.H. and Campochiaro, P.A., 1990. Growth factor responsiveness of human retinal pigment epithelial cells. *Invest. Ophthalmol. Vis. Sci.* **31**: 839-846.
12. McLaren, M.J., Holderby M., Brown, M.E. and Inana, G., 1992. Kinetics of ROS binding and ingestion by cultured RCS rat RPE cells: modulation by conditioned media bFGF. *Invest.Ophthalmol. Vis. Sci.* **33**:1027.
13. Hamill, O.P., Marty, A., Neher, E., Sakmann, B. and Sigworth, F.J., 1981. Improved patch-clamp techniques for high-resolution current recordings from cells and cell-free patches. *Pflüger's Arch.* **391**: 85-100.
14. Wen, R., Liu, G.M. and Steinberg, R.H., 1993. Whole-cell recordings in fresh and cultured cells of the human and the monkey retinal pigment epithelium. *J. Physiol. (Lond.)* **465**: 121-147.

15. Strauß, O., Richard, G. and Wienrich, M., 1993. Voltage-dependent potassium currents in cultured human retinal pigment epithelial cells. *Biochem. Biophys. Res. Commun.* **191**: 775-781.

16. Strauß, O., Weiser, T. and Wienrich, M., 1994. Potassium currents in cultured cells of the rat retinal pigment epithelium. *Comp. Biochem. Physiol.* **109A**: 975-983.

17. Tao, Q., Rafuse, P.E. and Kelly, M.E.M., 1994. Potassium currents in cultured rabbit retinal pigment epithelial cells. *J. Membrane Biol.* **141**: 123-138.

18. Hughes, B.A. and Steinberg, R.H., 1990. Voltage-dependent currents in isolated cells of the frog retinal pigment epithelium. *J. Physiol. (Lond.)* **428**: 273-297.

19. Wen, R., Liu, G.M. and Steinberg, R.H., 1994. Expression of a tetrodotoxin-sensitive Na^+-current in cultured human retinal pigment epithelial cells. *J. Physiol (Lond.)* **476**: 187-196.

20. Botchkin, L.M. and Matthews, G., 1994. Voltage-dependent sodium channels develop in rat retinal pigment epithelium cells in culture. *Proc. Natl. Acad. Sci. USA* **91**: 4564-4568.

21. Ueda, Y. and Steinberg; R.H., 1993. Voltage-operated calcium channels in fresh and cultured rat retinal pigment epithelial cells. *Invest. Ophthalmol. Vis. Sci.* **34**: 3408-3418.

22. Strauss, O. and Wienrich, M., 1994. Ca^{2+}-conductances in cultured rat retinal pigment epithelial cells. *J. Cell. Physiol.* **160**: 89-96.

23. Botchkin, L.M. and Matthews, G., 1993. Chloride current activated by swelling in retinal pigment epithelium cells. *Am. J. Physiol.* **265**: C1037-C1045.

24. Ueda, Y. and Steinberg, R.H., 1994. Chloride currents in freshly isolated rat retinal pigment epithelial cells. *Exp.Eye Res.* **58**: 331-342.

25. Strauss, O., Wiederholt, M. and Wienrich, M., 1996. Activation of Cl⁻ currents in cultured rat retinal pigment epithelial cells by intracellular applications of inositol-1,4,5-triphosphate: differences between rats with retinal dystrophy (RCS) and normal rats. *J. Membrane Biol.* **151**: 189-200.

26. Hughes, B.A. and Segawa, Y., 1993. c-AMP activated chloride currents in amphibian retinal pigment epithelial cells. *J. Physiol. (Lond.)* **466**: 749-766.

27. Strauss, O. and Wienrich, M., 1993. Cultured retinal pigment epithelial cells from RCS rats express an increased calcium conductance compared with cells from normal rats. *Pflüger's Arch.* **425**: 68-76.

28. McDonald, T.F., Pelzer, S., Trautwein, W. and Pelzer, D.J., 1994. Regulation and Modulation of calcium channels in cardiac, skeletal, and smooth muscle cells. *Physiological Rev.* **74**: 365-507.

29. Edwards, R.B., 1977. Culture of rat retinal pigment epithelium. *In Vitro* **13**: 301-304.

30. Herbert J.M., Augereau J.M. and Maffrand J.P., 1990. Chelerythrine is a potent and specific inhibitor of protein kinase C. *Biochem Biophys Res Commun.* **172**:993-999.

14

STRUCTURES OF THE OLIGOSACCHARIDES OF RHODOPSIN FROM NORMAL AND RCS RATS

Edward L. Kean,[1,2*] Tamao Endo,[3] Naiqian Niu,[1] Daniel T. Organisciak,[4] Yuji Sato,[3] and Akira Kobata[3]

[1]Center for Vision Research
Department of Ophthalmology
[2]Department of Biochemistry
Case Western Reserve University
Cleveland, Ohio
[3]Tokyo Metropolitan Institute of Gerontology
Tokyo, Japan
[4]The Petticrew Research Laboratory
Department of Biochemistry and Molecular Biology
Wright State University
Dayton, Ohio

ABSTRACT

A study of the structures of the oligosaccharides of rhodopsin from the rat revealed that the major oligosaccharide isomer was same as that from cow, human and frog. Although the site of the dystrophy in the RCS (Royal College of Surgeons) rat has been shown to be the retinal pigment epithelium, the possibility was examined that alterations in the glycosylation of rhodopsin in the dystrophic animal might also be present. No differences were observed, however, in either the amounts or structures of the rhodopsin oligosaccharide chains from young or adult control rats and RCS rats, although some differences were detected in the relative distribution of some oligosaccharide isomers.

INTRODUCTION

As shown by studies of the structure of rhodopsin from the cow (1–3), frog (4), and human (5), the structures of the major oligosaccharide from each of these species are identical, and consists only of the hexasaccharide core region common to other asparagine-linked, complex glycoproteins:

---

* For correspondence at the Center for Vision Research, Department of Ophthalmology, Room 653 Wearn Building, Case Western Reserve University, 11100 Euclid Ave., Cleveland, Ohio 44106 U.S.A.

Degenerative Retinal Diseases, edited by LaVail *et al.*
Plenum Press, New York, 1997

GlcNAcβ1 → 2Manα1 → 3(Manα1 → 6)Manβ1 → 4GlcNAcβ1 → 4GlcNAc

In view of the wide use of the rat as an experimental animal, and the desire to explore the extent of this unique motif in other species, it was of interest to determine whether this abridged oligosaccharide structure was also expressed in the rat.

A knowledge of the structure of the carbohydrate chains of the normal rat was of interest also in order to investigate possible abnormalities in the glycosylation of rhodopsin in the retinal dystrophies. In recent years many mutations have been observed in the rhodopsin gene in the human disorder of retinitis pigmentosa (RP) (5), one class of which are mutations in regions that code for the Asn-X-Ser/Thr sequon. This sequence has been widely observed to be a necessary, but insufficient requirement for the glycosylation of asparagine-linked glycoproteins (6). Such mutations would result in losses in glycosylation of rhodopsin, although this has not been determined experimentally as yet in RP patients.

Alterations or defects in glycosylation of glycoproteins have been observed in a variety of disease states (7) and abnormal glycosylation of rhodopsin has been observed in the retinal degeneration of *Drosophila* (8). Abnormalities have been reported in glycosyltransferases that would be required for glycoprotein biosynthesis in membrane preparations from the RCS rat (9). Although the primary site of the dystrophy in the RCS rat has been known for many years to be the retinal pigment epithelium (10, 11), we investigated the possibility that abnormal glycosylation might also be expressed in rhodopsin isolated from animals with this hereditary dystrophy. We examined for the presence or absence of glycan chains of rhodopsin from normal and RCS rats, and determined the detailed structure of the oligosaccharide chains of this glycoprotein (12).

MATERIALS

Animals

The albino RCS rats used in this study (13) were originally obtained from W.K. Noell. The young rats (normal albino Sprague-Dawley, 19–31 days of age; the RCS rats, 21–26 days of age) were born and reared in a weak cyclic light environment consisting of 12 hrs light per day. The adult rats were male Sprague-Dawley animals, 2–3 months of age. All rats were dark adapted overnight and then killed in CO_2-saturated chambers. Retinas were rapidly excised, frozen, and stored in liquid nitrogen until analysis. The use of animals in this investigation conforms to the ARVO resolution on the use of animals in ophthalmic research.

ROS Preparation, Regeneration, Isolation and Purification of Rhodopsin

The details of these procedures were described previously (12). In short, they involved the following. ROS were prepared in the light by modifications of procedures described for rat ROS (14). Rhodopsin was regenerated by incubating the ROS at 4°C in the dark with 11-cis retinaldehyde and then extracted from the ROS with 0.05 M Tris-HCl, pH 7.0 containing 1% Emulphogene BC-720, 5 mM EDTA, and 0.29 mM PMSF. Purification involved immunoaffinity chromatography using the monoclonal antibody, 1D4, bound to agarose. Rhodopsin was recovered (about 90%) from the gel by eluting with a solution of a decapeptide (50 nmol/ml in 0.3% Emulphogene-0.05 M Tris-HCl, pH 7.0) modeled after a region near the carboxyl terminus of bovine rhodopsin (Ac-Lys-Thr-Glu-Thr-Ser-Gln-Val-Ala-Pro-Ala-COO⁻) (15). After concentration, rhodopsin was precipitated by mixing with a 20-fold volume of cold chloroform/methanol (1:1,v/v). Subsequent analyses were performed on the delipidated, denatured product.

Isolation and Analysis of Rhodopsin. After release from the peptide by hydrazinolysis, the oligosaccharides were reduced with NaB^3H_4, as described previously (16). The latter step introduces tritium into the reducing end of the oligosaccharide. The radioactive oligosaccharides were then subjected to anion-exchange chromatography (Mono-Q HR5/5 column). This was followed by serial affinity chromatography. The latter step involved chromatography on columns of AAL-Sepharose and Con A-Sepharose. Gel permeation column chromatography was then performed using a Superdex Peptide HR10/30 column (Pharmacia Biotech, 60 cm x 1 cm i.d.). During operation, the column was kept at 60°C. Each analysis was internally standardized by co-chromatography of partially hydrolyzed dextran, the glucose oligomers of which were monitored by a differential refractometer.

The labeled oligosaccharides purified in this manner were digested with the following glycosidases obtained as described previously (12). (a) *A. ureafaciens* sialidase, (b) jack bean α-mannosidase (c) diplococcal β–galactosidase, (d) diplococcal β-*N*-acetylhexosaminidase (e) snail β-mannosidase (f) jack bean β-*N*-acetylhexosaminidase; (g) diplococcal endoglycosidase-D.

RESULTS

Oligosaccharide Fractionation

The ion exchange chromatography separated the samples into neutral (N) and acidic fractions (A). The latter were completely converted to neutral oligosaccharides by *A. ureafaciens* sialidase digestion indicating that their acidic nature was due to the presence of sialic acid. The neutral oligosaccharide fractions obtained from fractions A were named fractions AN.

Serial lectin chromatography was then carried out. The samples were first subjected to chromatography on an AAL-Sepharose column, which binds fucose-containing oligosaccharides (17). The fractions which did not bind to the column are referred to as AAL⁻. A small bound fraction was eluted with buffer containing 1 mM fucose and is referred to as AAL⁺. Fraction AAL⁻ was applied to a second lectin column of Con A-Sepharose. Oligosaccharides that bound to the latter lectin could be separated by eluting sequentially with 5 mM α-methylglucopyranoside (fraction Con A⁺), followed by 100 mM α–methylmanno-pyranoside (fraction Con A⁺⁺). The fucosylated oligosaccharides comprised only a small fraction (0.8%–3.6%) of rat rhodopsin oligosaccharides.

Structural Analysis

Information concerning the anomeric configuration, sequence, size and nature of the linkages of the oligosaccharides was obtained by a combination of sequential glycosidase treatment using specific, purified glycosidases, followed by Superdex Peptide gel filtration whose elution characteristics were calibrated with a series of purified oligosaccharides. In addition, the size of the eluted material was also determined by reference to the elution pattern of a series of internal standards of glucose oligomers. These studies were performed on fractions N (AAL⁻Con A⁺⁺), fraction N (AAL⁻Con A⁺), and fraction AN (AAL⁻) (12).

An example of this procedure is seen in Fig. 1, showing the elution pattern from the gel filtration column after sequential glycosidase treatments. Fig. 1A shows the pattern of

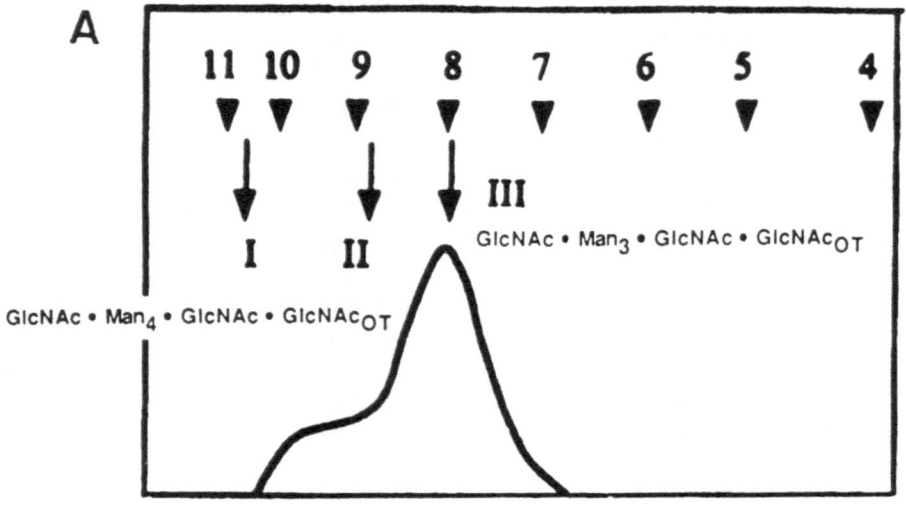

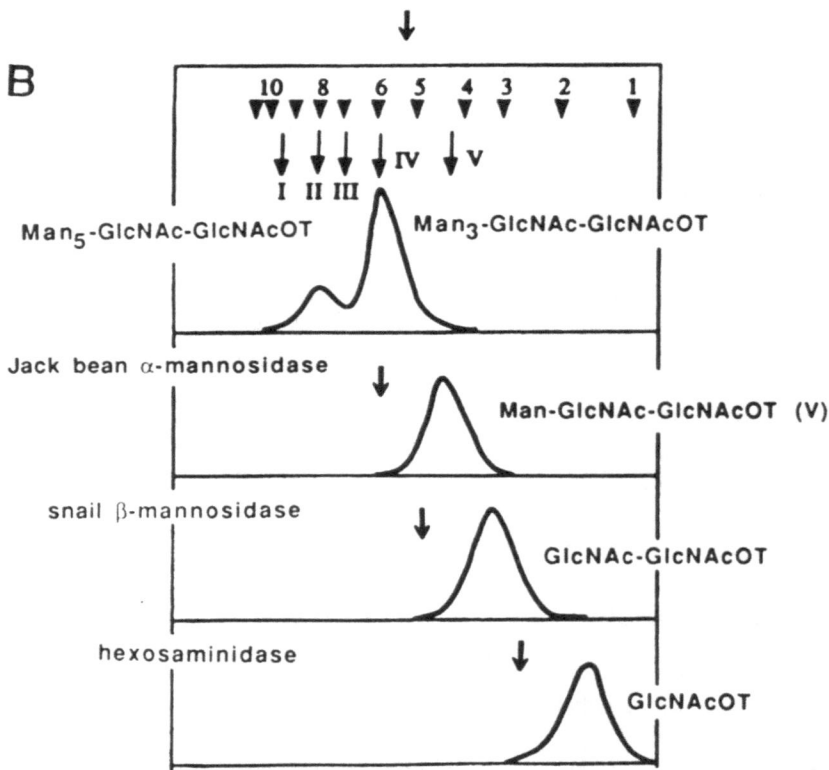

Figure 1. Superdex peptide column chromatography of the radioactive oligosaccharide obtained after serial lectin column affinity chromatography. (A), Fraction N (AAL⁻Con A⁺⁺); (B), Elution patterns of oligosaccharides after sequential treatment with the indicated purified glycosidases. Arrowheads indicate the elution positions of glucose oligomers added as internal standards (the numbers indicate glucose units). Arrows indicate the elution positions of authentic oligosaccharides.

Table 1. Proposed structures of the sugar chains of rat rhodopsin and their % molar ratios

	Molar ratio (%)		
Neutral fraction (N)	Adult	young normal	RCS

AAL⁻Con A⁺⁺

(1)

```
Man α1
        6
         Manβ1→ 4GlcNAcβ1→ 4GlcNAc—Asn
        3
GlcNAc β1→2 Man α1
```
78.3   67.7   58.4

(2)

```
Man α1
        6
         Man α1
Man α1       6
         3    Manβ1→ 4GlcNAcβ1→ 4GlcNAc—Asn
             3
GlcNAc β1→2 Man α1
```
9.4   14.9   25.1

AAL⁻Con A⁺

```
Man α1→6 (3) Man α1
                   6
                    Manβ1→ 4GlcNAcβ1→ 4GlcNAc—Asn
                   3
GlcNAc β1→2 Man α1
```
7.2   8.9   8.4

Acidic-neutral fraction (AN)

AAL⁻

```
Man α1
        6
         Man α1
Man α1       6
         3    Manβ1→ 4GlcNAcβ1→ 4GlcNAc—Asn
             3
Galβ1→4GlcNAc β1→2 Man α1
```
4.2   6.4   4.5

radioactivity from the Superdex Peptide column of material that had been recovered from Con A. One area corresponded to GlcNAc-Man$_3$GlcNAc-GlcNAc$_{OT}$ standard. Treatment of this material with diplococcal β-N-acetylhexosaminidase produced materials which eluted with the Man$_3$ and Man$_5$-GlcNAAc-GlcNAc$_{OT}$ standards, as seen in the upper panel of Fig. 1B. The subsequent panels in Fig. 1 B show the elution patterns after each preceding product was treated successively with the indicated purified exoglycosidase.

The structures proposed for the oligosaccharides of rhodopsin from the adult rat, young normal and RCS rats obtained in this manner are shown in Table 1.

As can be seen, the structure of the major oligosaccharide of rat rhodopsin is the same as found in cow, frog and human rhodopsins. The major oligosaccharide in the adult rat (Table 1, structure AAL⁻Con A⁺⁺, #1) accounted for almost 80% of the rhodopsin-oligosaccharides. Likewise, the structures of the oligosaccharide chains of rhodopsin from the normal and RCS rat were identical (12). However, differences were detected in the relative distribution of the two isomers which were recovered from Con A with 100 mM α-methylmannoside (fraction AAL⁻Con A⁺⁺) between normal and RCS. As seen in Table I there was about a 68% greater abundance in the RCS rat of the isomer containing two mannose residues attached to the Manα1→6 arm (structure AAL⁻Con A⁺⁺, #2) compared to the young normal rat, and a decrease of about 14% in the major isomer, structure #1. The significance of these differences and their relevance to the retinal dystrophy, if any, cannot be evaluated from this study.

Greater than 95% of the oligosaccharide chains of rhodopsin from the the adult rat are neutral, the remaining being sialic acid-containing. Acidic chains, i.e., containing sialic acid, have also been detected as minor constituents in human rhodopsin (16), but

Table 2. Extent of labeling of rhodopsin oligosaccharides from NaB^3H_4. Oligosaccharides were liberated from rhodopsins by hydrazinolysis, purified by ion exchange chromatography and paperchromatography, reduced with NaB^3H_4 and the extent of labeling determined

Source of rhodopsin	cpm
Normal young rat	1.40×106
RCS rat	1.34×106
Normal adult rat	1.34×106

comprise a larger complement in frog rhodopsin (4). Unlike the human, less than 1% of the rat are fucosylated.

Quantitation

Labeling the oligosaccharides with tritium during the reduction with NaB^3H_4 can be used as an index of the relative number of glycan chains present. As seen in Table 2, from the similarity in labeling, it can be concluded that the relative number of glycans attached to rhodopsin from the normal and dystrophic was essentially the same.

Other Studies

The amino acid content of control and RCS rat rhodopsin purified by rhodopsin antibody affinity chromatography was similar to that of bovine rhodopsin and human rhodopsin reported previously (12, 16). Likewise, SDS-PAGE patterns of rhodopsin of normal and RCS rats were the same and similar to bovine and human rhodopsins (12).

DISCUSSION

Previous structural studies of the oligosaccharide chains of rhodopsins of cow, frog, and human revealed a lack of, or limited extent of oligosaccharide chain extension and branching. This same motif has now been observed in the rat. This study represents the first detailed analysis of the structure of the rhodopsin oligosaccharides of this species. Previous studies have also described the limited activities of key glycosyltransferases which would be required for such modifications to have taken place. Chain extension would require the ability to galactosylate the oligosaccharide chains of rhodopsin. However, preparations from bovine, chick and human retinas showed little ability in this regard compared to rat liver (18, 19). Similarly, while the participation of GlcNAc-transferase II is required for oligosaccharide branching to occur, bovine and human retinas also showed little of this activity using rhodopsin/opsin as acceptors (20).

Unlike a previous report with the rat (21), neutral galactose-containing oligosaccharides were not detected in rat rhodopsin in the present experiments. Using in vivo kinetic tracer studies, a rapid, transient galactosylation of a small population of rat rhodopsin molecules was reported (21), a process which decreased with time. The present study, however, dealt with a static situation analyzing stable rhodopsin molecules. The differences in findings may reflect this different experimental design as well as differences in methods of ROS isolation and rhodopsin purification.

While various influences have been observed to be associated with the glycosylation of proteins (7), no generalized concept has emerged concerning the function of this post-translational modification. Thus, hard data, in contrast to speculation, is sparse concerning the influence of the sugar groups of rhodopsin on its physiological functions. As described in this report, with the exception of differences in the relative distribution of some of the isomers, our studies did not find abnormalities in the glycosylation of rhodopsin of the dystrophic rat as compared to the normal. Further studies on the function of glycosylation of this key molecule in the visual system are in process in this laboratory.

ACKNOWLEDGMENTS

This work was supported by NIH grants EY00393 (ELK), EY01959 (DTO), Grants-in Aid for Scientific Research from the Ministry of Education, Science and Culture of Japan (TE, YS), the Ohio Lions Eye Research Foundation and Research to Prevent Blindness, Inc.

ABBREVIATIONS USED

AAL, *Aleutia aurantia* lectin; Con A, Concanavalin A; endo-D, endo-β-*N*-acetylglucosaminidase D; RCS, Royal College of Surgeons; PMSF, phenylmethylsulfonyl fluoride. In structures, OT is used to indicate NaB^3H_4-reduced oligosaccharides.

REFERENCES

1. Liang, C.-J., Yamashita, K., Muellenberg, C.G., Shichi, H., and Kobata, A., 1979, Structure of the carbohydrate moieties of bovine rhodopsin, *J. Biol. Chem.* **254:** 6414–6418.
2. Fukuda, M.N., Papermaster, D.S., and Hargrave, P.A., 1979, Rhodopsin Carbohydrate. Structure of small oligosaccharides attached at two sites near the NH_2 terminus. *J. Biol. Chem.* **254:** 8201–8207.
3. Kean, E.L., Hara, S., Mizoguchi, A., Matsumoto, A., and Kobata, A, 1983, The enzymatic cleavage of rhodopsin by the retinal pigment epithelium. II. The carbohydrate composition of the glycopeptide cleavage product, Exp. Eye Res. **36:** 817–825.
4. Duffin, K.L., Lange, G.W., Welply, J.K., Florman, R., O'Brien, P.J., Dell, A., Reason, A.J., Morris, H.R. and Fliesler, S.J., 1993, Identification and oligosaccharide structure analysis of rhodopsin glycoforms containing galactose and sialic acid, *Glycobiol.* 3: 365–380
5. Dryja, T.P. and Li, T., 1995, Molecular genetics of retinitis pigmentosa, *Human Mol. Genetics* 4: 1739–1743.
6. Marshall, R.D.,1974, The nature and metabolism of the carbohydrate-peptide linkages of glycoproteins, *Biochem. Soc. Symp.* **40:** 17–26)
7. Varki, A.,1993, Biological roles of oligosaccharides: all of the theories are correct, *Glycobiol.* 3: 97–130.
8. Colley, N.J., Cassill, J.A., Baker, E.K., and Zuker, C.S., 1995, Defective intracellular transport is the molecular basis of rhodopsin-dependent dominant retinal degeneration, *Proc. Natl. Acad. Sci. USA,* **92:** 3070–3074
9. Mok, C and Matuk, Y., 1987, Effect of light on the transfer of sugars from sugar nucleotides to rod outer segment membranes of control and dystrophic rats, *Current Eye Res.* 6: 1173–1180.
10. Herron, W.L., Riegel, B.W., Myers, O.E., and Rubin, M.L., 1969, Retinal dystrophy in the rat-A pigment epithelial disease, *Invest. Ophthalmol.* 8: 595–604.
11. Mullen, R.J. and LaVail, M. M., 1976, Inherited retinal dystrophy: Primary defect in pigment epithelium determined with experimental rat chimeras, *Science* 192:799–801.
12. Endo, T., Niu, N., Organisciak, D.T., Sato, Y., Kobata, A., and Kean, E.L., Analysis of the oligosaccharide chains of rhodopsin from normal rats and those with hereditary retinal dystrophy, *Exp. Eye Res.* in press.

13. Delmelle, M., Noell, W.K., and Organisciak, D.T., 1975, Hereditary retinal dystrophy in the rat: Rhodopsin, retinol, vitamin A deficiency, *Exp. Eye Res.* **21:** 369–380.

14. Organisciak, D.T., Xie, A., Wang, H.-M., Jiang, Y.-L., Darrow, R.M.., and Donoso, L.A., 1991 Adaptive changes in visual cell transduction protein levels: Effect of light, *Exp. Eye Res.,* **53:** 773–779.

15. Molday, R.S., and MacKenzie, D., 1983, Monoclonal antibodies to rhodopsin: Characterization, cross-reactivity, and application as structural probes, *Biochem.* **22:** 653–660.

16. Fujita, S., Endo, T., Ju, J., Kean, E.L. and Kobata, A., 1994, Structural studies of the N-linked sugar chains of human rhodopsin, *Glycobiology.* **4:** 633–640.

17. Fukumori, F., Takeuchi, N., Hagiwara, T., Ohbayashi, H., Endo, T., Kochibe, N., Nagata, Y. and Kobata, A., 1990, Primary structure of a fucose-specific lectin obtained from a mushroom, *Aleuria aurantia, J. Biochem.,* **107:** 190–196.

18. Ju, J. and Kean, E.L., 1992, In vitro galactosylation of rhodopsin and opsin: kinetics, properties and characterization, *Exp. Eye Res.* **55:** 589–604

19. Kean, E.L., Ju, J., and Niu, N., 1995, Galactosylation of rhodopsin by the human retina, *Current Eye Res.* **14:** 413–419.

20. Ju, J., and Kean, E.L., 1994, Retinal GlcNAc-transferases and the glycosylation of rhodopsin, *Exp. Eye Res.* **59:** 565–576.

21. Smith, S.B., St Jules, R.S., and O'Brien, P.J., 1991, Transient hyperglycosylation of rhodopsin with galactose, *Exp. Eye Res.* **53:** 525–537.

15

INDOCYANINE GREEN VIDEOANGIOGRAPHY IN THE ROYAL COLLEGE OF SURGEONS RAT

Katsuhiro Yamaguchi, Keiko Yamaguchi, Takeo Satoh, and Shigeki Takahashi

Department of Ophthalmology
Yamagata University School of Medicine
2-2-2 Iidanishi, Yamagata, 990-23, Japan

SUMMARY

The ocular fundus of the Royal College of Surgeons rat was examined with a system for indocyanine green videoangiography (ICGV) with use of a scanning laser ophthalmoscope from Rodenstock. In 3-week-old RCS rats, a bright background fluorescence appeared gradually following the time course. In the late phase, by 5 minutes postinjection, the background fluorescence demonstrated a homogeneous brightness. However, in 5-week-old RCS rats, multiple spots of hyperfluorescence appeared over the dark background by 5 minutes postinjection. In 7-week-old RCS rats, multiple delimited, hyperfluorescent, irregularly round areas appeared over the dark background by 5 minutes postinjection. Such hyperfluorescent lesions increased in number and size in the RCS rats at age 9 weeks. The background fluorescence in the late phase in the RCS rats was abnormally dark, contributing to the mosaic appearance with bright and dark patches. The dark area detected during ICGV may be caused by the atrophy of the choriocapillaris. It is possible that the ICG molecule binds to the subretinal debris material and accounts for progressive increase in hyperfluorescence. Thus, ICGV enabled visualization of the chorioretinal abnormality in the RCS rat eye in vivo.

INTRODUCTION

Indocyanine green dye has been established as a useful adjunct in ophthalmic diagnosis. The dye has its emission and excitation peak in the near infrared region when bound to blood proteins. Current indocyanine green imaging technology includes digital video camera and scanning laser ophthalmoscopy (1,2). The system typically uses an excitation laser source with a wavelength of 780 nm. Thus, indocyanine green angiography enables visualization of the choroidal circulation, subretinal neovascular membranes, and other choroidal abnormalities (3,4). The choriocapillaris provides metabolic support to the outer neurosensory retina and retinal pigment epithelium. Several retinal dystrophic disorders, such as gyrate atrophy, choroideremia, and adult-onset vitelliform macular dystrophy are thought to be related to the choroidal abnormality. The Royal College of Surgeons (RCS) rat strain is characterized by a progressive degeneration of photoreceptor cells (5,6). In addition to degeneration of the photoreceptor cells, changes in the retinal vasculature have been described in this animal model for retinal degeneration, using intravenous histochemical demonstration with horseradish peroxidase (7,8). However, there is little information on the angiographical analysis in vivo. We have examined the ocular fundus of the RCS rat by indocyanine green videoangiography (ICGV).

MATERIALS AND METHODS

Dystrophic RCS rats were obtained from CLEA (Tokyo, Japan). Controls in this study were age-matched Wister-Kyoto rats. At least 5 rats of each group were examined at 3, 5, 7 and 9 weeks of age. All animals were maintained in a 12-hour light / 12 hour-dark light cycle provided by cool-white fluorescent lights at an intensity of 6 ft-c at cage level. All animals were maintained and treated in accordance with NIH guidelines. They were anesthetized with sodium pentobarbital (1mg/100g) intramuscularly, and their pupils were dilated with 0.5% tropicamide and 0.5% phenilephrine eyedrops. The system for ICGV with use of a scanning laser ophthalmoscope from Rodenstock was prepared. Then, ICGV was performed following the injection of the dye intravenously, yielding a dye dose of 2.5mg / 0.5ml.

RESULTS

In each RCS rats, ICGV demonstrated clearly visible choroidal and retinal circulations at the early phase. Then, a bright background fluorescence appeared gradually following the time course. In the late phase, the background fluorescence demonstrated a homogeneous brightness by 5 minutes postinjection, in the RCS rats at age 3 weeks (Fig.1A). However, in the RCS rats at age 5 weeks, multiple spots of hyperfluorescence appeared over the dark background by 5 minutes postinjection (Fig.1B). In the RCS rats at age 7 weeks, multiple delimited, hyperfluorescent, irregularly round areas appeared over the dark background by 5 minutes postinjection (Fig.1C). These lesions increased in number and size in the RCS rats at age 9 weeks (Fig.1D). In the control rats up to age 9 weeks, the choroidal and retinal circulations were clearly visible at the early phase of the ICGV (Fig.2A). Then, following the time course, a bright background fluorescence appeared gradually. In the late phase, by 5 minutes postinjection, the background fluorescence demonstrated a homogeneous brightness (Fig.2B).

DISCUSSION

The ICGV technology enabled visualization of the abnormality in the eye ground of the RCS rats after age 5 weeks. The background fluorescence of the RCS rats in the late phase was abnormally dark when compared to the control rats. Moreover, it become progressively darker following the postnatal weeks. This finding contributed to the mosaic appearance with bright patches observed in the ICGV images of the RCS rats. This abnormality was considered to have a strong relation to the retinal degeneration in this animal

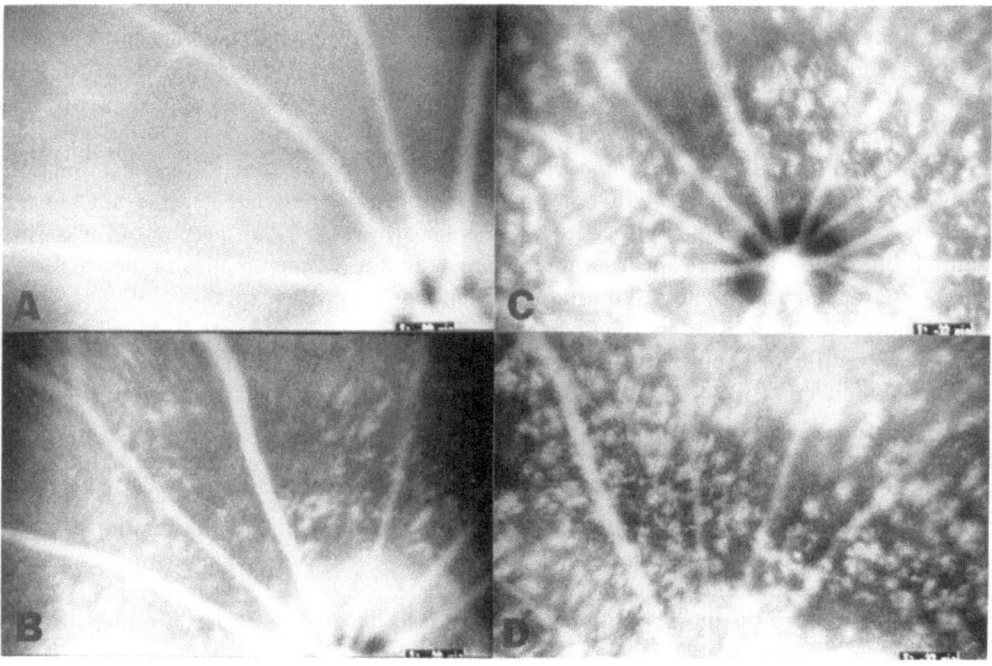

Figure 1. Indocyanine green videoangiography in the Royal College of Surgeons rats. (A) RCS rat aged 3 weeks: The background fluorescence demonstrates a homogeneous brightness. (B) RCS rat aged 5 weeks: Multiple spots of hyperfluorescence are seen over the dark background. (C) RCS rat aged 7 weeks: Multiple delimited, hyperfluorescent, irregularly round areas appeared over the dark background. (D) RCS rat aged 9 weeks: Hyperfluorescent lesions increase in number and size.

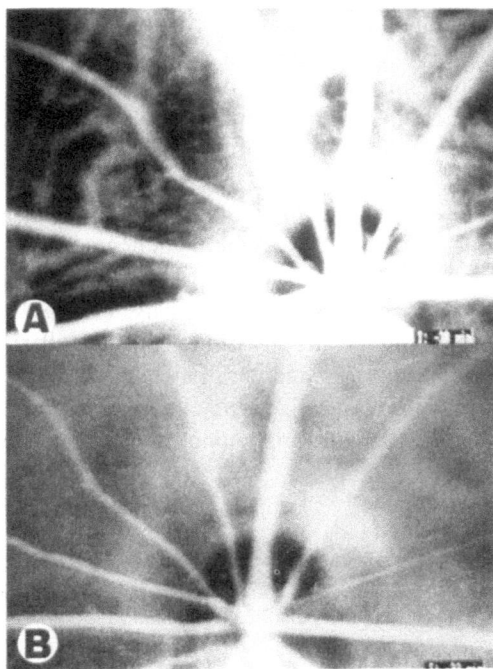

Figure 2. Indocyanine green videoangiography in the control Wister-Kyoto rats at age 9 weeks. (A) At the early phase of ICGV, the choroidal and retinal circulations are clearly visible. (B) In the late phase, by 5 *minutes postinjection:* The background fluorescence demonstrates a homogeneous brightness.

model. More specifically, in the RCS rats, the retina develops normally until about postnatal week 3. Then, photoreceptor cells degenerate progressively, and the outer nuclear layer becomes correspondingly thinner. Since retinal pigment epithelial (RPE) cells of RCS rat have defective phagocytosis of shed rod outer segments, the outer segment debris begins to accumulate at the surface of the RPE and continues to do so until age 2 months (9–11). In addition, RPE cells show certain morphological changes indicative of their dysfunction. The RPE become hyperplastic with abnormally flattened or convoluted surfaces. Caldwell et al (12) reported that the choriocapillaris was usually absent and the choroid was substantially thinner in the areas near large vascular vitreoretinal membranes in the RCS rats. Therefore, the dark area detected during ICGV could perhaps be explained by the atrophy of the choriocapillaris. A direct relationship between RPE cells and the choroidal vasculature has a significant importance in the ocular disease condition. Sarks (13) pointed out a correlation between RPE dropout and atrophy of corresponding areas of choriocapillaris in humans. Korte et al (14) have shown that by using sodium iodate, a toxin that destroys RPE cells in patches, areas of the choriocapillaris exposed to the degenerated clumps of RPE also undergo atrophy, while those corresponding areas of the choriocapillaris next to healthy patches of RPE are unaffected. Again, these reports support the relationship between the RPE and choriocapillaris. Therefore, our results of ICGV demonstrated an atrophic change in the choriocapillaris in the RCS rat eye.

With regard to the numerous hyperfluorescent lesions that increased in intensity and size in the time course of ICGV, maximum shape was generated by the postinjection time of 5 minutes. The lesion had become larger following the postnatal week. Also the number of lesions had increased with the age of the rats. Parodi et al (15) reported a correlation in location between the central yellow lesion visible in the red-free frame and the hyperfluorescent area detected by ICGV in patients with adult-onset foveomacular vitelliform dystrophy. Authors speculated that the hyperfluorescent area may represent progressive staining of the subretinal material. It was also hypothesized that ICG molecule binds to the eosinophilic material described histologically. In conjunction with this explanation based on the ICGV findings in human disease condition, the possibility arose that the subretinal debris material of the RCS rats could be a target to which the ICG molecule can bind. Since this debris continues to accumulate by postnatal age 2 month, it seemed reasonable that both the intensity of hyperfluorescence and size of the lesion had increased following the postnatal weeks.

For this ICG staining, a breakdown in permeability of the RPE seemed to play a role. In the RCS rats, the breakdown was shown as the result of dissolution of tight junctions between RPE cells (16,17). Therefore, the severity in the dissolution of tight junctions may reflect the intensity of the hyperfluorescence in the lesion. In addition, discontinuous atrophy of the choriocapillaris seemed to contribute to the formation of mosaic with bright and dark patches in the ICGV images in the RCS rats. Finally, the ICGV technology visualized the choroidal abnormality in the RCS rat eye ground, adding a new interest to this animal model for retinal degeneration.

REFERENCES

1. Guyer,D.R., Puliafito,C.A., Mones, J.M., Friedman, E., Chang, W., and Verdooner,S.R., 1992, Digital indocyanine green angiography in chorioretinal disorders, Ophthalmology 99: 287–291.
2. Bartsch,D.U., Weireb,R.N., Zinser,G., and Freeman,W.R., 1995, Confocal scanning infrared laser ophthalmoscopy for indocyanine green angiography, Am. J .Ophthalmol. 120: 642–651.

3. Manivannan,A., Kirkpatrick,N.J.P., Sharp,P.F., and Forrester,J.V. 1994, Clinical investigation of an infrared digital scanning laser ophthalmoscope, Br. J. Ophthalmol. 78: 84–90.

4. Woon,W.H., Fitzke,F.W., Bird,A.C., and Manshall,J., 1992, Confocal imaging of the fundus using a scanning laser ophthalmoscope, Br.J. Ophthalmol. 76: 470–474.

5. Dowling ,J.E., and Sidman ,R.L., 1962, Inherited retinal dystrophy in the rat, J. Cell Biol. 14: 73–109.

6. LaVail,M.M., Sidman,R.L., and Gerhardt,C.O., 1975, Congenic strains of RCS rats with inherited retinal dystrophy, J.Hered. 66: 242–244.

7. Mattes,M.T., and Bok, D., 1985, Blood vascular abnormalities in animal with inherited retinal degeneration, in: Retinal Degeneration: Experimental and Clinical Studies (M.M. LaVail, J.G.Hollyfield, and R.E.Anderson, eds), pp. 209–237, Alan R. Liss Inc., New York.

8. Seaton, A.D., and Turner,J.E., 1992, RPE transplants stabilize retinal vasculature and prevent neovascularization in the RCS rat, Invest. Ophthalmol.Vis.Sci. 33: 83–91.

9. Bok,D., and Hall, M.O., 1971, The role of the pigment epithelium in the etiology of inherited retinal dystrophy in the rat. J.Cell Biol. 49: 664–682.

10. LaVail,M.M., Sidman,R.L., and O'Neil,D., 1972, Photoreceptor-pigment epithelial cell relationships in rats with inherited retinal degeneration ; Radioautographic and electron microscope evidence for dual sourcre of extralamellar material, J.Cell Biol. 53: 185–209.

11. Tamai, M., and O'Brien, P.J., 1979, Retinal dystrophy in the RCS rat: in vivo and vitro studies of phagocytic action of the retinal pigment epithelium on the shed rod outer segments, Exp. Eye Res .28: 399–411.

12. Caldwell,R.B., Roque,R.S., and Solomon,S.W., 1989, Increased vascular density and vitreoreinal membranes accompany vascularization of the pigment epithelium in the dystrophic rat retina. Curr.Eye Res. 8: 923–927.

13. Sarks,S.H., 1990, Drusen and thier relationship to senile macular degeneration, Aust.J.Ophthalmol. 8: 117–130.

14. Korte,G.E., Reppucci,V., and Henkind,P., 1984, RPE destruction causes choriocapillary atrophy, Invest. Ophthalmol.Vis.Sci. 25: 1135–1145.

15. Parodi,M.B., Instulin,D., Russo,D., Ravalico,G., 1996, Adult onset foveomacular vitelliform dystrophy and indocyanine green videoangiography, Graefe's Arch.Clin. Exp. Ophthalmol. 234: 208–211.

16. Caldwell,R.B., McLaughlin,B.J., and Boykins,L.G., 1982, Intramembrane changes in retinal pigment epithelial cell junctions of the dystrophic rat retina, Invest. Ophthalmol. Vis.Sci. 23: 305–318.

17. Caldwell,R.B., and McLaughlin,B.J., 1983, Permeability of retinal pigment epithelial cell junctions in the dystrophic rat retina, Exp. Eye Res. 36: 415–427.

16

DEFECTIVE CHOROIDAL ANGIOGENESIS PRECEDES RETINAL PIGMENT EPITHELIAL PHAGOCYTIC DEFECT IN NEONATAL RCS RATS

Margaret J. McLaren

Department of Physiology and Biophysics
University of Miami School of Medicine
Miami, Florida

INTRODUCTION

The well known defect in phagocytosis of rod outer segment (ROS) membranes by the retinal pigment epithelium (RPE) of RCS rats, the subject of numerous investigations, has long been considered to be the primary pathology leading to retinal degeneration in this mutant (1–3). Light microscopic and ultrastructural studies of the developing photoreceptors of these animals have determined that the phagocytic defect is first evident around postnatal day 12, coincident with the onset of retinal phagocytosis in the normal animals (4). Recently, our laboratory has provided evidence that there is a second defect in dystrophic rat eyes that appears earlier in development than the phagocytic defect: this abnormality involves failed or delayed development of the choroidal vasculature, which normally proceeds rapidly between postnatal days 7–10 (5). In this chapter, I briefly review the data revealing the choroidal angiogenesis defect in mutant RCS rat pups, as well as our recent evidence indicating a reduction in the mRNA and protein for basic fibroblast growth factor (bFGF) in intact RPE of neonatal RCS rats (5). This new data is discussed in the context of previous conflicting investigations into bFGF expression in RCS rat retinas. Finally, the hypothesis is explored that the observed downregulation of bFGF in the mutant RPE during this period in postnatal ocular development may play a central role in the pathology, adversely affecting choroidal angiogenesis, RPE phagocytic function and trophic support of the photoreceptors.

METHODS

Experimental Animals

Breeding colonies were maintained by brother x sister matings of four RCS rat strains i.e. tan-hooded retinal dystrophic (RCS) and congenic control (RCS-rdy$^+$) and

Degenerative Retinal Diseases, edited by LaVail *et al.*
Plenum Press, New York, 1997

black-hooded (pigmented) dystrophic (RCS-p$^+$) and congenic control (RCS rdy$^+$p$^+$). Retinal phenotype of breeders and offspring was verified histologically on a routine basis. Pups were used for experiments on postnatal days (P) 7–14. All animal procedures conformed to the Resolution on the Use of Animals in Research of the Association for Research in Vision and Ophthalmology.

Light Microscopy of Developing Eyes of Neonatal Control and Mutant RCS Rats

Litters of pigmented control and dystrophic RCS rat pups, 7–11 days of age, were anaesthetized and fixed by intracardiac perfusion with mixed aldehydes (2.5% glutaraldehyde, 2.0% paraformaldehye in 0.1 M PO$_4$ buffer, pH 7.4). Fixed eyes were enucleated, bisected anteroposterially through the cornea and optic nerve head, embedded in paraffin and examined in H & E stained 8 :m-thick sections.

Preparation of Wholemounts of Choroid-Lined Eyecups

Choroid-lined eyecups were prepared from eyes of 7–11 day old pigmented control and mutant RCS rats following enzymatic removal of the RPE as described in Section 2.4. Eyecups were fixed in 2% paraformaldehyde and flatmounted between glass coverslips and microscope slides. The sclera is translucent in these preparations, enabling the intact choroidal vasculature to be observed by transmitted light and phase contrast microscopy. Choroid-lined eyecups were analyzed from 24 eyes sampled from six normal litters and 44 eyes from seven dystrophic litters. Counts of choroidal vessels were made using a 25x phase contrast objective of a Zeiss Photomicroscope III.

Preparation of Rat RPE-Lined Eyecups, Isolated RPE Sheets and Plated Primary RPE Cells

Eyecups used for immunohistochemistry were prepared from eyes of P7–14 tan-hooded normal and dystrophic rats. Under a dissecting microscope, eyes immersed in Hank's balanced salt solution were cut open at the limbus. The anterior segment and lens were discarded, the neural retina was gently removed from the eyecup, and the exposed RPE monolayer was fixed *in situ*. All other procedures using RPE involved, prior to dissection, incubation of eyes in Dispase; this single enzymatic step results in clean separation of the RPE monolayer from the underlying choroid, as previously described (5). Freshly isolated RPE sheets comprised of up to 100 attached cells were prepared for immunostaining by plating in 8-well glass chamber slides in Minimum Essential Medium with Earles's Salts (Gibco) + 20% fetal bovine serum (FBS) + 1% penicillin /strepto-mycin, and fixing them following attachment to the glass substrate (within 18 hr of plating).

Immunofluorescent Detection of bFGF in Normal and Dystrophic Rat RPE Cells

RPE-lined eyecups and sheets of RPE cells plated in multiwell slides were fixed at room temperature (RT) in 2% or 4% paraformaldehyde in 0.1M NaPO$_4$ buffer, pH 7.4, then washed extensively in buffer. For immunostaining, whole eyecups were processed through droplets of the various solutions, and plated RPE were processed on multiwell slides. All data were obtained from normal and dystrophic RPE samples prepared and immunostained concurrently. Both types of RPE preparation were first incubated for 30 min in PBS + 3% bovine serum albumin (BSA) + 1% FBS, then reacted for 1 hr at RT with primary antibody against bFGF diluted with PBS + 3% BSA, i.e. either 1) rabbit polyclonal antipeptide antibody raised against amino acids 1–24 of the human and bovine bFGF sequence (Sigma), 1:30 dilution or 2) mouse monoclonal antibody raised against purified bovine brain bFGF (Upstate Biotechnology Inc., type 1), 1:50–1:200 dilution. Following washing, bound antibody was detected by incubation with an appropriate FITC-labeled secondary antibody, as previously described (5). Controls included preadsorption of the antipeptide antibody with a tenfold excess of the corresponding bFGF synthetic peptide (Sigma) for 24 hr at 4C prior to use, and incubation of samples in all groups for 1 hr in PBS + 3% BSA alone, followed by secondary antibody. After the final washing step in PBS, all preparations were mounted in Fluoromount-G (Southern Biotechnology Associates Inc., Birmingham, AL), and slides were observed and photographed by phase contrast and fluorescence microscopy using a Zeiss Photomicroscope III.

Isolation of RNA from Uncultured P10 Rat RPE Cells

Immediately following isolation, pooled RPE sheets obtained from a litter of 10-day old pigmented normal or dystrophic rat pups (10–20 eyes; $0.5–1.0 \times 10^6$ RPE cells) were lysed by addition of a 1 ml volume of 4M guanidine thiocyanate solution, followed by vigorous vortexing and homogenization with a 7mm Polytron generator. Homogenates from each batch of cells were stored at -20C, then 5–7 batches, each obtained from pups of a different mother, were combined for RNA isolation. Low speed centrifugation was used to remove pigment granules and debris, then RNA was purified by CsCl centrifugation as described previously (5). Typically, 5–7 μg of purifed total RNA was obtained from 10^6 uncultured neonatal rat RPE cells. To produce RNA samples representative of RPE from a large number of rat pups, total RNA samples obtained from 100–125 individual eyes were combined for use on northern blots.

Northern Analysis of bFGF Expression in Uncultured Rat RPE Cells

An RT-PCR-amplified bFGF probe from rat RPE was prepared as previously described (5). RNA (10 μg/lane) isolated from uncultured P10 normal or dystrophic rat RPE or confluent cultures of BPEI-1 normal rat RPE cell line (6), grown with cobblestone morphology, was electrophoresed in formaldehyde gels, transferred to nylon filters, baked and used for hybridization with the rat RPE bFGF cDNA probe, and a probe for $-actin as described (5).

RESULTS

Analysis of Angiogenesis during Choroidal Differentiation in P7–11 Control and Mutant RCS Rats

Early morphological studies of fetal and neonatal rat eyes briefly described maturation of the choroidal vasculature occuring during the first postnatal week in the normal albino rat (7,8). To extend these findings and compare results in the control and mutant rat strains, we first examined paraffin sections of eyes of control and mutant pigmented rat pups, fixed by vascular perfusion on P7–11. In both normal and dystrophic eyes, by light microscopy, the choroid and sclera appeared essentially undifferentiated at P7 (Figs. 1a,b). In the normal pups, the choriocapillaris and the vortex venous system became apparent within the subsequent 1–3 days (Fig. 1c). By contrast, the choroidal tissue remained rela-

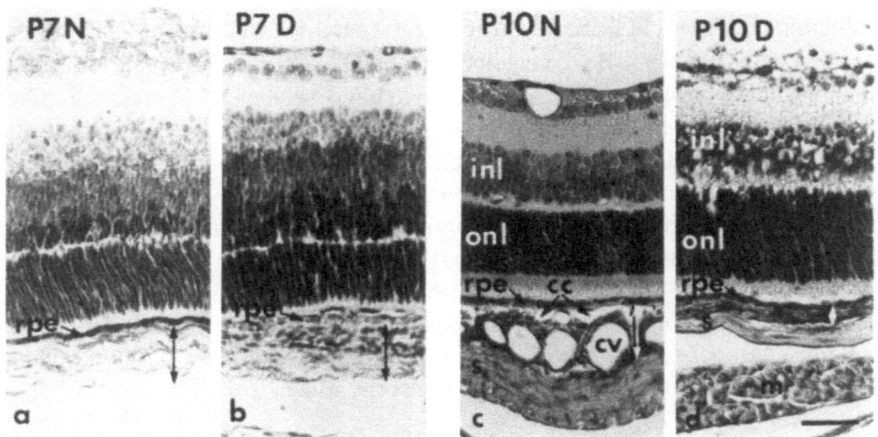

Figure 1. Analysis of choroidal angiogenesis in developing eyes of pigmented normal (N) and dystrophic (D) RCS rat pups. (a-d): H + E stained paraffin sections of eyes from normal and mutant rat pups perfusion-fixed on the indicated postnatal (P) day, showing the appearance of the posterior wall of the globe close to the optic nerve head. Black vertical arrows in (a,b) indicate extent of the undifferentiated choroid and sclera on P7; white vertical arrows in (c,d) show the width of the differentiating choroid on P10. rpe- retinal pigment epithelium; inl- inner nuclear layer; onl- outer nuclear layer; cc- choriocapillaris; cv- choroidal vessel; s- sclera; m- extraocular muscle. In P10 normal pups (c), but not mutants (d), an extensive network of choroidal vessels has developed, intimately surrounded by stroma containing abundant heavily pigmented melanocytes. Bar- 100 μm (a-d). Reprinted with permission from McLaren, M.J. et al., 1996, FEBS Lett. 87: 63–70.

tively undifferentiated in P10–11 dystrophic pups (Fig. 1d). In the developing choroid of both normal and dystrophic pigmented rats, the connective tissue spaces became progressively more populated with darkly pigmented melanocytes; in sections containing transversely cut vessels, some of the melanocytes were seen to intimately wrap themselves around the walls of the forming vasculature (Fig. 1c).

Cross sections of eyes afford an obviously limited view of the choroidal vessels; to examine the distribution of the developing choroidal vascular tree in its entirety, wholemounts of intact choroid-lined eyecups from 7–11 day old rats were examined. In the normal rats, the developing choroidal vessels arose primarily from one of three regularly spaced, radially arranged vessels (designated V1-V3) which were already present at P7 and seemed to form the primary centers of angiogenesis. Each center was surrounded by darkly pigmented melanocytes (Fig. 2a). Sprouts arose in parallel from the walls of the main vessels, and their distribution could be appreciated as clear streaks or tracks, coursing through the pigmented stroma (Fig. 2a,b). Growth of the sprouts, and continued accumulation of melanocytes progressed rapidly between P8 and P10. In the normal animals, the three main vessels and their sprouts had a regular, pennate appearance, which was very similar among all eyecups examined (Fig. 2b). In contrast to this, the pattern of vessel formation was very irregular in the eyecups from the mutant RCS rats. The placement of the three main vessels was highly variable from one eye to another, and in some cases one or even two of the main vessels were not present at P7 or P8. In most of the mutant eyes at all ages observed, the diameter of the forming vessels appeared reduced relative to those of age-matched controls, and most notably, the distribution of vessels was markedly disorganized and highly variable even among littermates (Figs. 2c,d).

The number of vessels sprouting from each of the main vessels was quantitated in choroids of 8 and 10-day normal and mutant rats. As expected, the results revealed increasing choroidal angiogenesis between P8 and P10 in both rat strains, but significantly,

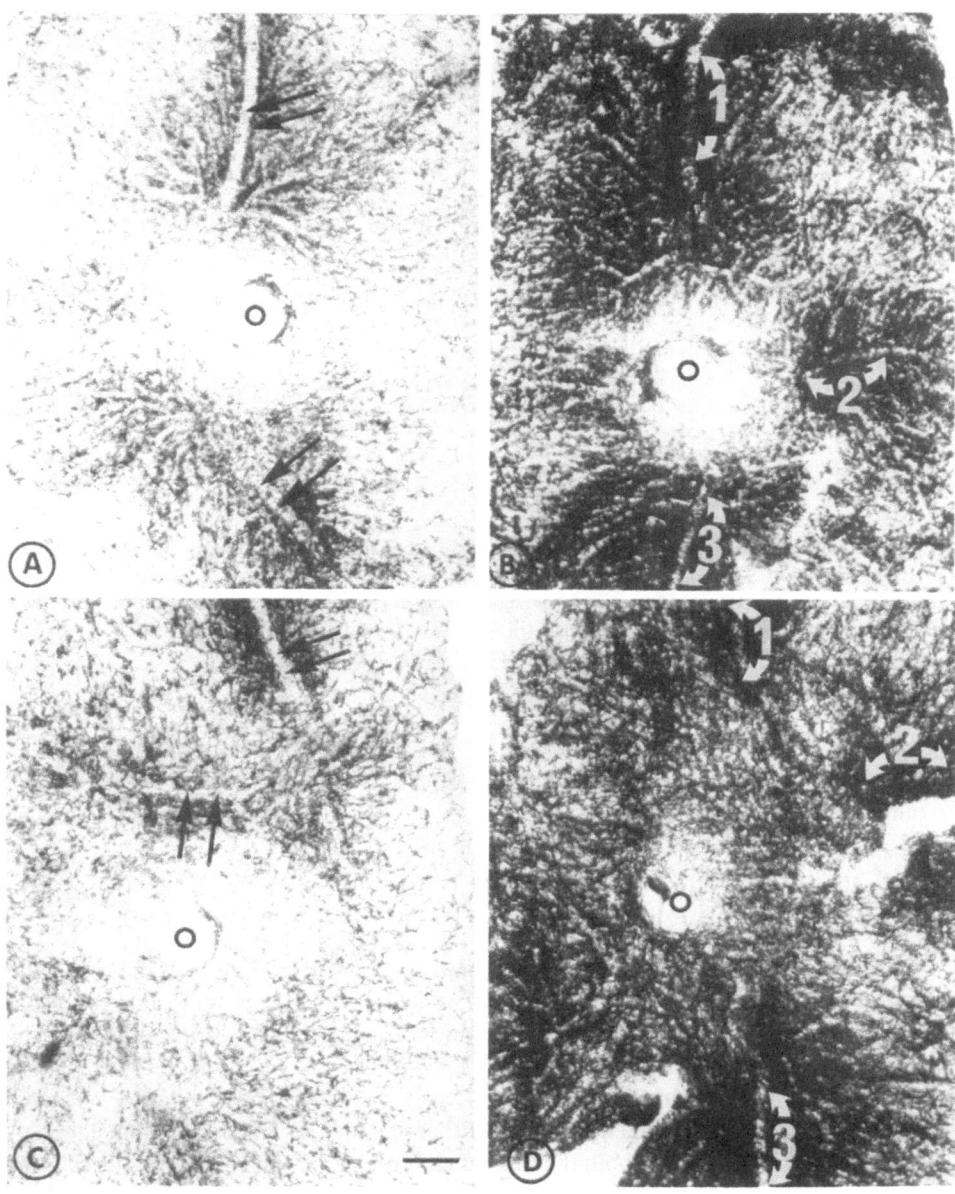

Figure 2. Views of representative wholemounts of choroid-lined eyecups, prepared following enzymatic removal of the RPE layer from eyes of P7 and P10 pigmented normal, (a,b) and P7 and P10 dystrophic (c,d) rat pups, respectively. Choroidal vessels develop as sprouts extending from two or three larger vessels present at P7 (double arrows in a,c). The distribution of the forming choroidal vessels is revealed as light lines and streaks coursing through heavily pigmented stroma (dark areas) containing abundant melanocytes. In normal eyes, the three main vessels (labeled 1–3) and their branches are evenly spaced and have a regular, pennate appearance (b). In the mutant eyes, the distribution of the main vessels is highly variable, and fewer branches are seen than in normal P10 eyecups (compare b,d). o- optic nerve head. Bar- 250 µm (a-d).

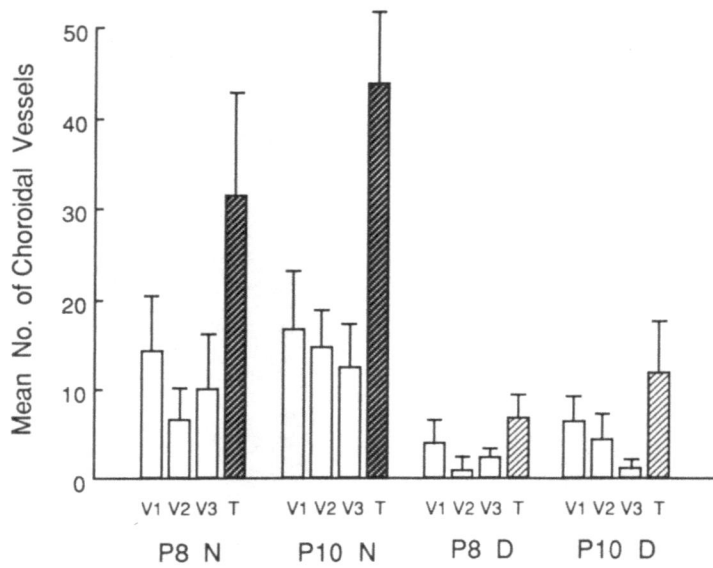

Figure 3. Quantitation of angiogenesis in wholemounts of choroid-lined eyecups from pigmented control, normal (N) and dystrophic (D) RCS rat pups. Counts were made, at the indicated postnatal ages, of vessels sprouting from the three main vessels (V1, V2, V3) forming foci of angiogenesis in the differentiating choroid. T- total counts of vessels. There is a striking reduction in angiogenesis in the mutant pups at both P8 and P10. ($P < 0.005$, Student's t-test). Reprinted with permission from McLaren, M.J. et al., 1996, FEBS Lett. 387: 63–70.

the numbers of vessels in choroids of mutant P8 and P10 rats were only 21% and 27% respectively, of control values (Fig. 3). Thus choroidal angiogenesis was markedly reduced in the mutant pups during this period.

Reduced Immunostaining of bFGF in Wholemounted RPE-Lined Eyecups and Isolated RPE Cells from P7-14 Dystrophic Rats

Following the observation that intraocular injection of bFGF on P23 can retard the photoreceptor degeneration in the RCS rat retina (9), several investigators including ourselves have examined the status of bFGF expression in normal and dystrophic retina and RPE using several approaches (5, 10–12). The defective *rdy* gene is known to be expressed in the RPE of these animals (13). In our study, we specifically wished to determine whether or not bFGF levels were the same in normal and dystrophic RPE during the critical developmental window from P7–14, which encompasses both the time of explosive choroidal angiogenesis (P7–10), and the normal time of onset of ROS phagocytosis in vivo (P12). Just as retinal sections provide a limited view of the choroidal vessels, they also permit analysis of an extremely limited sample of the whole RPE monolayer, which in neonatal rat eyes is comprised of about 60,000 cells. To overcome this problem, and to avoid potential loss of this soluble factor from sectioned RPE cells, we developed a simple new immunofluorescent technique to directly assess the distribution and staining intensity of positively labeled cells within the intact RPE monolayer, as described in the Methods. The RPE and choroidal melanocytes are amelanotic and hence essentially transparent in the pink-eyed, tan-hooded strains of normal and dystrophic RCS rats; taking advantage of this fact, it was possible to view the immunoreactivity of the entire RPE monolayer en face in wholemounts of RPE-lined eyecups (Fig. 4a).

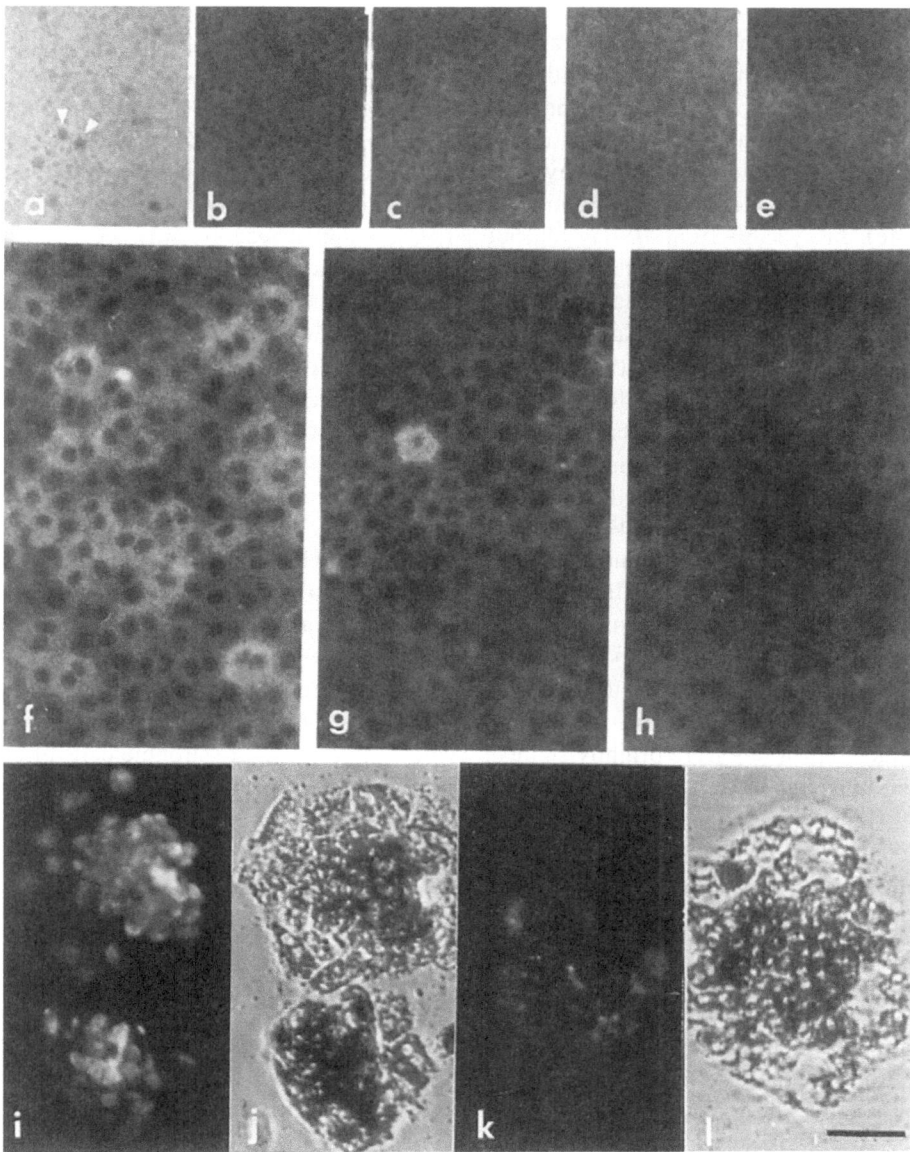

Figure 4. Immunofluorescent detection of bFGF in control and dystrophic rat RPE cells. a-e: Low power views of wholemounted RPE-lined eyecups from tan-hooded P8 normal (a-c) and P8 dystrophic (d,e) rats. Two eyecups from one normal animal (a,b) and two samples from a mutant pup (d,e) are compared. Samples in (a,d) reacted with anti-bFGF [1–24] antipeptide antiserum; controls in (b,e) reacted with the same antiserum preadsorbed with bFGF [1–24] peptide; control in (c) reacted with secondary antibody alone. Arrowheads in (a) indicate sites (dark holes) where individual RPE cells have fallen out of the otherwise brightly fluorescent epithelium, revealing the underlying unreactive choroid. Note the weak bFGF immunoreactivity in the mutant RPE compared with control (a,d). (f-h): Higher magnification of eyecups from P10 normal (f,h) and dystrophic (g) rats. (f,g)- preparations reacted with monoclonal antibody against bovine brain bFGF; (h) -control. (i-l): Clumps of isolated RPE cells from P7 pigmented normal (i,j) and dystrophic (k,l) rats. Cells in (i,k) reacted with monoclonal anti-bFGF antiserum; (j,l) show phase contrast images of (i,k), respectively. Note strong bFGF immunostaining only in normal RPE cells (i). Bar- 100 µm (a-e); 40 µm (f-h); 125 µm (i-l). Reprinted with permission from McLaren, M.J. et al., 1996, FEBS Lett. 387: 63–70.

Similar results were obtained with both bFGF antibodies. In eyecup preparations from 8-day old normal rats, most of the RPE cells in the monolayer were brightly stained following reaction with anti-bFGF antibodies (Fig. 4a). Specificity of the reaction was shown by loss of immunostaining with the bFGF [1–24] peptide antibody following preadsorption with the corresponding peptide (Fig. 4b). Controls from all age groups showed only weak background fluorescence upon omission of the primary antibody (Fig. 4c). In contrast, the RPE monolayer of 8-day old mutant rats, when reacted with the bFGF antisera, showed very weak immunostaining (Fig. 4d) which was barely above control levels (Fig. 4e). In eyecups from 10-day old normal rats, a mixed reaction was seen in which about 30–50% of the RPE cells showed a distinct positive reaction, whereas in adjacent cells within the epithelium, staining was closer to background levels (Fig. 1f; compare with control in Fig. 4h). In 12-day old normal rats, even fewer RPE cells were immunoreactive with bFGF antibodies (not shown). In all of the positively stained normal RPE cells, many of which are binucleate in rats, the immunoreactivity was diffusely distributed throughout the cytoplasm, revealing the hexagonal shape of the cells and the one or two unstained nuclei (Fig. 4f). In the dystrophic eyecups, as seen in the 7–8 day old animals, almost no strongly positive cells were observed in RPE monolayers from 10–14 day old RCS rats (P10 result shown in Fig. 4g).

To ensure that the immunostaining observed in the RPE in eyecup preparations was intrinsic to the RPE cells, bFGF immunostaining was also performed on small sheets of freshly isolated RPE cells from eyes of P7 pigmented normal and dystrophic rats. The cells were cultured overnight to allow for attachment to the glass slides, then fixed for immunostaining. Consistent with the result using intact RPE monolayers in eyecups from P8 tan-hooded control and mutant rats, clumps of isolated RPE cells from P7 pigmented congenic control rats were brightly stained, whereas sheets of P7 dystrophic cells of comparable appearance were minimally labeled following reaction with the anti-bFGF antibodies (Figs. 4i,j; compare with 4k,l).

Reduced Expression of bFGF mRNA in Northern Blots of RNA from Uncultured RPE of P10 Mutant Rat Pups

Finally, northern analysis was used to evaluate the level of expression of bFGF mRNA in RPE of neonatal rat eyes, using samples of pooled RNA from uncultured normal or dystrophic RPE (representative of approximately 125 eyes of each rat strain), and RNA from cultured BPEI-1 normal rat RPE cells for reference (Fig. 5a). The result revealed a faint but distinct 7.0 kb bFGF transcript in P10 uncultured RPE from both rat strains. A slightly stronger bFGF signal was seen in the cultured RPE cell line. All of the signals, however, were much weaker than $-actin, a very highly expressed transcript in both fresh and cultured RPE cells (Fig. 5a). Densitometric analysis of bFGF expression, normalized to actin, revealed an approximately twofold reduction in bFGF levels in uncultured P10 mutant RPE, as compared with age-matched congenic controls (Fig. 5b).

DISCUSSION

A major pursuit of our laboratory has been to understand the molecular mech-anisms involved in RPE phagocytosis and its failure in the mutant RCS rat RPE, and to elucidate the unknown *rdy* gene leading to this defect (4–6,14–17). The analysis of the differentiating choroidal vasculature was a new departure, prompted by the findings of the bFGF ex-

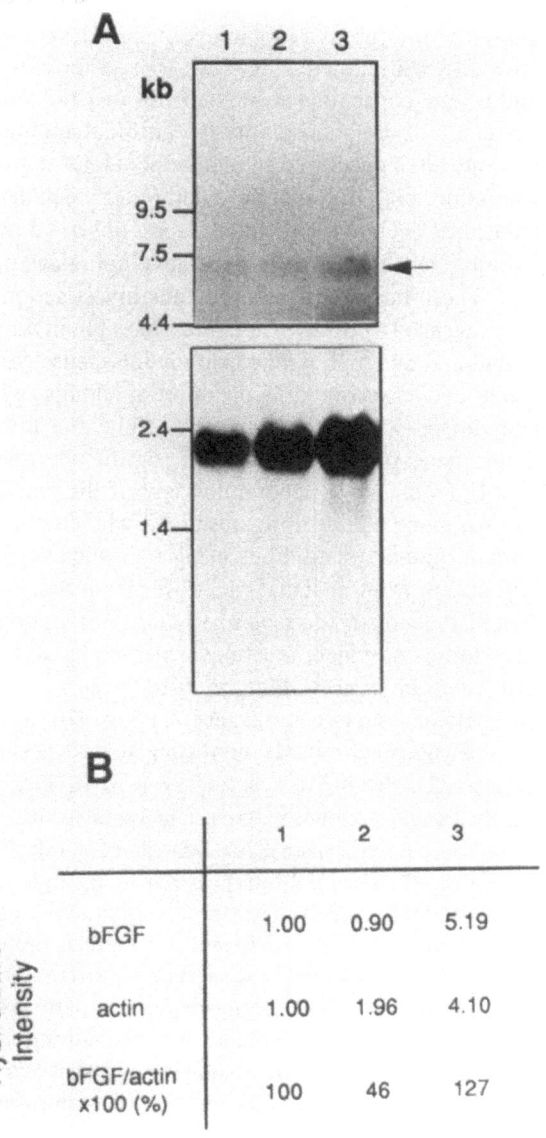

Figure 5. Northern blot analysis of bFGF expression in uncultured neonatal rat RPE and normal RPE cell line. (a): Ten μg of total RNA from freshly isolated RPE sheets from P10 RCS rdy$^+$p$^+$ control rats (lane 1), P10 RCS p$^+$ dystrophic rats (lane 2) or cultured BPEI-1 normal RPE cells (lane 3) was electrophoresed, blotted, and hybridized with ^{32}P-labeled RT-PCR-amplified bFGF probe (upper panel), or $-actin probe (lower panel) and autoradiographed. Arrow indicates a faint 7.0 kb bFGF message present in each lane, but most prominent in lane 3. (b): Densitometric analysis of relative hybridization intensity of bFGF and actin in autoradiographs shown in (a). Results show slightly more than twofold reduction in bFGF expression in lane 2 (P10 mutant RPE) vs. lane 1 (P10 normal RPE). Reprinted with permission from McLaren, M.J. et al., 1996, FEBS Lett. 387: 63–70.

	1	2	3
bFGF	1.00	0.90	5.19
actin	1.00	1.96	4.10
bFGF/actin x100 (%)	100	46	127

pression analysis in normal and dystrophic neonatal rat RPE, i.e. the markedly reduced bFGF immunostaining in RPE from 7–14 day old RCS rat pups, and the reduction in bFGF mRNA levels in sheets of freshly isolated RPE cells from 10 day old mutant pups. During embryogenesis, basic FGF mediates several fundamental processes including induction of mesoderm from endoderm, mitogenesis in mesoderm- and neuroectoderm-derived cells and angiogenesis (reviewed in ref. 18). Unlike the case in humans, differentiation and maturation of the retina, choroid and sclera in the rat occurs during the first postnatal month. Rudimentary photoreceptor outer segments first appear about P7. Subsequent to P12, the time of onset of ROS phagocytosis by the normal RPE (and of the RPE phagocytic defect in the RCS rat), the normal retina continues to mature for at least another two weeks, whereas overt signs of photoreceptor cell death become apparent in the mutant retinas by about one month of age (19). Thus the genetic defect is first expressed while the dystrophic eye is still developing (4). Because bFGF is a key factor in other differentiating systems, we looked for evidence of any developmental anomaly that might be a consequence of the observed reduction in bFGF expression. This led to our awareness of the abnormality in choroidal angiogenesis in the mutant rat pups. Past investigations have described retinal vessel abnormalities in RCS rats with advanced retinal degeneration, but choroidal vasculature has been described as appearing normal in aged RCS rats (20,21). Thus the differences we observed may indicate that the process of choroidal vasculogenesis is delayed in the developing eye of the mutant rats.

There have been two previous investigations, examining either bFGF immuno-staining patterns (10) or mRNA expression (11) in which no differences were detected between normal and dystrophic retinas in vivo. Because we have been primarily interested in the molecular events surrounding the onset of pathology in the mutant RPE, the site of the RCS genetic defect (13), we have so far confined our investigation of bFGF to study of expression of this gene product in the RPE, during the relevant early postnatal period in vivo, i.e. P7–14. To enable a panoramic view of the reaction in the intact RPE monolayer of each eye, we made use of immunostained wholemounts of RPE-lined eyecup preparations. Our immunofluorescent findings using both the eyecup preparations as well as freshly isolated RPE sheets from neonatal rats differ from those in a previous study using immunoperoxidase histochemistry to examine bFGF localization in retinal cross sections (10). In the former study, in which reactions were compared in normal and dystrophic retinas of 19 different ages from P1-P90, no differences were reported in staining of the neural retina or RPE between the two strains at P7, P8 or P10 (ages also observed in our study), or at any of the other ages examined. Interestingly, in that study, and in another using only normal rat retinas (22), the bFGF reaction seen in RPE cells in retinal sections was reported to be mainly in the nucleus, whereas our results with intact neonatal normal RPE cells demonstrated only a cytoplasmic localization of bFGF. It has been shown that multiple isoforms of bFGF (18, 22, 23 and 24 kD) arise from a single mRNA transcript by use of alternate translation start sites (23). The three higher MW species are localized to the cell nucleus, whereas only the 18kD isoform is found in the cytoplasm (24–26). Although lacking the signal sequence of a typical secretory protein, the 18kD protein is apparently exported extracellularly by a novel energy-dependent pathway independent of the ER/Golgi (27).

One possible explanation for the staining differences observed in our study of intact RPE and previous studies using retinal sections is that the soluble, cytoplasmic form of bFGF may be preferentially lost in processing, once the RPE cells are cut open upon sectioning. Additionally, it is conceivable that in our preparations, permeabilization of the cells with paraformaldehyde fixation alone may have been sufficient to allow the antibodies to gain access to cell cytoplasm but not the cell nuclei; if so, then the nuclear isoforms of bFGF, if present, may have gone undetected. Regardless of the status of the nuclear bFGF, our results seem to suggest that the level of the 18 kD secretory form of bFGF was greatly reduced in the dystrophic rat RPE cells, at least between P7-P12, when this protein was readily detected in the cytoplasm of the normal RPE cells. The finding of a more than two-fold reduction in the bFGF mRNA levels in RNA samples from uncultured RPE pooled from large numbers of P10 mutant rats provided evidence that the downregulation of bFGF may have been at the transcriptional level. This result differs from that of an earlier study using RT-PCR to detect bFGF mRNA levels in RPE of 21-day and 3-month old normal and RCS rats, in which no differences were observed between normal and dystrophic rats at either age (11). These conflicting results may have been due to technical differences in the methods used to collect RPE samples and measure bFGF mRNA (non-quantitative RT-PCR vs. northern analysis); alternatively, the decrease in bFGF observed in 7–12 day mutant rat RPE may be an occurrence unique to this particular time in development.

With the realization that the phagocytic defect is not the only manifestation of the mutant gene action in the dystrophic RPE, it now becomes necessary to seek a unifying explanation for a syndrome involving at the least, abnormal choroidal angiogenesis, failed RPE phagocytosis and photoreceptor cell death in the developing eye. Is it possible that expression of bFGF by the RPE is a common denominator, required for trophic support of the photoreceptors, normal choroidal angiogenesis and RPE phagocytosis? Let us examine separately the current evidence for each of these proposed functions of bFGF. From the chimeric rat experiment which first established the RPE as the site of action of the rdy gene (13), and from later studies showing photoreceptor rescue by transplantation of neo-natal normal RPE cells (28,29), it is clear that the mutant photoreceptors can be rescued by a diffusible factor coming from normal RPE. To my knowledge, bFGF remains the only neurotrophic factor tested to date which has the capability of sustaining the mutant photoreceptors when injected into the RCS eye. Basic FGF is also the only known factor produced by the RPE to have this effect. Given the observed cytoplasmic localization of bFGF in the RPE during P7-P12, it is reasonable to postulate that the 18kD secretory form of bFGF may be released extracellularly by the normal RPE cells of rats of this age. Some of this bFGF might be incorporated into the interphotoreceptor matrix, available for tro-phic support of the photoreceptors, as previously proposed (30). Indeed, in a preliminary study in which we injected bFGF into eyes of 9-day old RCS rats, we observed photore-ceptors with remarkably normal looking ultrastuctural appearance in the injected, but not control fellow eyes of these animals fixed on P28 (31). Thus exogenous bFGF appeared to have neurotrophic effects when administered not only on P23 (9), but also earlier, at the onset of photoreceptor differentiation.

Basic FGF is a known angiogenic factor in many systems (reviewed in 18,32,33). Direct infusion of bFGF into choroids of minipigs is reported to stimulate choroidal angio-genesis in this experimental system (34). The finding in our study of markedly reduced immunostaining of the 18kD form of bFGF in the dystrophic rat RPE, coincident with re-duced choroidal angiogenesis relative to controls, is suggestive that secretion of bFGF by the RPE may play a key role in the normal development of the choroidal vasculature (5). This hypothesis remains to be tested experimentally, however reults of several investiga-tions are consistent with it. It has been established in the mature eye that destruction of the RPE leads to degeneration of the choriocapillaris, leading to the concept that maintainence of this essential component of the choroidal circulation is under control of factors origi-nating in the RPE (35). More recently, it has been demonstrated that elongation of choroi-dal endothelial tubes in vitro is stimulated by co-culture with RPE, and inhibited by antibodies against bFGF and another angiogenic factor, VEGF (36). Thus it would not be surprising if bFGF production by the RPE were involved in choroidal angiogenesis in the neonatal rat eye.

Finally, we have obtained evidence both from in vitro (17) and in vivo (31) studies that exogenous bFGF stimulates ROS phagocytosis by RPE cells. For the in vitro studies, densely plated confluent primary cultures of P9–11 normal and dystrophic rat RPE cells were pretreated for 24 hr with MEM, or MEM + bFGF, aFGF, EGF or NGF, then ROS phagocytosis was assayed over patches of RPE cells with pigmented, cuboidal morphol-ogy most representative of RPE in vivo. In normal cultures, RPE cells with this phenotype are slow to bind ROS during the first 9–11 hr of exposure to them, then subsequently bind large numbers of ROS per cell, and ingest them synchronously over the next 1–2 hr (16).

In untreated dystrophic cultures, in which the ingestion phase of ROS phagocytosis is defective (37), cells with this phenotype bind equal numbers of ROS as controls, but ingest only 10–15% of control values after a 19 hr incubation (16). The results of the growth factor experiments showed that this defect in ROS ingestion could be reversed in mutant cultures by pretreatment with bFGF and to a lesser extent aFGF, but not by the unrelated factors EGF and NGF (17). The normal mechanism of ingestion, and the bFGF-mediated rescue of ingestion in the mutant RPE cells, but not ROS binding, were also shown to require new gene transcription, since both were inhibitable with actinomycin D (17). The specific genes which are transcribed under these conditions remain to be determined.

Is the bFGF-mediated phagocytic rescue simply a phenomenon of culture? When injected into the eyes of RCS rats on P23, bFGF does not seem to reverse the phagocytic defect in vivo, based on the continued presence of the debris zone in the subretinal space (9). However, in the experiment described above, in which we injected bFGF into eyes of RCS rat pups (on P9–11), we did observe reduction in the thickness of this layer on P28, in the eyes injected with bFGF, and the RPE in the treated eyes displayed several ultrastructural features including bodies resembling large phagosomes, that are observed in normal P28 RPE fixed during the daily peak of shedding and phagocytosis (31). Thus bFGF administered earlier in development, at the normal time of onset of phagocytosis in vivo, may have stimulated ROS phagocytosis, at least for several days in the treated eyes. Interestingly, photoreceptor survival does not seem to be irreversibly linked with ongoing phagocytosis, since photoreceptor rescue by intraocular bFGF injection on P23 continues for several months in apparent absence of phagocytosis (9).

In summary, there appears to be supportive, although not yet definitive, evidence for the hypothesis that bFGF from the RPE is involved in each of the mechanisms of choroidal angiogenesis, RPE phagocytosis and trophic support of the photoreceptors during normal development of the rat retina. In pursuit of the *rdy* gene, we have screened for mutations in the coding sequence of the bFGF gene and in reverse-transcribed mRNA from RCS rats, with negative results (5). Although a mutation in the bFGF promotor region has not been excluded, our current thinking is that the bFGF expression defect in the RPE is more likely to be downstream of a mutation in another gene, i.e. the as yet unidentified *rdy* gene in this model.

ACKNOWLEDGMENTS

Co-authors of the original work described in this paper were G. Inana, M.D., Ph.D., M.E. Brown, M.D., Ph.D. and W. An, M.D. This work was supported by Research to Prevent Blindness, Inc., Walter G. Ross Foundation and Foundation Fighting Blindness, Inc. Figures 1,3,4 and 5 are reprinted from FEBS Lett. 387: (McLaren, M.J., An. W., Brown, M.E., and Inana, G, "Analysis of basic fibroblast growth factor in rats with inherited retinal degeneration", pp. 63–70,1996), with kind permission of Elsevier Science-NL, Sara Burgerhartstraat 25, 1055 KV Amsterdam, The Netherlands.

REFERENCES

1. Herron, W.L., Reigel, B.W., Myers, O.E., and Rubin. M.L., 1969, Retinal dystrophy in the rat- a pigment epithelial disease, *Invest. Ophthalmol. Vis. Sci.* **8**: 595–604.
2. Bok, D., and Hall, M.O., 1971, The role of the pigment epithelium in the etiology of inherited retinal dystrophy in the rat, *J. Cell Biol.* **49**: 664–682.

3. Edwards, R.B., and Szamier, R.B., 1977, Defective phagocytosis of isolated rod outer segments by RCS rat retinal pigment epithelium in culture, *Science* **197**: 1001–1003.

4. Irons, M.J., 1989, Localization of 5'-nucleotidase activity in RCS rat retinas, in: *Inherited and Environmentally Induced Retinal Degenerations*, (M.M. LaVail, R.E. Anderson, and J.G. Hollyfield, eds.), pp. 301–314, Alan R. Liss, New York.

5. McLaren, M.J., An. W., Brown, M.E., and Inana, G, 1996, Analysis of basic fibroblast growth factor in rats with inherited retinal degeneration, *FEBS Lett.* **387**: 63–70.

6. McLaren, M.J., Sasabe, T., Li, C.-Y., Brown, M.E., and Inana, G., 1993, Spontaneously arising immortal cell line of rat retinal pigmented epithelial cells, *Exp. Cell Res.* **204**: 311–320.

7. Berson, D., 1965, The development of the choroid and sclera in the eye of the foetal rat with particular reference to their developmental interrelationship, *Exp. Eye Res.* **4**: 102–103.

8. Braekevelt, C.R., and Hollenberg, M.J., 1970, Development of the retinal pigment epithelium, choriocapillaris and bruch's membrane in the albino rat, *Exp. Eye Res.* **9**: 124–131.

9. Faktorovich, E.G., Steinberg, R.H., Yasamura, D., Matthes, M.T., and LaVail, M.M., 1990, Photoreceptor degeneration in inherited retinal dystrophy delayed by basic fibroblast growth factor, *Nature* **347**: 83–86.

10. Connolly, S.E., Hjelmeland, L.M., and LaVail, M.M, 1992, Immunohistochemical localization of basic fibroblast growth factor in mature and developing retinas of normal and RCS rats, *Curr. Eye Res.* **11**: 1005–1017.

11. Rakoczy, P.E., Humphrey, M.F., Cavaney, D.M., Chu, Y., and Constable, I.J., 1993, Expression of basic fibroblast growth factor and its receptor in the retina of Royal College of Surgeons rats. A comparative study, *Invest. Ophthamol. Vis. Sci.* **34**: 1845–1852.

12. Malecaze, F., Mascarelli, F., Bugra, K., Fuhrmann, G., Courtois, Y., and Hicks, D, 1993, Fibroblast growth factor receptor deficiency in dystrophic retinal pigment epithelium. *J. Cell. Physiol.* **154**: 631–642.

13. Mullen, R.J., and LaVail, M.M., 1976, Inherited retinal dystrophy. Primary defect in pigment epithelium determined with experimental rat chimeras. *Science* **192**: 799–801.

14. Irons, M.J., 1987, Redistribution of Mn^{++}-dependent pyrimidine 5'-nucleotidase (MDPNase) activity during shedding and phagocytosis, *Invest. Ophthalmol. Vis. Sci.* **28**: 83–91.

15. McLaren, M.J., Inana, G., and Li, C.-Y., 1993, Double fluorescent vital assay of phagocytosis by cultured retinal pigment epithelial cells. *Invest. Ophthalmol. Vis. Sci.* **34**: 317–326.

16. McLaren, M.J., 1996, Kinetics of rod outer segment phagocytosis by cultured retinal pigment epithelial cells. Relationship to cell morphology. *Invest. Ophthalmol. Vis. Sci.* **37**: 1213–1224.

17. McLaren, M.J., and Inana, G., 1995, ROS phagocytosis by cultured normal rat RPE and bFGF-treated dystrophic RPE is transcriptionally controlled. *Invest. Ophthalmol. Vis.* Sci. **36(4)**: S815.

18. Gospodarowicz, D., Ferrara, N., Schweigerer, L., and Neufeld, G, 1987, Structural characterization and biological functions of fibroblast growth factor. *Endocrine Rev.* **8**: 95–114.

19. LaVail, M.M., 1979, The retinal pigment epithelium in mice and rats with inherited retinal degeneration, in: *The Retinal Pigment Epithelium*, (K.M. Zinn, and M.F. Marmor, eds.), pp. 357–380, Harvard University Press, Cambridge, Massachusetts.

20. Roque, R.S., and Caldwell, R.B., 1991, Pigment epithelial cell changes precede vascular transformations in the dystrophic retina, *Exp. Eye Res.* **28**: 787–798.

21. Fitzgerald, M.E., Caldwell, R.B., and Reiner, A., 1992, Vasoactive intestinal polypeptide-containing nerve fibers are increased in abundance in the choroid of dystrophic RCS rats. Curr. Eye Res. **11**: 501–515.

22. Gao, H., Hollyfield, J.G., 1992, Basic fibroblast growth factor (bFGF) immunolocalization in the rodent outer retina demonstrated with an anti-rodent bFGF antibody. *Brain Res.* **585**: 355–360.

23. Florkiewicz, R.Z., and Sommer, A., 1989, Human fibroblast growth factor gene encodes four polypeptides: Three initiate translation from non-AUG codons, *Proc. Natl. Acad. Sci.* **86**: 3978–3981.

24. Renko, M., Quarto, N., Morimoto, T., and Rifkin, D.B., 1990, Nuclear and cytoplasmic localization of different fibroblast growth factor species. *J. Cell. Physiol.* **144**: 108–114.

25. Florkiewicz, R.Z., Baird, A., and Gonzalez, A.-M., 1991, Multiple forms of FGF-2: Differential nuclear and cell surface localization, *Growth Factors* **4**: 265–275.

26. Ishigooka, H., Aotaki-Keen, A.E., and L.M. Hjelmeland, 1992, Subcellular localization of bFGF in human retinal pigment epithelium in vitro, *Exp. Eye Res.* **55**: 203–214.

27. Florkiewicz, R.Z., Majack, R.A., Buechler, R.D., and E. Florkiewicz, 1995, Quantitative export of FGF-2 occurs through an alternative, energy-dependent, non-ER/Golgi pathway, *J. Cell. Physiol.* **162**: 388–399.

28. Li, L., and Turner, J.E., 1988, Inherited retinal dystrophy in the RCS rat: Prevention of photoreceptor degeneration by pigment epithelial cell transplantation. *Exp. Eye Res.* **47**: 911–917.

29. Li, L., and Turner, J.E., 1991, Optimal conditions for the long-term photoreceptor cell rescue in RCS rats: The necessity for healthy RPE transplants. *Exp. Eye Res.* **52**: 669–679. 30. Hageman,G.S., Kirchoff-

Rempe, M.A., Lewis, G.P., Fisher, S.K., and Anderson, D.H., Sequestration of basic fibroblast growth factor in the primate interphotoreceptor matrix. *Proc. Natl. Acad. Sci.* **88**: 6706–6710.

31. McLaren, M.J., Hernandez, E., and Inana, G., 1994, Effect of intraocular injection of bFGF in RCS rat pups, Invest. Ophthalmol. Vis. Sci. 34(4):Suppl., 1498.

32. Klagsbrun, M., and Folkman, J., 1990, Angiogenesis, in: *Peptide Growth Factors and Their Receptors, II* (M.B. Sporn, and A.B. Roberts, eds.), pp. 549–586, Springer-Verlag, New York.

33. Slavin, J., 1995, Fibroblast growth factors: at the heart of angiogenesis. *Cell Biol Int.* **19**: 431–444.

34. Soubrane, G., Cohen, S.Y., Delayre, T., Tassin, J., Hartmann, M.-P., Coscas, G.J., Courtois, Y., and Jeanny, J.-C., 1994, Basic fibroblast growth factor experimentally induced choroidal angiogenesis in the minipig. *Curr. Eye Res.* **13**: 183–195.

35. Korte, G.E., Reppucci, V., and Henkind, P., 1984, RPE destruction causes choriocapillary atrophy, *Invest. Ophthalmol. Vis. Sci.* **25**: 1135–1145.

36. Sakamoto, T., Sakamoto, H., Murphy, T.L., Spee, C., Soriano, D., Ishibashi. T., Hinton, D.R., and Ryan, S.J., 1995, Vessel formation by choroidal endothelial cells in vitro is modulated by retinal pigment epithelial cells, *Arch Ophthalmol.* **113**: 512–520.

37. Chaitin, M.H., and Hall, M.O., 1983, Defective ingestion of rod outer segments by cultured dystrophic rat pigment epithelial cells, *Invest. Ophthalmol. Vis. Sci.* **24**: 812–820.

17

COMPARATIVE BIOLOGY OF RETINOID DEPRIVATION AND REPLACEMENT IN FLIES AND RODENTS

William S. Stark

Department of Biology
Saint Louis University
3507 Laclede Avenue
St. Louis, Missouri 63103-2010

ABSTRACT

The purpose of this review is to compare the findings from recent deprivation and replacement studies in *Drosophila*, rats and mice. Opsin is decreased in rhabdomeres of retinoid-deprived flies. In *Drosophila*, the visual system recovers quickly upon retinoid replacement even after prolonged deprivation. Importantly, photoreceptors do not degenerate. Total deprivation of all retinoids including retinoic acid is possible in *Drosophila*. Thus deprivation of vitamin A chromophore precursors can be achieved with or without deprivation of retinoic acid, a vertebrate transcriptional activator and morphogen. Replacement or rearing with specific dietary supplements including retinoic acid suggest that opsin transcription is regulated transcriptionally by a nuclear receptor acting within the -701 to -488 region in the *ninaE* [opsin gene] promoter. Phospholipase C [PLC], a phototransduction protein acting downstream from rhodopsin, is decreased by retinoid deprivation only in that the rhabdomeres are smaller when deprivation decreases opsin. However, a *Drosophila* retinoid binding protein expressed in Semper cells is specifically eliminated by retinoid deprivation. Total retinoid deprivation is not possible in the vertebrate since retinoic acid is essential. However, it was deemed useful to re-examine the effects of vitamin A deprivation and to investigate replacement in the rodent. Extended periods of deprivation were necessary to decrease sensitivity in rats and mice. Retinol replacement effected full recoveries. Vitamin A deprivation decreases interphotoreceptor matrix retinoid binding protein in the rat and the mouse. In summary, photoreceptors in the vertebrate, as in *Drosophila*, are robust against degeneration even when rhodopsin is drastically reduced.

Degenerative Retinal Diseases, edited by LaVail *et al.*
Plenum Press, New York, 1997

INTRODUCTION

Rhodopsin and Its Chromophore in *Drosophila*

The compound eye of adult *Drosophila* is composed of about 800 ommatidia with three different types of retinula cells. The rhodopsin-containing organelle of each cell is the rhabdomere. R1–6, 6 cells per ommatidium, is the predominant receptor type, expressing Rh1, a blue absorbing rhodopsin (1, 2). Rhodopsin is a typical G-protein coupled receptor with the protein portion, opsin, spanning the membrane seven times. Opsin is covalently linked to a light absorbing pigment. In *Drosophila*, this chromophore, 11-*cis*-3-hydroxyretinal (3, 4, 5, 6, 7), is derived from dietary carotenoids.

Retinoid Deprivation in *Drosophila*

Investigation of vitamin A's effects on the visual system of the insect was initiated in classic studies of sensitivity in the fly (8) and structure in the moth (9). The present era of investigation of fly vision utilizing retinoid manipulations began about two decades ago (10, 11, 12, 13, 14, 15, 16).

Retinoic acid is not toxic nor is it essential for flies, unlike the case for vertebrates. Thus, rearing *Drosophila* from egg to adult on a totally retinoid deprivational medium [Sang's medium] is particularly straightforward. Deprivation greatly reduces P-face particles in fly rhabdomeric microvilli as assayed by freeze-fracture electron microscopy (10, 16, 17). The absence of the opsin band on western blots from carotenoid deficient *Drosophila* (18, 19) adds to the evidence that opsin synthesis is blocked. Deprivation decreases visual pigment measured by spectrophotometry (16) and microspectrophotometry [MSP] of living flies (20, 21). Electron microscopic [EM] morphometry showed that retinoid deprivation decreases rhabdomere size (21).

Sensitivity is decreased at least two log units in retinoid-deprived *Drosophila* (12, 22, 23). Deprivation also preferentially decreases ultraviolet [UV] sensitivity in the fruitfly *Drosophila* (7, 12, 15) and in the housefly *Musca* (8, 15) through elimination of a retinoid pigment which sensitizes the blue-absorbing rhodopsin (15, 24). Deprivation eliminates the inactivation in R1–6 photoreceptors induced by bright blue light and the associated prolonged depolarizing afterpotential [PDA] (13, 25). Inactivation and afterpotential are such fundamental properties of the normal fly's electroretinogram [ERG] that some visual mutants [and their genes] have been named *nina* [neither inactivation nor afterpotential], and R1–6's rhodopsin [Rh1] gene is called *ninaE* (e.g. 1).

RESULTS

Retinoid Replacement Mediates Rapid Recovery in *Drosophila*

Retinoid replacement is conveniently achieved by placing deprived adults into a vial containing only carrot juice, a readily-assimilated carotenoid source (21, 26, 27). Replacement increases visual pigment as measured by MSP (21, 26, 27), rhabdomere size (21), and EM immunogold labeling (21).

The rhabdomeric P-face particle density increases half-way to the replete level in the first 12 hr and is fully restored by 2 days (28). This recovery is quite quick, considering that the flies are drinking their carotenoids (28). On western blots, the opsin recovery is detectable at 4 hr into carrot juice replacement (29).

Most of the sensitivity restoration for retinoid replaced flies is achieved within about one day (23), even if replacement is withheld until 11 days into adult life. This necessarily implies that, in *Drosophila*, receptors do not degenerate as a result of retinoid deprivation.

Retinoids Including Retinoic Acid Control Opsin Gene Transcription in *Drosophila*

It is not surprising that the availability of chromophore controls rhodopsin synthesis in *Drosophila* at the translational level (19). If you don't have chromophore, how can you make the finished visual pigment? Immunogold decoration using an Rh1 monoclonal antibody is particularly high in the rough endoplasmic reticulum after one day of retinoid replacement, indicating rapid *de novo* opsin biosynthesis (30). Retinoids, including retinoic acid, control transcription of *ninaE* as shown by promoter-reporter (31) and northern analyses (29). Promoter deletions suggested that retinoids act at −701 to −488 (31) in the *ninaE* promoter. Preliminary footprinting analysis is consistent with this estimate, and preliminary gel retardation analysis has narrowed the 5′ boundary to −693 and the 3′ boundary to −554 [W. L. Picking and W. S. Stark, unpublished]. The recovery of *ninaE* mRNA levels was noted as early as 1 hr into replacement with carrot juice (29). In summary, retinoids control opsin gene transcription quickly in addition to serving as opsin's chromophore.

Chromophore Deprivation Differs from Retinoid Deprivation in Flies

Two challenges to the transcriptional control hypothesis were addressed: [1] Ozaki et al. (19) did not see decreased opsin mRNA with yeast food in *Drosophila*; and [2] Huber et al. (32) did not see decreased mRNA with beef heart reared *Calliphora*. Northern blots (29) replicated high mRNA with yeast food but also confirmed none with the defined deprivation food, Sang's medium. Furthermore, high *ninaE* mRNA was found when Beef Brain Heart Infusion, *trans*-retinoic acid, 9-*cis*-retinoic acid and β-carotene were added to Sang's medium. On western blots (29), we saw Rh1 in normal food flies, none in deprived, none in *ninaE^{oI17}* [Rh1 deletion mutant], none when Beef Brain Heart Infusion or *trans*-retinoic acid were added to Sang's medium, but clear Rh1 when β-carotene was added to Sang's.

Thus, thorough retinoid deprivation does indeed eliminate opsin gene transcription, and replacement activates it quickly. Treatments used in experiments which negated transcriptional control must leave transcription effectors which act on the *ninaE* promoter. Elimination of precursors for visual pigment synthesis is "chromophore deprivation" since other factors, including retinoic acid, can activate transcription.

A *Drosophila* Retinoid Binding Protein Is Decreased by Retinoid Deprivation

Retinoids control a *Drosophila* retinoid binding protein (33) referred to as a Retinoid and Fatty Acid Binding Glycoprotein [RFABG] which has homology to secreted insect carrier proteins called lipophorins (34, 35). This protein has also been called DRBP [*Drosophila* Retinoid Binding Protein] to emphasize its relevance to retinoids in *Drosophila*. It may be a functional homologue of vertebrate IRBP [interphotoreceptor matrix RBP]; in the vertebrate retina, IRBP is well-known as a transporter of retinoids between retinal pigment epithelium [RPE] and photoreceptors.

Western blots show that DRBP is high in retinoid replete flies and nearly zero in deprived flies (36, 37). Replicating some of the dietary manipulations utilized in the opsin studies, western blots indicated that DRBP is high in a wider range of supplemented diets than for rhodopsin. DRBP is high in retinoic acid supplemented Sang's medium, yeast-glucose food, β-carotene supplemented Sang's, Beef Brain Heart Infusion supplemented Sang's and $ninaE^{oll7}$ (36). While DRBP binds retinoids and fatty acids (33), retinoids, not fatty acids, are relevant to expression—DRBP is diminished by retinoid deprivation with Sang's medium which has sufficient fatty acids (38).

Since DRBP binds retinoids (33), it seems possible that DRBP functions in retinoid transport or metabolism for photoreceptor function, inferred from analogy to vertebrate IRBP. One might expect no DRBP mRNA in deprived flies which have no retinoids, suggesting that the mechanism of retinoid control of DRBP transcription could be the same as that proposed for opsin. Preliminary northern analyses undertaken in collaboration with R. K. Kutty and B. N. Wiggert are consistent with this hypothesis.

DRBP Is Expressed in Semper Cells in *Drosophila*

The DRBP-antiDRBP complex was visualized with FITC conjugated anti-IgG (36). DRBP is high in Semper [cone] cells, virtually zero rhabdomeres, and notable in the intraommatidial matrix. Pre-immune staining gave the expected result of a negative control. Although Semper cells are not receptor cells, they are adjacent to photoreceptors with the intraommatidial matrix confluent to both cell types. Thus Semper cells could process retinoids for photoreceptors, or the extracellular matrix could shuttle retinoids.

Replicating and extending western analyses, DRBP was high in Semper cells when flies were reared on Sang's medium, yeast-glucose food, β-carotene supplemented Sang's, Beef Brain Heart Infusion supplemented Sang's and mutants [$ninaE^{oll7}$ and *glass*, the later lacking photoreceptors]. Based on previous findings in the blowfly (39) and honeybee (40) which have shown that vitamin A and photoisomerase are localized distally, the Semper cells, also distal, are prime candidates for involvement in retinoid metabolism.

Retinoid Deprivation Decreases PLC [Phospholipase C] in flies

Retinoid control of a downstream molecule of the phototransduction cascade was tested. Biochemical (41) and molecular (42) approaches showed that *Drosophila*'s *norpA* [no receptor potential A] gene encodes a phosphatidylinositol-specific phospholipase C [PLC]. *NorpA* PLC functions directly downstream from the G-protein in the *Drosophila* phototransduction cascade. Its role was proved in rescue experiments (43). An antiserum against the major product of *norpA* (44) was used. Western blots revealed the expected 130-kDa band (43) and showed that retinoid deprivation reduced PLC by about half (37, 45).

With 1 μm LR White sections and FITC conjugated anti-rabbit-IgG, PLC was localized using fluorescence and confocal microscopes (45). Immunocytochemistry confirmed rhabdomeric localization of PLC (43). As expected, rhabdomeres did not fluoresce in *norpA* mutants. Rhabdomeres showed bright fluorescent emission in retinoid replete and deprived flies, but rhabdomeres were smaller in deprived flies, in agreement with earlier observations (21). Thus, the PLC reduction is probably based on reduction in size of the organelle, suggesting that PLC reduction results as a consequence of opsin reduction.

Rods Are Small but Do Not Degenerate after Half a Year of Vitamin A Degeneration in the Rat

In the fruitfly *Drosophila*, it seemed amazing to see healthy photoreceptors devoid of opsin (16). In rodents, by contrast, the literature had been dominated by a widely-accepted report that vitamin A deprivation led to retinal degeneration (46, 47). Rats deprived of vitamin A [but with the necessary retinoic acid] were used to address the similarities and differences in vertebrate and invertebrate opsin regulation (48). Rhodopsin in rat rods was decreased to 14% by 26 weeks of deprivation without photoreceptor cell loss. Although the volume of the outer segment was decreased to 42% of the normal size, the density of opsin in the disk membrane was not reduced, as assessed by P-face particle counts [confirming earlier findings on *Xenopus* (49)] and EM immunogold counts. This is in stark contrast with the fly situation. It is thus quite clear that half a year of vitamin A deprivation in the rat does not lead to retinal degeneration.

Retinoid Replacement Effects a Complete Recovery in Vitamin A Deprived Rats

Visual recovery by vitamin A replacement was tested with the ERG (50). An immediate [6 hr to 1 day] phase of recovery suggested that the opsin which had developed without chromophore could accept chromophore to form functional rhodopsin. Recovery to near the replete controls' level by 14 days showed that outer segments could recover in a time course approximating the turnover of that organelle.

Vitamin A Deprivation and Replacement Decrease and Recover Vision in the Mouse

Because of increased utilization of transgenic mice, the rat studies were replicated in mice (51). Sensitivity was decreased by 2 log units after 60 weeks of deprivation. Also the c-wave of the ERG was decreased. Complete recoveries of sensitivity and the c-wave are seen in mice 7 days after retinol administration. Thus, in the mouse, as in the rat, recovery after extended vitamin A deprivation suggests that the receptors do not degenerate.

There is a UV peak in the dark adapted spectral sensitivity of vitamin A replete control mice. In an interesting parallel with the fly, this UV sensitivity is preferentially decreased in deprived mice and recovered upon replacement. However, unlike the situation in the fly [the sensitizing pigment hypothesis, see above], chromatic adaptation suggested that this UV sensitivity peak is mediated by a UV cone mechanism [known to be present in the rodent retina (52)].

The Nature of Vitamin A Control of IRBP in the Rodent

The decrease in DRBP by retinoid deprivation in *Drosophila* shows a striking parallel with the rat (53) and the mouse (51) where vitamin A deprivation *vs* replacement decreases and recovers IRBP respectively. IRBP recovery may actually be the prerequisite for photoreceptor recovery. Northern analyses in mice failed to demonstrate retinoid control of IRBP gene expression (51), but it should be remembered that mice cannot be deprived of retinoic acid.

DISCUSSION

Retinoid Deprivation and Replacement Facilitate the Study of Photoreceptor Renewal and Maintenance

Visual proteins are maintained to keep photoreceptors healthy. Visual membranes are in jeopardy because reactive substances are bombarded by light in a highly-metabolic system. A major thrust of vision research has been to understand this turnover process in order to elucidate causes and prevention of blindness. Retinoid deprivation contributes to understanding membrane renewal especially in that replenishment synchronizes a massive synthesis and deployment of rhodopsin. Recoveries after extended deprivation in *Drosophila*, rats and mice imply that vitamin A deprivation does not necessarily lead to degeneration.

Vitamin A, Retinoic Acid and Opsin Transcription in the Fly and the Vertebrate

The discovery that retinoids, including retinoic acid, stimulate opsin gene transcription in *Drosophila* (29, 31) show that retinoid deprivation and repletion are powerful manipulations. This research now addresses mechanisms of action of receptors like the retinoic acid receptor, and regulation of tissue-specific gene expression in eukaryotic cells. In the vertebrate, retinoic acid is pivotal in developmental regulation. By contrast, the conventional wisdom for the fly had previously been that retinoids serve only as precursors of photopigment chromophore.

The work in *Drosophila* now relates to new evidence that vitamin A may activate vision gene transcription in the vertebrate. Steady state mRNA measurements from protection assays in bovine retina showed that opsin, arrestin and cone opsin [but not IRBP] gene activation follows 11-*cis* retinylester production (54).

Retinoic Acid and the Visual System

Deciphering the relevance of retinoic acid in vertebrate visual systems has been more difficult than in the fly since it is essential for development. However, it now appears that retinoic acid is relevant to visual development in rat photoreceptors (55), rod development in zebrafish (56), UV cones in the trout (57), rescue of rat (58) and goldfish (59) photoreceptors from light damage, and in specification of dorsoventral visual axis in zebrafish (60). Retinoids are converted to retinoic acid in cultured retinal cells of rabbit (61) and chick (62).

Why Might Retinoic Acid Be Used to Report Chromophore Availability for Control of Opsin and DRBP in *Drosophila*?

The above issues converge around the question - How could the receptor superfamily for steroid hormones and retinoic acid be relevant to vision? It seems strange that opsin, with its retinaldehyde chromophore, would be regulated by a messenger, possibly retinoic acid, oxidized beyond chromophore availability. Perhaps the system borrows the already existing mechanism of a transcription factor like the retinoic acid receptor with a DNA binding site plus a retinoic acid binding site. Retinoic acid production is an unavoidable byproduct of vision when light acts on chromophores in tissues with dehydrogenases (63). Thus, retinoic acid would be a reliable messenger for regulation relevant to presence of chromophore.

ACKNOWLEDGEMENTS

Supported by NIH grant RO1 EY07192. I thank my laboratory coworkers D.-M. Chen for electrophysiology, W. L. Picking for molecular biology, and K. Shim for immunocytochemistry. S. J. Fliesler and M. Richards [Saint Louis University Anheuser-Busch Eye Institute] assisted in the preparation of the *Drosophila* opsin antibody. For collaboration in recent studies on the mouse, I thank several colleagues, especially G. I. Liou [Medical College of Georgia, Department of Ophthalmology] and M. L. Katz [University of Missouri - Columbia, Department of Ophthalmology]. For collaboration in recent studies on DRBP, I thank several colleagues, especially B. N. Wiggert and R. K. Kutty [National Eye Institute]. For collaboration in recent studies on PLC, I thank several colleagues, especially R. D. Shortridge and S. Kim [State University of New York - Buffalo, Department of Biology]. I thank R. H. White [University of Massachusetts - Boston, Department of Biology] for his collaboration in freeze-fracture electron microscopy. I thank C. F. Thomas [University of Wisconsin - Madison, Integrated Microscopy Resource, NIH grant DRR-570] for assistance with confocal microscopy.

REFERENCES

1. O'Tousa, J.E., Baehr, W., Martin, R.L., Hirsh, J., Pak, W.L., and Applebury, M.L., 1985, The *Drosophila ninaE* gene encodes an opsin, *Cell* **40**: 839–850.
2. Zuker, C.S., Cowman, A.F., and Rubin, G.M., 1985, Isolation and structure of a rhodopsin gene from D. melanogaster, *Cell* **40**: 851–858.
3. Goldsmith, T.H., Marks, B.C., and Bernard, G.D., 1986, Separation and identification of geometric isomers of 3-hydroxyretinoids and occurrence in the eyes of insects, *Vis. Res.* **26**: 1763–1769.
4. Seki, T., Fujishita, S., Ito, M., Matsuoka, N., Kobayashi, C., and Tsukida, D., 1986, A fly, *Drosophila melanogaster* forms 11-cis 3-hydroxyretinal in the dark, *Vis. Res.* **26**: 255–258.
5. Tanimura, T., Isono, K., and Tsukahara, Y., 1986, 3-hydroxyretinal as a chromophore of *Drosophila melanogaster* visual pigment analysed by high-pressure liquid chromatography, *Photochem. Photobiol.* **34**: 255–228.
6. Isono, K., Tanimura, T., Oda, Y., and Tsukahara, Y., 1988, Dependency on light and vitamin A derivatives of the biogenesis of 3-hydroxyretinal and visual pigment in the compound eyes of *Drosophila melanogaster*, *J. Gen. Physiol.* **92**: 587–600.
7. Stark, W.S., Schilly, D., Christianson, J.S., Bone, R.A., and Landrum, J.T., 1990, Photoreceptor-specific efficiencies of β-carotene, zeaxanthin and lutein for photopigment formation deduced from receptor mutant *Drosophila melanogaster*, *J. Comp. Physiol.* **166**: 429–436.
8. Goldsmith, T.H., and Fernandez, H.R., 1966, Some photochemical and physiological aspects of visual excitation in compound eyes, in *The functional organization of the compound eye*, (C.G. Bernhard, ed.), pp. 125–143, Pergamon Press, Oxford-New York.
9. Carlson, S.D., Steeves, H.R.I., Vandenberg, J.S., and Robbins, W.E., 1967, Vitamin A deficiency: Effect on retinal structure in the moth *Manduca sexta*, *Science* **158**: 268–270.
10. Boschek, C.B., and Hamdorf, K., 1976, Rhodopsin particles in the photoreceptor membrane of an insect, *Z. Naturforsch.* **31C**: 763.
11. Razmjoo, S., and Hamdorf, K., 1976, Visual sensitivity and the variation of total photopigment content in the blowfly photoreceptor membrane, *J. Comp. Physiol.* **105**: 279–286.
12. Stark, W.S., Ivanyshyn, A.M., and Hu, K.G., 1976, Spectral sensitivities and photopigments in adaptation of fly visual receptors, *Naturwissen.* **63**: 513–518.
13. Stark, W.S., and Zitzmann, W.G., 1976, Isolation of adaptation mechanisms and photopigment spectra by vitamin A deprivation in *Drosophila*, *J. Comp. Physiol.* **105**: 15–27.
14. Stark, W.S., 1977, Diet, vitamin A and vision in *Drosophila*, *Drosoph. Inform. Serv.* **52**: 47.
15. Stark, W.S., Ivanyshyn, A.M., and Greenberg, R.M., 1977, Sensitivity and photopigments of R1-6, a two-peaked photoreceptor, in *Drosophila, Calliphora* and *Musca*, *J. Comp. Physiol.* **121**: 289–305.
16. Harris, W.A., Ready, D.F., Lipson, E.D., Hudspeth, A.J., and Stark, W.S., 1977, Vitamin A deprivation and *Drosophila* photopigments, *Nature (Lond.)* **266**: 648–650.
17. Schinz, R.H., Lo, M.V.C., Larrivee, D.C., and Pak, W.L., 1982, Freeze-fracture study of the *Drosophila* photoreceptor membrane: Mutations affecting membrane particle density, *J. Cell Biol.* **93**: 961–969.
18. deCouet, H.G., and Tanimura, T., 1987, Monoclonal antibodies provide evidence that rhodopsin in the outer rhabdomeres of *Drosophila melanogaster* is not glycosylated, *Euro. J. Cell Biol.* **44**: 50–56.

19. Ozaki, K., Nagatani, H., Ozaki, M., and Tokunaga, F., 1993, Maturation of major *Drosophila* rhodopsin, ninaE, requires chromophore 3-hydroxyretinal, *Neuron* **10**: 1113–1119.

20. Stark, W.S., and Johnson, M.A., 1980, Microspectrophotometry of *Drosophila* visual pigments: Determinations of conversion efficiency in R1–6 receptors, *J. Comp. Physiol.* **140**: 275–286.

21. Sapp, R.J., Christianson, J.S., Maier, L., Studer, K., and Stark, W.S., 1991a, Carotenoid replacement therapy in *Drosophila*: recovery of membrane, opsin and visual pigment., *Exp. Eye Res.* **53**: 73–79.

22. Zimmerman, W.F., and Goldsmith, T.H., 1971, Photosensitivity of the circadian rhythm and of visual receptors in carotenoid-depleted *Drosophila*, *Science* **171**: 1167–1169.

23. Chen, D.M., and Stark, W.S., 1992, Electrophysiological sensitivity of carotenoid deficient and replaced *Drosophila*, *Vis. Neurosci.* **9**: 461–469.

24. Kirschfeld, K., Franceschini, N., and Minke, B., 1977, Evidence for a sensitising pigment in fly photoreceptors, *Nature (Lond.)* **269**: 386–390.

25. Stark, W.S., Frayer, K.L., and Johnson, M.A., 1979, Photopigment and receptor properties in *Drosophila* compound eye and ocellar receptors, *Biophys. Struct. Mechanism* **5**: 197–209.

26. Stark, W.S., Hartman, C.R., Sapp, R.J., Carlson, S.D., Claude, P., and Bhattacharyya, A., 1987, Vitamin A replacement therapy in *Drosophila*, *Dros. Inf. Serv.* **66**: 136–137.

27. Stark, W.S., Sapp, R.J., and Schilly, D., 1988, Rhabdomere turnover and rhodopsin cycle: maintenance of retinula cells in *Drosophila melanogaster*, *J. Neurocytol.* **17**: 499–509.

28. Stark, W.S., and White, R.H., 1996, Carotenoid replacement in *Drosophila*: Freeze-fracture electron microscopy. *J. Neurocytol.* **25**: 233–241.

29. Picking, W.L., Chen, D.-M., Lee, R.D., Vogt, M.E., Polizzi, J.L., Marietta, R.G., and Stark, W.S., 1996, Control of *Drosophila* opsin gene expression by carotenoids and retinoic acid: Northern and Western analyses, *Exp. Eye Res.* in press.

30. Sapp, R.J., Christianson, J.S., and Stark, W.S., 1991b, Turnover of membrane and opsin in visual receptors of normal and mutant *Drosophila*, *J. Neurocytol.* **20**: 597–608.

31. Sun, D., Chen, D.M., Harrelson, A., and Stark, W.S., 1993, Increased expression of chloramphenicol acetyltransferase by carotenoid and retinoid replacement in *Drosophila* opsin promoter fusion stocks, *Exp. Eye Res.* **57**: 177–187.

32. Huber, A., Wolfrum, U., and Paulsen, R., 1994, Opsin maturation and targeting to rhabdomeral photoreceptor membranes requires the retinal chromophore, *Eur. J. Cell Biol.* **63**: 219–229.

33. Duncan, T., Geetha, K., Chader, G.J., and Wiggert, B., 1994, A glycoprotein binding retinoids and fatty acids is present in *Drosophila*, *Arch. Biochem. Biophys.* **312**: 158–166.

34. Kutty, R.K., Kutty, G., Kamadur, R., Duncan, T., Koonin, E.V., Rodriguez, I.R., Odenwald, W.F., and Wiggert, B., 1996, Molecular characterization and developmental expression of a retinoid- and fatty acid-binding glycoprotein from *Drosophila*: A putative lipophorin, *J. Biol. Chem.* **271**: 20641–20649.

35. Kutty, R.K., Kutty, G., Duncan, T., Rodriguez, I.R., Shim, K., Stark, W.S., and Wiggert, B., 1996, Molecular cloning of a novel gene encoding a *Drosophila* retinoid- and fatty-acid-binding glycoprotein expressed in cone (Semper) cells in the compound eye, *Invest. Ophthalmol. Vis. Sci.* **37**: S802.

36. Shim, K., Picking, W.L., Kutty, R.K., Thomas, B.N., and Stark, W.S., 1997, Control of *Drosophila* retinoid binding protein expression by retinoids and retinoic acid: Northern, western and immunocytochemical analyses, *Exp. Eye Res.* in preparation.

37. Lee, R.D., 1994, The effects of retinoids and carotenoids upon *Drosophila* photoreceptors: Regulation of the expression of opsin, phospholipase C, *Drosophila* retinal binding protein, and membrane buildup, Saint Louis University, MS Thesis.

38. Stark, W.S., Lin, T.N., Brackhahn, D., Christianson, J.S., and Sun, G.Y., 1993, Fatty acids in the lipids of *Drosophila* heads: Effects of visual mutants, carotenoid deprivation and dietary fatty acids, *Lipids* **28**: 345–350.

39. Schwemer, J., 1993, Visual pigment renewal and the cycle of chromophore in the compound eye of the blowfly, in *Sensory Systems of Arthropods*, (K. Wiese, ed.), pp. 54–68, Birkhauser Verlag, Basel.

40. Smith, W.C., and Goldsmith, T.H., 1991, Localization of retinal photoisomerase in the compound eye of the honeybee, *Vis. Neurosci.* **7**: 237–249.

41. Yoshioka, T., Inoue, H., and Hotta, Y., 1985, Absence of phosphotidylinositol phosphodiesterase in the head of a *Drosophila* visual mutant, norpA (no receptor potential A), *J. Biochem.* **97**: 1251–1254.

42. Bloomquist, B.T., Shortridge, R.D., Schnewly, S., Perdew, M., Montell, C., Steller, H., Rubin, G., and Pak, W.L., 1988, Isolation of a putative phospholipase C gene of *Drosophila, norpA*, and its role in phototransduction, *Cell* **54**: 723–739.

43. McKay, R.R., Chen, D.M., Miller, K., Kim, S., Stark, W.S., and Shrotridge, R.D., 1995, Phospholipase C rescues visual defect in *norpA* mutant of *Drosophila melanogaster*, *J. Biol. Chem.* **270**: 13271–13276.

44. Zhu, L., McKay, R.R., and Shortridge, R.D., 1993, Tissue-specific expression of phospholipase C encoded by the *norpA* gene of *Drosophila melanogaster*, *J. Biol. Chem.* **268**: 15994–16001.
45. Stark, W.S., Shim, K., Thomas, C.F., Lee, R.D., and Shortridge, R.D., 1996, Retinoid deprivation reduces phospholipase C (PLC-β), product of the *Drosophila norpA* gene, *Invest. Ophthalmol. Vis. Sci.* **37**: S803.
46. Dowling, J.E., and Wald, G., 1958, Vitamin A deficiency and night blindness, *Proc. Natl. Acad. Sci. USA* **44**: 648–661.
47. Dowling, J.E., and Wald, G., 1960, The biological activity of vitamin A acid, *Proc. Nat. Acad. Sci. USA* **46**: 587–608.
48. Katz, M.L., Kutryb, M., Norberg, N., Gao, C.L., White, R.H., and Stark, W.S., 1991, Maintenance of opsin density in photoreceptor outer segments of retinoid-deprived rats, *Invest. Ophthalmol. Vis. Sci.* **32**: 1968–1980.
49. Engbretson, G.A., Hassin, G., and Witkovsky, P., 1979, The threshold versus pigment relation of *Xenopus* rods, *Vision Res.* **19**: 367–374.
50. Katz, M.L., Chen, D.-M., Stientjes, H.J., and Stark, W.S., 1993, Photoreceptor recovery in retinoid-deprived rats after vitamin A replenishment, *Exp. Eye Res.* **56**: 671–682.
51. Liou, G.I., Matragoon, S., Chen, D.-M., Gao, C.-L., Fei, Y., Katz, M.L., and Stark, W.S., 1997, Retinol-dependent visual sensitivity and interphotoreceptor retinoid-binding protein (IRBP) content in the mouse, *Exp. Eye Res.* submitted.
52. Jacobs, G.H., Neitz, J., and Deegan, J.F.I., 1991, Retinal receptors in rodents maximally sensitive to ultraviolet light, *Nature* **353**: 655–656.
53. Katz, M.L., Gao, C.L., and Stientjes, H.J., 1993, Regulation of the interphotoreceptor retinoid-binding protein content of the retina by vitamin A, *Exp. Eye Res.* **57**: 393–401.
54. Timmers, A.M., Wintjes, E.T., and DeGrip, W.J., 1996, Is vitamin A metabolism in RPE involved in the regulation of photoreceptor specific gene expression during retinal development?, *Invest. Ophthalmol. Vis. Sci.* **37**: S693.
55. Kelley, M.W., Turner, J.K., and Reh, T.A., 1994, Retinoic acid promotes differentiation of photoreceptors *in vitro*, *Development* **120**: 2091–2102.
56. Hyatt, G.A., Schmitt, E.A., and Dowling, J.A., 1995, Retinoic acid induces the expansion of rod photoreceptors in the developing zebrafish retina, *Neurosci. Abstr.* **21**: 1557.
57. Browman, H.I., and Hawryshyn, C.W., 1994, Retinoic acid modulates retinal development in the juveniles of a teleost fish, *J. Exp. Biol.* **193**: 191–207.
58. Unoki, K., Ohba, N., Arimura, H., Muramatsu, H., and Muramatsu, T., 1994, Rescue of photoreceptors from the damaging effects of constant light by midkine, a retinoic acid-responsive gene product, *Invest. Ophthalmol. Vis. Sci.* **35**: 4063–4068.
59. Chen, D.-M., Dong, G., and Stark, W.S., 1996, Ultraviolet light damage and reversal by retinoic acid in the goldfish: Microscopy and electrophysiology, *Invest. Ophthalmol. Vis. Sci.* **37**: S1057.
60. Marsh-Armstrong, N., McCaffery, P., Gilbert, W., Dowling, J.E., and Drager, U.C., 1994, Retinoic acid is necessary for development of the ventral retina in zebrafish, *Proc. Natl. Acad. Sci. USA* **91**: 7296–7290.
61. Edwards, R.B., Adler, A.J., Dev, S., and Claycomb, R.C., 1992, Synthesis of retinoic acid from retinol by cultured Muller cells, *Exp. Eye Res.* **54**: 481–490.
62. Stenkamp, D.L., and Adler, R., 1993, Biological effects of retinoids and retinoid metabolism in cultures of chick embryo retina neurons and photoreceptors, in *Retinal Degeneration Clinical and Laboratory Applications*, (J.G. Hollyfield, R.E. Anderson and M.M. LaVail, ed)., pp. Plenum Press, New York.
63. Wagner, E., McCaffery, P., Mey, J., and Drager, U., 1995, Light-mediated retinoic acid production, *Neurosci. Abstr.* **21**: 1173.

18

RHODOPSIN-DEPENDENT MODELS OF
Drosophila PHOTORECEPTOR DEGENERATION

David R. Hyde, Scott Milligan, and Troy Zars

Department of Biological Sciences
University of Notre Dame
Notre Dame, Indiana 46556

INTRODUCTION

Many model systems have been exploited to examine the various mechanisms and molecular defects that can lead to retinal degeneration. While humans and mice are the two major vertebrate organisms for these analyses, the invertebrate *Drosophila melanogaster* provides a compelling alternative model system to study hereditary retinal degeneration. The invertebrate and vertebrate photoreceptor cells are dramatically different in both structure and organization, their visual transduction second messenger systems, and the direction of change in membrane polarity as a result of light excitation (1, 2). However, the processes that lead to retinal degeneration may be conserved between vertebrates and invertebrates. This is exemplified by mutations in analogous molecules, most notably both dominant and recessive rhodopsin mutations. *Drosophila* is ideal for a genetic analysis of retinal degeneration due to its small size, rapid life cycle and ease of mutant generation. *Drosophila* has provided a large number of mutants that undergo light-dependent, light-enhanced, and light-independent retinal degeneration (3). These mutants are defective in both expected and novel components that are required for visual transduction and photoreceptor cell structure. Because there has not been an exhaustive search for *Drosophila* retinal degeneration mutants, we expect that many mechanisms and molecular details remain to be identified. Recent results have provided a wealth of information and new ideas about photoreceptor cell structure and physiology.

Dominant and recessive rhodopsin mutations can lead to retinal degeneration by several mechanisms. Proposed mechanisms for dominant retinal degeneration include improper folding or reduced stability of rhodopsin, blocking or misrouting the wild-type rhodopsin protein, a dominantly active rhodopsin, a constitutively phosphorylated rhodopsin, and a class of mutations with undefined mechanisms (4–13). Recessive rhodopsin mutations also cause retinal degeneration several ways. Null mutations have been identified in both vertebrates and invertebrates, although the best characterized degeneration mechanism is in *Drosophila* (14–16). The *Drosophila ninaE^{117}* null mutant exhibits a developmental defect in photoreceptor cell rhabdomeres, which leads to a rapid retinal

Degenerative Retinal Diseases, edited by LaVail *et al.*
Plenum Press, New York, 1997

degeneration (15, 16). This absence of rhodopsin protein is thought to cause an imbalance in cohesive forces in the compact microvillar environment, which causes the rhabdomeres to flatten into long sheets of membrane. Expression of low rhodopsin levels slows the retinal degeneration process relative to null mutant flies. This slower degeneration mechanism may be a perturbation in the relative membrane biosynthesis/turnover rate or a defect in protein maturation/traffic (15, 16).

We report here our recent work on two different genes that, when mutated, lead to retinal degeneration *via* a rhodopsin-mediated pathway. First, flies defective in the *retinal degeneration B* (*rdgB*) gene, which encodes a novel integral membrane form of a phosphatidylinositol transfer protein, exhibit a recessive light-enhanced retinal degeneration and electrophysiological defect (17). We generated a dominant *rdgB* mutation that also causes retinal degeneration and exhibits a defect in the electrophysiological light-response. Unlike the recessive *rdgB* mutation, this dominant mutation exerts these phenotypes by blocking the expression of the mature rhodopsin protein. Second, we isolated two dominant rhodopsin mutations. Unlike the previously reported *Drosophila* dominant rhodopsin mutations that block the maturation of the wild-type rhodopsin (5, 7), these two mutations permit the expression, maturation, and correct localization of the wild-type rhodopsin to the photoreceptor rhabdomeres. This suggests that the dominant degeneration phenotype is likely associated with the expression and localization of the mutant rhodopsin in the photoreceptor cell.

MATERIALS AND METHODS

Electrophysiology

The electrophysiological light-response was recorded by electroretinograms (ERGs) essentially as described elsewhere (18, 19). The orange and blue light intensities were 3×10^1 mW / cm^2 and 5×10^2 mW/cm^2, respectively. ERG traces were recorded using a MacAdios II analog to digital converter using the SuperScope II software program (GW Instruments) on a Macintosh IIx machine.

Immunoblot Analyses

Heads from newly eclosed (less than one day old), dark raised flies were collected and homogenized in 20 µl extraction buffer (2% SDS, 2 mM KCL, 3% urea, 10 mM Tris pH 8.0, 2 mM EDTA, 2 mM EGTA, 5 mM DTT) (20). The homogenate was incubated in a boiling water bath for five minutes followed by cooling on ice for two minutes. Debris was pelleted at 16250 *xg* for ten minutes and 15 µl of supernatant was transferred to a new tube containing 5 µl of 4x SDS loading buffer (250 mM Tris pH 8.0, 8% SDS, 40% glycerol, 0.04% bromophenol blue, 20 mM DTT). This mixture was boiled five minutes, 2 µl of 10X iodoacetamide (92 mg/ml) was added, and 10 µl was electrophoresed on a 12.5% SDS-polyacrylamide gel (21). Proteins were transferred to nitrocellulose (Nitro ME, Micro Separations, Inc.) using a semi-dry transfer apparatus (Bio-Rad) at 20 volts for one hour. The membrane was blocked in 5% blotto (5% non-fat dry milk in TBS (20 mM Tris pH 7.5, 500 mM NaCl) for two hours with shaking. The membrane was washed for 20 minutes in TTBS (0.05% Tween-20 in TBS). The membrane was incubated overnight in rabbit anti-rhodopsin polyclonal antiserum (provided by J. O'Tousa) diluted 1:1000 in 2% blotto, followed by washing three times with TTBS for ten minutes each. The membrane was incubated in goat anti-rabbit alkaline phosphatase conjugated secondary antibody diluted 1:15000 in 2% blotto for two hours. The membrane was washed two times with TTBS for five minutes each. A final five minute wash with 0.1 M Tris pH 9.5 preceded detection with the Bio-Rad alkaline phosphatase colorimetric development kit. Levels of mature R1–6 opsin in different samples were determined by laser scanning densitometry of four identical immunoblots with a LKB Bromma Ultroscan XL Enhanced Laser Densitometer. Absorbance values were between 1.0 and 2.0, within the linear range of the laser densitometer. The area under the curve generated from the scan was measured and used in the comparisons between different mutants. All values are given as fraction of wild-type, with the students t-test used to determine significant differences ($p < 0.05$).

Scoring Retinal Degeneration

Different *ninaE* allelic combinations and control flies were collected daily and subjected to one of three light regimens; constant light, 12 hour light:dark cycle, or constant darkness. Flies exposed to light were scored daily for the deep pseudopupil. The flies from constant dark were aged for a given length of time, scored once for a deep pseudopupil, and discarded. The percent of flies retaining their deep pseudopupil for a given day was calculated from the total number of flies isolated on day one. The percent of flies with a deep pseudopupil from different trials were grouped by age and averaged. The differences at each day between different light regimens were tested for significance ($p < 0.05$) using the students unpaired t-test.

Immunocytochemistry

One day old wild-type, *ninaE^{pp100}*, *ninaE^{pp36}*, and *ninaE^{D1}* heterozygous flies (*ninaE* dominant/*ninaE^{117}*; P[rh1-hsv]) expressing a wild-type rhodopsin cDNA with a HSV epitope tag appended to the C-terminus were fixed and embedded. Flies were anesthetized and their eyes surgically removed into PBS-buffered 2% formaldehyde and 2% gluteraldehyde. The eyes were washed three times in PBS and dehydrated in an ethanol series (50, 70, 80, and 90%) for ten minutes each. The eyes were transferred through two changes of LR White (Sigma Chemical Co.) for 30 minutes at room temperature. The eyes were added to a final aliquot of LR White in gelatin capsules and incubated at 55°C for at least 12 hours. One micron sections were cut using an ultramicrotome and glass knives into double distilled water. The sections were incubated with 100 µl of a 1:100 dilution of an anti-HSV monoclonal antibody (Novagen) overnight at 4°C in 1X PBS containing 0.1% BSA (PBS-BSA). The section were washed for five minutes in 100 µl of PBS-BSA then incubated for two hours with 100 µl of 1:100 dilution of FITC-conjugated goat-anti-mouse (Southern Biotechnology Associates, Inc., Birmingham, AL) in PBS-BSA at room temperature. The sections were washed for five minutes in 100 µl of PBS-BSA, drained, and 20 µl of Vectashield (Vector) was added. The sections were kept in relative darkness until viewed with either a Nikon Microphot -FXA or Axiovert 135 TV inverted microscope. The images of the rhodopsin-HSV signal in the rhabdomeres was gathered using the Metamorph software on a Dell computer.

RESULTS

The Retinal Degeneration B (*rdgB*) Gene

The *retinal degeneration B (rdgB)* gene was identified as a recessive light-enhanced photoreceptor cell degeneration mutation (22). Very young *rdgB* mutant flies rapidly lose their electrophysiological light-response prior to any obvious signs of photoreceptor degeneration (17). A variety of genetic experiments revealed that the rdgB protein acts within the light-initiated phosphoinositide cascade, downstream of rhodopsin, DGqα (G-protein α subunit), phospholipase C, and protein kinase C (17, 23–25). The 1054 amino acid rdgB protein possesses six hydrophobic regions that likely serve as membrane spanning segments (26). This is consistent with the rdgB protein segregating with the membrane fraction of homogenized heads, which is not removed by alkaline washes (20). The absence of a cleavable signal peptide in the rdgB sequence prior to the first hydrophobic domain suggests that the amino-terminus is cytosolic (26, 27). The amino-terminal 281 amino acids of the rdgB protein share 42% identity to the entire rat brain phosphatidylinositol transfer protein (PI-TP; (28)). This high degree of homology is surprising because all other PI-TP molecules are soluble proteins of approximately 35 kDa (29, 30), yet rdgB is an integral membrane protein of 160 kilodaltons (20). We expressed rdgB's PI-TP domain as a soluble protein in *E. coli* and demonstrated that this truncated protein catalyzes the transfer of phosphatidylinositol between two different membrane bilayers *in vitro* (29, 30). Therefore, rdgB's PI-TP domain possesses the same biochemical activity as the soluble PI-TPs.

To further characterize the importance of the PI-TP domain in rdgB, we generated several *in vitro* missense mutations at the threonine 59 residue. PI-TPs are capable of transferring both phosphatidylinositol (PI) and phosphatidylcholine (PC) between membrane bilayers *in vitro*. Several mutations at threonine 59 in the rat brain PI-TP (including threonine to glutamic acid, T59E) selectively abolished PI transfer, without affecting PC transfer (31). Because threonine 59 is conserved in rdgB, we created some of the corresponding mutations, expressed the mutated soluble rdgB PI-TP domains and assayed for PI and PC transfer *in vitro*. Surprisingly, the T59E mutation had no effect on either the PI or PC transfer rates of the soluble rdgB PI-TP domain (data not shown). However, when we introduced this mutation into the full-length rdgB molecule and expressed it in *rdgB²* null mutants (*rdgB²*; P[rdgB-T59E]), the retinal degeneration was not suppressed, while the wild-type full-length rdgB molecule fully rescued the retinal degeneration phenotype (data not shown). Additionally, the ERG light-responses of *rdgB²*; P[rdgB-T59E] flies were improved relative to the *rdgB²* null flies, but were still not wild-type. Primarily, these flies could not generate the wild-type ERG light-response amplitude for multiple flashes of light (Figure 1). Therefore, the T59E mutation did not affect PI-TP activity *in vitro*, yet was ineffective in restoring wild-type rdgB activity *in vivo*.

We examined the rdgB-T59E mutation in *rdgB⁺* flies (*rdgB⁺*; P[rdgB-T59E]) to examine if it affected the wild-type *rdgB* activity. The *rdgB⁺*; P[rdgB-T59E] flies lost their deep pseudopupil in a dosage-dependent and light-enhanced manner (data not shown). We observed that the ERG light-response of young white-eyed *rdgB⁺*; P[rdgB-T59E] flies had a nina phenotype (Figure 1). The nina phenotype (neither inactivation nor afterpotential) was identified in a group of *Drosophila* mutants that exhibited a common ERG phenotype (32, 33). In white-eyed wild-type flies, a high intensity blue light will elicit a light-coincident receptor potential (LCRP), which is followed by a prolonged depolarizing afterpotential (PDA; (34, 35)). While subsequent blue light stimuli during this PDA fail to generate responses from photoreceptor cells R1–6, small responses from photoreceptors R7 and R8 are observed (wild-type in Figure 1). The inactivation phenomenon is due to the photoconversion of a significant amount of rhodopsin to the stable metarhodopsin state. An orange light stimulus during the PDA will photoconvert the meta-rhodopsin back to rhodopsin and reactivate the cells, such that they can fully respond to another light stimulus. As shown in Figure 1, the white-eyed *rdgB⁺*; P[rdgB-T59E] flies exhibited an ERG response

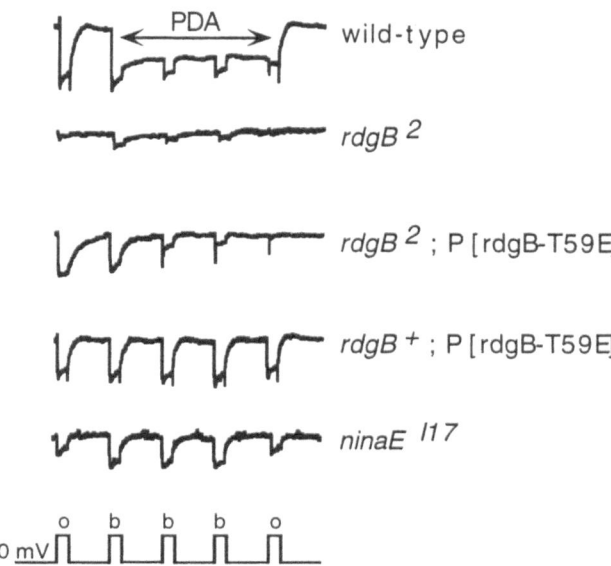

Figure 1. The electroretinogram (ERG) light-responses of white-eyed flies expressing various *rdgB* alleles were measured. We compared wild-type flies (wild-type), the *rdgB²* null mutant (*rdgB²*), the *rdgB²* null mutant expressing a germline transformed copy of the T59E mutation (*rdgB²*; P[rdgB-T59E]), and a wild-type fly expressing two copies of the germline transformed T59E mutation (*rdgB⁺*; P[rdgB-T59E]). For comparison, the ERG light-response of a white-eyed *ninaE¹¹⁷* rhodopsin null mutant (*ninaE¹¹⁷*) was also determined. The flies were less than eight hours old and raised under normal 12 hour light:dark conditions. The flies were prepared for the ERG recordings under dim red light and then dark adapted for 30 minutes prior to the first recording. The ERG light-responses were recorded from a pattern of five second flashes of either orange (o) or blue (b) light, with 20 seconds of dark adaptation between flashes. The PDA generated by blue light stimulation in white-eyed wild-type flies is shown (PDA in wild-type). The scales for time and response amplitude (in seconds and millivolts, respectively), along with the pattern of orange and blue stimuli are shown at the bottom. The *cinnabar* and *brown* mutations were both introduced into each of the above genotypes to block production of the eye pigments and generate the white-eye phenotype.

that was similar to the *ninaE¹¹⁷* mutant phenotype. Because all the *nina* mutants reduce the wild-type rhodopsin levels either directly or indirectly (33), it is likely that the nina ERG light response in the *rdgB⁺*; P[rdgB-T59E] flies is due to low rhodopsin levels. Additionally, the loss of the deep pseudopupil in the *rdgB⁺*; P[rdgB-T59E] flies was similar to the loss of the outer rhabdomeres in *ninaE* mutants using optical neutralization (33).

We examined the steady-state rhodopsin levels using immunoblots of fly head homogenates. We examined wild-type, *rdgB²* null mutants, *rdgB²*; P[rdgB-T59E], and *rdgB⁺*; P[rdgB-T59E] flies (Figure 2). The *rdgB²* null mutants expressed only 4% of the wild-type levels of rhodopsin, which was expected because the *rdgB* mutants lack the subrhabdomeric cisternae (SRC), which is an extension of the endoplasmic reticulum that is adjacent to the rhabdomere (36). Rhodopsin, which is localized in the rhabdomere and the SRC (37), presumably traffics through the SRC. The absence of the SRC in *rdgB²* flies may disrupt rhodopsin trafficking such that it is no longer stable and is drastically reduced in abundance in the photoreceptor cell. Expression of one copy of the rdgB-T59E mutant protein in a *rdgB²* fly (*rdgB²*; P[rdgB-T59E]) increased the stable rhodopsin expression to 11% of the wild-type level. Surprisingly, we found that *rdgB⁺* flies expressing the P[rdgB-T59E] transgene expressed less than 1% of the wild-type rhodopsin level. This dominant

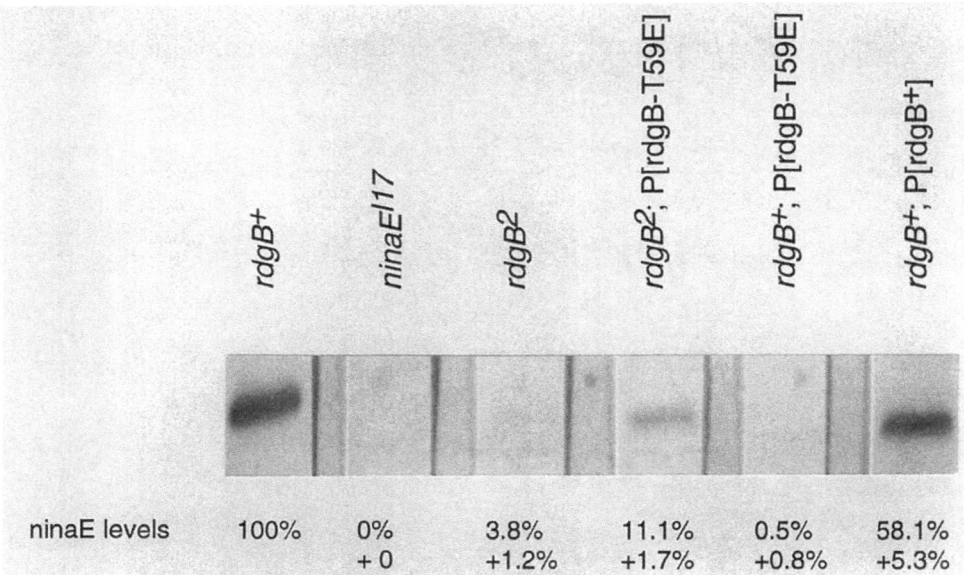

	rdgB+	ninaEll7	rdgB2	rdgB2; P[rdgB-T59E]	rdgB+; P[rdgB-T59E]	rdgB+; P[rdgB+]
ninaE levels	100%	0% + 0	3.8% +1.2%	11.1% +1.7%	0.5% +0.8%	58.1% +5.3%

Figure 2. The rdgB-T59E mutation dominantly blocks the stable expression of the photoreceptor R1–6 opsin. Rhodopsin levels were determined by immunoblots of fly head protein extracts from various *rdgB* alleles. The homogenate of two head equivalents was electrophoresed through SDS-polyacrylamide gels and electrotransferred to nitrocellulose. The rhodopsin protein was detected with a rabbit anti-rhodopsin polyclonal antiserum, followed by incubation with goat anti-rabbit conjugated to alkaline phosphatase. We compared wild-type flies (*rdgB+*), the *rdgB2* null mutant (*rdgB2*), the *rdgB2* null mutant expressing one copy of a germline transformed T59E mutation (*rdgB2*; P[rdgB-T59E]), wild-type flies expressing two copies of the germline transformed T59E mutation (*rdgB+*; P[rdgB-T59E]), and wild-type flies expressing two copies of a germline transformed wild-type rdgB cDNA (*rdgB+*; P[rdgB+]). We used the *ninaE^{l17}* null mutant (*ninaE^{l17}*) as a control for the complete absence of rhodopsin protein. The relative abundance of rhodopsin in fly head homogenates was calculated from four gels. The blots were quantitated with a scanning laser densitometer and the relative amount of rhodopsin was determined. The average percentage of mature rhodopsin in various *rdgB* backgrounds relative to wild-type flies and the standard deviation are shown.

effect is specific for rhodopsin. The *trp*-encoded Ca^{2+} channel, which is also expressed throughout the photoreceptor rhabdomeres (38), is not quantitatively reduced in the *rdgB+*; P[rdgB-T59E] flies (data not shown). Additionally, the *rdgB+*; P[rdgB-T59E] flies express significant levels of rdgB protein (data not shown), which suggests that the SRC is intact. Therefore, the rdgB-T59E mutant protein is likely disrupting a critical and specific function of rdgB+ in the maturation and/or transport of rhodopsin in the *Drosophila* photoreceptor cell.

Dominant Mutations in the Major Rhodopsin Gene, *ninaE*

One form of autosomal dominant retinitis pigmentosa (ADRP) in humans is due to a group of missense mutations in the rod opsin gene (13). Similarly, two groups have independently identified a number of dominant retinal degeneration mutations in *Drosophila* that are due to missense mutations in the major rhodopsin gene (5, 7). One of the stronger *Drosophila* rhodopsin mutations that was identified is the *ninaE^{D1}* allele (7). These dominant mutations appear to interfere with the maturation of the wild-type rhodopsin, which results in very low functional rhodopsin levels in the rhabdomere. While this appears to be similar to one subset of the human ADRPs, it does not account for the additional mechanistic changes caused by the other dominant human opsin mutations. The identification of additional dominant rhodopsin mutations in *Drosophila*, which mimic the other mechanisms of ADRP, could prove useful in elucidating how these dominant mutations lead to loss of photoreceptor cells and eventually could be beneficial in devising therapy for this condition.

We identified two new missense mutations in the *Drosophila* rhodopsin gene (*ninaE*), which cause dominant retinal degeneration. These mutations (*ninaE^{pp36}* and *ninaE^{pp100}*) were shown to be *ninaE* alleles based on the following criteria. First, these mutations failed to complement *ninaE^{117}*, which is a null allele (39). Young *ninaE^{pp36}*/*ninaE^+* and *ninaE^{pp100}*/*ninaE^+* flies exhibited a wild-type ERG light-response, which included both the prolonged depolarizing afterpotential (PDA) and inactivation of the R1–6 response (Figure 3). However, these dominant mutations exhibited a nina ERG light-response phenotype when placed *in trans* with *ninaE^{117}* (Figure 3). This nina ERG phenotype suggests these mutants express reduced levels of functional rhodopsin. Notice the *ninaE^{pp36}*/*ninaE^{117}* and *ninaE^{pp100}*/*ninaE^{117}* trans-heterozygotes exhibited a much larger light-response than the *ninaE^{D1}*/*ninaE^{117}* flies (Figure 3), which indicates that they possess a higher rhodopsin level than *ninaE^{D1}*/*ninaE^{117}* flies. Second, we sequenced the *ninaE* genes from these dominant mutant flies and identified missense mutations. The *ninaE^{pp36}* allele is an alanine to valine change at amino acid 93, which is in the first intracellular loop and adjacent to the second membrane spanning domain. The *ninaE^{pp100}* mutation is a

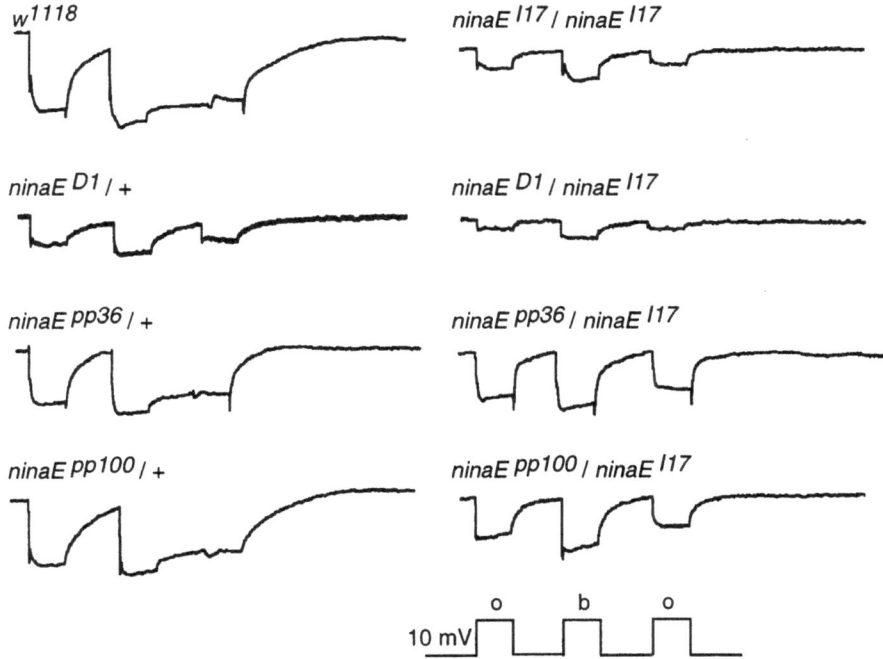

Figure 3. One day old white-eyed flies expressing various *ninaE* alleles were examined for their light-response as measured by the electroretinogram (ERG). We compared white-eyed wild-type flies (wild-type), the *ninaE^{117}* null mutant, the dominant *ninaE^{D1}* mutant, the dominant *ninaE^{pp36}* mutant, the dominant *ninaE^{pp100}* mutant, and each of the above *in trans* with the *ninaE^{117}* mutation. The ERG light-responses were recorded with a pattern of orange (o), blue (b), orange light with five seconds of darkness between stimuli. The scales for time and response amplitude (in seconds and millivolts, respectively), along with the pattern of orange and blue stimuli are shown at the bottom. The *w^{1118}* mutation was introduced into each of the above genotypes to generate the white-eye phenotype.

glycine to arginine change at position 299, which immediately follows the sixth membrane spanning domain and is in the third extracellular loop. These two lines of data demonstrate that these dominant mutations are *ninaE* alleles.

Unlike the previously characterized dominant *ninaE* alleles, the *ninaE^{pp36}* and *ninaE^{pp100}* alleles exhibited a remarkably different time course for degeneration. We measured retinal degeneration by the loss of the deep pseudopupil, which is a virtual image of the rhabdomeres from approximately 20 adjacent ommatidia (40). The absence of the deep pseudopupil is an indication that either the rhabdomeres were lost or that the highly structured ommatidial arrangement was disrupted. We examined the loss of the deep pseudopupil in flies carrying one copy of either the *ninaE^{pp36}* or *ninaE^{pp100}* allele. Both the *ninaE^{pp36}* and *ninaE^{pp100}* mutants lost their deep pseudopupil significantly earlier than the *ninaE^{D1}* mutants (Figure 4). However, there was a marked difference between the *ninaE^{pp36}* and *ninaE^{pp100}* degenerations. The *ninaE^{pp36}* mutants only degenerated under constant light conditions, with no loss of the deep pseudopupil when raised for up to 15 days in either a 12 hour light:dark cycle (Figure 4) or constant darkness (data not shown). In contrast, the *ninaE^{pp100}* mutants showed a light-independent degeneration, with nearly identical onset and rates of degeneration regardless of the light conditions (Figure 4). This suggests that the *ninaE^{pp36}* and *ninaE^{pp100}* mutants are utilizing different mechanisms of degeneration relative to each other. We also found that flies carrying two copies of either the *ninaE^{pp36}* or *ninaE^{pp100}* alleles lost their deep pseudopupil earlier than heterozygotes (data not shown). This is in contrast to the *ninaE^{D1}* allele, in which approximately 50% of the flies with one copy of the dominant allele lost the deep pseudopupil in 15 days and greater

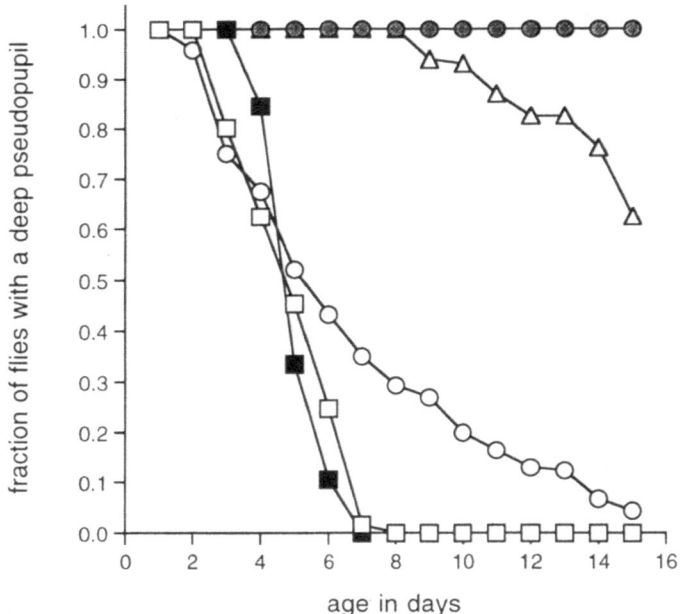

Figure 4. Heterozygous dominant *ninaE* mutants (*ninaE/ninaE^+*) were raised under different light conditions and assayed for retinal degeneration by the loss of the deep pseudopupil. The *ninaE^{pp36}/ninaE^+* flies were raised in either constant light or a 12 hour light:dark cycle (open circles and filled circles, respectively). The *ninaE^{pp100}/ninaE^+* flies were raised in either constant light or constant darkness (open and filled squares, respectively). The *ninaE^{D1}/ninaE^+* flies were raised in constant light (open triangles). The fraction of flies that retained a deep pseudopupil was plotted against the age of the flies.

than 25 days when two copies of the dominant allele were present (7). We confirmed that the loss of the $ninaE^{pp36}$ and $ninaE^{pp100}$ deep pseudopupils was due to degeneration by examining sections of retinal tissue. In all cases, the loss of the deep pseudopupil correlated with the loss of rhabdomeres (data not shown). The rapid degeneration observed with the $ninaE^{pp36}$ and $ninaE^{pp100}$ alleles and the observation that the flies carrying two mutant alleles degenerated faster than flies with only one dominant mutant allele, clearly demonstrates that these mutations cause degeneration by a mechanism different than the previously characterized dominant mutations.

We examined if the dominant mutations affected rhodopsin protein levels in photoreceptors R1–6 using immunoblots and rhodopsin protein immunolocalization. The previously characterized dominant $ninaE$ alleles appeared to exert their effect by significantly reducing the steady state level of rhodopsin in the photoreceptor cells by blocking the maturation of both the mutant and wild-type proteins (5, 7). A different situation occurs with the $ninaE^{pp36}$ and $ninaE^{pp100}$ alleles. The $ninaE^{pp36}/ninaE^+$ and $ninaE^{pp100}/ninaE^+$ flies had approximately 29% and 39% of the amount of rhodopsin as wild-type flies, respectively (data not shown). In contrast, $ninaE^{D1}/ninaE^+$ and $ninaE^{117}/ninaE^+$ flies had only 9% and 29% of the rhodopsin as wild-type flies, respectively (7). Unlike $ninaE^{D1}$, the $ninaE^{pp36}$ and $ninaE^{pp100}$ mutations did not block production of the wild-type rhodopsin relative to the $ninaE^{117}$ null allele. In addition, the $ninaE^{pp36}/ninaE^{pp36}$ and $ninaE^{pp100}/ninaE^{pp100}$ flies expressed 6% and 18% of the amount of rhodopsin as wild-type flies, respectively (data not shown). The comparable $ninaE^{117}$ and $ninaE^{D1}$ homozygotes both lacked detectable rhodopsin on immunoblots. Therefore, the $ninaE^{pp36}$ and $ninaE^{pp100}$ alleles differ from the previously characterized dominant rhodopsin mutations by stably expressing rhodopsin protein. We further demonstrated this point by immunolocalization of a HSV-tagged rhodopsin protein in several dominant mutant backgrounds. The HSV-tagged rhodopsin protein is clearly detected in the R1–6 rhabdomeres by immunofluorescence in the $ninaE^+$, $ninaE^{pp36}$, and $ninaE^{pp100}$ background (Figure 5, Panels A, C, and D, respectively). However, using the equivalent exposure setting, the HSV-tagged rhodopsin protein is weakly detected in the $ninaE^{D1}$ background. Therefore, the $ninaE^{pp100}$ and $ninaE^{pp36}$ degeneration mechanisms do not significantly block the maturation of rhodopsin as described for the other dominant rhodopsin mutations.

DISCUSSION

We described three different pathways that rhodopsin can induce photoreceptor degeneration. First, the absence of rhodopsin in the rhabdomere leads to loss of the rhabdomere and eventually, the loss of the photoreceptor cell. This pathway, which was shown with the dominant rdgB-T59E mutation, is similar to the recessive $ninaE^{117}$ null mutant phenotype (15, 16). Because rhodopsin is the most abundant protein in the rhabdomere, its absence could destabilize the microvillar membrane and cause rhabdomere loss. Second, the dominant rhodopsin mutation $ninaE^{pp36}$ exhibits photoreceptor degeneration only under constant light conditions. Third, the dominant rhodopsin mutation $ninaE^{pp100}$ causes photoreceptor degeneration regardless of the light conditions, even constant darkness. These last two mechanisms are very different than the previously characterized dominant rhodopsin mutations in *Drosophila* for two major reasons (5, 7). First, these mutations do not block the maturation and/or transport of rhodopsin to the rhabdomere. Second, these mutations exhibit a stronger mutant phenotype when two copies of the mutant allele are present than when only one copy is present. Because these two dominant mutations do not

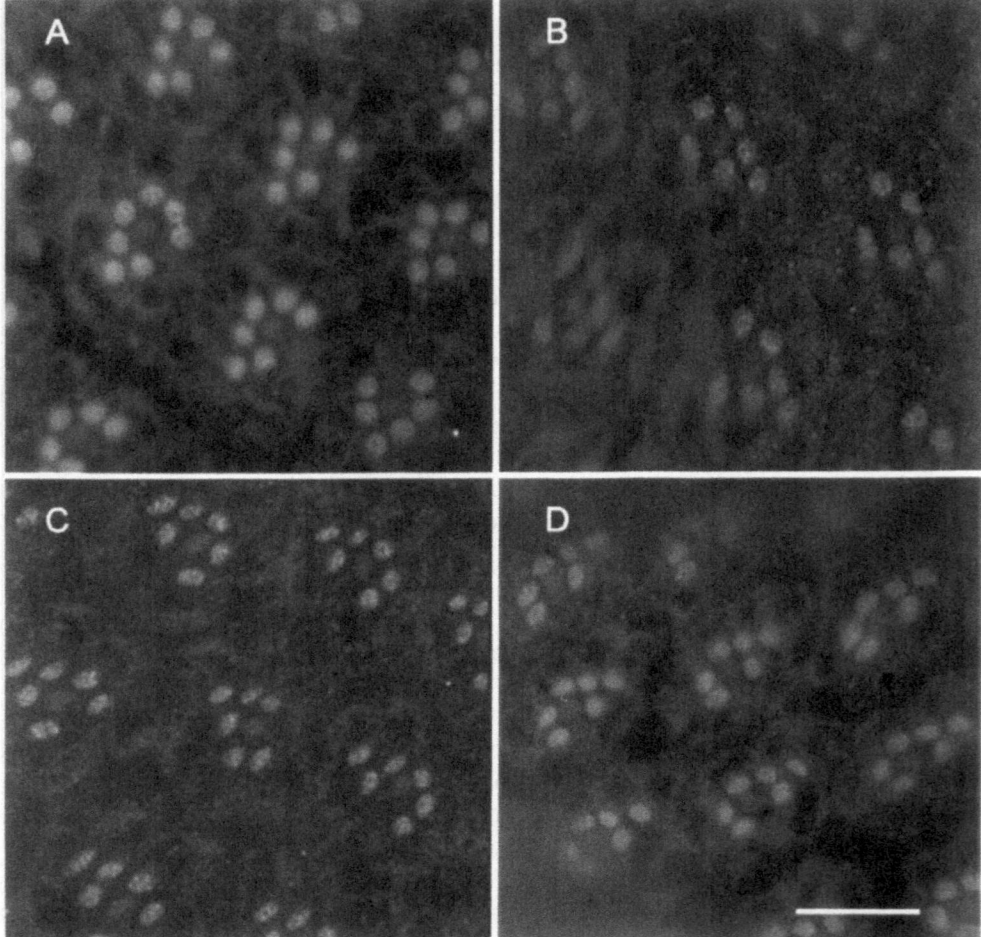

Figure 5. Immunolocalization of a HSV-tagged ninaE protein in various *ninaE* mutant backgrounds. *Drosophila* heads containing a HSV tagged germline transformed *ninaE* gene and either a *ninaE⁺* (A), *ninaE^{D1}* (B), *ninaE^{pp36}* (C) or *ninaE^{pp100}* (D) allele were embedded in plastic and one micron sections were cut. The sections were stained with an anti-HSV monoclonal IgG and detected with FITC-conjugated goat-anti-mouse IgG. The staining was restricted to the R1–6 rhabdomeres, which is consistent with the *ninaE* expression in the retina. Control sections from the *ninaE⁺* fly expressing the HSV-tagged rhodopsin, which were treated identically with the single exception that the anti-HSV monoclonal IgG was omitted, revealed no significant signal in the R1–6 rhabdomeres relative to the R7 rhabdomere (data not shown). The indirect fluorescence was recorded with a cooled slow scan CCD camera on an Axiovert 135TV microscope. The images were processed with a Dell computer using the Meta-Morph and Adobe Photoshop software. The scale bar in Panel D corresponds to 6.3 microns.

block the maturation and/or transport of the wild-type rhodopsin protein, they must cause the degeneration by affecting the stability and/or physiology of the rhabdomere/photoreceptor cell.

The identification of two different dominant rhodopsin mutations, one that exhibits a light-dependent degeneration and the other a light-independent degeneration, suggests some interesting models for their degeneration mechanisms. The light-dependent *ninaE^{pp36}* degeneration suggests that there is a defect in the physiological response to the light stimulus. The *norpA* and *rdgC* mutations exhibit similar light-dependent degeneration, although these mutations are recessive. In both *norpA* and *rdgC*, the degeneration appears to be due to an inability to dephosphorylate the phosphorylated metarhodopsin. The *rdgC* gene encodes a Ca^{2+}-dependent serine/threonine protein phosphatase (41), which recognizes the phosphorylated meta-rhodopsin as a substrate (42). The *norpA*-encoded phospholipase C is the visual transduction effector molecule and is required to hydrolyze PIP_2 to DAG and IP_3, which releases Ca^{2+} from intracellular stores (43). This calcium is required to activate the rdgC protein (42). The *ninaE^{pp36}* mutation may affect rhodopsin conformation such that the phosphorylated meta-rhodopsin is not recognized by the rdgC protein. The light-independent *ninaE^{pp100}* degeneration could be due to a constitutively activated rhodopsin protein. The continual stimulation of the visual transduction cascade, in the absence of light, could lead to degeneration. Alternatively, the *ninaE^{pp100}* mutation could affect rhodopsin conformation in the rhabdomere such that it destabilizes the rhabdomere and causes degeneration. Further genetic and molecular experiments should elucidate the degeneration mechanisms for these two dominant mutations. Regardless of the molecular details, it is clear that these mutations exhibit a degeneration process that does not affect the maturation and/or transport of rhodopsin to the rhabdomere as has been previously described for other dominant opsin mutations.

The identification of the dominant rdgB-T59E mutant phenotype affecting the steady-state levels of rhodopsin may help elucidate a potential role of the wild-type rdgB protein in the photoreceptor cell. Previously, the rdgB protein was suggested to be involved in vesicular trafficking from the SRC to the rhabdomere (20). This model was based on the analogy of the yeast phosphatidylinositol transfer protein, SEC14, and its requirement for vesicular movement from the Golgi (44, 45). Several lines of evidence also implicated mammalian PI-TPs as regulators of membrane trafficking events (reviewed in (46)), such as the priming of preformed secretory granules for Ca^{2+}-activated secretion (47) and the budding of both constitutive secretory vesicles and dense core secretory granules from the *trans*-Golgi network (48). It is known that the *rdgB* recessive mutations lose the SRC compartment coincident with the loss of the rhabdomeric microvilli (36). Rhodopsin has been suggested to traffic to the rhabdomere through the SRC based on immunolocalization of rhodopsin protein and the coincident loss of rhodopsin immunoreactivity in the rhabdomere and the loss of the SRC structure (37). The absence of the SRC could adversely affect rhodopsin stability, due to the lack of the correct compartment to move through on its way to the rhabdomere. Consistent with this hypothesis, *rdgB^2* null mutants possess very low steady-state levels of rhodopsin (4% of wild-type levels, Figure 2). Because the flies analyzed on the immunoblots were less than 12 hours old and lacked any obvious signs of degeneration at that time, the low rhodopsin levels are likely not due to the degeneration process. Additionally, the dominant *rdgB* mutant exhibits wild-type levels of rdgB protein on immunoblots, although the precise subcellular localization of the rdgB protein has not been determined. This suggests that the SRC is present in newly eclosed dominant *rdgB* mutant flies. Therefore, the absence of rhodopsin protein cannot be due to the lack of the SRC. Because the steady-state level of the *trp*-encoded calcium channel is unaffected by the dominant *rdgB* mutation, the rdgB-T59E protein is likely not perturbing a general process affecting all integral membrane proteins in the rhabdomere. Based on rdgB's subcellular localization in the SRC, it is likely that the rdgB-T59E protein is affecting the maturation and/or transport of rhodopsin between the SRC and the rhabdomere. This provides the strongest evidence to date that the rdgB protein is required in some aspect of vesicular trafficking between the SRC and rhabdomere.

Several lines of genetic and biochemical data implicate the rdgB protein as functioning downstream of protein kinase C in the *Drosophila* visual transduction cascade (17,

23–25). The role of rdgB in the visual transduction cascade appears to contradict its function in the maturation/transport of rhodopsin to the rhabdomere. These models can be reconciled if a component of the visual transduction cascade (for example IP_3, Ca^{2+}, or PKC) regulates the $rdgB^+$ activity involved in rhodopsin maturation/transport. Alternatively, the $rdgB^+$ protein may perform multiple roles in the photoreceptor cell and its role in rhodopsin trafficking, the recessive mutant retinal degeneration, and ERG phenotypes may not be directly related. We are currently performing additional *in vitro* mutagenesis experiments with rdgB to dissect its activity. In addition, we isolated several genetic suppressors of the *rdgB* retinal degeneration and ERG phenotypes, which may correspond to molecules that interact with rdgB and would help elucidate the function of the rdgB protein.

ACKNOWLEDGMENTS

We thank Joseph O'Tousa for providing us with the HSV-tagged *ninaE* construct and the rabbit anti-rhodopsin polyclonal antiserum, Yan Cheng for performing the P-element germline transformations and tissue sectioning. We are indebted to Douglas Fishkind for training us on the use of the MetaMorph software and use of his Axiovert. We thank Douglas McAbee and Joseph O'Tousa for their critical reading and comments of this manuscript. This work was supported by the National Institutes of Health (EY08058) and The Foundation Fighting Blindness (#93–152).

REFERENCES

1. Fein, A. and Szuts, E. Z. "Photoreceptor structure." Photoreceptors: Their role in vision. Hutchinson, Fuller, Mullins and Villegas ed. 1982 Cambridge University Press. Cambridge.
2. Yarfitz, S. and Hurley, J. B., 1994, Transduction mechanisms of vertebrate and invertebrate photoreceptors, *J. Biol. Chem.* **269**: 14329–14332.
3. Smith, D. P., Stamnes, M. A. and Zuker, C. S., 1991, Signal Transduction in the Visual System of *Drosophila*, *Annu. Rev. Cell Biol.* **7**: 161–190.
4. Nathans, J., Merbs, S. L., Sung, C. H., Weitz, C. J. and Wang, Y., 1992, Molecular genetics of human visual pigments, *Annu. Rev. Genet.* **26**: 403–424.
5. Colley, N. J., Cassill, J. A., Baker, E. K. and Zuker, C. S., 1995, Defective intracellular transport is the molecular basis of rhodopsin-dependent dominant retinal degeneration, *Proc. Natl. Acad. Sci. USA* **92**: 3070–3074.
6. Goto, Y., Peachey, N. S., Ripps, H. and Naash, M. I., 1995, Functional abnormalities in transgenic mice expressing a mutant rhodopsin gene, *Invest. Ophthalmol. Vis. Sci.* **36**: 62–71.
7. Kurada, P. and O'Tousa, J. E., 1995, Retinal degeneration caused by dominant rhodopsin mutations in *Drosophila*, *Neuron* **14**: 571–579.
8. Li, T. S., Franson, W. K., Gordon, J. W., Berson, E. L. and Dryja, T. P., 1995, Constitutive activation of phototransduction by K296E opsin is not a cause of photoreceptor degeneration, *Proc. Natl. Acad. Sci. USA* **92**: 3551–3555.
9. Robinson, P. R., Cohen, G. B., Zhukovsky, E. A. and Oprian, D. D., 1992, Constitutively active mutants of rhodopsin, *Neuron* **9**: 719–725.
10. Robinson, P. R., Buczylko, J., Ohguro, H. and Palczewski, K., 1994, Opsins with mutations at the site of chromophore attachment constitutively activate transducin but are not phosphorylated by rhodopsin kinase, *Proc. Natl. Acad. Sci. USA* **91**: 5411–5415.
11. Roof, D. J., Adamian, M. and Hayes, A., 1994, Rhodopsin accumulation at abnormal sites in retinas of mice with a human P23H rhodopsin transgene, *Invest. Ophthalmol. Vis. Sci.* **35**: 4049–4062.
12. Sung, C. H., Makino, C., Baylor, D. and Nathans, J., 1994, A rhodopsin gene mutation responsible for autosomal dominant retinitis pigmentosa results in a protein that is defective in localization to the photoreceptor outer segment, *J. Neurosci.* **14**: 5818–5833.
13. Nathans, J., 1994, In the eye of the beholder: visual pigments and inherited variation in human vision, *Cell* **78**: 357–360.

14. Rosenfeld, P. J., Cowley, G. S., Mcgee, T. L., Sandberg, M. A., Berson, E. L. and Dryja, T. P., 1992, A null mutation in the rhodopsin gene causes rod photoreceptor dysfunction and autosomal recessive retinitis-pigmentosa, *Nat. Genet.* **1**: 209–213.

15. Leonard, D. S., Bowman, V. D., Ready, D. F. and Pak, W. L., 1992, Degeneration of photoreceptors in rhodopsin mutants of *Drosophila*, *J. Neurobiol.* **23**: 605–626.

16. Kumar, J. P. and Ready, D. F., 1995, Rhodopsin plays an essential structural role in *Drosophila* photoreceptor development, *Development* **121**: 4359–4370.

17. Harris, W. A. and Stark, W. S., 1977, Hereditary retinal degeneration in *Drosophila melanogaster*: a mutant defect associated with the phototransduction process, *J. Gen. Physiol.* **69**: 261–291.

18. Larrivee, D. C., Conrad, S. K., Stephenson, R. S. and Pak, W. L., 1981, Mutation that selectively affects rhodopsin concentration in the peripheral photoreceptors of *Drosophila melanogaster*, *J. Gen. Physiol.* **78**: 521–545.

19. Blake, A., Bender, L. B., O'Day, P. M., Lonergan, M. and Venkatesh, T. R., 1991, Fused rhabdomeres (*fur*) in *Drosophila*: An eye mutation that alters rhabdomere morphology and retinal function, *J. Neurogenetics* **7**: 213–228.

20. Vihtelic, T. S., Goebl, M., Milligan, S., O'Tousa, J. E. and Hyde, D. R., 1993, Localization of *Drosophila retinal degeneration B*, a membrane-associated phosphatidylinositol transfer protein, *J. Cell Biol.* **122**: 1013–1022.

21. Laemmli, U. K., 1970, Cleavage of structural proteins during assembly of the head of bacteriophage T4, *Nature (London)* **227**: 680–685.

22. Hotta, Y. and Benzer, S., 1970, Genetic dissection of the *Drosophila* nervous system by means of mosaics, *Proc. Natl. Acad. Sci. USA* **67**: 1156–1163.

23. Stark, W. S., Chen, D.-M., Johnson, M. A. and Frayer, K. L., 1983, The *rdgB* gene of Drosophila: retinal degeneration in different alleles and inhibition by *norpA*, *J. Insect Physiol.* **29**: 123–131.

24. Lee, Y. J., Shah, S., Suzuki, E., Zars, T., O'Day, P. M. and Hyde, D. R., 1994, The *Drosophila dgq* gene encodes a Gα protein that mediates phototransduction, *Neuron* **13**: 1143–1157.

25. Smith, D. P., Ranganathan, R., Hardy, R. W., Marx, J., Tsuchida, T. and Zuker, C. S., 1991, Photoreceptor deactivation and retinal degeneration mediated by a photoreceptor-specific protein kinase C, *Science* **254**: 1478–1484.

26. Vihtelic, T. S., Hyde, D. R. and O'Tousa, J. E., 1991, Isolation and characterization of the *Drosophila retinal degeneration B* (*rdgB*) gene, *Genetics* **127**: 761–768.

27. Engelman, D. M. and Steitz, T. A., 1981, The spontaneous insertion of proteins into and across membranes: the helical hairpin hypothesis, *Cell* **23**: 411–422.

28. Dickeson, S. K., Lim, C. N., Schuyler, G. T., Dalton, T. P., Helmkamp, G. M. J. and Yarbrough, L. R., 1989, Isolation and sequence of cDNA clones encoding rat phosphatidylinositol transfer protein, *J. Biol. Chem.* **264**: 16557–16564.

29. Wirtz, K. W. A., 1991, Phospholipid transfer proteins, *Annu. Rev. Biochem.* **60**: 73–99.

30. Cleves, A., McGee, T. and Bankaitis, V., 1991, Phospholipid transfer proteins: a biological debut, *Trends in Cell Biology* **1**: 30–34.

31. Alb, J. G., Gedvilaite, A., Cartee, R. T., Skinner, H. B. and Bankaitis, V. A., 1995, Mutant rat phosphatidylinositol/phosphatidylcholine transfer proteins specifically defective in phosphatidylinositol transfer: Implications for the regulation of phospholipid transfer activity, *Proc. Natl. Acad. Sci. USA* **92**: 8826–8830.

32. Pak, W. L. "Study of photoreceptor function using Drosophila mutants." Neurogenetics: Genetic Approaches to the Nervous System. Breakfield ed. 1979 Elsevier North-Holland. New York.

33. Stephenson, R. S., O'Tousa, J., Scavarda, N. J., Randall, L. L. and Pak, W. L. "*Drosophila* mutants with reduced rhodopsin content." The Biology of Photoreception. Cosens and Vince-Price ed. 1983 Cambridge University Press. Cambridge.

34. Minke, B., Wu, C.-F. and Pak, W. L., 1975, Isolation of light-induced response of the central retinula cells from the electroretinogram of *Drosophila*, *J. Comp. Physiol.* **98**: 345–355.

35. Cosens, D. and Wright, R., 1975, Light elicited isolation of the complementary visual input systems in white-eye *Drosophila*, *J. Insect Physiol.* **21**: 1111–1120.

36. Matsumoto-Suzuki, E., Hirosawa, K. and Hotta, Y., 1989, Structure of the subrhabdomeric cisternae in the photoreceptor cells of *D. melanogaster*, *J. Neurocyto.* **18**: 87–93.

37. Suzuki, E. and Hirosawa, K., 1991, Immunoelectron microscopic study of the opsin distribution in the photoreceptor cell of *Drosophila melanogaster*, *J. Electron Microsc.* **40**: 187–192.

38. Niemeyer, B. A., Suzuki, E., Scott, K., Jalink, K. and Zuker, C. S., 1996, The *Drosophila* light-activated conductance is composed of the two channels TRP and TRPL, *Cell* **85**: 651–659.

39. O'Tousa, J. E., Baehr, W., Martin, R. L., Hirsh, J., Pak, W. L. and Applebury, M. L., 1985, The *Drosophila ninaE* gene encodes an opsin, *Cell* **40**: 839–850.

40. Franceschini, N. "Pupil and Pseudopupil in the Compound Eye of *Drosophila*." Information Processing in the Visual Systems of Arthropods. Wehner ed. 1972 Springer-Verlag, Berlin.

41. Steele, F. R., Washburn, T., Rieger, R. and O'Tousa, J. E., 1992, *Drosophila retinal degeneration C (rdgC)* encodes a novel serine/threonine protein phosphatase, *Cell* **69**: 669–676.

42. Byk, T., Baryaacov, M., Doza, Y. N., Minke, B. and Selinger, Z., 1993, Regulatory arrestin cycle secures the fidelity and maintenance of the fly photoreceptor cell, *Proc. Natl. Acad. Sci. USA* **90**: 1907–1911.

43. Bloomquist, B. T., Shortridge, R. D., Schneuwly, S., Perdew, M., Montell, C., Stellar, H., Rubin, G. and Pak, W. L., 1988, Isolation of a putative phospholipase C gene of *Drosophila, norpA*, and its role in phototransduction, *Cell* **54**: 723–733.

44. Bankaitis, V. A., Malehorn, D. E., Emr, S. D. and Greene, R., 1989, The *Saccharomyces cerevisiae SEC14* gene encodes a cytosolic factor that is required for transport of secretory proteins from the yeast Golgi complex, *J. Cell Biol.* **108**: 1271–1281.

45. Bankaitis, V. A., Aitken, J. F., Cleves, A. E. and Dowhan, W., 1990, An essential role for a phospholipid transfer protein in yeast Golgi function, *Nature* **347**: 561–562.

46. Alb, J. G., Jr., Kearns, M. A. and Bankaitis, V. A., 1996, Phospholipid metabolism and membrane dynamics, *Curr. Opin. Cell Biol.* **8**: 534–541.

47. Hay, J. C. and Martin, T. F. J., 1995, ATP-dependent inositide phosphorylation required for Ca2+-activated secretion, *Nature* **374**: 173–177.

48. Ohashi, M., De Vries, K. J., Frank, R., Snoek, G., Bankaitis, V., Wirtz, K. and Huttner, W. B., 1995, A role for phosphatidylinositol transfer protein in secretory vesicle formation, *Nature* **377**: 544–547.

19

RECESSIVE DEGENERATION OF PHOTORECEPTOR CELLS CAUSED BY POINT MUTATIONS IN THE CYTOPLASMIC DOMAINS OF *Drosophila* RHODOPSIN

Joachim Bentrop,[1] Karin Schwab,[1] William L. Pak,[2] and Reinhard Paulsen[*]

[1]Zoologisches Institut
Universität Karlsruhe (T.H.)
Lehrstuhl 1, Kornblumenstraße 13, D-76128 Karlsruhe
[2]Department of Biological Sciences
Lilly Hall of Life Sciences
Purdue University
W. Lafayette, Indiana 47907

ABSTRACT

The cellular mechanism of photoreceptor cell death in inherited retinal degeneration is not yet understood. Mutations in photoreceptor-specifically expressed genes, for example the rhodopsin gene, have been identified as primary genetic defects. Using transgenic *Drosophila* as a model system, we investigated whether mutations in distinct amino acids which are conserved within the cytoplasmic domains throughout the rhodopsin family, namely Leu81, Asn86 or Glu271, may cause inherited retinal degeneration. Substitutions at these sites are shown to interfere with two rhodopsin functions: (i) Rhodopsin biosynthesis is partially blocked, leading to a lowered amount of functional rhodopsin in photoreceptor cells. (ii) Photoreceptor cells expressing mutant rhodopsins undergo age-dependent degeneration in a recessive manner. Degeneration starts with a deterioration of the membranes constituting the photoreceptive cell compartment. In later states of degeneration, photoreceptor cell bodies and the extracellular interphotoreceptor space are filled with remnants of the photoreceptive membrane. Retinal degeneration is interpreted to result from alterations in rhodopsin's cytoplasmic surface which destabilize protein-protein interactions required in maintaining the architecture of the photoreceptive membrane compartment.

INTRODUCTION

More that 60 rhodopsin mutations have been identified causing autosomal dominant *Retinitis pigmentosa*, while only three recessive alleles have been reported to date (1–3).

Degenerative Retinal Diseases, edited by LaVail *et al.*
Plenum Press, New York, 1997

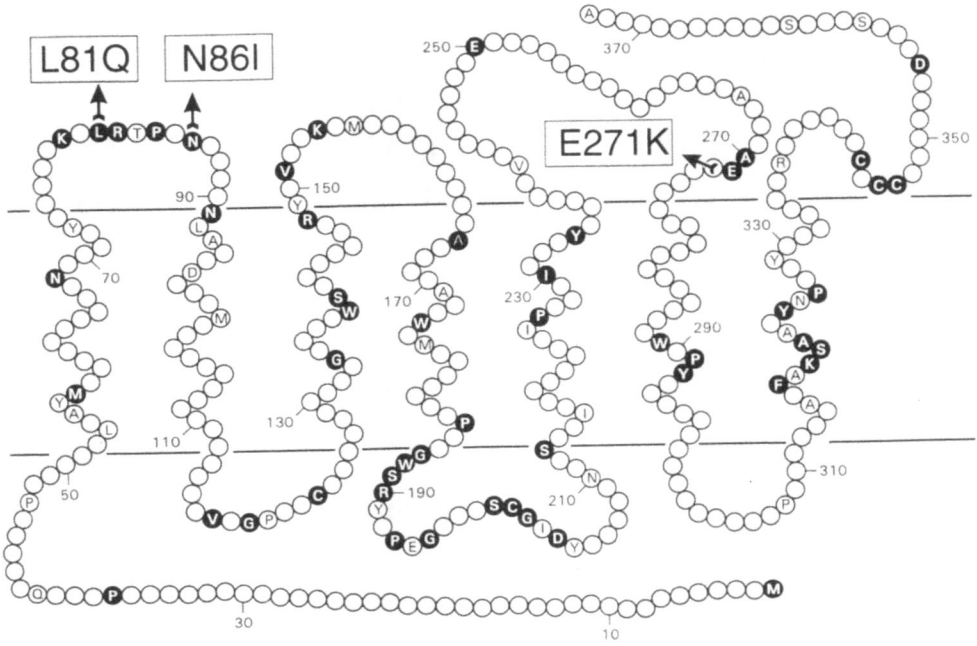

Figure 1. Structure model of *Drosophila* Rh1 rhodopsin. Letters indicate amino acids conserved in human rod and *Drosophila* Rh1 rhodopsins, solid circles indicate amino acids conserved in human rod, blue, red, and green sensitive cone rhodopsins and in *Drosophila* Rh1. Also shown are the rhodopsin mutations analyzed in the present study.

Moreover, the cellular mechanisms by which these mutations lead to photoreceptor cell degeneration have not yet been established.

The eye is exceptional in the extent of conservation of both development and function between vertebrates and invertebrate in general, and humans and *Drosophila* in particular (4). For many years, *Drosophila* has been a well established model system for the study of inherited retinal degeneration (5,6), and it was here, that rhodopsin mutations were first shown to induce photoreceptor cell degeneration (7–10). Rhodopsin mutant phenotypes are characterized by a reduced amount of visual pigment in the eye and by a loss of the photoreceptive membrane compartment, in an age-dependent manner. Like in humans, the great majority of retinal degeneration-causing rhodopsin mutations isolated to date display a dominant degeneration phenotype (11, 12), and the affected amino acids are localized in the transmembrane helices or on the extracellular surface of rhodopsin.

Sequence comparisons of human and *Drosophila* rhodopsins (13–16; Fig. 1) show an amazingly high degree of conservation on the amino acid level, also in the cytoplasmic domains of the proteins. This leads to assume that this region serves a function which is conserved across species and which is strongly dependent on the presence of specific amino acid residues. In the current view, the cytoplasmic surface of rhodopsin is thought to provide the domain for interactions with other proteins of the phototransduction machinery or with proteins involved in rhodopsin targeting. An aspect neglected so far are protein-protein interactions involved in stabilizing the rhodopsin-containing membrane compartments, the rhabdomeric microvilli of invertebrate or the disc membrane stacks of vertebrate photoreceptor cells.

We, therefore, designed the current study, exploiting the power of *Drosophila* genetics, to analyze whether substitutions of amino acids in the cytoplasmic domains of rhodopsin are candidate causes for retinal degeneration. Transgenic animals were generated, which express mutant rhodopsin in a rhodpsin-null mutant background, and they were studied for the effect of rhodopsin mutations on photoreceptor function and structure maintenance.

MATERIALS AND METHODS

Construction of Mutants, Fly Stocks

Oligonucleotide-directed *in vitro* mutagenesis was carried out to induce amino acid mutations L81Q, N86I, or E271K, respectively (listing indicates original amino acid, location, mutation) (10). The mutant Rh1 gene was then cloned into P-element transformation vector Carnegie3 rosy2 (17). P-element mediated transformation into host strain *ninaE^{oII7}* was carried out as described (10), and transformant lines were made homozygous for the P-element insert and the X chromosome mutation *white*. *NinaE^{oII7}* was used as Rh1-Null mutant control, these flies contain a large deletion in the 5'-region of the gene and make no detectable Rh1-transcript (13). To obtain flies heterozygous for one copy of mutant Rh1 gene and one copy of wild-type rhodopsin, homozygous mutant males were crossed to wild-type virgins, and the F1 off-spring was analyzed. Flies were raised on a standard corn meal diet and were kept under a 12-h light/12-h dark cycle.

Transmission Electron Microscopy and Immunolabeling of Ultrathin Sections

For conventional transmission electron microscopy, flies were perfused with fixative, eyes were dissected, fixed, embedded, and sectioned following the protocoll of Leonard et al (9). Sections were stained with 2% uranyl acetate and lead citrate. For immunolabelling of ultrathin sections, flies were fixed with 0.1% glutaraldehyde/ 3% paraformaldehyde in PB (0.1 M sodium phosphate buffer, pH 7.2) for 1 hour at room temperature and for 2 hours at 4°C, dehydrated in a graded ethanol series, infiltrated and then embedded in LR-White. Ultrathin sections were cut with a Reichert Ultracut microtome and were collected on formvar-coated nickel grids. Immunogold-labeling with anti-opsin antibodies was carried out according to Wolfrum (18), gold particles were silver-enhanced, and sections were then stained with 2% uranyl acetate. All sections were examined with a Zeiss EM 912 electron microscope.

Western Blot Detection of Opsin

For the detection of opsin protein in mutant flies, total membranes from compound eyes (1–2 days post-eclosion) were dissected, buffer washed and extracted in SDS sample buffer (19). Samples containing 10 μg of protein (equivalent of 10 heads) were separated by SDS-PAGE. For immunoblotting, proteins were electrophoretically transferred onto

PVDF-membranes. Membranes were incubated with polyclonal antibodies directed against peptide I237-K258 of *Drosophila* Rh1-opsin followed by binding of alkaline phosphatase-conjugated Protein A and visualization through a chromogenic reaction with 5-bromo-4-chloro-3-indolyl phosphate/4-nitro-blue-tetrazolium chloride.

Spectrophotometry

Heads of 150 flies were dissected under red light, total head membranes were isolated, and visual pigment was extracted in 4% Digitonin, 100 mM phosphate buffer (Na/K), pH 6.2 as described (19). Spectrophotometric measurements as carried out according to Paulsen (20).

Tangential sections through the distal half of adult fly ommatidia, ages = post-eclosion. (A-E) Flies were homozygous for the Rh1 mutation, (A) day 0, (B) 4 weeks (C) 8 weeks, (D) 1 week, (E) 6 weeks; (F) fly hetereozygous for the Rh1 mutation over one copy of the wild type gene, 8 weeks. Inset in (A): Immunogold-localization of opsin in a rhabdomere representative for R1–6, fly homozygous for Rh1 L81Q, 1 day. Numbers indicate the identity of photoreceptor cells; the long arrow points at the subrhabdomeral catacombs; arrow heads mark vesicles and sheets of apposed membranes filling the rhabdomeral stalk in degenerating photoreceptor cells; short arrows indicate membrane shedding into the extracellular cavity. Scale bars = 1 μm.

RESULTS

Retinal Degeneration in Rhodopsin Mutants

To assess the functional importance of amino acids conserved in the cytoplasmic domains of human and *Drosphila* opsins, we generated *Drosphila* transgenics expressing mutant Rh1 rhodopsin genes which result in substitutions L81 → Q, N86 → I, or E217 → K respectively. Main focus of the present study was to elucidate whether these point mutations cause retinal degeneration. Therefore, we investigated photoreceptor morphology as a function of age. Figs. 2–4 display cross sections through ommatidia of mutant flies at different ages. The central rhabdomeres (R7) is not affected by the mutation.

At eclosion, L81Q mutant flies have intact rhabdomeres (Fig. 2A) and normal catacomb-like structures at the base of the microvilli. Microvilli remain largely intact until about four weeks post-eclosion, at which time, however, the catacomb-like structures deteriorate (Fig. 2B). By eight weeks post-eclosion, about half of the R1–6 rhabdomeres have degenerated (Fig. 2C). In the course of photoreceptor degeneration, two types of degenerative processes are obvious in addition to the changes visible within the subrhabdomeric catacombs: As rhabdomeres begin to deteriorate, the microvillar membrane starts protruding into the cell body as sheets of apposed membranes (Fig. 2D) eventually filling the whole rhabdomeric stalk with membrane vesicles (Fig. 2E). Also, in later phases of degeneration, vesiculated rhabdomeric membranes are shedded into the central extracellular, interphotoreceptor cavity (Fig. 2C). This membrane shedding is interpreted to reflect an abnormality in membrane degradation, since such extracellular membrane whorls do not occur in the wild type.

In Rh1 N86I mutants, photoreceptor degeneration proceeds much faster than in L81Q. Although the rhabdomeres do not look much different from those of wild type at eclosion (Fig. 3A), by four weeks post-eclosion, three or four R1–6 rhabdomeres are missing in each ommatidium (Fig. 3B). All R1–6 rhabdomeres have disappeared by six week post-eclosion (not shown), and degeneration of the photoreceptor cell bodies is clearly visible by eight

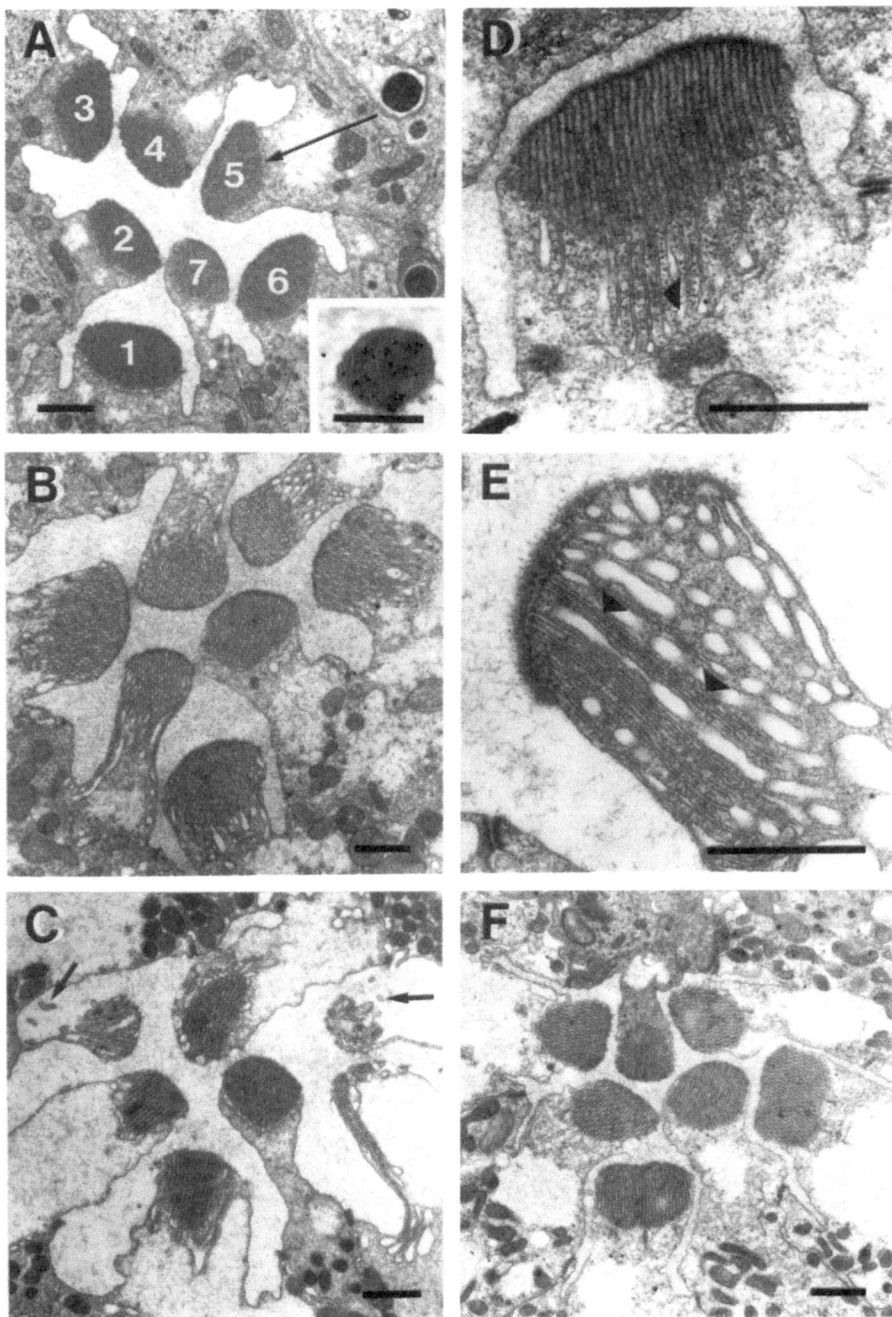

Figure 2. Photoreceptor cell degeneration in Rh1 L81Q mutant flies.

weeks post-eclosion, at which time only the central rhabdomere (R7) is still intact (Fig. 3C). Obviously, membrane breakdown is very fast in this mutant. Membrane whorls representing degraded microvilli are few, and, instead, from early on, small vesicles containing electron dense material can be found in the rhabdomeric region (Fig. 3D). Multi-vesicular body-like structures begin filling the extracellular cavity in later states of degeneration (Fig. 3E).

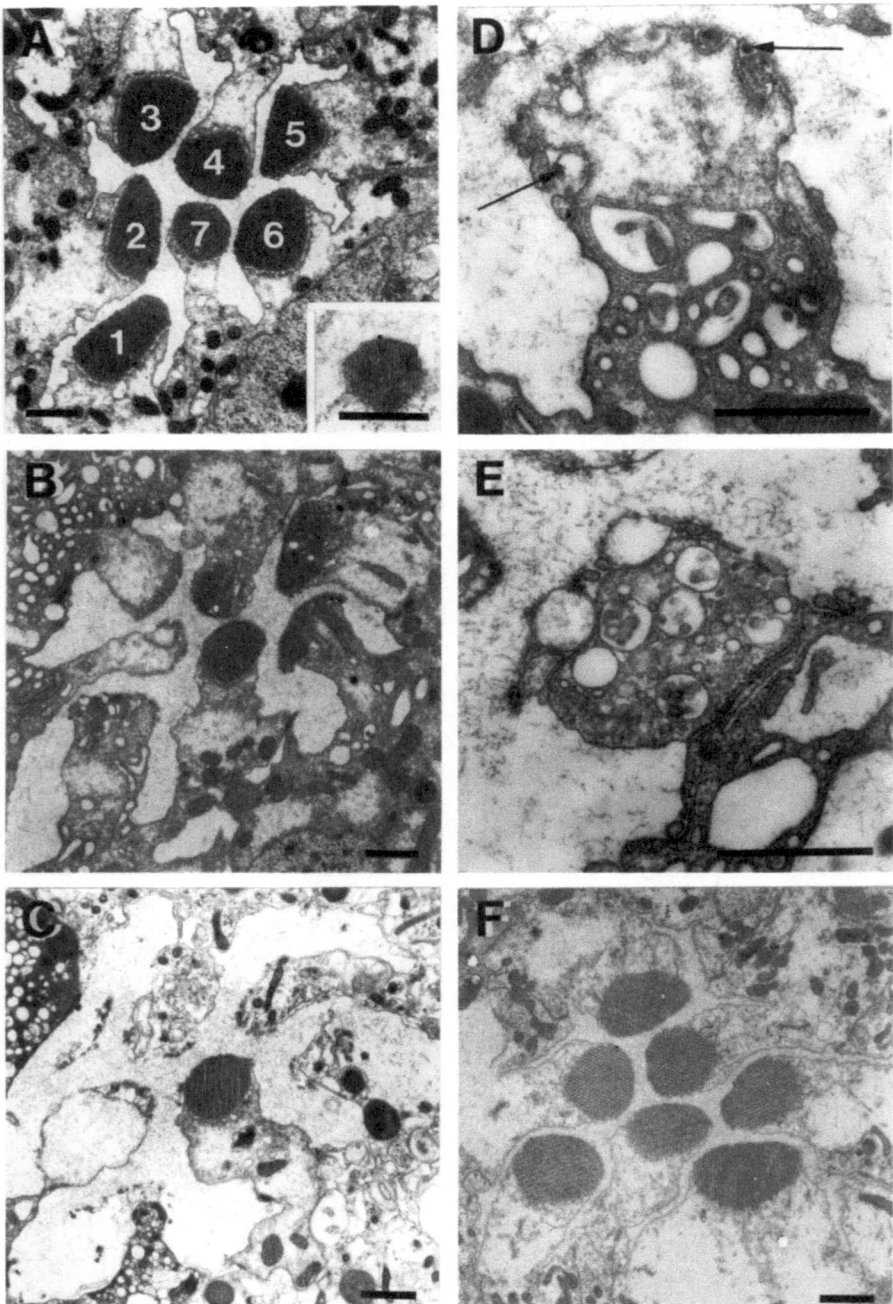

Figure 3. Photoreceptor cell degeneration in Rh1 N86I mutant flies.

Tangential sections through the distal half of adult fly ommatidia, ages = post-eclosion. (A-E) Flies were homozygous for the Rh1 mutation, (A) day 0, (B) 4 weeks, (C) 8 weeks, (D) electron dense granula (arrow) in the rhabdomeric stalk, 6 weeks, (E) multi-vesicular body-like structure in the extracellular cavity, 6 weeks; (F) fly heterozygous for the Rh1 mutation over one copy of the wild type gene, 8 weeks. Inset in (A): Immunogold-lo-

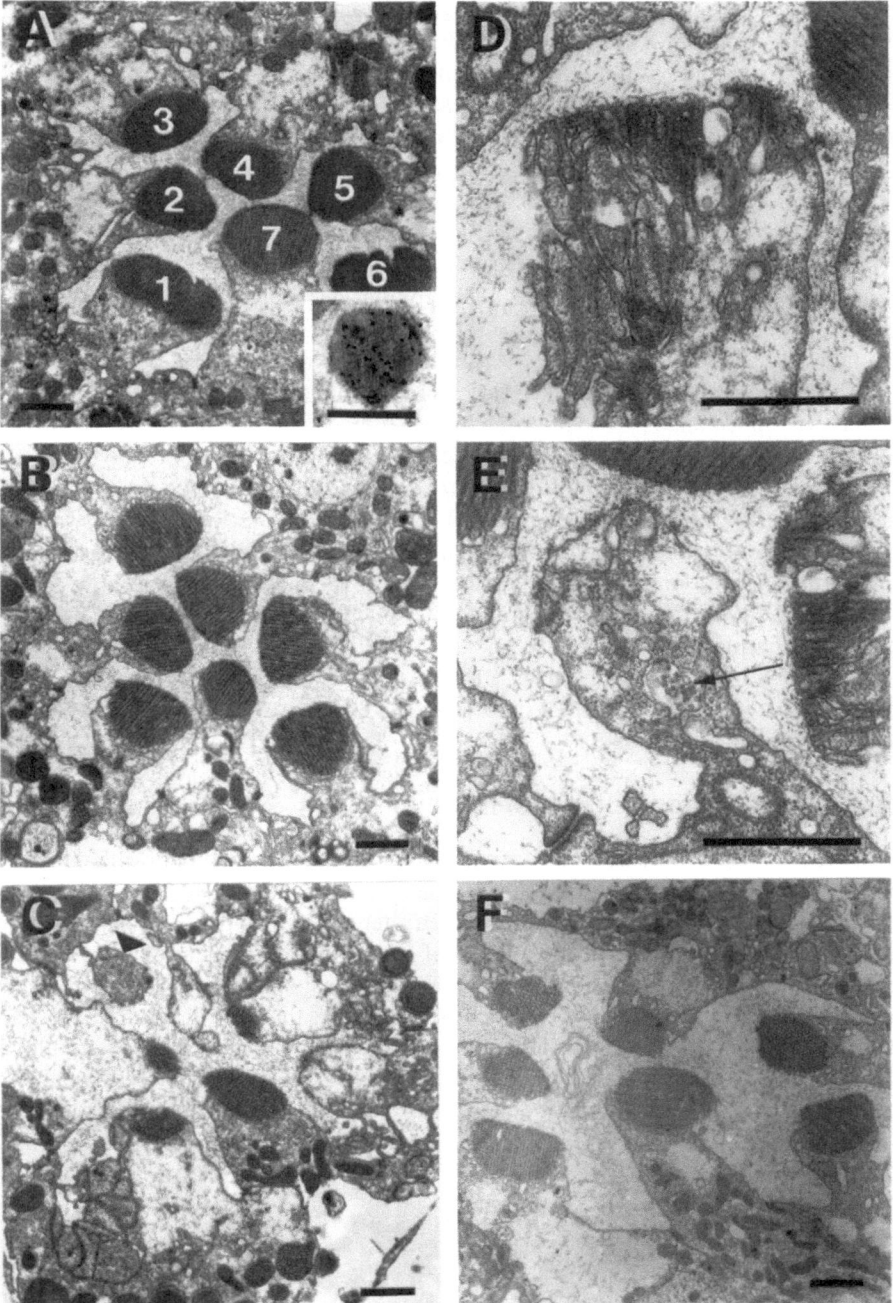

Figure 4. Photoreceptor cell degeneration in Rh1 E271K mutant flies.

calization of opsin in a rhabdomere representative for R1–6, fly homozygous for the Rh1 N86I, 1 day. Numbers indicate the identity of photoreceptor cells. Scale bars = 1 μm.

Tangential sections through the distal half of adult fly ommatidia, ages = post-eclosion. (A-E) Flies were homozygous for the Rh1 mutation, (A) day 0, (B) 4 weeks (C) 8 weeks, arrowhead points at cell debris in the extracellular cavity; (D) representative R1–6 rhabdomere showing irregular membrane invaginations, 6 weeks; (E) multivesicular body (arrow) in representative R1–6 photoreceptor, 6 weeks; (F) fly hetereozygous for the Rh1 mutation over one copy of the wild type gene, 8 weeks. Inset in (A): Immunogold-localization of opsin in a rhabdomere representative for R1–6, fly homozygous for the Rh1 E271K, 1 day. Numbers indicate the identity of photoreceptor cells. Scale bars = 1 μm.

Rh1 E271K mutants show a slower degeneration time course. All R1–6 rhabdomeres still look intact four weeks post-eclosion, and most of the catacomb-like structure looks still intact (Fig. 4B). One or two rhabdomeres per ommatidium are missing at 6 weeks post-eclosion (not shown), and photoreceptor cells still display partial rhabdomeres at 8 weeks post-eclosion (Fig. 4C). However, first signs of microvillar degeneration are visible as early as 1 day post-eclosion, as some microvilli are shortened within many rhabdomeres (Fig. 4A). In this mutant, however, microvilli remnants protruding into the rhabdomeric stalk do not form regular sheets of parallel membranes as in Rh1 L81Q mutants (Fig. 4D). Degeneration coincides with the appearance of big multi-vesicular body-like structures in the photoreceptor-cell (Fig. 4E) and cellular debris in the extracellular, interphotoreceptor cavity (Fig. 4C).

Relevant for an assessment of the mutations studied here as candidates for *Retinitis pigmentosa*-causing mutations in the human rhodopsin gene is the question, whether the degeneration phenotype is dominamt or not. In neither mutant, we found evidence for rhabdomere degeneration in eight week-old heterozygotes carrying one copy of mutant and one of wild-type rhodopsin, (Figs. 2G, 3G & 4G), indicating that the degeneration phenotype is completely recessive.

Immunoblots after separation by SDS-PAGE of protein extracts. All samples contained 10 μg protein, except for ΔAsn20 (20 μg). The blot was probed with an anti-Rh1 opsin antiserum.

Expression of Mutant Opsin Genes

In order to relate ultrastructural changes in mutant flies to the rhodopsin mutation introduced, we first checked for the expression of the mutant opsin genes. RNA analyses showed that the expression of the opsin gene has not been altered as a result of the transformation procedure (19). Thus, any changes in the amount of opsin or in ultrastructure maintenance in the mutants must result from alterations in post-transcriptional steps of rhodopsin biogenesis or from misfunctions of the mutant rhodopsin molecules. At the post-transcriptional level, we investigated the content of opsin protein in the mutant photoreceptors by Western blot analysis, using an anti-Rh1 opsin antiserum. Photoreceptors of Rh1 L81Q mutants contain an estimated one fifth of wild-type opsin (Fig. 5). In Rh1 N86I mutants there is no evidence for the

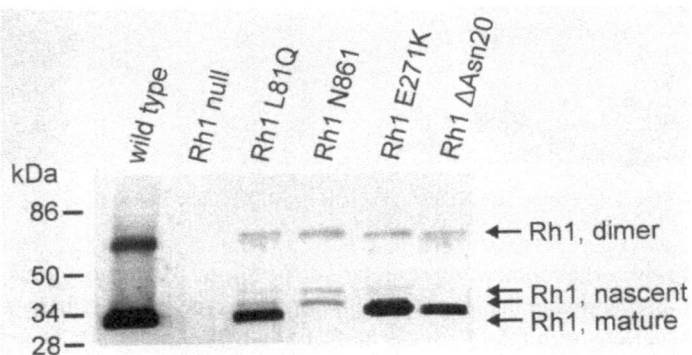

Figure 5. Opsin levels in Rh1 wild type and mutant eyes.

synthesis of a protein corresponding to mature opsin. In Rh1 E271K mutants an opsin band is visible with about 1/3 of the intensity of that in wild-type samples. Interestingly, in all three mutants, two fainter bands of lower electrophoretic mobility are visible. These higher molecular weight species of opsin represent the nascent, glycosylated form of the protein (21,22,19). The opsin band pattern in ΔAsn20 mutant flies, which lack one of the two potential glycosylation sites in the extracellular domains of the opsin molecule (Asn20 → Ile, (10)), shows no higher molecular weight band, indicating that the two glycosylated forms of opsin detected in the mutants studied here represent different processing intermediates of a polysaccharide chain linked to Asn20; thus, we find no evidence for a glycosylation of Asn196. The amount of nascent opsin in photoreceptors of all three mutants is higher than in wild-type photoreceptors (all lanes in Fig. 5 represent the same amount of protein), indicating that mutant opsins is to accumulating in the glycosylated form.

Rhodopsin contents were measured by spectrophotometry of digitonin extracts obtained from fly head membranes. Flies were homozygous for the mutant rhodopsin gene, or heterozygous (one copy of mutant Rh1 gene/one copy of wild-type Rh1 gene) as indicated. Values are expressed relative to the wild-type rhodopsin content of 0.55 pmol per head (n = 5).

Rhodopsin Synthesis and Targeting

For mutants Rh1 L81Q and Rh1 E271K we could demonstrate that the mutant opsin, once deglycosylated, becomes attached to the chromophore (11-cis-3hydroxyretinal) and forms rhodopsin with the same spectral properties as the wild type (19). The amount of rhodopsin in homozygous mutants, measured by spectrophotometry, roughly reflects the amount of opsin detected by Western Blot analysis (Table 1). No visual pigment is detectable in Rh1 N86I flies, indicating that the amount is below the sensitivity limits of the method, i.e., less than about 2% of the amount in wild type. Electrophysiological measurements show, however, that some functional Rh1 rhodopsin is also formed in Rh1 N86I, where is capable of triggering the signal transduction cascade (19). The amount of spectrally intact rhodopsin in heterozygotes is above 50% of the wild-type level, indicative of a gene-dosage dependance of the amount of rhodopsin, i.e. all mutations studied are semidominant with respect to the amount of rhodopsin.

To evaluate whether the rhodopsin mutations affect the transport of the rhodopsin molecule, an immunocytochemical analysis was carried out to subcellularly localize the opsin. Opsin molecules in Rh1 L81Q and Rh1 E271K mutants are correctly targeted to the rhabdomeral microvilli (Fig. 2E & 4E). No opsin can be detected in Rh1 N86I rhabdomeres (Fig. 3E), labelling is indistinguishable from the background in negative controls. The difference in labelling intensity between mutant and wild-type rhabdomeres (not shown) roughly correlates to the ratio in the opsin amounts measured by spectrophotometry.

Table 1. Relative rhodopsin content in mutant fly retinae

	Relative rhodopsin content	
Rh1 mutation	Homozygous	Heterozygous (mutant/wild type)
Rh1 wild type	100 ± 14	—
Rh1 null	0	52.5 ± 5.8
Rh1 L81Q	19.1 ± 1.4	59.2 ± 6.5
Rh1 N86I	<2.0	51.5 ± 6.8
Rh1 E271K	31.2 ± 2.4	68.3 ± 4.2

DISCUSSION

The rhodopsin molecule harbours sequence informations for protein-protein interactions involved in a variety of processes: folding, post-translational modifications and targeting of the rhodopsin molecule, activation of the phototransduction cascade, and maintenance of the ultrastructure of the light-sensitive membrane compartment of photoreceptor cells. Rhodopsin mutations interfering with any of those functions may eventually lead to degeneration of the photoreceptor cell. Based on the hypothesis, that amino acids highly conserved among rhodopsins are extremely cruical for these rhodopsin functions, we studied the effects of mutations of such amino acids within the cytoplasmic domains of the protein. Site-directed mutagenesis was carried out leading to the substitution of single amino acids conserved between the human and *Drosophila* Rh1-opsin, and transgenic flies were generated which express the mutant rhodopsin genes in a rhodopsin-null background.

All mutations studied, L81Q, N86I, or E271K, resepctively, lead to a decrease in rhodopsin expression. Interestingly, in all mutants, more of the glycosylated form of opsin can be detected than in the wild type. It has been demonstrated for *Drosophila* that the deglycosylation of rhodopsin during its biosynthesis is a consequence of chromophore binding, and it is a prerequisit for rhodopsin integration into the rhabdomeric membrane. These considerations suggest that mutations L81Q, N86I, and E271K interfere with normal opsin processing at the chromophore binding and/or deglycosylation step. Which of these steps is blocked is still to be determined in future experiments. No drastic pile-up of nascent opsin occurs in either mutant which is consistent with findings of Huber *et al.* (22) that glycosylated opsin is unstable and subjected to degradation unless it is rapidly processed to chromophore-bound, deglycosylated, mature form of rhodopsin. We conclude that the cytoplasmic surface of rhodopsin forms a binding surface for proteins involved in opsin processing, in which the conservation of single amino acid is of high importance for the correct formation of that interface. Interestingly, the blockage of opsin processing by either mutations is not complete. A small fraction of mutant molecules are processed correctly to form spectrally intact rhodopsin, which is capable of triggering the visual transduction cascade. Moreover, the mutant molecules, once processed, are transported to the rhabdomere. Evidently, the structural changes resulting from correct processing provide the rhodopsin molecule with a tag for proper targeting to the rhabdomeric membrane. The few mutant molecules forming intact rhodopsin may account for molecules which are misread by the enzymes involved in opsin processing because of a structure similar enough to that of wild type.

In addition to their effect on rhodopsin expression, all three mutations induce a slow degeneration of the photoreceptor cells. During pupal development normal looking photoreceptor cells with intact rhabdomeres are formed, the rhabdomeral architecture is indistinguishable from that of wildtype. The catacomb-like structures which separate individual microvilli at their bases (23) are well defined. As a function of time, at first these catacombs disappear, then microvilli deteriorate and involutions of microvillar membranes fill the cell body. In later phases of degeneration, signs of abnormal membrane brakedown become obvious, electron dense granules and multivesicular body-like structures within the cell body and in the extracellular, interphotoreceptor cavity. Accordingly, degeneration starts with the disintegration of the photoreceptive membrane compartment, imposing onto the photoreceptor cell a need for the brakedown of abnormal quantities of rhabdomeric membrane, which then eventually leads to degeneration of the whole cell. As Kumar and Ready (23) have shown, the formation of normal rhabdomeres requires the rhodopsin-induced separation of membranes at the microvillar neck which defines the length of the microvilli. Even the low amount of rhodopsin formed in Rh1 N86I mutants is sufficient for the initial formation of intact microvilli; however, all three mutations studied, eliminate some structural feature required for the long-term integrity of the rhabdomere. Defects in the phototransduction and adaptation cascades can be largely excluded as the cause of photoreceptor cell degeneration in these mutants, scince the electrophysiological analysis shows no differences in the mutant ERGs from wild-type ERGs other than those resulting from a reduced rhodopsin content. We conclude from this that degeneration in mutants results from alterations in rhodopsins cytoplasmic surface destabilizing protein-protein interactions required for the maintenance of the microvillar architecture.

Mutations L81Q, N86I and E271K are recessive for the degeneration phenotype, as no obvious photoreceptor degeneration is observed in heterozygotes, and they are semi-dominant with regard to the rhodopsin content in photoreceptor cells. Accordingly, if rhodopsin maturation or the structural stabilization of the rhabdomeral microvilli should involve the formation of protein complexes in which several opsin molecules are associated (11), the alterations introduces by L81Q, N86I, or E271K mutations must be minor enough not to affect the maturation and transport of wild-type opsin molecules.

Taken together, we show that substitutions of single, highly conserved amino acids in cytoplasmic domains of rhodopsin affect two steps in rhodopsin function: Firstly, the correct receptor protein processing is blocked. Blockage however is not complete, and a few mutant molecules are processed to form spectrally intact rhodopsin capable of triggering the phototransduction cascade. Secondly, mutations L81Q, N86I, or E271K elimit some structural feature required for long-term photoreceptor cell maintainance, leading to a brakedown of the photoreceptive membrane compartment. Since L81, N86 and E271 are conserved in human rhodopsin, mutations of these amino acids must be regarded as possible candidates for recessive forms of *Retinitis pigmentosa* in humans as well.

ACKNOWLEDGMENTS

The authors thank K. Weber for expert technical assistance and L.L.Randall for recording the ERGs. We are indepted to J.E. O'Tousa for the gift of the Rh1 DAsn20 mutant of *Drosophila*. Supported by the Deutscher Akademischer Austauschdienst (Prog. 312, to J.B.), the Deutsche Forschungsgemeinschaft (Pa 274/5-1 to R.P. and J.B.) and NEI grant no. EY00033 (to W.L.P.).

REFERENCES

1. Nathans, J., Merbs, S.L., Sung, C.-H., Weitz, C.J. and Wang, Y., 1992, Molecular genetics of human visual pigments, *Annu. Rev. Genet.* **26**: 403–426.
2. Berson, E.L., 1993, Retinitis Pigmentosa, *Invest. Ophthalmol. Vis. Sci.* **34**: 1659–1676.
3. Dryja, T.P. and Li, T., 1995, Molecular genetics of retinitis pigmentosa, *Hum. Mol. Genet.* **4**: 1739–1743.
4. Halder, G., Callaerts, P. and Gehring, W.J., 1995, Induction of ectopic eyes by targeted expression of the *eyeless* gene in *Drosophila*, *Science* **276**: 1788–1792.
5. Pak, W.L., 1994, Retinal degeneration mutants of *Drosophila*, in: *Molecular Genetics Of Inherited Eye Disorders* (A.F. Wright, and B. Jay, eds.), pp. 29–52, Harwood Academic Publishers, Chur.
6. Stark, W.S., Hunnius, D., Mertz, J. and Chen, D.-M., 1995, *Drosophila* as a model for photoreceptor dystrophies and cell death, in: *Degenerative Deseases Of The Retina* (R.E. Anderson, ed.), pp. 217–226, Plenum Press, New York.
7. Leonard, D.S. and Pak, W.L., 1984, Photoreceptor degeneration associated with mutations in presumptive opsin structural gene in Drosophila, *Soc. Neurosci. Abstr.* **10**: 1032.

8. O'Tousa, J.E., Leonard, D.S. and Pak, W.L., 1989, Morphological defects in oraJK84 photoreceptors caused by mutations in R1–6 opsin gene of *Drosophila*, *J. Neurogenetics* **6**: 41–52.

9. Leonard, D.S., Bowman, V.D., Ready, D.F. and Pak, W.L, 1992, Degeneration of photoreceptors in rhodopsin mutants of *Drosophila*, *J. Neurobiology* **23**: 605–626.

10. O'Tousa, J.E., 1992, Requirement of N-linked glycosylation site in *Drosophila* rhodopsin, *Vis. Neurosci.* **8**: 385–390.

11. Colley, N.J., Cassill, J.A., Baker, E.K. and Zuker, C.S, 1995, Defective intracellular transport is the molecular basis of rhodopsin-dependent dominant retinal degeneration, *Proc. Natl. Acad. Sci. USA* **92**: 3070–3074.

12. Kurada, P. and O'Tousa, J.E., 1995, Retinal degeneration caused by dominant rhodopsin mutations in *Drosophila*, *Neuron* **14**: 571–579.

13. O'Tousa, J.E., Baehr, W., Martin, R.L., Hirsh, J., Pak, W.L. and Applebury, M.L., 1985, The *Drosophila ninaE* gene encodes an opsin, *Cell* **40**: 839–850.

14. Zuker, C.S., Cowman, A.F. and Rubin, G.H., 1985, Isolation and structure of a rhodopsin gene from *Drosophila melanogaster*, *Cell* **40**: 851–858.

15. Nathans, J. and Hogness, D.S., 1984, Isolation and nucleotide sequence of the gene encoding human rhodopsin, *Proc. Natl. Acad. Sci. USA* **81**: 4851–4855.

16. Nathans, J., Thomas, D. and Hogness, D.S., 1986, Molecular genetics of human color vision: The genes encoding blue, green and red pigments, *Science* **232**: 971–975.

17. Rubin, G.M. and Spradling, A.C., 1983, Vectors for P element-mediated gene transfer in *Drosophila, Nucleic Acids Res.* **11**: 6341–6351.

18. Wolfrum, U., 1995, Centrin in the photoreceptor cells of mammalian retinae, *Cell Motil. Cytoskeleton* **32**: 55–64.

19. Bentrop, J., Schwab, K., Pak, W.L. and Paulsen, R., 1996, Site-directed mutagenesis of highly conserved amino acids in the first cytoplasmic loop of *Drosophila* Rh1 opsin blocks rhodopsin synthesis in the nascent state, *EMBO J.*, submitted.

20. Paulsen, R., 1984, Spectral characteristics of isolated blowfly rhabdoms, *J. Comp. Physiol. A* **155**: 47–55.

21. Ozaki, K., Nagatani, H., Ozaki, M. and Tokunaga, F., 1993, Maturation of major *Drosophila* rhodopsin, *ninaE*, requires chromophore 3-hydroxyretinal, *Neuron* **10**: 1113–1119.

22. Huber, A., Wolfrum, U. and Paulsen, R., 1994, Opsin maturation and targeting to rhabdomeral photoreceptor membranes requires the retinal chromophore, *Eur. J. Cell Biol.* **63**: 219–229.

23. Kumar, J.P. and Ready, D.F., 1996, Rhodopsin plays an essential structural role in *Drosophila* photoreceptor development, *Development* **121**: 4359–4370.

20

NITRIC OXIDE-INDUCED INCREASES IN RETINAL cGMP

A Role in Photoreceptor Degenerations

I. G. Morgan[*] and J. W. Wellard

Visual Sciences Group
Centre for Visual Science and Research School of Biological Sciences
Australian National University
Canberra, ACT 2601, Australia

INTRODUCTION

Although nitric oxide (NO) has only recently been recognised as a messenger within the central nervous system, there is already an extensive body of literature on the presence of an NO system within the vertebrate retina. The available evidence implicates NO in the regulation of retinal and choroidal blood flow, the light responses of photoreceptors, post-receptoral processing within the retina, in light-adaptation (for a recent comprehensive review see ref.1), and in the control of eye growth (2). A major problem with these postulated functions is that there is, as yet, no evidence of any influence of light over the rate of production of NO. Indeed, studies in our laboratory have systematically explored the effects of short-term and long-term exposure to light and dark on NOS activity, measured by following the rate of production of citrulline from exogenous arginine, and have found that the variations used have had no effect on NOS activity. NO has also been implicated in the pathological responses of the retina, where there is better evidence for an increased production of NO and a role for cGMP.

The evidence for a role in light adaptation comes largely from the ability of NO donors to promote reponses normally associated with light-adaptation, such as the formation of horizontal cell spinules (3) and the uncoupling of horizontal cells (4). Further evidence comes from the interaction of the NO system with the retinal dopaminergic and melatonin systems, since both of these messengers are believed to play an important role in the control of light-adaptive phenomena (for reviews see refs. 5,6). NO-donors have been re-

---

[*] Address for correspondence: Dr I.G. Morgan, Visual Sciences Group, Centre for Visual Science and Research School of Biological Sciences, Australian National University, GPO Box 475, Canberra, ACT 2601, Australia. Phone: 616-249-4671; fax: 616-249-3808; Email: ian.morgan@anu.edu.au.

ported to inhibit dopamine release (7,8) and melatonin, which has an inhibitory effect on dopamine release (9), has been reported to inhibit NO synthase (NOS) activity (10). A role of NO in the control of eye growth may be related to this interaction, since dopamine has also been implicated in the control of eye growth (11).

We have recently described a retinal micro-circuit composed of the photoreceptors, the dopaminergic amacrine cells and the enkephalin-, neurotensin-, and somatostatin-like immunoreactive (ENSLI) amacrine cells, which is based on reciprocal inhibitory interactions. This micro-circuit functions as a retinal dark-light switch, with the photoreceptors releasing melatonin and the ENSLI amacrine cells releasing enkephalins, somatostatin and neurotensin in the dark, and the dopaminergic amacrine cells releasing dopamine in the light (12,13). One part of this circuit involves dopaminergic feedback inhibition of the rate of synthesis and release of melatonin by the photoreceptors. Dopamine, released in response to light (14) interacts with D2/D4-dopamine receptors (15,16) on the photoreceptors (17,18), inhibiting adenylate cyclase activity (19) and decreasing cAMP levels within the photoreceptors. The lower cAMP levels appear to destabilise the rate-limiting enzyme in melatonin synthesis (serotonin N-acetyltransferase; NAT), increasing the rate of degradation or inactivation of NAT (20), thus depressing the rate of melatonin synthesis and release.

Because of the evidence for a role for NO in light-adaptation and the interactions between NO, dopamine and melatonin which have been described, we decided to determine whether NO-donors had an effect on NAT activity.

NO-DONORS SUPPRESS RETINAL NAT ACTIVITY

As shown in Table 1, the NO-donor sodium nitroprusside (SNP) depressed NAT activity in retinas maintained in the dark, as though the retinas had been exposed to light or to the non-specific dopamine agonist 6,7-ADTN. The depression was dependent on the concentration of SNP in a dose-dependent manner. The suppressive effects of SNP were blocked by the NO scavenger mannitol, and were not produced by the spectator ion ferricyanide. These results clearly implicate NO in the suppression of NAT activity by SNP. However, it should be noted that, since NO-donors are reported to inhibit dopamine release (7,8), it would be difficult to explain the result in terms of an increase in the dopaminergic feedback.

Table 1. Effect of NO-donors and scavengers on retinal NAT activity (% dark control)

Light control	22.0
6,7-ADTN	18.6
SNP (125uM)	61.6
SNP (250uM)	46.5
SNP (500uM)	10.5
SNP (500uM) plus mannitol (2mM)	104.5
Potassium ferricyanide (500uM)	114.7

All experiments were carried out by incubating dark-adapted retinas in the dark for 60 min in a medium consisting of 120mM NaCl, 25mM NaHCO$_3$, 25mM D-glucose, 5mM KCl, 3mM MgCl$_2$, 1.8mM CaCl$_2$, with a pH of 7.4 and saturated with 95% O$_2$/5%CO$_2$. NAT activity was determined radiochemically as previously described (63).

Table 2. Effects of inhibitors of soluble
guanylate cyclase and cGMP analogues
on retinal NAT activity (% dark control)

SNP (500uM)	30.7
LY 83583 (50uM)	84.1
Dibutyryl cGMP (10mM)	41.6
8-Bromo-cGMP (10mM)	43.1

NO-DONORS SUPPRESS RETINAL NAT BY A cGMP-DEPENDENT MECHANISM

NO is known to affect the operation of at least three enzymes, soluble guanylate cyclase (21), ADP-ribosyltransferase (22) and cyclooxygenase (23). The best-characterised effects of NO are produced by activation of soluble guanylate cyclase activity, and increases in intracellular cGMP levels. In the dark-maintained retina, 500 uM SNP, which suppresses NAT activity maximally, caused an approximately three-fold increase in cGMP levels. These NO-induced increases in cGMP appear to be causally related to the suppression of NAT, since the ability of SNP to suppress NAT activity was blocked by the inhibitor of soluble guanylate cyclase activity LY-83583 (24). Moreover, the membrane-permeable cGMP analogues dibutyryl-cGMP and 8-bromo-cGMP also suppressed NAT activity (Table 2). In this case, a maximum of just over 50% suppression was obtained, probably due to limitations on the solubility of the analogues.

PHOSPHODIESTERASE INHIBITORS ALSO SUPPRESS NAT ACTIVITY

Consistent with the involvement of a cGMP-dependent mechanism, the specific inhibitor of cGMP-phosphodiesterase, zaprinast (25), produced a dose-dependent suppression of NAT activity (Table 3). The non-specific phosphodiesterase inhibitor IBMX, also produced a dose-dependent inhibition of NAT activity, although it was less potent on a molar basis than zaprinast. This result is somewhat surprising, since although IBMX will produce accumulations of cGMP, it will also produce increases in cAMP levels, which are believed to increase or at least stabilise NAT activity (20). This suggests that the protective effects of cAMP can be overcome by increased levels of cGMP. The molecular mechanisms by which increased cGMP levels suppress NAT activity, over-riding the protective effects of increased levels of cGMP are worth further investigation.

Table 3. Effects of phosphodiesterase
inhibitors on retinal NAT activity
(% dark control)

Zaprinast (1uM)	67.1
Zaprinast (100uM)	49.4
Zaprinast (10mM)	16.5
IBMX (3uM)	90.4
IBMX (300uM)	47.0
IBMX (30mM)	25.3

SOME FUNDAMENTAL PARADOXES

As noted above, the suppression of NAT activity by SNP was not expected on the basis of the reported ability of NO-donors to decrease dopamine release (7,8), although it is consistent with the reported light-adaptive effects of NO-donors (3,4). It needs to be stressed that light-adaptive effects of NO-donors are inconsistent with a depression of dopamine release, since light-induced dopamine release is believed to be a major trigger of light-adaptive responses (5,6).

The results we have obtained are further paradoxical, in that SNP suppresses NAT activity within the photoreceptors by a mechanism which involves increased levels of retinal cGMP. Light, which also suppresses NAT activity within the photoreceptors, is known however to decrease cGMP levels within the photoreceptors (26). However, there is one known site within the retina, where light and NO-donors have the same effect on cGMP levels, that is in the ON-bipolar cells (27,28). If the NO-donors were acting at that site to depolarise the ON-bipolar cells, then an increased rate of dopamine release would be anticipated, which could lead to the suppression of NAT activity by dopaminergic feedback onto the D2/D4-dopamine receptors of the photoreceptors. Despite the evidence for a suppression, rather than a stimulation, of dopamine release by NO-donors (7,8), we therefore tested whether the observed effects could be due to dopaminergic feedback from the inner retina.

FEEDBACK FROM THE INNER RETINA IS NOT INVOLVED IN THE SUPPRESSION OF NAT ACTIVITY BY NO-DONORS

The first hypothesis to be tested was that depolarisation of the ON-bipolar cells by SNP produced stimulation of the dopaminergic amacrine cells, enhanced release of dopamine and activation of D2/D4-dopamine receptors on the photoreceptors. This possibility was tested in two ways. Firstly, the non-specific excitatory amino acid antagonist kynurenic acid was used to block glutamate-mediated transmission between ON-bipolar cells and dopaminergic amacrine cells. Kynurenic acid at concentrations as high as 5mM had no effect on the suppression of NAT activity by SNP (Table 4). Secondly, the D2-dopamine antagonist sulpiride was used to block the effects of dopamine on the D2/D4-dopamine receptors on the photoreceptors. Once again, sulpiride at concentrations as high as 50mM was without effect. These two observations appear to rule out a role for dopaminergic feedback.

However, it remained possible that some other unknown mediator was released from the inner retina as a result of a SNP-induced depolarisation of the ON-bipolar cells. To test for this possibility, experiments were carried out on chicken retinas which had been previously lesioned with kainic acid. These lesions eliminate all of the OFF-bipolar cells and most of the amacrine cells (29), yet SNP was still able to suppress retinal NAT activity to the same extent.

Table 4. Effects of kynurenic acid and sulpiride on retinal NAT activity (% dark control)

SNP (500uM)	19.4
SNP plus kynurenic acid (5mM)	8.9
SNP plus sulpiride (50mM)	16.1

These results, taken together, rule out a role for dopaminergic feedback in the suppression of NAT activity by SNP, and make it unlikely that any other messenger released from the inner retina is involved. These results therefore suggest that the suppression of NAT activity by SNP is likely to be intrinsic to the photoreceptors.

A RETURN TO THE ORIGINAL PARADOX

Thus the original paradox remains unresolved. Both SNP and light suppress NAT activity, SNP by a pathway that involves stimulation of soluble guanylate cyclase and increases in cGMP levels within the photoreceptors. In contrast, light increases the activity of cGMP-phosphodiesterase (30) and decreases cGMP levels (26). The hypothesis of a direct effect of NO on the photoreceptors is made more viable by the demonstration of soluble guanylate cyclase activity in rods (31), and by the demonstration that cGMP immunoreactivity in photoreceptors is increased by SNP (32.). SNP normalises the dark current and light responses in isolated frog rods (33), as does co-application of arginine and NADPH, a cofactor of NOS (34). Given the lack of evidence of light-regulated synthesis of NO in the retina, it is unclear whether these effects are physiologically relevant, but they nevertheless demonstrate that photoreceptors contain the physiological substrate for SNP to increase cGMP levels, even if it is not clear how the increased levels suppress NAT activity.

NO DOES NOT APPEAR TO PLAY A ROLE IN PHYSIOLOGICAL REGULATION OF NAT ACTIVITY OR LIGHT-ADAPTATION

The original paradox also indicates that NO does not play a role in the physiological process of light-induced suppression of NAT activity, since this is associated with a decrease in cGMP levels. How increases in cGMP levels lead to suppression of NAT is not known, but it could involve competition with some step in the pathway which links cAMP levels to NAT activity, or to a separate suppressive effect directly stimulated by cGMP-dependent activities such as protein kinases or phosphodiesterases.

Equally, NO does not seem likely to play a role in physiological light adaptation, which appears to be involve an increase in dopamine release (5,6), whereas NO-donors suppress dopamine release (7,8). This simply generates a further unresolved question, namely how can light and NO-donors produce the same effects, when they produce differing effects on dopamine release. One possibility is that dopamine stimulates NO synthesis and release, and that the feedback inhibition of dopamine release by NO-donors is irrelevant to the feedforward effects of NO. However, this assumes that light stimulates NO synthesis, for which there is no evidence, although in preliminary experiments, we have detected a small stimulatory effect of dopamine agonists on NOS activity.

Another possibility is more complex, and depends on the way in which melatonin and dopamine combine to exert effects on a number of physiological light-adaptive responses (35). In analysing the distribution of receptors for dopamine (D1-type) and enkephalins, we have noticed that there is a striking parallel, although not identity, in their distribution (36,37,38). This implies that many cells express both D1-dopamine receptors and enkephalin receptors. Since D1-dopamine receptors activate adenylate cyclase (39), while enekphalin receptors inhibit it (40), the counter-phase release of dopamine in the light and enkephalins in the dark (12,13) could be paralled by antagonistic regulation of adenylate cyclase activity and cAMP levels in the target cells which coexpress the receptors. Melatonin is also known to inhibit adenylate cyclase (41), but at present little is known in detail about the distribution of melatonin receptors in the retina. However, it does appear that at least one cell type, the displaced ganglion cells which project to the nucleus of the basal optic root, express D1-dopamine (37,38), enkephalin (36) and melatonin (42) receptors. Gan and Iuvone (43) have shown that melatonin and D1-dopamine agonists exert antagonistic effects upon adenylate cyclase in neural cells prepared from embryonic day 8 chicken retinas. This suggests that co-expression of D1-dopamine and melatonin receptors may be quite common in retinal cells, as co-expression of D1-dopamine and enkephalin receptors appears to be from our immunohistochemical studies (36–38). If this is the case, the light-adaptive effects of NO-donors could in part be mediated by the removal of melatonin-mediated inhibition of dopamine release and of adenylate cyclase activity. The regulatory pathways involved are potentially quite complex, since melatonin appears to exert complex, circadian-time-dependent effects on glutamate uptake and release (44), cGMP synthesis and degradation (45), and cAMP levels (46) in the retina.

A ROLE FOR NO AND cGMP IN RETINAL PATHOLOGY

While there is no evidence for a role for NO and NO-induced increases in cGMP levels in normal retinal physiology, there is evidence for roles for both in retinal pathology.

Responses to Excitotoxins and Ischaemia

Exposure of the retina to excitotoxins such as kainic acid and NMDA results in substantial destruction of retinal cells, with species-specific patterns of cell loss (47). NMDA and other glutamate analogues are able to stimulate cGMP synthesis (48), and NMDA-toxicity is blocked by NOS inhibitors (49,50). NMDA receptors also seem to be involved in destruction of retinal cells in response to ischaemia (51), and ischaemic cell loss can be blocked by NOS inhibitors (50,52). These results suggest that the NO-cGMP pathway may be involved in excitotoxin- and ischaemia-induced cell death.

Photoreceptor Degeneration in Retinitis Pigmentosa and Some Animal Models

Some forms of retinitis pigmentosa, and some animal models of this disease, are known to involve production of a defective cGMP-phosphodiesterase (53). The response of retinal melatonin levels or NAT activity over these periods do not seem to have been reported, but it seems possible that increases in cGMP levels would lead to suppression of NAT activity and hence to suppression of retinal melatonin levels.

Any link of these changes to photoreceptor degeneration is of course speculative, although Ulshafer et al. (54) have reported that incubation of isolated Xenopus retina with cGMP analogues for 8h results in massive death of photoreceptors, particularly of rods. In contrast, we have found that exposure of the retina to SNP for periods of up to 1h does not seem to produce significant changes in the morphology of the photoreceptors.

The link between increased cGMP levels and photoreceptor cell death is currently unknown. Constant dark, which maintains cGMP levels at the high end of the normal physiological range does not lead to photoreceptor loss. But the increases beyond the normal range which are produced in retinitis pigmentosa and in the animal models are associated with cell death. It is possible that suppression of NAT activity is part of the process. Melatonin and dopamine are known to interact antagonistically in the control of the shedding of photoreceptor disks (35). Under conditions of chronically suppressed melatonin synthesis, dopamine release may be abnormally high, which may push the cells towards apoptosis. In addition, low melatonin levels will affect the uptake and release of glutamate (44), synthesis and degradation of cGMP and cAMP (45,46) and possibly NOS activity (10). Further studies are clearly needed to investigate the link.

Cell Death in Uveitis

In the two cases outlined above, increased production of NO is produced by exposure to excitotoxins or ischaemia, or increased levels of cGMP are produced by defects in cGMP phosphodiesterase. Retinal cell death also occurs in various forms of pathological and experimentally induced uveitis and uveoretinitis, which are often associated with destruction of the outer layers of the retina. In these cases, there is marked induction of iNOS in the retinal pigment epithelium, the choroid and the Muller cells (55,56). NOS inhibitors appear to be able to reduce the cell loss which occurs (57–59). Thus NO seems to be involved, but whether the effect of NO on the photoreceptors, increased levels of cGMP and the down-regulation of NAT activity is part of the causal chain is still to be established.

Cell Death in Constant Light

It has also been reported that NOS inhibitors reduce photoreceptor loss in rats maintained in constant light (60). This indicates that activation of NOS and increased production of NO have a role in the pathogenic process. What leads to the activation of NOS is unclear, and whether the NO-cGMP-NAT-melatonin link is involved is also unclear. Under conditions of constant light, NAT activity and melatonin synthesis are in any case suppressed. Given that melatonin has been reported to inhibit NOS activity (10), increased production of NO might occur as a result. While there are many other potential pathways for light-induced damage (61)), some involvement of this pathway is supported by the protective effects of dopamine agonists and the potentiating effects of melatonin on light-induced damage (62).

CONCLUSIONS

We have shown that the NO-donor SNP suppresses retinal NAT activity by a cGMP-dependent mechanism, probably by acting directly on the photoreceptors. However, NO does not seems to be involved in the normal physiological suppression of NAT activity by light. Nor does NO seem to be involved in physiological light-adaptation, although it evokes some light-adaptive responses. We suggest that it may mimic the effects of light because it suppresses NAT activity and melatonin synthesis, as does light, and that light-adaptive responses which are regulated by the interaction of dopamine and melatonin may respond to the chronic reduction in melatonin production. The fundamental problem with implicating NO in any physiological part of visual processing is that there is no evidence that changes in visual stimulation regulate NOS activity and the production of NO.

In contrast, the ability of NO to suppress NAT activity by a cGMP-dependent mechanism may be important in a number of pathological processes. Increased output of

NO by excitotoxin- or ischaemia-induced stimulation of nNOS may be important for cell loss under these conditions. Increased production of NO by iNOS may be important in photoreceptor loss in uveitis and uveoretinitis. Increased production of NO by mechanisms which are as yet uncertain also seems to be involved in constant light-induced damage to the retina. Whether these processes depend in any way on the ability of NO to increase cGMP levels in the photoreceptors is not clear, but increased levels of cGMP are clearly pathogenic, since some forms of retinitis pigmentosa involve defective cGMP-phosphodiesterase, and incubation of isolated retinas with cGMP analogues are known to cause the destruction of photoreceptors.

It is not clear if the pathogenic effects of increased cGMP levels depend in any way on the suppression of NAT activity and melatonin synthesis, but there are plausible links to mechanisms which would cause cell disruption, particularly of the photoreceptors. The interaction of dopamine and melatonin regulates the turn-over of photoreceptor outer segments and phagocytosis by the retinal pigment epithelium. Depressed melatonin levels could lead to increased NOS activity. If the complete pathway of NO-cGMP-NAT-melatonin is involved in any of these pathological responses, then there are potential sites of pharmacological intervention at each of these steps.

REFERENCES

1. Goldstein, I.M., Ostwald, P., and Roth, S., 1996, Nitric oxide: a review of its role in retinal function and disease, Vis.Res. 36: 2979–2994.
2. Kawasaki, Y., Ohmi, G., Fujikado, T., Taniguchi, N., and Tano, Y., 1996, Role of nitric oxide in experimental myopia, Exptl.Eye Res.63: S142.
3. Greenstreet, E.H., and Djamgoz, M.B., 1994, Nitric oxide induces light-adaptive morphological changes in retinal neurones, NeuroReport, 6: 310–317.
4. Miyachi, E., Murakami, M., and Nakaki, T., 1990, Arginine blocks gap junctions between retinal horizontal cells, NeuroReport, 1: 107–110.
5. Djamgoz, M.B., and Wagner, H.J.,1992, Localization and function of dopamine in the adult vertebrate retina, Neurochem.Int.20: 139–91.
6. Witkovsky, P., and Dearry, A.,1992, Functional roles of dopamine in the vertebrate retina, Prog.Retinal Res.11: 247–292.
7. Bugnon,O., Schaad, N.C., and Schorderet, M., 1994, Nitric oxide modulates endogenous dopamine release in bovine retina, NeuroReport 5: 401–404.
8. Djamgoz, M.B., Cunningham, J.R., Davenport, S.L., and Neal, M.J., 1995, Neurosci.Letts. 198: 33–36.
9. Dubocovich, M.L.,1983, Melatonin is a potent modulator of dopamine release in the retina, Nature 306: 782–4.
10. Pozo, D., Reiter, R.J., Calvo, J.R., and Guerrero, J.M., 1994, Physiological concentrations of melatonin inhibit nitric oxide synthase in rat cerebellum, Life Sci. 55: 455–460.
11. Stone, R.A., Lin, T., Laties, A.M., Iuvone, P.M.,1989, Retinal dopamine and form-deprivation myopia, Proc.Natl.Acad.Sci.USA 86: 704–6.
12. Morgan, I.G., and Boelen, M.K., 1996, A retinal dark-light switch: a review of the evidence, Vis.Neurosci. 13: 399–409.
13. Morgan, I.G., and Boelen, M.K., 1997, The retinal dark-light switch: a neural flip-flop device, Prog.Retinal Res., in press.
14. Kramer, S.G.,1971, Dopamine: A retinal neurotransmitter. I. Retinal uptake, storage, and light-stimulated release of H3-dopamine in vivo, Invest.Ophthalmol.vis.Sci. 10: 438–452.
15. Iuvone, P.M., Boatright, J.H., and Bloom, M.M.,1987, Dopamine mediates the light-evoked suppression of serotonin N-acetyltransferase activity in retina, Brain Res. 418: 314–324.
16. Zawilska, J.B., and Nowak, J.Z.,1994, Does D4 dopamine receptor mediate the inhibitory effect of light on melatonin biosynthesis in chick retina?, Neurosci.Letts.166: 203–206.
17. Muresan, Z., and Besharse, J.C.,1993, D2-like dopamine receptors in amphibian retina: localization with fluorescent ligands, J.Comp.Neurol.331: 149–160.

18. Wagner, H.J., Luo, B.G., Ariano, M.A., Sibley, D.R., and Stell, W.K.,1993, Localization of D2 dopamine receptors in vertebrate retinae with anti-peptide antibodies, J.Comp.Neurol.331: 469–481.

19. Iuvone, P.M.,1986, Evidence for a D2 dopamine receptor in frog retina that decreases cyclic AMP accumulation and serotonin N-acetyltransferase activity, Life Sci. 38: 331–342.

20. Alonso-Gomez, A.L., and Iuvone, P.M.,1995, Melatonin biosynthesis in cultured chick retinal photoreceptor cells: calcium and cyclic AMP protect serotonin N-acetyltransferase from inactivation in cycloheximide-treated cells, J.Neurochem. 65: 1054–1060.

21. Katsuki, S., Arnold, W., Mittal, C., and Murad, F., 1977, Stimulation of guanylate cyclase by sodium nitroprusside, nitroglycerin and nitric oxide in various tissue preparations and comparison to the effects of sodium azide and hydroxylamine, J.Cyclic Nucleotide Res. 3: 23–35.

22. Brune, B., and Lapetina, E.G., 1980, Activation of a cytosolic ADP-ribosyltransferase by nitric oxide-generating agents, J.Biol.Chem. 264: 8455–8458.

23. Salvemini, D., Misko, T.P., Masferrer, J.L., Seibert, K., Currie, M.G., and Needleman, P., 1993, Nitric oxide activates cyclooxygenase enzymes, Proc.Natl.Acad.Sci.USA 90: 7240–7244.

24. Kontos, H.A., and Wei, E.P., Hydroxyl radical-dependent inactivation of guanylate cyclase in cerebral arterioles by methylene blue and by LY83583, Stroke 24: 427–434.

25. Lugnier, C., Schoeffter, P., Le Bec, A., Strouthou, E., and Stoclet, J.C., 1986, Selective inhibition of cyclic nucleotide phosphodiesterases of human, bovine and rat aorta, Biochem.Pharmacol. 35: 1743–1751.

26. Goridis, C., and Virmaux, N., 1974, Light-regulated guanosine 3',5'-monophosphate phosphodiesterase of bovine retina, Nature 248: 57–58.

27. Shiells, R.A., and Falk, G., 1990, Glutamate receptors of rod bipolar cells are linked to a cyclic GMP cascade via a G-protein, Proc.Roy.Soc.London B 242: 91–94.

28. Nawy, S., and Jahr, C.E., 1990, Suppression by glutamate of cGMP-activated conductance in retinal bipolar cells, Nature 346: 269–271.

29. Ingham, C.A., and Morgan, I.G., 1983, Dose-dependent effects of intravitreal kainic acid on specific cell types in the chicken retina, Neurosci. 9: 165–181.

30. Stryer, L., 1986, Cyclic GMP cascade of vision, Annu.Rev.Neurosci. 9: 87–119.

31. Margulis, A., Sharma, R.K., and Sitaramayya, A., 1992, Nitroprusside-sensitive and insensitive guanylate cyclases in retinal outer rod segments, Biochem.Biophys.Res.Comm. 185: 909–914.

32. Berkelmans, H.S., Schipper, J., Hudson, L., Steinbusch, H.W.M., and Da Vente, J., 1989, cGMP immunocytochemistry in aorta, kidney, retina and brain tissues of the rat after perfusion with nitroprusside. Histochem. 93: 143–148.

33. Schmidt, K.F., Noll, G.N., and Yamamoto, Y., 1992, Sodium nitroprusside alters dark voltage and light responses in isolated retinal rods during whole-cell recording, Vis.Neurosci. 9: 205–209.

34. Tsuyama, Y., Noll, G.N., and Schmidt, K.F., 1993, L-Arginine and nicotinamide adenine dinucleotide phosphate alter dark voltage and accelerate light response recovery in isolated retinal rods of the frog (Rana temporaria), Neurosci.Letts. 149: 95–98.

35. Besharse, J.C., Iuvone, P.M., and Pierce, M.E.,1988, Regulation of rhythmic photoreceptor metabolism: a role for post-receptoral neurons, Prog.Retinal Res.7: 21–61.

36. Morgan, I.G., Yang, D.S., Li, Z.K., and Maderspach, K.,1994, Localisation of enkephalin receptors in the chicken retina, Proc.Aust.Neurosci.Soc. 5: 207.

37. Morgan, I.G., and Yang, D.S., 1996, Overlapping distributions of D1-dopamine and enkephalin receptors in the chicken retina, Proc.Aust.Neurosci.Soc. 7: 218.

38. Firth, S.I., Morgan, I.G., and Boelen, M.K. 1997, Localisation of D1-dopamine receptors in the chicken retina, ANZ J.Ophthalmol., in press.

39. Kebabian, J.W., and Calne, D.B.,1979, Multiple receptors for dopamine, Nature 277: 93–96.

40. Walczak, S.A., Wilkening, D., and Makman, M.H., 1979, Interaction of morphine, etorphine and enkephalins with dopamine-stimulated adenylate cyclase of monkey amygdala, Brain Res. 160: 105–116.

41. Iuvone, P.M., and Gan, J.,1994, Melatonin-receptor-mediated inhibition of cyclic AMP accumulation in chick retinal cultures, J.Neurochem. 63: 118–124.

42. Krause, D.N., Siuciak, J.A., and Dubocovich, M.L.,1994, Unilateral optic nerve transection decreases 2-[125I]-iodomelatonin binding in retinorecipient areas and visual pathways of chick brain, Brain Res. 654: 63–74.

43. Iuvone, P.M., and Gan, J.,1995, Functional interaction of melatonin receptors and D1 dopamine receptors in cultured chick retinal neurons, J.Neurosci. 15: 2179–2185.

44. Faillace, M.P., Sarmiento, M.I., and Rosenstein, R.E.,1996, Melatonin effect on [3H]glutamate uptake and release in the golden hamster retina, J.Neurochem. 67: 623–628.

45. Faillace, M.P., Keller-Sarmiento, M.I., and Rosenstein, R.E.,1996, Melatonin effect on the cyclic GMP system in the golden hamster retina, Brain Res. 711: 112–117.

46. Faillace, M.P., Sarmiento, M.I., Siri, L.N., and Rosenstein, R.E.,1994, Diurnal variations in cyclic AMP and melatonin content of golden hamster retina, J.Neurochem. 62: 1995–2000.

47. Morgan, I.G.,1983, Kainic acid as a tool in retinal research, Prog.Retinal Res.2: 249–266.

48. Zeevalk, G.D., Hyndman, A.G., and Nicklas, W.J,1989, Excitatory amino acid-induced toxicity in chick retina: amino acid release, histology and the effects of chloride channel blockers, J.Neurochem. 53: 1610–1619.

49. Zeewalk, G.D., and Nicklas, W.J.,1994, Nitric oxide in retina: relation to excitatory amino acids and excitotoxicity, Exptl.Eye Res. 58: 343–350.

50. Akaike, A., Kashii, S., and Honda, Y.,1996, Role of nitric oxide on NMDA receptor-mediated neuronal death in the retina, Exptl.Eye Res.63: S152.

51. Mosinger, J.L., Price, M.T., Bai, H.Y., Xiao, H., Wozniak, D.F., and Oleny, J.W.,1991, Blockade of both NMDA and non-NMDA receptors is required for optimal protection against ischemic neuronal degeneration in the in vivo adult monkey retina, Exp.Neurol. 113: 10–17.

52. Geyer, O., Almog, J., Lupu-Meiri, M., Lazar, M., and Oron, Y.,1995, Nitric oxide synthase inhibitors protect rat retina against ischemic injury, FEBS Letts. 374: 399–402.

53. McLaughlin, M.E., Sandberg, M.A., Berson, E.L., and Dryja, T.P.,1993, Recessive mutations in the gene encoding the beta-subunit of rod phosphodiesterase in patients with retinitis pigmentosa, Nature Genet 4: 130–134.

54. Ulshafer, R.J., Garcia, C.A., and Hollyfield, J.G.,1980, Sensitivity of photoreceptors to elevated levels of cGMP in the human retina, Invest.Ophthalmol.vis.Sci. 19: 1236–1241.

55. Goureau, O., Bellot, J., Thillaye, B., Courtois, Y., and de Kozak, Y.,1995, Increased nitric oxide production in endotoxin-induced uveitis. Reduction of uveitis by an inhibitor of nitric oxide synthase, J.Immunol. 154: 6518–6523.

56. Jacquemin, E., de Kozak, Y., Thillaye, B., Courtois, Y., and Goureau, O.,1996, Expression of inducible nitric oxide synthase in the eye from endotoxin-induced uveitis rats, Invest.Ophthalmol.vis.Sci. 37: 1187–1196.

57. Parks, D.J., Cheung, M.K., Chan, C.C., and Roberge, F.G.,1994, The role of nitric oxide in uveitis, Arch.Ophthalmol. 112: 544–546.

58. Tilton, R.G., Chang, K., Corbett, J.A., Misko, T.P., Currie, M.G., Bora, N.S., Kaplan, H.J., and Williamson, J.R.,1994, Endotoxin-induced uveitis in the rat is attenuated by inhibition of nitric oxide production, Invest.Ophthalmol.vis.Sci.35: 3278–3288.

59. Mandai, M., Yoshimura, N., Yoshida, M., Iwaki, M., and Honda, Y.,1994, The role of nitric oxide synthase in endotoxin-induced uveitis: effects of NG-nitro-L-arginine, Invest.Ophthalmol.vis.Sci. 35: 3673–3680.

60. Goureau, O., Jeanny, J.C., Becquet, F., Hartmann, M.P., and Courtois, Y.,1993, Protection against light-induced degeneration by an inhibitor of NO synthase, NeuroReport 5: 233–236.

61. Organisciak, D.T., and Winkler, B.S., Retinal light damage: practical and theoretical considerations,1994, Prog.Retinal Res. 13: 1–29.

62. Bubenik, G.A., and Purtill, R.A.,1980, The role of melatonin and dopamine in retinal physiology, Can.J.Pharmacol. 58: 1457–1462.

63. Morgan, I.G., Boelen, M.K., and Miethke, P.,1995, Pineal activity is under the control of retinal d1-dopaminergic pathways, NeuroReport 6: 446–448.

21

GLYCOHISTOCHEMICAL STUDY OF LIGHT-INDUCED RETINAL DEGENERATION

Removal System of Apoptotic Cells

Fumiyuki Uehara,[1] Norio Ohba,[1] Toyoko Yanagita,[1] Munefumi Sameshima,[1] Naoto Iwakiri,[1] Akiko Okubo,[1] Yoshiko Maeda,[1] Kazuhiko Unoki,[1] and Taeko Miyagi[2]

[1]Department of Ophthalmology
Kagoshima University Faculty of Medicine
8-35-1 Sakuragaoka, Kagoshima-shi 890, Japan
[2]Research Institute
Miyagi Cancer Center
Natori-shi, Miyagi 981-12, Japan

INTRODUCTION

We are interested in the physiological roles of the sialoglycoconjugates associated with rod photoreceptor cells, which specifically metabolize sialic acids by balancing sialyltransferase with sialidase (1, 2). Defects of O- and N-linked sialoglycoconjugates may cause retinal dysplasia and retinal degeneration, respectively, in humans (3). Retinal degeneration can be experimentally induced in rats by modification of sialic acids of retinal sialoglycoconjugates (4,5), but the pathomechanism underlying this degenerative process remains to be clarified.

It has been shown that vertebrate photoreceptor cells are damaged by excessive light exposure (6,7). Ki 67 antigen is a nuclear protein, which is expressed in the stage of the cell cycle (8). This antigen-expression has been widely used as a marker of cellular proliferation. On the other hand, the TdT-mediated dUTP-X nick end labeling (TUNEL) technique is now widely used for the detection of apoptotic cell death at the single cell level (9). Apoptosis is considered to be a common process in the late stages of many types of retinal degeneration. In the present work, therefore, these two methods with opposite targets were combined with glycohistochemical methods to examine the mechanism underlying light-induced retinal degeneration, particular paying attention to the change in retinal sialoglycoconjugates.

Degenerative Retinal Diseases, edited by LaVail *et al.*
Plenum Press, New York, 1997

MATERIALS AND METHODS

Tissue Preparation

Albino Wistar rats were reared under cyclic lighting conditions (12 hours on/12 hours off) with an in-cage illumination level of 10 lux at room temperature of 22 1 . At the age of 8–10 weeks they were kept in a constant-intensity light, which was produced with two 36-watt fluorescent lamps placed 50 cm above the floor of the cage. The illumination level in the cage was maintained at 1,100 to 1,500 lux. The experimental animals were sacrificed after 2, 3 and 5 days of constant exposure to the light with an overdose of carbon dioxide at between 1000 and 1100 hours. Control animals were sacrificed at the same times. The eyes were enucleated from the rats, immersion-fixed in buffered formalin, incubated in phosphate-buffered saline (PBS), dehydrated in an ethanol series, and then embedded in paraffin. All animal procedures conformed to the Guidelines of the Kagoshima University Faculty of Medicine for Animal Experiments, and the ARVO Resolution on the Use of Animals in Research.

Histochemistry

Five-μm sections of the specimens were cut, mounted on poly-L-lysine-coated glass slides, deparaffinized in xylene, and then rehydrated in a graded ethanol series. The tissue sections for immunohistochemistry were immersed in 0.01 M sodium citrate buffer, pH 6.0, and then boiled for 4 min in a pressure cooker to unmask antigens. Endogenous peroxidase activity was blocked by treatment of the tissue sections with 0.2% H_2O_2 in methanol for 30 minutes. After the slides had been immersed in PBS for 30 minutes, nonspecific binding was blocked by incubating the sections with 3% horse serum (for immunohistochemistry) or 2% bovine serum albumin (for lectin histochemistry) for 30 minutes at room temperature. The tissue sections were then overlaid with either an antiserum (1/100 diluted with PBS) to cytosolic sialidase {produced by one of the authors (Miyagi, 10)}, a monoclonal antibody to Ki 67 antigen (Novocastra Laboratories, UK), or biotinylated lectins {*Maackia amurensis* lectin II (MAL II, 11) or jacalin (12)} in a humidified chamber for 1 hour at room temperature. After the slides had been washed in PBS for 30 minutes, the tissue sections for immunohistochemistry were incubated with biotinylated anti-rabbit IgG (for the antiserum to cytosolic sialidase) or biotinylated anti-mouse IgG (for the antibody to Ki 67 antigen; 1/100 diluted with PBS; Vector Laboratories) for 1 hour at room temperature. After the slides had been washed in PBS for 30 minutes, binding was visualized with the reagents of a Vectastain Elite ABC kit and diaminobenzidine as the peroxidase substrate following the standard protocol for a Vectastain kit (Vector Laboratories). The sections were dehydrated, coverslipped, and then examined for the location of the brown reaction product. TUNEL-positive cells (apoptotic cells) were immunohistochemically detected using an in situ cell death detection kit (Boehringer Mannheim, Germany) following the standard protocol of Boehringer Mannheim.

RESULTS

Figure 1 shows the sites of binding of the antibody (Ab) to cytosolic sialidase in rat retinas. In the control retina from rats reared under cyclic lighting conditions, the Ab bound diffusely to the photoreceptor inner segments (Fig. 1A) as reported previously (2). The retinas from the experimental rats exposed to constant light for 2 days remained almost intact, with well retained outer and inner segments of photoreceptor cells consisting

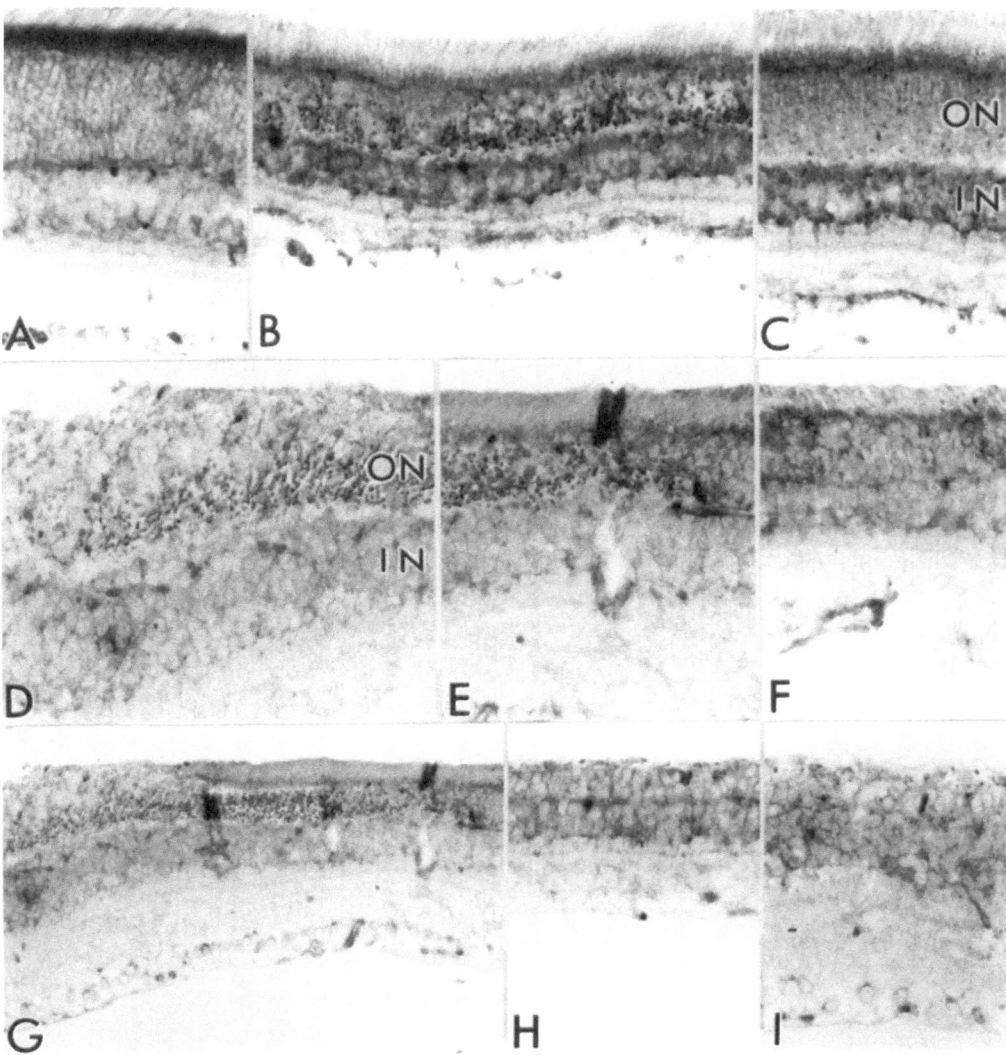

Figure 1. Immunohistochemical localization of cytosolic sialidase in rat retinas. A: Control retina from a rat reared under cyclic lighting conditions; B: superior hemisphere of a retina exposed to constant light for 2 days; C: inferior hemisphere of a retina exposed to constant light for 2 days; D, E, G: superior hemisphere of a retina exposed to constant light for 3 days; F: inferior hemisphere of a retina exposed to constant light for 3 days; H: superior hemisphere of a retina exposed to constant light for 5 days; I: inferior hemispere of a retina exposed to constant light for 5 days. ON: outer nuclear layer; IN: inner nuclear layer.

of a row of 8 to 10 nuclei in the outer nuclear layer (ONL). However, at this stage, the distribution of the Ab to cytosolic sialidase had already changed. Particle-like materials were detected with Ab relatively diffusely (intenser on the inner side) and sparsely in the ONL of the superior and inferior hemispheres, respectively (Fig, 1B, C). The retinas from the rats exposed to constant light for 3 days (Fig. 1D-G) also remained relatively intact except for the equatorial portion of the superior hemisphere (Fig. 1D, E, G). The inferior hemisphere at this stage still showed a little binding of the Ab (Fig. 1F). Figure 1E and G show the transitional zone between the intense and sparse binding regions of the Ab on the superior hemisphere with well retained outer and inner segments of photoreceptor cells. Figure 1D shows that as the morphology was destroyed, the distribution of the Ab-binding decreased. Following 5 days of constant light exposure, the outer and inner segments of photoreceptors had decreased in height to a considerable extent, and the thickness of the ONL was decreased to a row of 2–4 nuclei (Fig. 1H, 1I). At this stage, the distribution of sialidase was observed only sparsely on both superior and inferior hemispheres (Fig. 1H, I). These observations suggest that the increase in the distribution of sialidase may precede the degenerative change of the photoreceptors.

Figure 2 shows the binding sites of MAL II in the control (Fig. 2A) and experimental rat retinas (Fig. 2B-E). In the control retinas, MAL II bound intensely and weakly to the photoreceptor outer segments and the other layers including the ONL, respectively (Fig. 2A) as reported previously (13). Following 2 days of constant light exposure, the intensity of MAL II binding to the ONL of the superior hemisphere (Fig. 2B) was slightly weaker than that of the inferior one (Fig. 2C). The superior equatorial region of the retinas from the rats exposed to constant light for 3 days showed a distinct decrease in MAL II binding (in the left and right areas from the points indicated by arrows in Fig. 2D and Fig. 2E, respectively) in comparison with the surrounding regions (in the right and left areas from the points indicated by arrows in Fig. 2D and E, respectively).

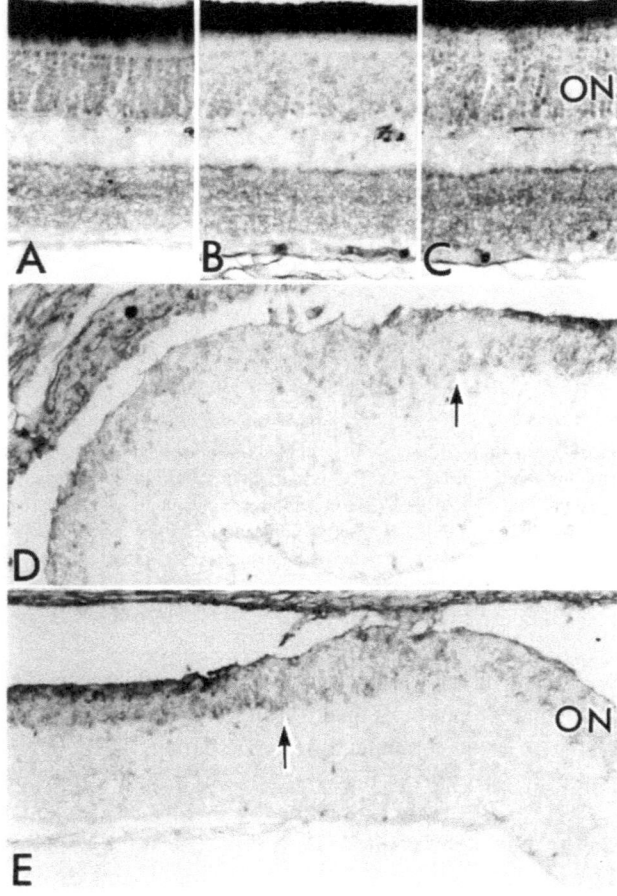

Figure 2. *Maackia amurensis* lectin II binding in rat retinas. A: Control retina from a rat reared under cyclic lighting conditions; B: superior hemisphere of a retina exposed to constant light for 2 days; C: inferior hemisphere of a retina exposed to constant light for 2 days; D, E: superior hemisphere of a retina exposed to constant light for 3 days. ON: outer nuclear layer.

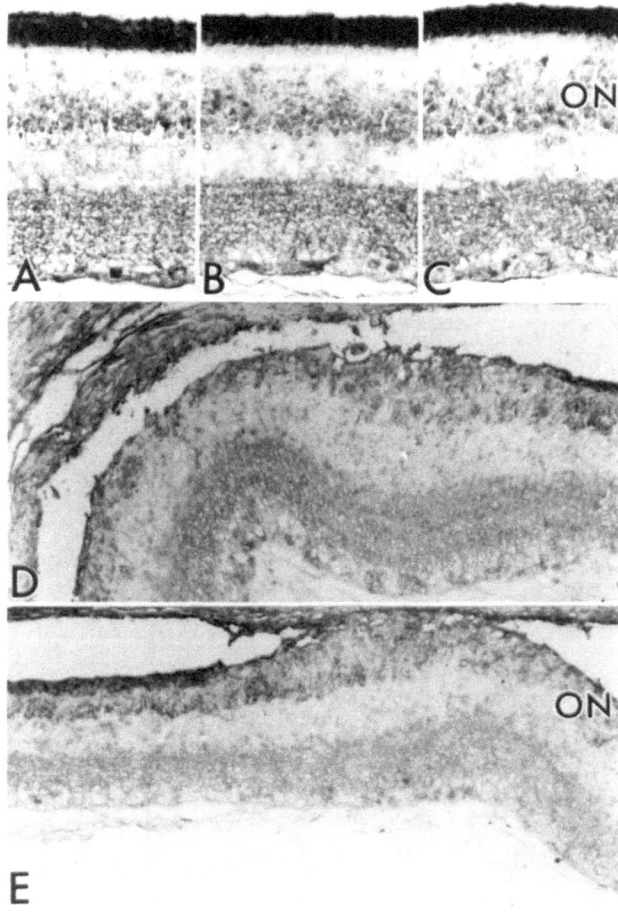

Figure 3. Jacalin binding in rat retinas. A: Control retina from a rat reared under cyclic lighting conditions; B: superior hemisphere of a retina exposed to constant light for 2 days; C: inferior hemisphere of a retina exposed to constant light for 2 days; D, E: superior hemisphere of a retina exposed to constant light for 3 days. ON: outer nuclear layer.

Figure 3 shows the binding sites of jacalin in the control (Fig. 3A) and experimental rat retinas (Fig. 3B-E). In the control retinas (Fig. 3A) and the retinas exposed to constant light for 2 days (Fig. 3B, C), jacalin bound intensely and weakly to the photoreceptor outer segments and the other layers including the ONL, respectively. The ONL on the superior equatorial region of the retinas from the rats exposed to constant light for 3 days did not show a decrease in jacalin binding in comparison with the surrounding regions, although the binding to the photoreceptors was decreased (Fig. 3D, E).

Figure 4 shows TUNEL-positive cells in the control (Fig. 4A) and experimental rat retinas (Fig. 4B-G). In the control retinas, no positive cell was detected (Fig. 4A). After 2 days of constant light exposure, TUNEL-positive cells were detected sparsely in the ONL of both superior and inferior hemispheres (Fig. 4B, C). Following 3 days of constant light exposure, entirely and relatively diffuse labelings were observed in the ONL of the superior and inferior hemispheres, respectively (Fig. 4D, E). After 5 days of constant light exposure, TUNEL-positive cells were still observed diffusely in the ONL of both superior and inferior hemispheres (Fig. 4F, G).

Figure 5 shows Ki 67 antigen-positive cells in the control (Fig. 5A) and experimental rat retinas (Fig. 5B-G). In the control retinas (Fig. 5A) and the retinas exposed to constant light for 2 days (Fig. 5B, 5C), positive cells were detected on the outer and inner

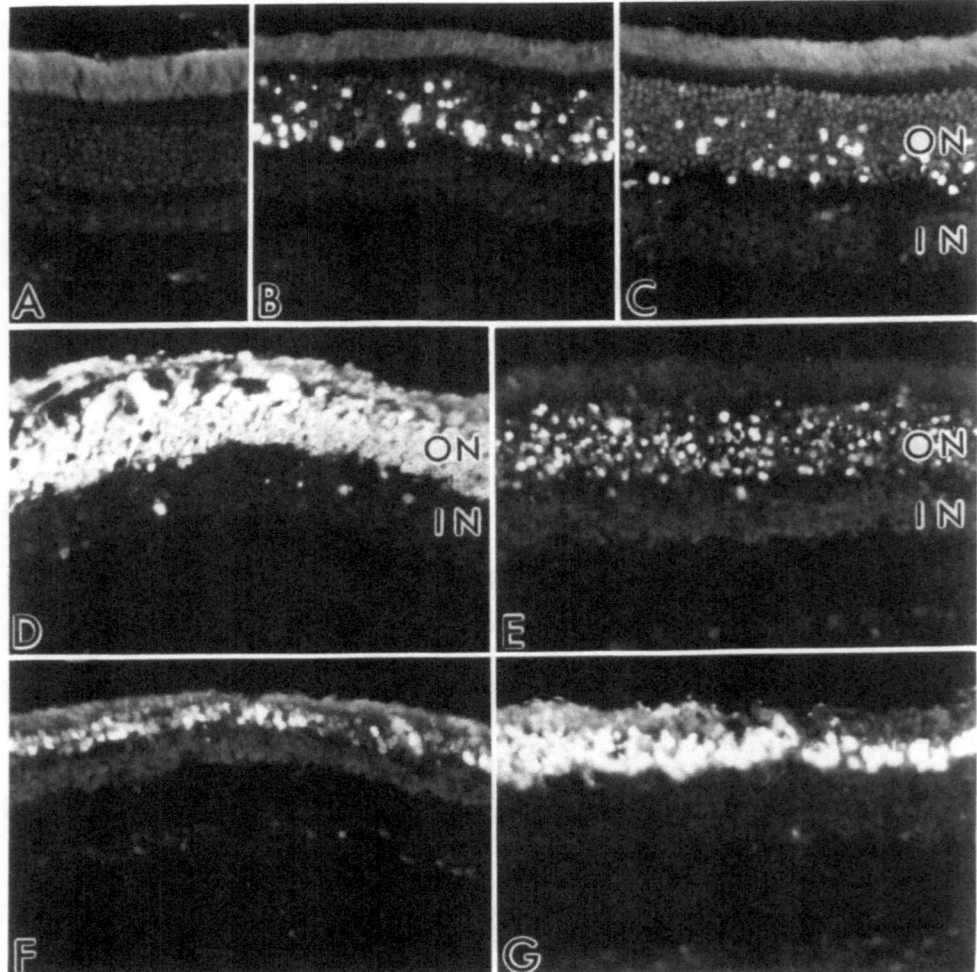

Figure 4. TdT-mediated dUTP-X nick end labeling of rat retinas. A: Control retina from a rat reared under cyclic lighting condition; B: superior hemisphere of a retina exposed to constant light for 2 days; C: inferior hemisphere of a retina exposed to constant light for 2 days; D: superior hemisphere of a retina exposed to constant light for 3 days; E: inferior hemisphere of a retina exposed to constant light for 3 days; F: superior hemisphere of a retina exposed to constant light for 5 days; G: inferior hemispere of a retina exposed to constant light for 5 days. ON: outer nuclear layer; IN: inner nuclear layer.

sides of the inner nuclear layer (INL) sparsely and intensely, respectively. Ki 67 antigen-positive cells increased in the INL and diffusely appeared in the ONL of the superior equatorial region after 3 days of constant light exposure (Fig. 5D). In the ONL of the inferior hemisphere at this stage, positive cells were sparsely detected (Fig. 5E). Following 5 days of constant light-exposure, the distribution of the antigen was still observed sparsely on both superior and inferior hemispheres (Fig. 5F, 5G).

Figure 6 consists of 5 lower magnification figures of retinas from rats exposed to constant light for 3 days to show the relation between different histochemical stainings (Fig. 6A-E). Figure 6A, B and C present the distributions of cytosolic sialidase, MAL II binding and jacalin binding, respectively. These figures show that the sialidase digested the terminal sialyl residues, which are required for MAL II binding but not for jacalin binding. Figure

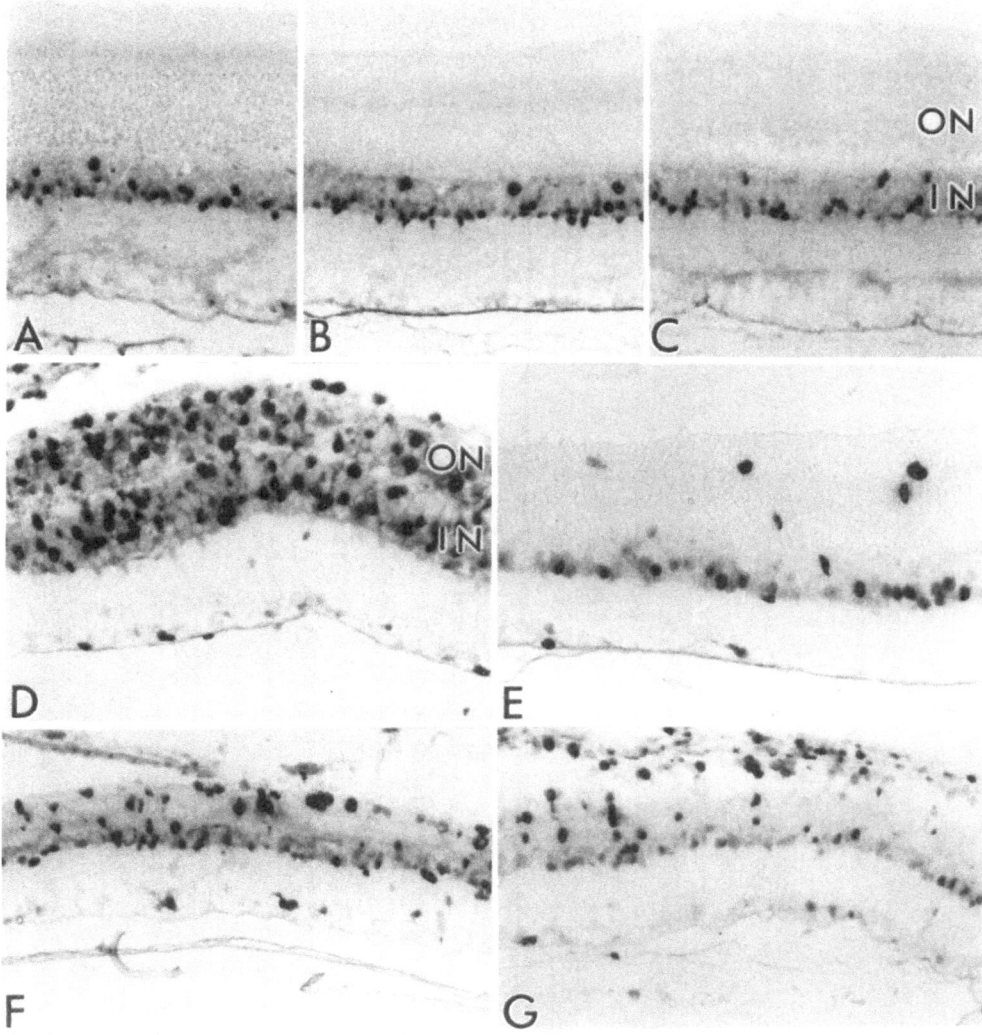

Figure 5. Ki 67 antigen-distribution of rat retinas. A: Control retina from a rat reared under cyclic lighting condition; B: superior hemisphere of a retina exposed to constant light for 2 days; C: inferior hemisphere of a retina exposed to constant light for 2 days; D: superior hemisphere of a retina exposed to constant light for 3 days; E: inferior hemisphere of a retina exposed to constant light for 3 days; F: superior hemisphere of a retina exposed to constant light for 5 days; G: inferior hemisphere of a retina exposed to constant light for 5 days. ON: outer nuclear layer; IN: inner nuclear layer.

6D shows that TUNEL-positive cells were present not only in the ONL of the equatorial region with destroyed photoreceptors, but also in that of the surrounding regions with relatively preserved photoreceptors. Figure 6E shows that Ki 67 antigen-positive cells were localized in the ONL and INL of the equatorial region with shortened photoreceptors.

DISCUSSION

The present histochemical findings show that the increase in the distribution of sialidase and the apoptotic change precede the proliferative response in the ONL. Although the

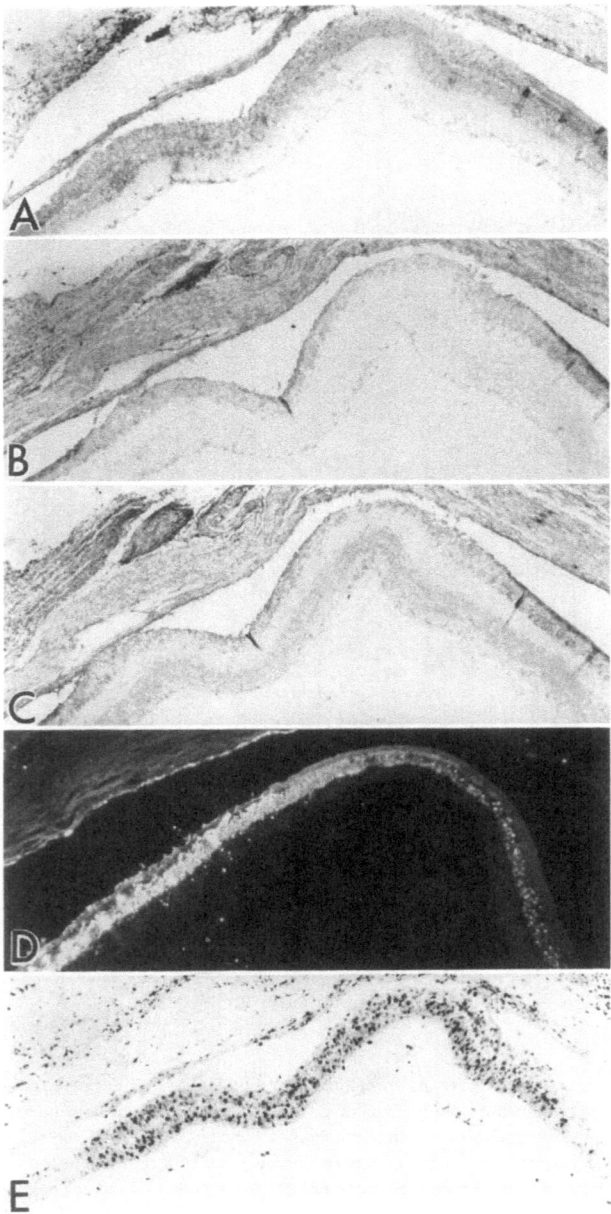

Figure 6. Histochemistry of the rat retinas following 3 days of constant light exposure. A: Immunohistochemical localization of cytosolic sialidase; B: *Maackia amurensis* lectin II binding; C: jacalin binding; D: TdT-mediated dUTP-X nick end labeling; E: immunohistochemical localization of Ki 67 antigen.

increase in sialidase and that in TUNEL-positive cells were simultaneously observed, it is natural to consider that the apoptotic change may induce the increase in sialidase as well as those of other enzymes including DNA endonuclease and cysteine proteases. In fact, the particle-like materials which were detected with the Ab to sialidase may be apoptotic bodies from their shrunk appearance. It is also noteworthy that these changes of the inner side are intenser than those of the outer side in the ONL. This finding may be due to the anatomical feature that rod photoreceptor cells are more densely present on the inner side of the ONL since cytosolic sialidase is selectively expressed in the rod cells (2). This may be one of the reasons why rod photoreceptor cells are more sensitive than cones to destruction by constant light (14). Cone cells may survive following long-term exposure to constant light due to lack of the removal system rmediated by sialidase. The superior hemisphere of the retina was more sensitive to the effects of light than the inferior hemisphere, also consistent with the previous observations (7). The increased sialidase may digest and remove terminal sialic acids from the glycoconjugates on the surfaces of the apoptotic cells, resulting in the decrease in their staining with MAL II, which recognizes the sialic acid α2,3 galactose sequence (11). The preservation of their staining in the ONL with jacalin, which is specific for galactose β1,3 N-acetylgalactosamine with or without terminal sialic acids (12), suggests that the galactosyl residues of the glycoconjugates on their surfaces may be intact. These sugar residues may be required for the next step of apoptosis described below.

Our preliminary immunohistochemical examination revealed that some of the Ki 67 antigen-positive cells accumulated in the ONL of retinas exposed to constant light for 3 days are macrophages. Since the surface of macrophages is usually negatively charged due to the distribution of sialoglycoconjugates, cells on which sialic acids are present are not recognized by macrophages (15). The mechanism by which apoptotic cells are rapidly removed by macrophages from a lesion has not yet been clarified. The present findings suggest that the sugar residues exposed on digestion with intrinsic increased sialidase may stimulate the accumulation of Ki 67 antigen-positive cells including macrophages. Apoptotic cells without sialic acids on their surfaces may be recognized by macrophages on which a C-type lectin specific for galactose/N-acetylgalactosamine is present (16). Through this process, the apoptotic cells may be removed from the lesion, resulting in the thinness of the ONL following 5 days of constant light exposure. Based on these findings, a model for the mechanism for the removal of apoptotic cells is proposed, as shown in Figure 7.

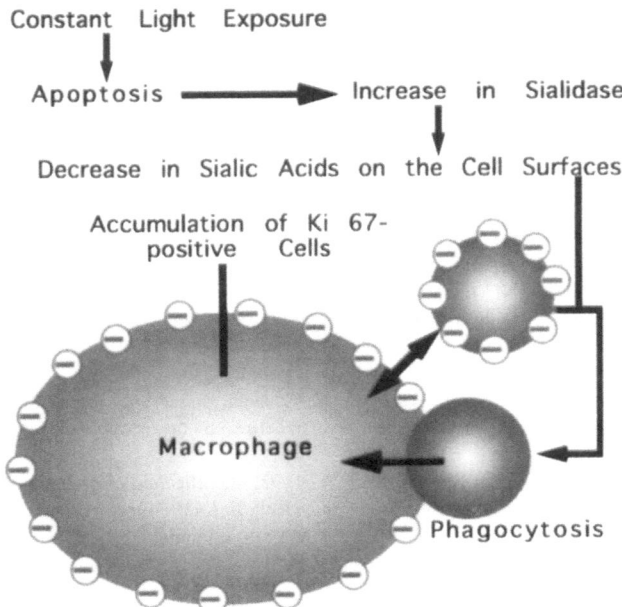

Figure 7. A model for the removal system of apoptotic cellsGeller et al. (17) showed that Ki 67 antigen-positive cells are increased in cat and rabbit retinas at 3 days after retinal detachment. Yi et al. (18) showed that macrophages are abundant in laser lesions of rat retinas on the 3rd day after intense laser photocoagulation. A similar biological response, of which the peak is observed around 3 days after the treatment, may occur in these different experimental disease conditions, including the present constant light exposure.

We also previously reported that intravitreous injection of sialidase induces a loss of the ONL with preservation of the inner retina (5). In the rat retina, we observed, on electron microscopy, that apoptotic changes are present in the ONL and phagosomes increase in RPE cells (19). This observation is interesting in that extrinsic sialidase may also induce photoreceptor cell loss. Both intrinsic increased- and extrinsic administered-sialidase may induce photoreceptor cell-degeneration through similar mechanism. It is possible that this process mediated by sialidase may be common not only to various types of photoreceptor cell degeneration of different causes, but also to apoptosis in other tissues and organs. It is also possible that some type of retinal degeneration may be directly induced by abnormal metabolism of sialic acids, an increase in sialidase or a decrease in sialyltransferase, as proposed previously (1, 3).

ACKNOWLEDGMENTS

This work was supported by a Grant-in-Aid for Scientific Research from the Japanese Ministry of Health and Welfare and a Grant-in-Aid from the Japan Medical Association.

REFERENCES

1. Uehara, F., 1993, Molecular cell glycobiology of the retina, *J. Jpn. Ophthalmol. Soc.* 97: 1370–1393.
2. Uehara, F., Ohba, N., Sameshima, M., Yanagita, T., Iwakiri, N., Ozawa, M. and Miyagi, T., 1996, Immunohistochemical localization of cytosolic sialidase in photoreceptor cells, *Jpn. J. Ophthalmol.* 40: 187–191.
3. Uehara, F., Ohba, N., Sameshima, M., Yanagita, T., Iwakiri, N. and Ozawa, M., 1996, Sialoglycoconjugates and retinal degeneration, in: *Retinal Degeneration and Regeneration* (S. Kato, N.N. Osborne and M. Tamai, eds.), pp. 73–80, Kugler Publications, Amsterdam.
4. Uehara, F., Ohba, N., Sameshima, M., Takumi, K., Unoki, K., Muramatsu, T., Yasumura, D. and LaVail, M.M., 1989, Nucleotide-induced retinal changes, in: *Inherited and Environmentally Induced Retinal Degenerations* (M.M. LaVail, R.E. Anderson and J.G. Hollyfield, eds.), pp. 577–584, Alan R Liss, New York.
5. Uehara, F., Ohba, N., Yasumura, D. and LaVail. M.M., 1990, Neuraminidase induced retinal changes, *J. Eye (Atarashii Ganka)* 7: 1701–1704.
6. Noell, W.K., Walker, V.S., Kang, B.S. and Berman, S., 1966, Retinal damage by light in rats, *Invest. Ophthalmol. Vis. Sci.* 5: 450–473.
7. LaVail, M.M., Gorrin, G.M., Repaci, M.A. and Yasumura, D., 1987, Light-induced retinal degeneration in albino mice and rats: Strain and species differences, in: *Degenerative Retinal Disorders, Clinical and Laboratory Investigations* (J.G. Hollyfield, R.E. Anderson and M.M. LaVail, eds.), pp. 430–454, Alan R Liss, New Yoyk.
8. Gerdes, J., Lemke, H., Baisch, H., Wacker, H.H., Schwab, U. and Stein, H., 1984, Cell cycle analysis of a cell proliferation-associated human nuclear antigen defined by the monoclonal antibody Ki-67, *J. Immunol.* 133: 1710–1715.
9. Gavrieli, Y., Sherman, Y. and Ben-Sasson, S.A., 1992, Identification of programmed cell death in situ via specific labeling of nuclear DNA fragmentation, *J. Cell Biol.* 119: 493–501.
10. Miyagi, T., Sagawa, J., Konno, K. and Tsuiki, S., 1990, Immunological discrimination of intralysosomal, cytosolic, and two membrane sialidase present in rat tissues, *J. Biochem.* 107: 794–798.
11. Sata, T., Lackie, P.M., Taatjes, D.J., Peumans, W. and Roth, J., 1989, Detection of the Neu5Ac (α2,3)Gal(β1,4)GlcNAc sequence with the leukoagglutinin from *Maackia amurensis*: Light and electron microscopic demonstration of differential tissue expression of terminal sialic acid in α2,3- and α-2,6-linkage, *J. Histochem. Cytochem.* 37: 1577–1588.
12. Hortin, G.L. and Trimpe, B.L., 1990, Lectin affinity chromatography of proteins bearing O-linked oligosaccharides: Application of jacalin-agarose, *Analytical Biochemistry.* 188: 271–277.
13. Uehara, F., Sameshima, M., Unoki, K., Okubo, A., Yanagita, T., Sugata, M., Iwakiri, N. and Ohba, N., 1994, *Maackia amurensis* lectin binding in developing rat retina, *Jpn. J. Ophthalmol.* 38: 364–367.

14. LaVail, M.M., 1976, Survival of some photoreceptor cells in albino rats following long-term exposure to constant light, *Invest. Ophthalmol.* 15: 64–70.

15. Soejima, T., Dantsuji, Y. and Nagayama, A., 1996, Function of macrophages, *Seitai no Kagaku.* 47: 248–252.

16. Sato, M., Kawakami, K., Osawa, T. and Toyoshima, S., 1992, Molecular cloning and expression of cDNA encoding a galactose/N-acetylgalactosamine-specific lectin on mouse tumoricidal macrophages, *J. Biochem.* 111: 331–336.

17. Geller, S.F., Lewis, G.P., Anderson, D.H. and Fisher, S.K., 1995, Use of the MIB-1 antibody for detecting proliferating cells in the retina, *Invest. Ophthalmol. Vis. Sci.* 36: 737–744.

18. Yi, X., Takahashi, K., Ogata, N. and Uyama, M., 1996, Immunohistochemical proof of origin of macrophages in laser photocoagulation lesion in the retina, *Jpn. J. Ophthalmol.* 40: 192–201.

19. Sameshima, M., Okubo, A., Uehara, F., Nakashima, Y., Sugata, M. and Ohba, N., 1994, Electron microscopic study of neuraminidase-induced retinal degeneration, *J. Jpn. Ophthalmol. Soc.* Suppl. 98: 129.

22

LIGHT-INDUCED RETINAL DEGENERATION IS PREVENTED IN MICE LACKING *c-fos*

Farhad Hafezi,[1][*] Andreas Marti,[1] Joachim P. Steinbach,[2] Kurt Munz,[2] Adriano Aguzzi,[2] and Charlotte E. Reme[1]

[1]Department of Ophthalmology
[2]Department of Neuropathology
University Hospital Zurich, Switzerland

SUMMARY

Although there is evidence that the protooncogene *c-fos* may be a mediator of apoptosis, its precise role is unclear (16). In the retina, *c-fos* is physiologically expressed in a diurnal manner and is inducible by light pulses (10,20). We observed light-elicited apoptosis in photoreceptors of pigmented $c\text{-}fos^{+/+}$ and $c\text{-}fos^{+/-}$ mice in a dose-dependent manner. To determine whether *c-fos* is essential in the chain of events culminating in apoptosis, we have studied light-induced retinal damage in mice lacking *c-fos*. After dark adaptation, $c\text{-}fos^{+/+}$, $c\text{-}fos^{+/-}$ and $c\text{-}fos^{-/-}$ mice show normal retinal morphology. After 2h of light exposure, all groups exhibited vesiculations of photoreceptor outer segments but very little apoptosis. After 12 additional hours in darkness, the rate of apoptosis increased dramatically in control mice but not in $c\text{-}fos^{-/-}$ mice. Therefore, *c-fos* is essential for delayed light-induced apoptosis of photoreceptors.

INTRODUCTION

A disorder of the apoptotic pathway is nowadays seen as an important factor contributing to a variety of diseases: an upregulation of apoptosis is observed in neurodegenerative diseases (AIDS, M. Alzheimer, M. Parkinson), a downregulation is a pathogenetic mechanism in malignant tumors. In contrast to necrosis, apoptosis is a genetically regulated form of cell death that does not damage adjacent tissue. A specific, yet largely unknown cascade of protein/protein interactions leads to the death of the cell. Morphologically, apoptosis appears as a condensation of nucleus and cytoplasm with a subsequent breakdown into the so called apoptotic bodies. Internucleosomal fragmentation of DNA is

---

* F.H. and A.M. contributed equally to the presented work.

a prominent feature which can be detected histochemically (TUNEL method) or by means of DNA isolation and agarose gel electrophoresis (19).

A growing number of gene products involved in the regulation of apoptosis has been identified in the past years (9,19). Recent evidence indicates that the protooncogene *c-fos* may be a mediator of apoptosis (4,6,8,11,16). *c-fos* upregulation was observed concomitant with apoptosis in several systems (2,8,14). The *c-fos* gene encodes a nuclear phosphoprotein that forms a heterodimeric complex with members of the Jun family of proteins to constitute the transcription factor AP-1 (activator protein 1) (5). Besides its potential role in apoptosis, a variety of other biological functions is also ascribed to *c-fos* such as regulation of proliferation and differentiation in a variety of cell types. Although several *in vitro* studies imply a role for *c-fos* during apoptosis, *in vivo* studies failed to prove that this factor is essential (6,11,15). In the retina, c-fos is expressed in a diurnal manner and is inducible by light (10,20).

Our lab and others have shown recently that diffuse, bright light induces apoptosis of retinal photoreceptors in a very synchronous manner in the rat retina (1,13) (and Hafezi et al., Exp Eye Res, submitted). Threshold intensities lead to a vesiculation of rod outer segments whereas light exposure for 2 h and sacrifice after additional 12h in darkness leads to an abundant apoptosis of photoreceptors and a disorganization of the outer retina.

To test whether *c-fos* is involved in the pathway leading to light-induced apoptosis of photoreceptors we used our model of retinal light damage in mice lacking the *c-fos* gene and compared them to control littermates (*c-fos*$^{+/+}$ and $^{+/-}$ mice).

MATERIALS AND METHODS

All procedures concerning animals in this study adhered to the ARVO resolution for the care and use of animals in Vision Research. Male *c-fos*$^{-/-}$ mice and controls were bred on a mixed C57BL/6 x SV129 background and maintained in a 12:12 light-dark cycle (lights on at 6 a.m.) with 10–20 lux within the cages. *c-fos*$^{-/-}$ and controls were dark adapted for 36h and killed or exposed to diffuse, cool white fluorescent light for 2h. At the end of light exposure, mice were killed immediately or following additional 12h in darkness. Eyes were rapidly enucleated under dim red light. For light microscopy, eyes were fixed in 2.5% glutaraldehyde and the upper and lower central retinae were trimmed under a dissecting microscope equipped with a red filter. For *in situ* detection of DNA strand breaks, eyes were fixed in 2% paraformaldehyde and embedded in paraffine. Tdt-mediated dUTP nick-end labeling (TUNEL) was performed with modifications using an "*in situ* cell death detection kit" (Boehringer Mannheim, Germany). DNA strand breaks were labeled with fluorescein-tagged dUTP and visualized with a FITC filter. For DNA fragmentation analysis, retinas of two animals were pooled. Extraction of total retinal DNA was performed as described (17). 5μg of total DNA was analyzed on a 1.8% agarose gel. DNA was visualized at 254 nm by staining with SYBR GREEN (Molecular Probes, The Netherlands) and compared to a 100 bp ladder molecular weight marker (Pharmacia Biotech, Uppsala, Sweden).

RESULTS

In a first step, we established a dose-response relationship of retinal light damage in control mice (data not shown). The threshold light dose to induce apoptosis of photoreceptors was 5'000 lux for 2 hours. We measured the transparency of the ocular optical media

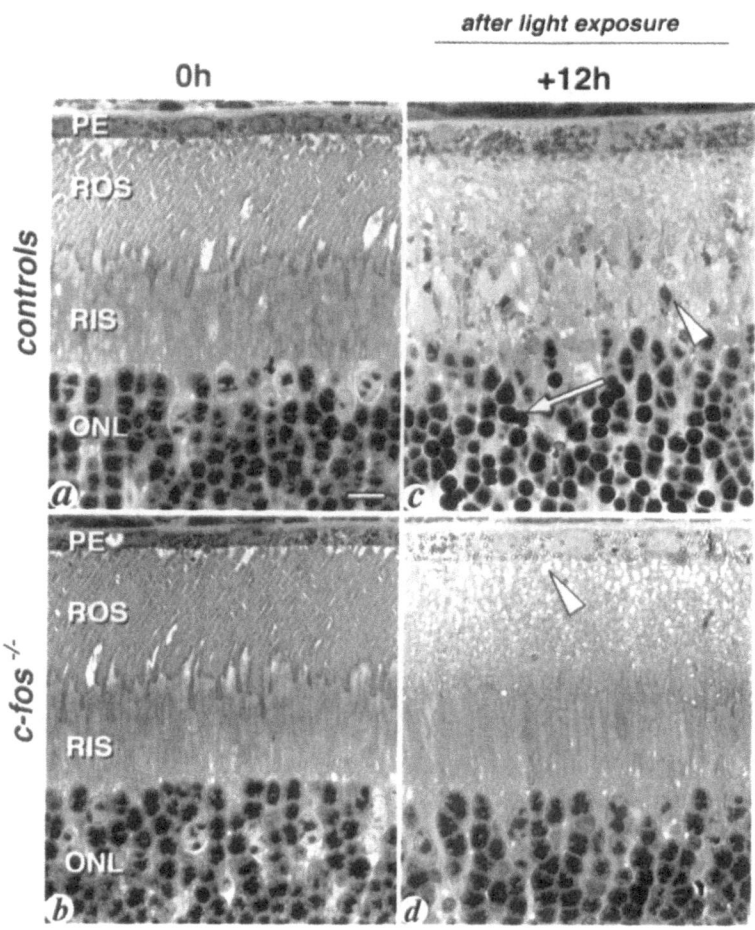

Figure 1. Light microscopy of retinal apoptosis in *c-fos*$^{-/-}$ mice (b,d) and controls (a,c). a,b) Dark adapted (0h) control mice (a) and *c-fos*$^{-/-}$ mice (b) show normal retinal morphology. c) Retina of a control mouse at 12h in darkness following the 2h light exposure. The great majority of the photoreceptor nuclei is condensed (arrow), the RIS displays condensed cytoplasm (arrowhead). ROS are deteriorated. d) Retina of a *c-fos*$^{-/-}$ mouse at 12h in darkness following the 2h light exposure. No apoptotic photoreceptor nuclei can be observed (representative fields were chosen). ROS display vesiculations of tips (arrowhead). ROS: rod outer segments. RIS: rod inner segments. ONL: outer nuclear layer. Scale bar 10µm.

and the light transmission of homogenized crystalline lenses measured by spectro-photometry. Both were found to be similar *c-fos*$^{-/-}$ mice and controls (*c-fos*$^{+/+}$ and *c-fos*$^{+/-}$ littermates) (data not shown). In all our experiments, no differences in the extent of light damage were observed between *c-fos*$^{+/+}$ and *c-fos*$^{+/-}$ control mice.

Dark adapted animals displayed regular retinal morphology (Fig. 1 a,b). We exposed unanesthetized, free moving *c-fos*$^{-/-}$ mice and control littermates to our light regimen of 5'000 lux for 2h. After 12 additional hours in darkness striking differences between knockouts and controls were observed. Apoptosis was abundant in controls (Fig. 1 c) but almost no apoptosis was observed in *c-fos*$^{-/-}$ mice (Fig. 1 d, table). The equal extent of ROS vesiculations confirmed equal retinal irradiance levels (Fig. 1 c,d).

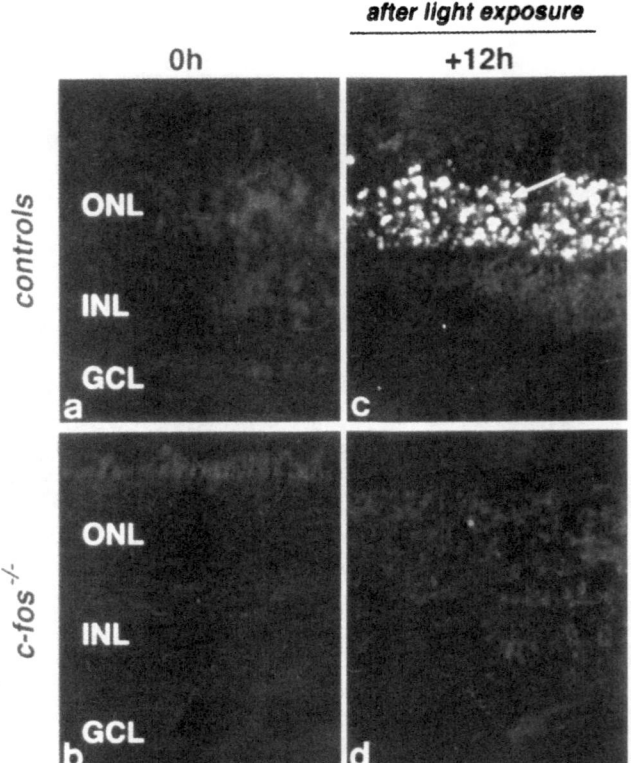

Figure 2. Detection of DNA strand breaks in photoreceptor nuclei in light microscopic sections by *in situ* nick end-labeling (TUNEL). (a,b) Dark adapted (0h) control mice (a) and *c-fos*$^{-/-}$ mice (b) show no TUNEL positive photoreceptor nuclei. (c,d) Analysis after 12h in darkness following the 2h light exposure. Retina of a control mouse (c) shows numerous positively stained photoreceptor nuclei in the ONL. (d) TUNEL positive photoreceptor nuclei are virtually absent in *c-fos*$^{-/-}$ mice (representative fields were chosen). ONL: outer nuclear layer. Scale bar 10μm.

TUNEL staining corroborated the morphological analysis. Retinas of dark-adapted animals of both groups remained negative (Fig. 2 a,b). Controls showed positively labeled photoreceptor nuclei carrying fragmented DNA after 12 h in darkness (Fig. 2 c) whereas retinas of *c-fos*$^{-/-}$ mice showed no staining (Fig. 2 d).

Quantitative analysis of apoptotic nuclei revealed a statistically highly significant difference between littermate controls and *c-fos* knockouts after 12 h in darkness (Table 1).

DISCUSSION

In this study we showed that c-Fos is an essential component for the regulation of delayed light-induced apoptosis of retinal photoreceptors *in vivo*. Alterations of ROS tips

Table 1. Counts of apoptotic photoreceptor nuclei in *c-fos*$^{-/-}$ and control mice 12h after light exposure

	c-fos$^{-/-}$	controls
After 12h	1.3 ± 1.0*	48.5 ± 11.4*

*p.001. Student's unpaired t-Test. Values are means ± s.d.

represent early and reversible signs of moderate light damage as described (12,18). They may be due to photochemically induced free radical formation with lesions of phospholipid membranes.

c-Fos and its family members (FosB, FosδB, Fra-1, Fra-2) need the simultaneous expression of Jun family members to constitute transcription factor AP-1 (5). However, Jun proteins do not depend on Fos expression. Depending on their composition, different AP-1 complexes may or may not show functional redundancy for one another (7). Our data suggest that, in the adult retina, the lack of the AP-1 constituent c-Fos cannot be compensated for by other AP-1 members during light-induced apoptosis. In contrast, redundancy may explain why apoptosis occur in the developing retina during normal histogenesis (21). Even though lacking the *c-fos* gene, *c-fos*$^{-/-}$ mice showed normal retinal morphology (Fig. 1 b). Similar findings were described during the morphogenesis of other organs showing where it was observed that developmental apoptosis occurs normally in *c-fos*$^{-/-}$ mice (15). To determine whether the lack of redundancy is a key factor for the observed protection in the absence of c-Fos, a detailed analysis of the AP-1 complex in the retina is needed.

CONCLUSION

In conclusion we report that c-Fos is an essential mediator of delayed light-induced apoptosis of photoreceptor cells in the adult retina. These findings provide a first step towards the understanding of mechanisms of light-induced apoptosis.

Furthermore, this is the first time that light damage could almost completely be suppressed *in vivo*. This finding may also be of particular interest for the pathogenesis of human degenerative diseases such as retinitis pigmentosa. This hereditary disorder is characterized by a dystrophy of photoreceptors with apoptosis as the final common pathway of cell death(3). Intriguingly, *c-fos* is induced concomitant with photoreceptor death in various mouse strains affected with retinal dystrophies (2,14). Suppression of *c-fos* might therefore be a means to prevent retinal light damage and prolong the time course of retinal dystrophies.

ACKNOWLEDGMENTS

We thank L. Vogel and C. Imsand for skilled technical assistance, T.P. Williams for critical reading of the manuscript and B. Gloor for continuous support. This work was supported by the Swiss National Science Foundation, grant no. 31–40791.94, Sandoz Foundation, Basel, Switzerland, Bruppacher Foundation, Zürich, Switzerland and Caroline and Ian Leaf and family.

REFERENCES

1. Abler, A.S., Chang, C.-J., Ful, J., Tso, M.O.M. and Lam, T.T., 1996, Photic injury triggers apoptosis of photoreceptor cells, *Res Comm Mol Pathol Pharmacol*, 92:177–189.
2. Agarwal, N., Patel, H., Brun, A.-M. and Nir, I., 1995, Alteration of c-fos by light/dark in rds mouse retina: possible involvement in apoptosis of photoreceptors, *Invest Ophthalmol Vis Sci*, 36:2921.
3. Chang, G.Q., Hao, Y. and Wong, F., 1993, Apoptosis: final common pathway of photoreceptor death in rd, rds, and rhodopsin mutant mice, *Neuron*, 11:595–605.
4. Colotta, F., Polentarutti, N., Sironi, M. and Mantovani, A., 1992, Expression and involvement of c-fos and c-jun protooncogenes in programmed cell death induced by growth factor deprivation in lymphoid cell lines, *J Biol Chem*, 267:18278–18283.

5. Curran, T. and Franza, B.R., Jr., 1988, Fos and Jun: the AP-1 connection, *Cell*, 55:395–397.
6. Estus, S., Zaks, W.J., Freeman, R.S., Gruda, M., Bravo, R. and Johnson, E.M., 1994, Altered gene expression in neurons during programmed cell death: identification of c-jun as necessary for neuronal apoptosis, *J Cell Biol*, 127:1717–1727.
7. Lord, K.A., Abdollahi, A., Hoffman Liebermann, B. and Liebermann, D.A., 1993, Proto-oncogenes of the fos/jun family of transcription factors are positive regulators of myeloid differentiation, *Mol Cell Biol*, 13:841–851.
8. Marti, A., Jehn, B., Costello, E., Keon, N., Ke, G., Martin, F. and Jaggi, R., 1994, Protein kinase A and AP-1 (c-Fos/JunD) are induced during apoptosis of mouse mammary epithelial cells, *Oncogene*, 9:1213–1223.
9. Morgan, J.I. and Curran, T., 1995, Immediate-early genes: ten years on, *TINS*, 18:66–67.
10. Nir, I. and Agarwal, N., 1993, Diurnal expression of c-fos in the mouse retina, *Brain Res Mol Brain Res*, 19:47–54.
11. Preston, G.A., Lyon, T.T., Yin, Y., Lang, J.E., Solomon, G., Annab, L., Srinivasan, D.G., Alcorta, D.A. and Barrett, J.C., 1996, Induction of apoptosis by c-Fos protein, *Mol Cell Biol*, 16:211–218.
12. Remé, C.E., Malnoë, A., Jung, H.H., Wei, Q. and Munz, K., 1994, Effect of dietary fish oil on acute light-induced photoreceptor damage in the rat retina, *Invest Ophthalmol Vis Sci*, 35:78–90.
13. Remé, C.E., Weller, M., Szczesny, P., Munz, K., Hafezi, F., Reinboth, J.J. and Clausen, M., Light-induced apoptosis in the rat retina in vivo: Morphological features, threshold and time course. In Anderson, R.E., LaVail, M.M. and Hollyfield, J.G. (Eds.), *Retinal Degeneration*, Plenum Press, New York, London, 1995, pp. 19–25.
14. Rich, K.A., Zhan, Y. and Blanks, J.C., 1994, Aberrant expression of c-fos accompagnies photoreceptor cell death in the rd mouse, *Invest Ophthalmol Vus Sci*, 35:1833.
15. Roffler-Tarlov, S., Gibson Brown, J.J., Tarlov, E., Storalov, J., Chapman, D.L., Alexiou, M. and Papaiannou, V.E., 1996, Programmed cell death in the absence of c-fos and c-jun, *Development*, 122:1–9.
16. Smeyne, R.J., Vendrell, M., Hayward, M., Baker, S.J., Miao, G.G., Schilling, K., Robertson, L.M., Curran, T. and Morgan, J.I., 1993, Continuous c-fos expression precedes programmed cell death in vivo, *Nature*, 363:166–169.
17. Strange, R., Li, F., Saurer, S., Burkhardt, A. and Friis, R.R., 1992, Apoptotic cell death and tissue remodelling during mouse mammary gland involution, *Development*, 115:49–58.
18. Szczesny, P.J., Munz, K. and Remé, C.E., Light damage in the rat retina: patterns of acute lesions and recovery. In Pleyer, U., Schmidt, K. and Thiel, H.J. (Eds.), *Cell and Tissue Protection in Ophthalmology*, Hippokrates Verlag, Stuttgart, 1995, pp. 163 - 175.
19. Wyllie, A.H., 1995, The genetic regulation of apoptosis, *Curr Opin Genet Dev*, 5:97–104.
20. Yoshida, K., Kawamura, K. and Imaki, J., 1993, Differential expression of c-fos mRNA in rat retinal cells:regulation by light/dark cycle, *Neuron*, 10:1049–1054.
21. Young, R.W., 1984, Cell death during differentiation of the retina in the mouse, *J Comp Neurol*, 229:362–373.

23

ISCHEMIC NEURONAL DEATH IN THE FISH RETINA

In Respect of Oxygen Radicals and Glutathione

S. Kato,[1][*] Z.-Y., Zhou,[1] K. Sugawara,[1] Y. Yasui,[1] N. Takizawa,[1] K. Sugitani,[2] and K. Mawatari[2]

[1]Department of Neurobiology
NIRI
[2]Department of Laboratory Sciences University of Kanazawa
Faculty of Medicine
Kanazawa 920, Japan

INTRODUCTION

It has been postulated that ischemia leads to a loss of cellular homeostasis and a shortage of available adenosine triphosphate (ATP) (1–3). This causes a release of glutamate and other mediators, which results in a rapid cellular efflux of potassium and influxes of sodium, calcium, and chloride with obligated water. The neuronal damage is probably secondary to the influx and mobilization of calcium and activation of a variety of enzymes, such as protein kinases, proteases, and lipases (1,4,5). Glutamate N-methyl-D-aspartate (NMDA) receptors are thought to play a major role in the damaging effects elicited by an ischemic insult because the activation of these receptors results in an influx of calcium (1,3,6). Recent works have reported the involvement of oxygen radicals in the NMDA or ischemia induced neurotoxicity (7–9). However, such a direct evidence involving oxygen radicals during ischemia in vivo has not been studied. In the present study we first tested an availability of isolated fish retina to the ischemia in vitro. The nature of the ischemic insult to the retina in vitro can be precisely controlled by eliminating the glucose (hypoglycemia), oxygen (anoxia) or glucose and oxygen (ischemia) from the medium. Moreover, drugs can be added to the medium to reach the retina without having to cross barriers. We have used fish retinas in our neurobiological study (10–12). In spite of the useful and unique advantages (10,12), very few study of ischemia with fish retina has been conducted. In such a line, we measured

---

* Correspondence should be addressed to S. Kato at Department of Neurobiology, NIRI, University of Kanazawa, 13-1 Takara-machi, Kanazawa, Ishikawa 920, Japan. TEL/FAX: +81 762 34 4235; E-mail: satoru@med.kanazawa-u.ac.jp.

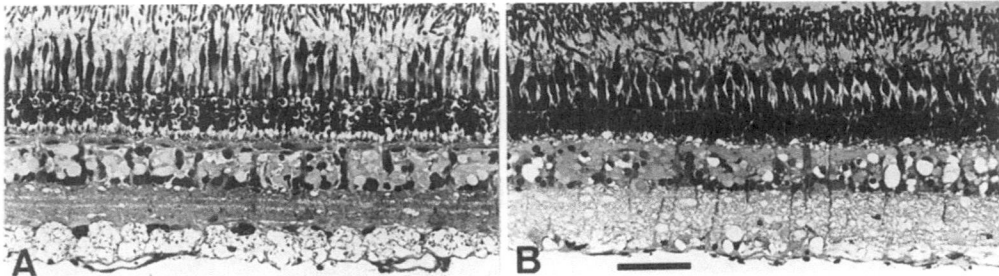

Figure 1. Morphological changes of isolated carp retina after ischemia. A: Retina perfused in normal fish Ringer solution with glucose and oxygen for 5 hrs. B: Retina perfused in ischemic (hypoglycemia+anoxia) condition for 2 hrs. Note neuronal damages (edema and vacuolization) of inner retina. Scale bar=100 mm.

an accumulation of reactive oxygen species (ROS) and a cellular level of glutathione (GSH), which is a major antioxidant against oxidative stress, after ischemia in vitro with the isolated carp retina. We further investigated some metabolic respects of GSH, particularly the synthetic pathway of GSH from a precursor cysteine.

ISCHEMIC INSULT IN VITRO TO THE ISOLATED FISH RETINA

We first tested an availability of our experimental condition as an in vitro model of retinal ischemia. We estimated it by using both morphological and biochemical (ATP) parameters during ischemic treatment.

Morphological Changes during Retinal Ischemia

When isolated carp retina was treated with ischemic condition (hypoglycemia+anoxia) for 2 hrs, a characteristic neuronal damage could be seen. Edematous changes of inner retina with multiple vacuolizations were conspicuous in the ischemic condition (Fig. 1B and Table 1A) as compared to the normal retina which was perfused in normal fish Ringer solution with oxygen for 5 hrs (cf. Fig. 1A). The similar morphological features were induced at a similar time course of ischemia in vivo by occlusion of retinal vessels of mammal (13,14).

Biochemical Changes during Retinal Ischemia

The ATP depletion was an important biochemical sign of ischemia. We measured cellular levels of ATP after ischemic condition. The cellular level of ATP was completely depleted less than one tenth of control 3 hrs after ischemic condition (Table 1B). A sig-

Table 1. Ischemic insults to the isolated carp retina

A.	Morphology	Neuronal damages of inner retina
B.	Adenosine triphosphate (ATP)	Depletion of cellular level of ATP
C.	Reactive oxygen species (ROS)	Increase of ROS accumulation
D.	Glutathione (GSH) content	Decrease of cellular level of GSH
E.	GSH synthesis	Inhibition of GSH synthesis from cysteine
	Localization of [35S] cysteine uptake	Neurons of inner retina

nificant decrease of concentration of ATP was initiated in early (1 hr) exposure of ischemic condition. These data with the isolated fish retina guaranteed a valuability of our retinal preparation as an in vitro model of ischemia.

GENERATION OF REACTIVE OXYGEN SPECIES DURING ISCHEMIA

To clarify the involvement of oxygen radicals in the ischemic insults to the fish retina, we measured an accumulation of reactive oxygen species (ROS) after ischemia. The ROS accumulation was estimated utilizing a converting reaction of 2,7-dichlorofluoresin diacetate (DCFH-DA) to 2,7-dichlorofluorescein (DCF) oxidized by ROS (15). The DCF formation in the fish retina was increased 2.2 fold 4 hrs after ischemia (Table 1C). The rate of DCF formation was further enhanced 3.2 fold by reperfusion of normal Ringer solution with oxygen.

GLUTATHIONE METABOLISM DURING ISCHEMIA

Glutathione (GSH) is a major non protein SH compound and plays a key role in the redox state of cells (16). As the above data indicated a generation of oxygen radicals during ischemia, we next examined cellular level of GSH and its synthesis during ischemia.

GSH Content

When isolated carp retina was treated with ischemia for 5 hrs, a significant or large reduction of cellular level of GSH occurs. The retinal GSH was reduced to 70% or 60% of control retina at 2 hrs or 5 hrs after ischemia, respectively (Table 1D). In the normal retina, no significant reduction of GSH could be detected even in a 5 hr-incubation of normal Ringer solution with oxygen.

GSH Synthesis from Cysteine

The synthesis of GSH is largely dependent on the cellular level of cysteine. The cysteine is introduced into the cells through two pathways. One source is cysteine directly taken up via ASC transport system and the other source is cystine taken up via x$\bar{\text{c}}$ antiporter system (17). The cystine taken up is readily reduced to cysteine. Our previous autoradiographic study suggests that the localization of x$\bar{\text{c}}$ antiporter in the carp retina is limited to glial Müller cells with [^{35}S]cystine (18). In such a similar line, we examined incorporating cells in the carp retina with [^{35}S]cysteine. Some neurons of inner retina incorporated with [^{35}S]cysteine could be seen in autoradiography (Table 1E). We further examined GSH synthesis from cysteine. Carp retinas were incubated with 5 mM [^{35}S]cysteine for 30 min to 4 hrs. The separation of cysteine and GSH was performed by a HPLC method with sulfhydryl group specific fluorogenic reagents (19). The incorporation of [^{35}S]cysteine into GSH was measured by counting the radioactivity. A significant synthesis of GSH from [^{35}S]cysteine could be seen 1 hr after incubation.

GSH Synthesis during Ischemia

In the morphological features after ischemic treatment, neuronal damages (edematous changes with vacuolizations) of inner retina were characteristic (Fig. 1). The autoradiographic study of [^{35}S]cysteine uptake also showed active incorporating neurons of inner retina (Table 1E). Therefore, we investigated effects of cysteine on GSH content during ischemia. Adding 1 mM cysteine did not restore the cellular level of GSH after ischemia (Table 1E). On the other hand, cellular level of GSH was recovered to control by adding 1 mM cysteine in the presence of glucose.

DISCUSSION

In Vitro Study of Ischemia

In this study, we aimed to answer two questions whether the ischemic insults in vitro to the retina have an equivalency with those in vivo to the retina and whether free radicals involve in the neuronal death induced by ischemia. Moreover, how does the antioxidant glutathione behave during ischemia? We used the isolated fish retina throughout this experiment. The fish retina remains viable for a longer time period than that of mammals (10,20). The amplitude of electroretinographic b-wave did not deteriorate for 5–6 hrs in the isolated carp retina in a similar normal condition (10). The morphological and biochemical (ATP) data showed typical neuronal damage of inner retina and ATP depletion in a short time (1–3 hrs) of ischemic treatment. Therefore, we concluded the ischemic condition with isolated carp retina as a sensitive and useful in vitro model of ischemia.

Accumulation of Oxygen Radicals during Ischemia

The increase rate of formation of DCF suggests a direct evidence of accumulation of reactive oxygen species (ROS) after ischemic treatment. Three kinds of ROS, O_2, H_2O_2 and $\cdot$ OH, are generated from the four-steps reduction of molecular oxygen. These reactions are catalyzed by transition metal ions such as iron, which provide electrons at each step (21). This type of chain reaction and sequent generation of cytotoxic ROS may promote the ischemic damage of fish retina. The reperfusion (reoxygenation) experiment with further increase of DCF formation in this study can warrant a useful model of ischemia or reperfusion. The study of toxic and protective mechanism for retinal ischemia is now in progress using this in vitro preparation (21–24).

Glutathione Synthesis from Cysteine during Ischemia

The cellular level of glutathione (GSH) was significantly reduced to 70% of control 2 hrs after ischemia. Although GSH is a ubiquitous molecule in the living cells, the concentration gradients in the CNS including the retina has not yet been studied. A recent work has reported in the rabbit retina that glial Müller cells and some neurons of inner retina have a high concentration of GSH (25). The reduction of cellular level of GSH during ischemia in this study is supposed to mainly reflect a reduction of neuronal GSH because of the neuronal target of ischemia. Furthermore, we clarified for the first time two dissociative glial and neuronal synthetic pathways of GSH in the retina with radioactive cystine and cysteine, respectively. The former is predominantly utilized cystine as a precursor of GSH synthesis, while the latter cysteine. Glial Müller cells have a rich expression of cystine/glutamate antiporter (x$\bar{c}$) system (18), whereas neurons have an only uptake system for cysteine via ASC transport system in this study (Fig. 2) (26). The retinal GSH synthesis from cysteine was completely inhibited by ischemic condition. A full restoration of cellular level of GSH in the presence of glucose suggests that the complete inhibition of

Glial cells   Neuronal cells

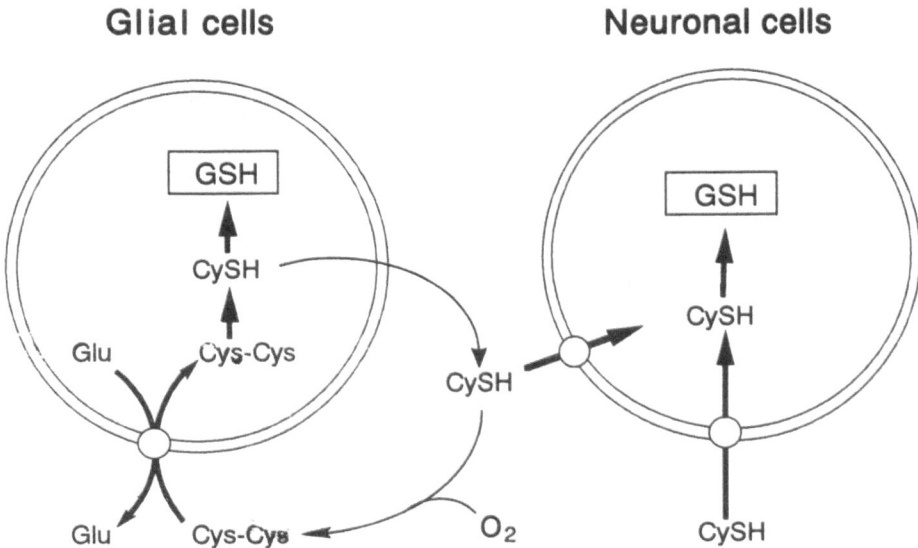

Figure 2. A dissociation of glutathione (GSH) synthesis from cystine (Cys-Cys) or cysteine (CySH) in the retinal glial or neuronal cells. The cystine is taken up to the glial cells through xc antiporter system, while the cysteine is taken up to the neuronal cells through ASC transport system. Details in text.

GSH synthesis from cysteine during ischemia ascribes to ATP depletion by ischemia. Both the accumulation of oxygen radicals (ROS) and inhibition of GSH synthesis from precursors by ischemia may further accelerate consumption of GSH and promotion of neuronal death during ischemia.

ACKNOWLEDGMENT

We thank Mr. S. Ishita and Mrs. Tami Urano for his technical and her secretarial assistance. This work was supported in part by research grants to SK from the Ministry of Education, Science and Culture of Japan.

REFERENCES

1. Choi, D.W., and Rothman, S.M., 1990, The role of glutamate neurotoxicity in hypoxic ischemic death, Ann. Rev. Neurosci. 13: 171–182.
2. Rothman, S.M., and Olney, J.W., 1986, Glutamate and the pathophysiology of hypoxic/ischemic brain damage, Ann. Neurol. 19: 105–111.
3. Sieso, B.K., 1992, Pathophysiology and treatment of focal cerebral ischemia; II. Mechanisms of damage and treatment, J. Neurosurg. 77: 337–354.
4. Meldrum, B., and Garthwaite, J., 1991, Excitatory amino acid neurotoxicity and neurodegenerative disease, in: The Pharmacology of Excitatory Amino Acids, (O. Lodge, and G. Collingridge, eds.), pp.54–62, Elsevier, Cambridge.
5. Hara, H., Sukamoto, T. and Kogure, K., 1993, Mechanism and pathogenesis of ischemia-induced neuronal damage, Prog. Neurobiol. 40: 645–670.
6. McBain, G.J, and Mayer, M.I., 1994, N-methyl-D-aspartate acid receptor structure and function, Physiol. Rev. 74: 723–759.
7. Lafon-Cazal, M., Pietri, S., Culcast, M. and Bockaert, J., 1993, NMDA-dependent superoxide production and neurotoxicity, Nature 364: 535–537.

8. Yue, T.L., Gu, J.L., Lysko, P.G., Cheng, H.Y., Barone, F.C., and Fenerstein, G., 1992, Neuroprotective effects of phenyl-t-butyl-nitrone in gerbil global brain ischemia and in cultured rat cerebellar neurons, Brain Res. 574: 193–197.

9. Olanow, C.W., 1994, A radical hypothesis for neurodegeneration, Trends Neurosci. 16: 439–444.

10. Kato, S., and Negishi, K., 1978, Effects of variations in the perfusate on the ERG and discharge of ganglion cells in carp retina, Exp. Eye Res. 26: 363–376.

11. Negishi, K., Teranishi, T., and Kato, S., 1989, The dopamine system of the teleost fish retina, Prog. Retinal Res. 9: 1–48.

12. Kato, S., Negishi, K., Teranishi, T., and Ishita, S., 1991, The use of the carp retina in neurobiology, Prog. Neurobiol. 37: 287–327.

13. Szabo, M.E., Droy-Lefax, M.T., Doly, M., and Braquet, P., 1991, Free radical-mediated effects in reperfusion injury: a histologic study with superoxide dismutase and EGB761 in rat retina, Ophthalm. Res. 23: 225–234.

14. Faberowski, N., Stefansson, E., and Davidson, R.C., 1989, Local hypothermia protects the retina from ischemia: a quantitative study in the rat, Invest. Ophthalmol. Vis. Sci. 30: 2309–2313.

15. Lebel, C.P., and Bondy, S.C., 1990, Sensitive and rapid quantitation of oxygen reactive species formation in rat synaptosomes, Neurochem. Int. 17: 435–440.

16. Meister, A., 1988, Glutathione metabolism and its selective modification, J. Biol. Chem. 263: 17205–17208.

17. Bannai, S., 1986, Exchange of cystine and glutamate across plasma membrane of human fibroblasts, J. Biol. Chem. 261: 2256–2263.

18. Kato, S., Ishita, S., Sugawara, K., and Mawatari, K., 1993, Cystine/glutamate antiporter expression in retinal Müller glial cells: implications for DL-a-aminoadipate toxicity, Neuroscience 57: 473–482.

19. Kato, S., Negishi, K., Mawatari, K., and Kuo, C.H., 1992, A mechanism for glutamate toxicity in the C6 glioma cells involving inhibition of cystine uptake leading to glutathione depletion, Neuroscience 48: 903–914.

20. Osborne, N.N., and Herrera, A.J., 1994, The effect of experimental ischemia and excitatory amino acid agonists on the GABA and serotonin immunoreactivity in the rabbit retina, Neuroscience 59: 1071–1081.

21. Olanow, C.W., 1992, An introduction to the free radical in Parkinson disease, Ann. Neurol. 32: Suppl. S2–S9.

22. Shinagawa, S., 1994, Serotonin protects C6 glioma cells from glutamate toxicity, Neuroscience 59: 1043–1050.

23. Mawatari, K., Yasui, Y., Sugitani, K., Takadera, T., and Kato, S., 1996, Reactive oxygen species involved in the glutamate toxicity of C6 glioma cells via xc antiporter system, Neuroscience 73: 201–208.

24. Kato, S., Mawatari, K., Sugitani, K., and Yasui, Y., 1996, DL-a-Aminoadipate is a toxin to Müller cells, Prog. Retinal and Eye Res. in press.

25. Pow, D.V., and Crook, D.K., 1995, Immunocytochemical evidence for the presence of high levels of reduced glutathione in radial glial cells and horizontal cells in the rabbit retina, Neurosci. Lett. 193: 25–28.

26. Sagara, J., Miura, K., and Bannai, S., 1993, Maintenance of neuronal glutathione by glial cells, J. Neurochem. 61: 1672–1676.

24

ISOLATION OF CANDIDATE GENES FOR RETINAL DEGENERATIONS

George Inana,[1] Akira Murakami,[1] Hitoshi Sakuma,[1] Tomomi Higashide,[1] Toshihiro Yajima,[1] and Margaret J. McLaren[2]

[1]Bascom Palmer Eye Institute
Department of Ophthalmology
[2]Department of Physiology and Biophysics
University of Miami School of Medicine
Miami, Florida

INTRODUCTION

The indentification of the molecular etiology is the first key step in understanding inherited retinal degenerations. Knowledge of the causative gene allows one to begin to investigate the mechanism by which the gene defect leads to the retinal degeneration. Understanding the pathophysiological mechanism of the disease may lead to finding the best treatment or cure for the retinal degeneration.

Isolation and characterization of retina-expressed genes have contributed greatly towards the identification of genes that can cause retinal degenerations when mutated. In fact, knowledge on isolated and characterized retinal genes plays a key role in the "positional candidate" approach to identifying genes responsible for inherited retinal degenerations. The positional candidate approach is the most successful strategy at this time to identify disease-causing genes, and it combines the results of mapping of genetic diseases and characterized genes to specific chromosomal loci. If a retina-expressed gene, especially one encoding a protein that is most likely important for the function of the retina, maps to the same chromosomal locus as an inherited retinal degeneration, the gene is considered a candidate causative gene for the disease and screened. This strategy has led to successful identification of a number of pathogenic genes for retinal degenerations, including rhodopsin for autosomal dominant retinitis pigmentosa (ADRP) (1), α and β subunits of cGMP phosphodiesterase for autosomal recessive retinitis pigmentosa (ARRP) and autosomal dominant stationary blindness (2, 3), TIMP3 for Sorsby's fundus dystrophy (4), type VII myosin for Usher syndrome type 1B (5), S-antigen for Oguchi disease (6), and guanylate cyclase for Leber's congenital amaurosis (Kaplan et al., VIIth International Symposium on Retinal Degeneration, October 1996, Sendai, Japan).

As successful as the positional candidate approach has been in identifying retinopathy-causing genes, the approach is heavily dependent on the availability of information

Degenerative Retinal Diseases, edited by LaVail *et al.*
Plenum Press, New York, 1997

on well-characterized retinal genes. The tremendous genetic heterogeneity of inherited retinal degenerations, well domonstrated in the cases of RP and Usher syndrome, also dictates that there are many different retinal genes that can cause these diseases when mutated. Yet, the number of isolated, characterized retina-specific genes is limited, such that investigators are running out of retinal genes to screen as candidate genes. Isolation and study of new retinal genes would also help us to further understand the biology and physiology of the retina, in addition to providing additional candidate genes for retinal degenerations. For these reasons, we have been isolating retina-enriched and retina-specific genes using a differential cDNA cloning strategy and studying them, including considering them as candidate genes for inherited retinal degenerations. Described below are the results on the first four human retinal genes we have isolated: rom-1, recoverin, X-arrestin, and HRG4.

METHODS

Preparation of Retina-Enriched cDNA Library

A retina-enriched cDNA library was prepared as previously described (7). Briefly, mRNA was isolated from the human retina, converted to double-stranded cDNA, subtracted with biotinylated fibroblast cDNA several times using streptavidin in combination with polymerase chain reaction (PCR) amplification, and cloned into pBluescript (Statagene, La Jolla, CA). Approximately three hundred recombinant clones were isolated. The present retina-specific clones were isolated in the initial analysis of 30 clones.

Southern and Northern Hybridization

High-molecular-weight genomic DNA was digested with restriction enzymes, electrophoresed in 0.8% agarose gel, transferred onto nitrocellulose or nylon filters by the Southern method (8), and hybridized to [^{32}P]-labeled cDNA probe. The hybridized filters were washed and autoradiographed as described (9). RNA was isolated from tissue or cells using guanidine thiocyanate (10). Human brain, liver, and lung RNA were obtained from Clonetech (Palo Alto, CA). The RNA was electrophoresed in denaturing agarose gel (11), transferred onto nylon membranes by the Southern blotting method (8) and hybridized with [^{32}P]-labeled DNA probe. The hybridized blots were washed and autoradiographed as described before (9). The quantity and quality of RNA on the blot were checked by hybridization with actin cDNA probe.

Screening of cDNA Library

A human retinal λgt11 cDNA library (12) was screened with [^{32}P]-labeled cDNA originally obtained from the retina-enriched library in order to isolate a full-length clone by standard procedure (9).

Isolation and Characterization of Genomic Clones

Approximately 10^6 clones from a cosmid (pWE 15) and a phage (Lambda Fix II) human placenta genomic library (Stratagene, La Jolla, CA) were screened with the human cDNA probe labeled with [^{32}P] (13), and positive clones were isolated by standard procedure (9). The gene clones were analyzed by restriction mapping with various restriction enzymes. Specific gene fragments were subcloned into pBluescript KS for sequencing and further analysis.

Chromosomal Mapping

DNAs from hamster-human somatic cell hybrids containing specific human chromosomes were used in genomic Southern analysis with the [^{32}P]-labeled cDNA probe (Bios, New Haven, CT). The hybridization pattern observed in a panel of somatic cell DNAs indicated the chromosomal location of the gene.

DNA Sequencing

DNA was sequenced by the dideoxy chain termination method using the Sequenase DNA sequencing kit (14) (United States Biochemical, Cleveland, OH). Sequences were analyzed by the IntelliGenetics (Mountain View, CA) and Genetic Computer Group (Madison, WI) software packages.

Sublocalization of X-Arrestin on the X Chromosome

High-molecular-weight genomic DNA from rodent-human somatic hybrids containing different segments of the human X chromosome (gift of Dr. T. Mohandas) was digested with Eco RI, electrophoresed in 0.8% agarose gel, transferred onto nylon filter by the Southern method (8), and hybridized to [^{32}P]-labeled cDNA probe. The hybridized filters were washed and autoradiographed as described (9). The accuracy of the DNA panel was checked by the hybridization of a known X chromosome marker, TIMP (ATCC, Rockville, MD), to the blot which showed the correct localization to the Xp region (data not shown).

Preparation of X-Arrestin Anti-Peptide Antibody

The human X-arrestin sequence was analyzed for a unique region showing high antigenicity by GenAlign and Plotstructure computer programs, respectively (GCG, Madison, WI). A 13 amino acid sequence including an arbitrarily-added cysteine (QKAVEAEGDEGSC, residues 377–388) was chosen from the carboxy terminus, and the peptide was synthesized and used to prepare an anti-peptide antibody as described (9). Briefly, 5 mg of peptide was coupled to keyhole limpet hemocyanin using m-maleimidobenzoic acid N-hydroxysuccinimide ester (MBS), purified through Sephadex G-25, mixed with Freund's complete adjuvant, and used for immunization of a New Zealand white rabbit. Every 2–3 weeks thereafter, the rabbit was immunized with a half dose of antigen in incomplete Freund's adjuvant, and serum was obtained. The antiserum, along with preimmune serum, was tested for antibody activity by Western blot analysis. All use of animals (rabbits and rats) was in accordance with the guidelines established in the ARVO Statement for the Use of Animals in Ophthalmic and Vision Research.

The antiserum was purified by affinity chromatography as described (9). Briefly, an affinity column was prepared by coupling bovine serum albumin to the synthetic peptide using MBS, purifying it through Sephadex G-25, then coupling it to CNBr-activated Sepahrose 4B (Pharmacia Biotech, Piscataway NJ). The antiserum was passed through the affinity column, and the bound antibody was eluted with glycine after washing. IgG concentration was estimated by spectrophotometry (OD$_{280}$).

Western Blot Analysis

Retinas were obtained from human Eye Bank eyes, homogenized in single-detergent-lysis-buffer (50mM Tris.HCl, pH8.0, 150mM NaCl, 0.02% sodium azide, 100μg/ml phenylmethylsulfonylfluoride, 1µg/ml aprotinin, 1% Triton X-100), and centrifuged to prepare retinal extract. Protein concentration was determined with Micro Protein Determination (Sigma, St. Louis, MO). Ten µg of retinal extract was subjected to sodium dodecyl sulfate-polyacrylamide gel electrophoresis (SDS-PAGE), and electro-blotted onto an Immobilon membrane (Millipore, Bedford, MA). The membrane was blocked with non-fat dry milk, reacted with the whole antiserum or affinity-purified antibody (50–1000x dilution, 0.2- 2.0 µg/ml, respectively), washed and reacted with biotinylated goat anti-rabbit IgG antibody. Bound antibody was visualized by reaction with streptavidin-alkaline phosphatase conjugate (GIBCO BRL, Gaithersburg, MD), followed by washing and treatment with nitroblue tetrazolium chloride and 5-bromo-4-chloro-3-indolylphosphate-p-toluidine-salt for color development.

Immunohistochemistry

Human Eye Bank eyes and pigmented (Long Evans) rat eyes were fixed in 4% paraformaldehyde in 0.1 M phosphate buffer, pH 7.4, and dissected portions of retina-lined eyecups were embedded in O.C.T. compound (Miles, Inc., Elkhart, IN). Cross sections of retina (6 µm thickness) were prepared on a cryostat and used for immunostaining. Sections were first post-fixed in 4% paraformaldehyde, washed in phosphate buffered saline (PBS) and blocked with 10% goat serum in PBS, then incubated for 1 hr at room temperature with one of the following antibodies: 1) affinity-purified rabbit anti-human X-arrestin peptide antibody (5 µg/ml); 2) rabbit antibody generated against a 22-residue peptide in the N-terminal extension of human red and green cone opsins (15, 16), 1:100 dilution; 3) MAbA9-C6 anti-bovine S-antigen monoclonal antibody (17), 1:100 dilution, or 4) G26c anti-bovine rhodopsin monoclonal antibody (18), 1:10 dilution. Following washing, sections were incubated with FITC-labeled goat anti-rabbit IgG, FITC-GARG (50x dilution; Organon Tecknika, Durham, NC) for the X-arrestin and red/green cone opsin antibodies, and FITC-labeled goat anti-mouse IgG, FITC-GAMG (50x dilution; BRL, Gaithersburg, MD) for the rhodopsin and S-antigen monoclonal antibodies. Slides were then washed and mounted in glycerol/PBS.

For double immunofluorescent histochemistry, human retinal sections were first processed for X-arrestin immunoreactivity as described above, except with the use of rhodamine labeled goat anti-rabbit IgG (Rh-GARG) (50x dilution; Boehringer, Indianapolis, IN) as the initial secondary antibody. Sections were then washed, post-fixed in cold methanol for 10 min. (only for subsequent reaction with the red/green cone opsin antibody, as recommended by John C. Saari, Ph.D.) and further reacted with the second specific antibody, i.e. either anti-human red/green cone opsin or anti-bovine S-antigen antibody. Following the wash step, sections were finally reacted with the appropriate FITC-tagged secondary antibody, i.e. FITC-GARG (for opsin) or FITC-GAMG (for S-antigen). Control sections for the double staining experiments were reacted with both secondary antibodies i.e. Rh-GARG and either FITC-GARG or FITC-GAMG. All slides were examined in a Zeiss Photomicroscope III by phase contrast and fluorescence microscopy using filter sets selective for rhodamine and fluorescein.

In Vitro Transcription and Translation

The HRG4 cDNA clone was linearized in the 3' non-coding region with Stu I and transcribed into RNA using the T3 promoter located upstream of the cloning site of the pBluescript vector (Stratagene, La Jolla, CA). A transcription reaction using one µg of DNA template was carried out with T3 RNA polymerase at 37°C for 45 minutes. The RNA transcript was extracted with phenol, precipitated with ethanol, and 1~2 µg of the

RNA template was translated in vitro with rabbit reticulocyte lysate (Stratagene, La Jolla, CA) with incorporation of [^{35}S]methionine (Du Pont NEN, Boston, MA) at 30°C for 1 hour. The translation products were analyzed in 12% SDS polyacrylamide gels and visualized by autoradiography. The rat cDNA clone was linearized with Stu I or Mae III, transcribed with T7 RNA polymerase and translated as above. The translation reaction with no RNA was also carried out as a negative control.

"Zoo" Southern Blot Analysis

Genomic DNA from mouse, rabbit, pig, calf and monkey was obtained from Clontech (Palo Alto, CA). Human genomic DNA was extracted from normal skin fibroblasts (19). Ten micrograms of each DNA sample was digested with EcoRI, phenol/chloroform extracted, ethanol precipitated, electrophoresed in a 0.8% agarose gel and transferred onto a nylon membrane by the Southern blotting method (8). The blot was hybridized with a full-length HRG4 cDNA probe as described for the northern analysis. Final washing after hybridization was performed at 42°C followed by autoradiography as described above.

In Situ Hybridization

In situ hybridization of HRG4/RRG4 riboprobes was performed in human or rat retina as described before (20) with some modifications. After fixation, frozen sectioning, and Proteinase K treatment, the sections were acetylated by immersion in 0.25% acetic anhydride in 0.1 M triethanolamine (pH 8.0) for 10 minutes at room temperature, followed by dehydration and hybridization. The sense (negative control) and antisense riboprobes were transcribed in vitro from the T3 or T7 promoter located on either side of the cloning site in the Bluescript vector with incorporation of [α-^{35}S]rCTP (Du Pont NEN, Boston, MA). Riboprobes (1×10^4 cpm/ml) for the human or rat tissue corresponded to a 569 bp fragment in the 3' non-coding region of the HRG4 cDNA, or a 360 bp fragment in the 3' non-coding region of the RRG4 cDNA, respectively. After hybridization, the slides were washed three times in 0.1X SSC for 20 minutes at 65°C and subjected to liquid emulsion autoradiography. After exposure for 4 days (8 days for rat retina of P0 and P10) at 4°C, slides were developed and counter-stained lightly with hematoxylin and eosin.

RESULTS AND DISCUSSION

Preparation of Retina-Enriched, -Specific cDNA Library

Human retinal mRNA was isolated and used in a subtractive cDNA cloning strategy (Fig. 1). Fibroblast mRNA population was used as a driver, and PCR amplification was used to rescue the small amount of subtracted material so that multiple subtractions could be performed in order to obtain a highly retina-enriched population of cDNAs. The final subtracted library contained approximately 300 clones, and approximately 100 appeared to represent true retinal genes. We began to analyze a subset of 30 clones, and the first four are described below.

rom-1

The first cDNA clone showed a retina-specific pattern of expression by Northern blot analysis (Fig. 2), and its encoded amino acid sequence showed a 35% overall homology to peripherin/rds, the photoreceptor disk membrane protein that is mutated in the rds

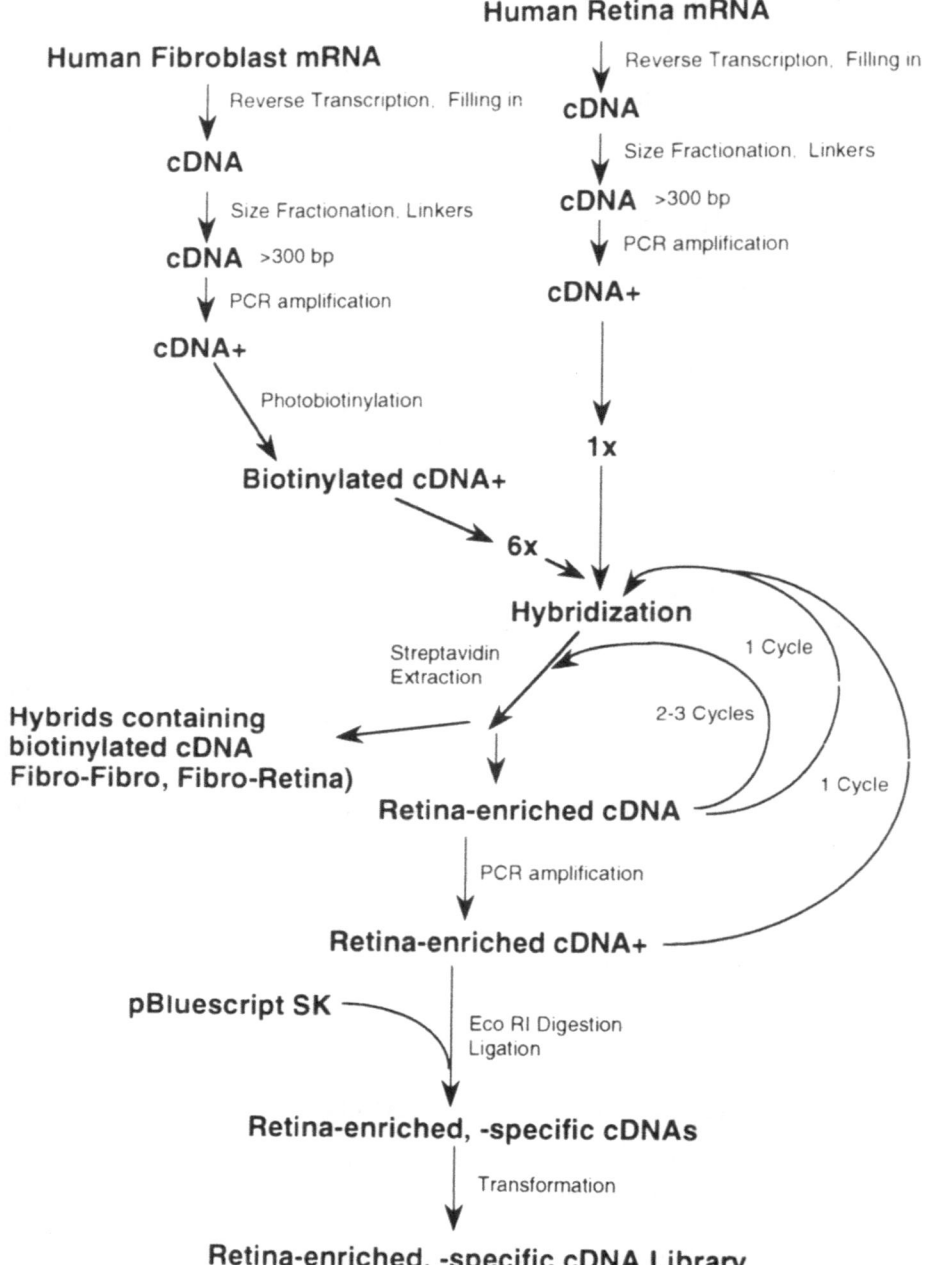

Figure 1. Preparation of retina-enriched, -specific cDNA library.

retinal degeneration mouse and some cases of ADRP in human (21–25). Bascom et al. had also cloned this new retinal cDNA which they named rom-1 (26), and they showed that the encoded protein is localized to the rod photoreceptor disk membrane, and it most likely interacts with peripherin. The function of rom-1 was speculated to be related to maintenance of the structural and possibly functional integrity of the rod photoreceptor disks. We cloned and characterized the gene for rom-1 which consisted of at least 3 exons,

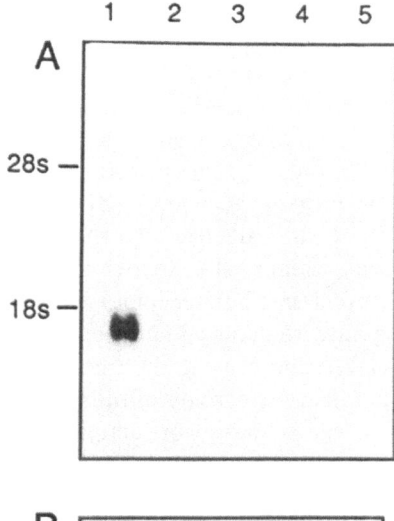

Figure 2. Northern blot analysis of Clone 1 (rom-1). A. RNA from human 1. retina; 2. brain; 3. lung; 4. liver; 5. fibroblast was hybridized with ^{32}P-labeled Clone 1 cDNA. B. Actin hybridization.

covering only 2 kbp of genomic sequence, and mapped the gene to chromosome 11. Determination of the exon/intron borders set the stage for screening this gene as a candidate gene for retinal degenerations (Fig. 3, see below). Thus, the first retinal gene we isolated by our subtractive cloning strategy turned out to be a novel retinal gene, coding for a protein that is most likely important for photoreceptor function.

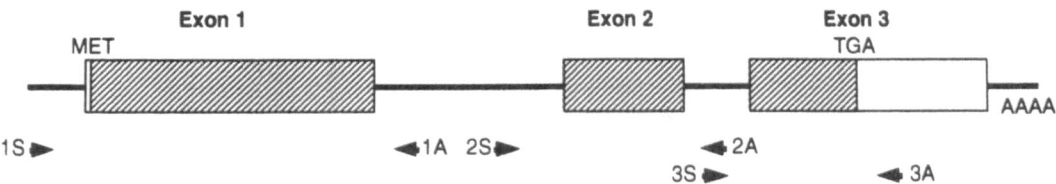

exon		PCR primers	size(bp) of product
1	1S	5' GC-clamp--ATCCCTGACACCTCTGCACT 3'	731
	1A	5' AAGGCAGCAGGAGGGAGCCT 3'	
2	2S	5' GC-clamp--AGACATCCTAACCCCTCTGT 3'	344
	2A	5' GTGTCAGATGCTTTGCTGAC 3'	
3	3S	5' GC-clamp--TCCAACCCTGGGCCTCTTGG 3'	365
	3A	5' TCAGACTTGTCCATCCCTC 3'	

Figure 3. Gene structure of rom-1. The exon/intron structure of the rom-1 gene is shown. PCR primers designed for amplification of the exonic sequences are shown.

Recoverin

The second cDNA clone from the subtracted library also corresponded to a gene with retina-specific expression as determined by Northern blot hybridization (Fig. 4). The 1.4 kb message encoded a 200 amino acid protein containing three calcium-bindiding EF-hand sites (7). The protein sequence showed 87% homology with the bovine photorecep-tor protein, recoverin (27), identifying it as human recoverin (Fig. 5). The human recoverin sequence also showed 58% homology to the calcium-binding chick cone pro-tein, visinin (28). Recoverin was originally thought to mediate the recovery of the dark current in photoreceptors after photoactivation by stimulating guanylate cyclase in re-sponse to reduced intracellular calcium concentration, thereby increasing the level of cGMP and causing the cation channel to open. More recently, recoverin has been shown to affect the activity of rhodopsin kinase, therefore to regulate rhodopsin phosphorylation, i.e., quenching of the activated state of rhodopsin (29). We cloned and characterized the recoverin gene which consisted of at least three exons encompassing approximately 10 kbp of genomic sequence (Fig. 6). The entire coding sequence was contained in the three exons, and all of the calcium-binding EF-hand regions were encoded by exon 1. The gene was mapped to chromosome 17 by hybridization analysis of a DNA panel of human-ro-dent somatic hybrids containing different human chromosomes. The localization of the re-coverin gene to chromosome 17 was of high interest because a number of human retinopathies have been mapped to 17, including Leber's congenital amaurosis and at least two ADRP loci (30–32). The characterization of the gene set the stage for screening of re-coverin as a candidate gene in various human retinal degenerations (see below).

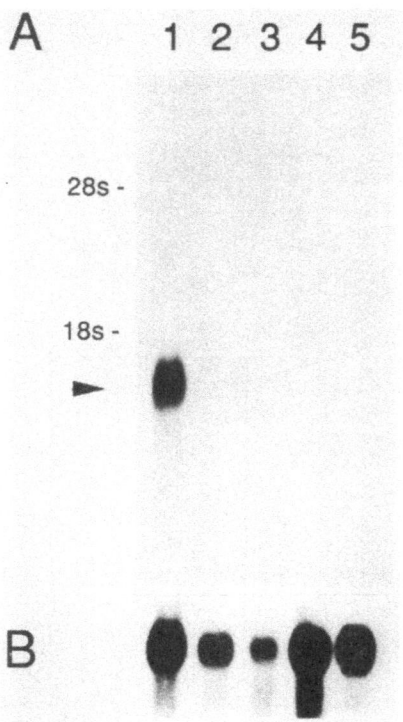

Figure 4. Northern blot analysis with the human recoverin cDNA probe. Ten μg of total RNA from various human tissues was applied in each lane: 1. retina; 2. brain; 3. lung; 4. liver; 5. skin fibroblasts. A, Hybridized with [^{32}P]-labeled cDNA clone; arrow points to the transcript of approximately 1.4 kb seen only in the retina. B, Hybridized with the human β-actin probe to check on the quantity and quality of RNA present in each lane. Positions of the 28s and 18s ribosomal RNA are shown.

```
RECO$CHICK     1 MGNSrSsALSrEvLqELrasTrytEEELsrWYegFqrqCsdGRIrcdeFerIYgnFFPnsePqgYArHVF
                 |||| | ||| | || |     |||| || |   | ||| |     || ||| | || |||
RECO$HUM       1 MGNSKSGALSKEILEELQLNTKFsEEELcSWYQSFLKdCPtGRITqQqFQsIYaKFFPdtDPKAYAQHVF
                 |||||||||||||||||||||| |||| ||||||| || |||| | || || |||| ||||||||||||
RECO$BOVIN     1 MGNSKSGALSKEILEELQLNTKFtEEELsSWYQSFLKeCPsGRITrQeFQtIYsKFFPeaDPKAYAQHVF

consensus        MGNSkSgALSkEiLeELqlnTkftEEELssWYqsFlk-Cp-GRIt-qeFq-IY-kFFP--dPkaYAqHVF


RECO$CHICK    72 SFDtNdDGTLDFrEYiIALHlTssGKThlKLEWAFSLfDVDrNGevSKsEVLEIitAIFKMIpeEerlqL
                 ||| | |||||| || ||||| || ||| |   ||| |||||||||| ||| || || ||||| |  |
RECO$HUM      72 SFDsNlDGTLDFKEYVIALHMTtAGKTNQKLEWAFSLYDVDGNGTISKNEVLEIVmAIFKMItPEDvKlL
                 ||| | |||||||||||||||| ||||||||||||||||||||||||||||||||| ||||| ||| | |
RECO$BOVIN    72 SFDaNsDGTLDFKEYVIALHMTsAGKTNQKLEWAFSLYDVDGNGTISKNEVLEIVtAIFKMIsPEDtKhL

consensus        SFD-N-DGTLDFkEYvIALHmTsaGKTnqKLEWAFSLyDVDgNGtiSKnEVLEIvtAIFKMI-pEd-k-L


RECO$CHICK   143 eDENsPqKRAdKlWaYFnKgenDKiaEgEFIdGvmkNdaImRLIQyEP        K K
                 ||| | ||| | ||| |     || | ||| |     | |||| ||            | |
RECO$HUM     143 dDENTPEKRAEKIWkYFGKnDDDKLTEKEFIEGTLANKEILRLIQFEPQKVKE   KmKna
                 |||||||||||||| ||| ||||||||||||||||||||||||||||||||   | |
RECO$BOVIN   143 eDENTPEKRAEKIWgfFGKkDDDKLTEKEFIEGTLANKEILRLIQFEPQKVKEklKeKkl

consensus        eDENtPeKRAeKiW-yFgK-ddDKltEkEFIeGtlaNkeIlRLIQfEPqkvkeklK-K--
```

Figure 5. Alignment of human, bovine, and chick recoverin sequences. The human (RECO$HUM), bovine (RECO$BOVIN), and chick (RECO$CHICK) recoverin (the chick protein originally called visinin) amino acid sequences were aligned by the GENALIGN program (GENALIGN is a copyrighted software product of IntelliGenetics, Inc.; the program was developed by Dr. Hugo Martinez of the Univestiy of California at San Francisco). Bars represent identical residues. A consensus sequence is shown at the bottom with uppercase letters indicating completely conserved residues. The three calcium-binding regions are overlined.

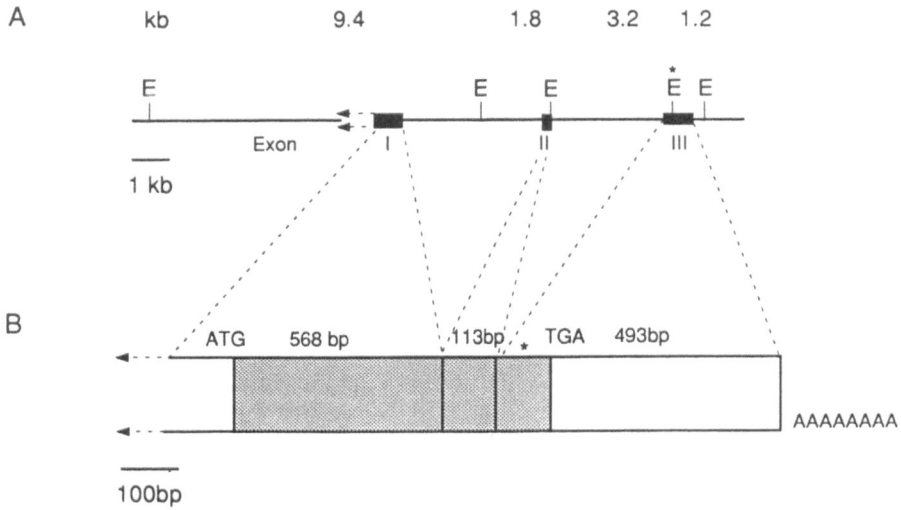

Figure 6. Human recoverin gene and cDNA. A, Recoverin gene. The putative exons I, II, and III are represented as filled boxes. The arrows at the left of exon I indicate that this exon may be longer upstream or interrupted by another intron. Eco RI sites are designated by E (asterisk indicates an RI site within an exon), and the numbers refer to the distances between the sites in kilobases. B, cDNA. ATG is the translational initiation codon, and TGA is the termination codon. Coding region is shaded. The exonic divisions of the sequence are indicated, with the arrows at the left end representing continuation of the exon upstream. The sizes of the exons are shown in basepairs.

X-Arrestin

The third retinal cDNA clone we analyzed from the subtracted library corresponded to another novel retinal gene. The 388 amino acid protein encoded by its message, whose expression was shown to be retina-specific by Northern blot analysis (Fig. 7), showed homology to arrestin (20). Arrestins are signal transduction proteins that are involved in receptor-mediated homologous desensitization. The first arrestin to be characterized was S-antigen, the rod photoreceptor arrestin (33), followed by β-arrestin, the arrestin for β-adrenergic receptor (34). Our arrestin appeared to be a new arrestin, distinct from S-antigen and β-arrestin, since its homology to the known arrestins was in the range of 49–58% while the homologies among the S-antigens and β-arrestins were all in the 80–90% range (20). Alignment of the new arrestin sequence with those of S-antigen and β-arrestin from various species confirmed this point, showing a unique carboxy-terminal region for the new arrestin (Fig. 8). The alignment also demonstrated the presence of transducin-like sequences in the new arrestin, typical of arrestins which are thought to compete with transducin/G-protein in interacting with the activated receptor. The gene for the new arrestin was mapped to Xcen-Xq22 by hybridization to a panel of DNA from somatic hybrids containing different parts of the human X chromosome (data not shown), and hence, the new arrestin was named X-arrestin.

In order to determine the localization of the X-arrestin protein within the retina, a peptide sequence unique to X-arrestin was chosen from the carboxy-terminal region and used to prepare an antibody (37). The antipeptide X-arrestin antibody specifically recognized a 47 kDa protein on a Western blot analysis of human retinal extract (Fig. 9). The

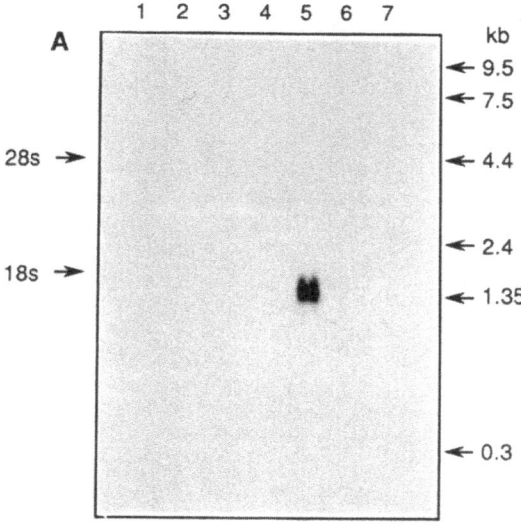

Figure 7. Northern blot analysis of human mRNA with the X-arrestin cDNA probe. RNA from different human tissues are present in lanes 1–7. Lane 1, lung; 2, liver; 3, brain; 4, retinal pigment epithelium; 5, neuroretina; 6 and 7, skin fibroblast. A, Hybridized with [^{32}P]-labeled cDNA clone; arrow points to the transcript of approximately 1.35 kb seen only in the retina. B, Hybridized with the human β-actin probe to check on the quantity and quality of RNA present in each lane. Positions of the 28s and 18s ribosomal RNA and RNA standards are shown.

Figure 8. Alignment of multiple arrestin sequences. The protein sequences of the new X-arrestin (XARRESTIN), bovine β-arrestin (BOVARRB), human thyroid arrestin (HUMARRESTM), bovine retinal arrestin (BOVANTS, S-antigen), and human retinal arrestin (HUMRETSA, S-antigen) were multiply aligned by the GENALIGN computer program (IntelliGenetics, Inc.). D–G and G/AXXXGK designate three regions showing homology to GTP phosphoryl binding sites, and the shadowed residues in the BOVANTS and HUMRETSA sequences are homologous to the pertussis toxin ADP-ribosylation site (35, 36). The regions designated 1–5 have been shown to be similar to transducin α in bovine β-arrestin (34).

antipeptide antibody, along with antibodies against the red and green opsin (generous gift of J. Saari), S-antigen (generous gift of L. Donoso), and rhodopsin (generous gift of P. Hargrave), was used in immunofluorescence analysis of human retinal sections. The X-arrestin immunofluorescence was demonstrated in the outer segment of cone photoreceptors, similar to that of the red- and green-sensitive cones (Fig. 10). Double immunofluorescence analysis co-localized X-arrestin and red and green opsin in the outer segment of the same cone photoreceptors (Fig. 11). The use of the A9C6 S-antigen antibody, which has been demonstrated to react with blue-sensitive cones in addition to rods (38, 39), revealed the presence of X-arrestin also in the blue-sensitive cones (Figs. 12). Thus, X-arrestin was shown to be present specifically in the outer segment of red-, green-,

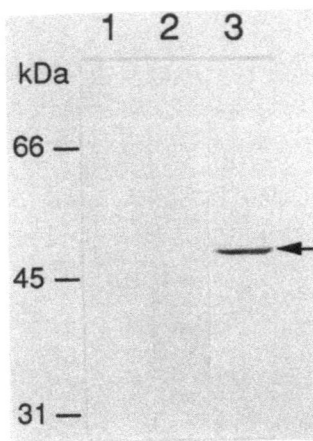

Figure 9. Western blot analysis of human retinal homogenate proteins. Human retinal extracts were electrophoresed in 10% SDS-PAGE, transferred to a nylon filter and reacted with preimmune serum from the rabbit used for antibody production (lane 1), pass-through fraction from the affinity purification (lane 2), and affinity-purified, human X-arrestin antipeptide antibody (lane 3). Bound antibody was visualized by the alkaline phosphatase color reaction. The X-arrestin anti-peptide antibody recognizes a protein in human retina of ~47 kDa (arrow). The size markers are bovine serum albumin (66 kDa), ovalbumin (45 kDa), and carbonic anhydrase (31 kDa).

and blue-sensitive cone photoreceptors. Considering the known function of arrestins, it is very likely that X-arrestin is involved in the desensitization of photoactivated red, green and blue cone opsins. These findings, along with the X-chromosome localization of the gene, made X-arrestin a prime candidate gene for X-linked cone diseases (see below).

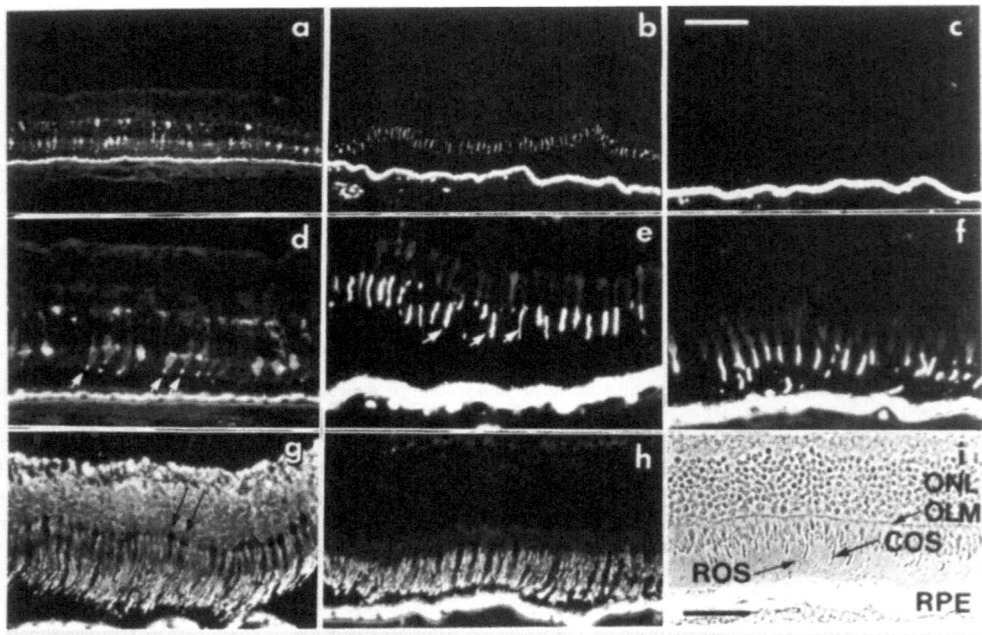

Figure 10. Localization of X-arrestin in human retina by indirect immunofluorescence. Frozen sections of peripheral (a) and perifoveal (b) retina from two human donor eyes reacted with affinity-purified human X-arrestin antipeptide antibody (a,b) or rabbit preimmune serum (c). The wavy horizontal line seen at low power in (a-c) is due to autofluorescence of lipofuscin in the aged retinal pigment epithelium. Images in (d,e) are higher magnifications of sections shown in (a,b), respectively. Arrows in (d,e) point to cone outer segments. There is a slight detachment of the retina in most areas of the perifoveal sections. Adjacent perifoveal sections (f-h) show immunostaining patterns with anti-human red/green cone opsin antibody (f), anti-bovine S-antigen antibody (g), and anti-bovine rhodopsin antibody (h). The phase contrast image of the field in (h) is shown in (i). ONL, outer nuclear layer; OLM, outer limiting membrane; COS, cone outer segment; ROS, rod outer segment; RPE, retinal pigment epithelium. Magnification (a-c) x60; (d-i) x150. Bar = 165 μm in (a-c) and 65 μm in (d-i).

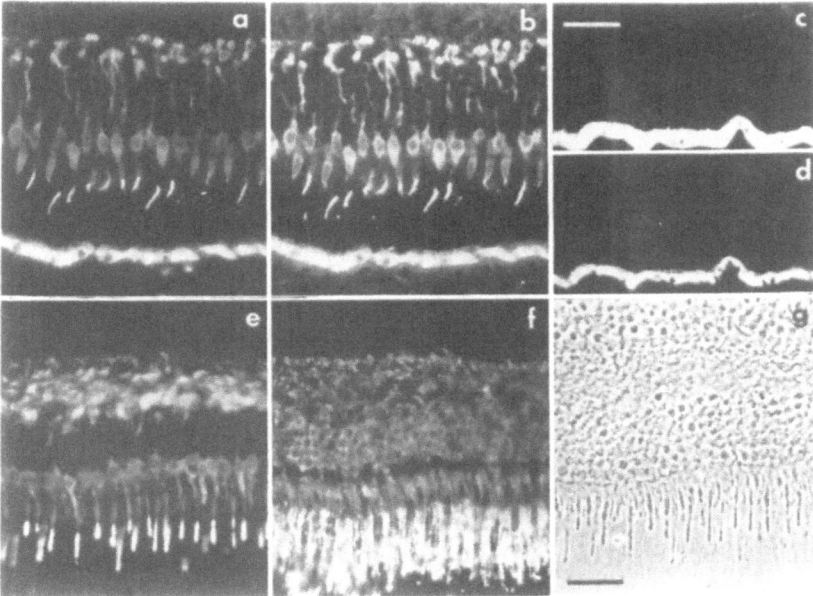

Figure 11. Double immunofluorescent localization of X-arrestin in human retina. (a,b) Double immunolabel with X-arrestin and red-green cone opsin antibodies. A section of perifoveal human retina was sequentially reacted with rabbit anti-X-arrestin antibody, Rh-GARG, rabbit anti-red/green cone opsin antibody and FITC-GARG. Note the precise correspondence of positively stained cones, doubly labeled with the X-arrestin antibody (viewed with the Rh filter set in a) and with the red/green cone opsin antibody (viewed through the FITC filter set in b). (c,d) Adjacent control section from retina shown in (a,b) doubly reacted with Rh-GARG (c) and FITC-GARG (d) alone, viewed with corresponding filter sets. (e,f) Double immunolabel with X-arrestin and S-antigen antibodies. Specific localization of X-arrestin to Rh-tagged cones is seen in (e), whereas S-Ag distribution is confined to the FITC-tagged rods in the section, shown in (f). (g) Phase contrast image of the section shown in (e,f). Magnification (a,b,e-g) x230; (c,d) x150. Bar = 165 µm in (c,d) and 43 µm in (a,b,e-g).

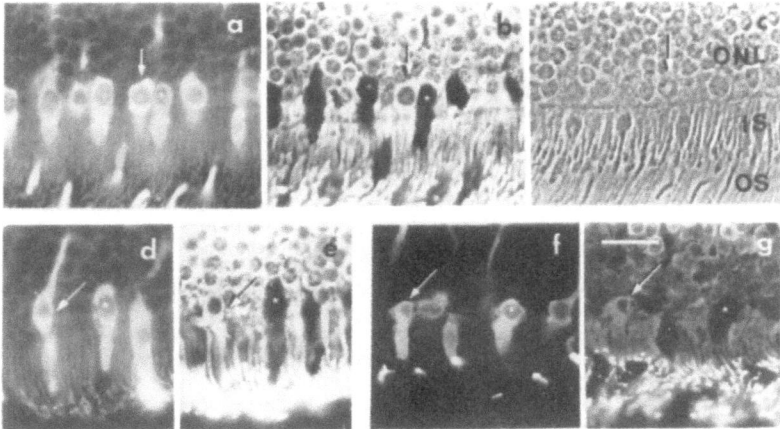

Figure 12. Localization of X-arrestin in blue cones. High magnification micrographs of the section shown in Fig. 11e-g, double stained with X-arrestin (a,d,f) and A9C6 S-Ag (b,e,g) antibodies. The panels show three examples of X-arrestin positive cones (arrows) showing immunoreactivity with A9C6, thereby identifying them as blue cones. Adjacent X-arrestin positive cone cells, negtive for A9C6 (asterisks) are red/green cones. Magnification, all panels x375. Bar = 27 µm.

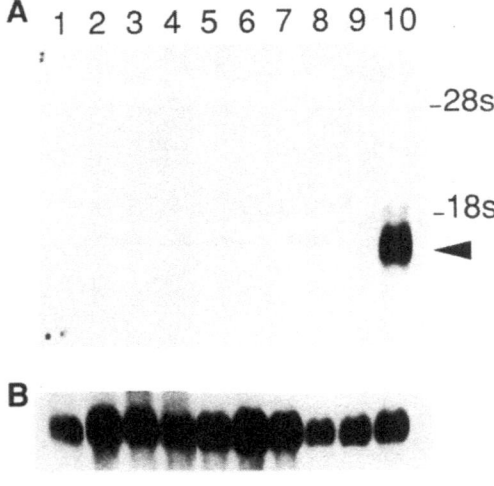

Figure 13. Northern blot analysis of HRG4 in human tissues. Ten µg of total RNA from various human tissues was applied in each lane: 1. kidney; 2. lung; 3. fetal brain; 4. adult brain; 5. skin fibroblast; 6. infant RPE; 7. adult RPE; 8. iris; 9. cornea; 10. retina. A. Hybridized with ³²P-labeled full-length cDNA probe of HRG4; arrowhead points to the transcript of approximately 1.4 kb seen only in the retina. B. Hybridized with the human β-actin probe to check on the quantity and quality of RNA present in each lane. Positions of the 28s and 18s ribosomal RNA are shown.

HRG4

The fourth and the last retinal cDNA to be discussed here also represented a novel retinal gene and showed a retina-specific pattern of expression by Northern blot analysis (40) (Fig. 13). The 1.4 kb message encoded a 240 amino acid protein which seemed to show a loose similarity to collagen but no definite match to any known sequence. A "zoo" genomic Southern blot analysis of DNA from various species demonstrated conservation of this gene sequence, suggesting an important function of its protein (data not

```
    HUMAN     1 MKVKKGGGGaGtatEsaPGpSgqSVaPipqPpAEsESGSESEPdaGPGPRpGPLQrKQPIG
                 |||||||||| |    |   || |   || |    | || |||||||| ||||| |||| |||||
     RAT      1 MKVKKGGGGtGpgaEpvPGaSnrSVePtrePgAEaESGSESEPepGPGPR1GPLQgKQPIG

  consensus     MKVKKGGGG-G---E--PG-S--SV-P---P-AE-ESGSESEP--GPGPR-GPLQ-KQPIG


    HUMAN    62 PEDVLGLQRITGDYLCSPEENIYKIDFVRFKIRDMDSGTVLFEIKKPPVSERLPINRRDLD
                 ||||||||||||||||||||||||||||||||||||||||||||||||||||||||||||
     RAT     62 PEDVLGLQRITGDYLCSPEENIYKIDFVRFKIRDMDSGTVLFEIKKPPVSERLPINRRDLD

  consensus     PEDVLGLQRITGDYLCSPEENIYKIDFVRFKIRDMDSGTVLFEIKKPPVSERLPINRRDLD


    HUMAN   123 PNAGRFVRYQFTPAFLRLRQVGATVEFTVGDKPVNNFRMIERHYFRNQLLKSFDFHFGFCI
                 ||||||||||||||||||||||||||||||||||||||||||||||||||||||||||||
     RAT    123 PNAGRFVRYQFTPAFLRLRQVGATVEFTVGDKPVNNFRMIERHYFRNQLLKSFDFHFGFCI

  consensus     PNAGRFVRYQFTPAFLRLRQVGATVEFTVGDKPVNNFRMIERHYFRNQLLKSFDFHFGFCI


    HUMAN   184 PSSKNTCEHIYDFPPLSEELISEMIRHPYETQSDSFYFVDDRLVMHNKADYSYSGTP
                 ||||||||||||||||||||||||||||||||||||||||||||||||||||||||
     RAT    184 PSSKNTCEHIYDFPPLSEELISEMIRHPYETQSDSFYFVDDRLVMHNKADYSYSGTP

  consensus     PSSKNTCEHIYDFPPLSEELISEMIRHPYETQSDSFYFVDDRLVMHNKADYSYSGTP
```

Figure 14. Alignment of HRG4 and RRG4. HRG4 (human) and RRG4 (rat) amino acid sequences were aligned by the GENALIGN program (GENALIGN is a copyrighted software product of IntelliGenetics, Inc.). Bars represent identical residues. A consensus sequence is shown at the bottom with uppercase letters indicating completely conserved residues.

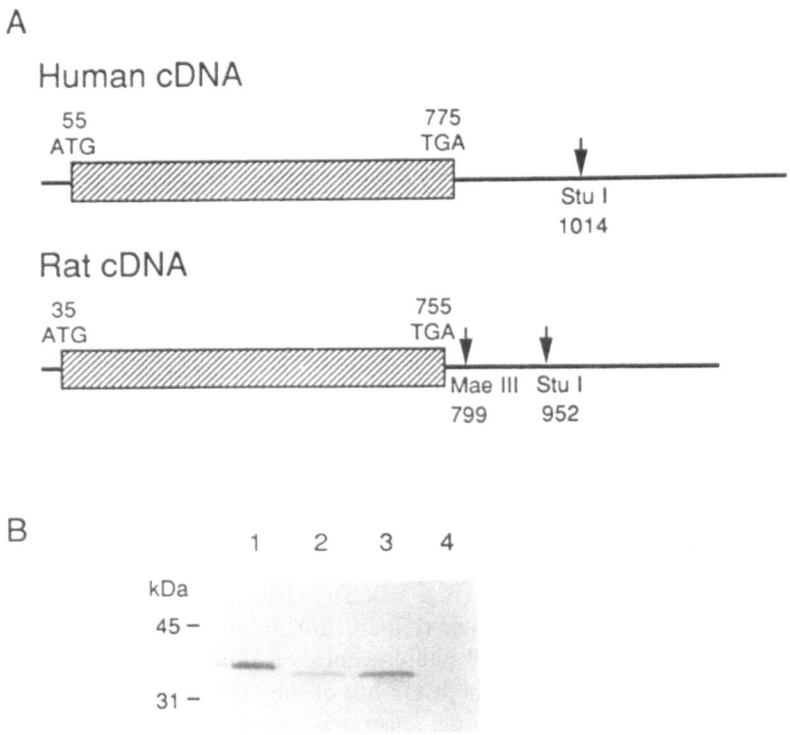

Figure 15. In vitro transcription and translation. A. The human cDNA clone was linearized with StuI in the 3′ non-coding region, as indicated by arrow, and transcribed into RNA using T3 RNA polymerase. The rat clone was cut with Mae III and transcribed with T7 RNA polymerase. RNA transcripts were then translated in vitro using the rabbit reticulocyte lysate with incorporation of [^{35}S]methionine. Hatched boxes are the open reading frames, and the numbers refer to their positions from the first base of the cDNAs. B. Autoradiograph of SDS-PAGE of the translation products with different template RNAs. Lanes: RNA templates made from 1. human cDNA digested with StuI; 2. human cDNA digested with PvuII; 3. rat cDNA digested with StuI; 4. no RNA.

shown). The novel gene was named Human Retinal Gene 4 (HRG4). Its rat homologue cDNA (RRG4) was cloned and shown to be also retina-specific in expression by Northern blot analysis (data not shown) and 92% homologous to the human HRG4 protein sequence (Fig. 14). The comparison of the human and rat sequence demonstrated the presence of a two-domain structure in the protein: the proximal 1/4 of the protein that is rich in proline and glycine and only 67% homologous, and the distal 3/4 of the protein that is 100% homologous. The two-domain structure is most likely significant functionally. The validity of the coding sequences present in the cDNA was tested by in vitro expression of the mRNA from the cDNA, translation, and SDS-PAGE analysis of the products (Fig. 15). The expected products from the putative coding seqences were obtained after digestion of the cDNA clones downstream of the putative termination codons, confirming the presence of the coding regions. The site of expression of HRG4 within the retina was determined by in situ hybridization to be the photoreceptors (Fig. 16). Signal was present in both the rod and cone inner segments, indicating expression in both types of photoreceptors. The expression of HRG4 during the development of the retina was analyzed by a developmental Northern blot hybridization of RNA from rat ret-

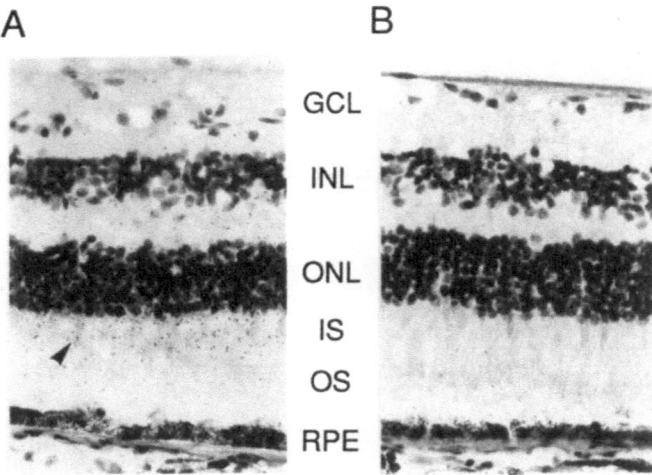

Figure 16. In situ hybridization of HRG4 in human retina. The result of liquid emulsion autoradiography of the retinal sections counterstained lightly with hematoxylin and eosin are shown. Hybridization was performed with antisense (A) or sense (B) riboprobe. GCL, ganglion cell layer; INL, inner nuclear layer; ONL, outer nuclear layer; IS, inner segment; OS, outer segment; RPE, retinal pigment epithelium. In addition to rods, some cones (arrowhead) show hybridized signals in their inner segments.

ina at different ages, starting at birth (Fig. 17). The HRG4 expression was detectable at P5 when the differentiation of the outer retina begins, and maximal expression was attained by P23 when the maturation of photoreceptors is complete. Thereafter, this level of expression was maintained throughout the life of the animal. The developmental pattern of expression of HRG4 was also demonstrated by in situ hybridization which confirmed the result of the developmental Northern analysis (Fig. 18). Thus, expression of HRG4 was temporally and spatially correlated with photoreceptor development and function. Since some photoreceptor proteins, especially those involved in phototransduction, show a diurnal variation in expression, the expression level of HRG4 during the diurnal cycle was examined. No change in the level of HRG4 expression was detected during the diurnal cycle while the expression of S-antigen was maximal during the light period and minimal during the dark period (data not shown). Interestingly, recently HRG4 was found to be homologous to a new neuroprotein, unc119, cloned in C. elegans (41). This protein appears to be important for chemosensation in the worm, such that a mutant worm manifests various defects including in coordination and feeding. Thus, HRG4 is most likely also a neuroprotein that is vital to the function of photoreceptors, making it a good candidate gene for retinal degeneration.

Screening of Candidate Genes

Work is in progress to screen the new retinal genes described above as candidate genes in various retinal diseases.

Rom-1 appeared to be a good candidate gene for retinopathies in that it is related to peripherin/rds, a gene that has already been related to retinal degeneration in the rds mouse and some cases of human ADRP and macular pattern dystrophies (22–25, 42–44). In fact, a mutation in rom-1 was shown to cause RP when present with a second mutation in peripherin/rds in the "digenic RP" (45). In our screening we uncovered a putative null mutation in rom-1 in a family with RP without a mutation in peripherin/rds (46). No mutation was found in two other genes we also tested, rhodopsin and cGMP-phosphodiesterase β-subunit. A variable pattern of the disease phenotype was observed in the family mem-

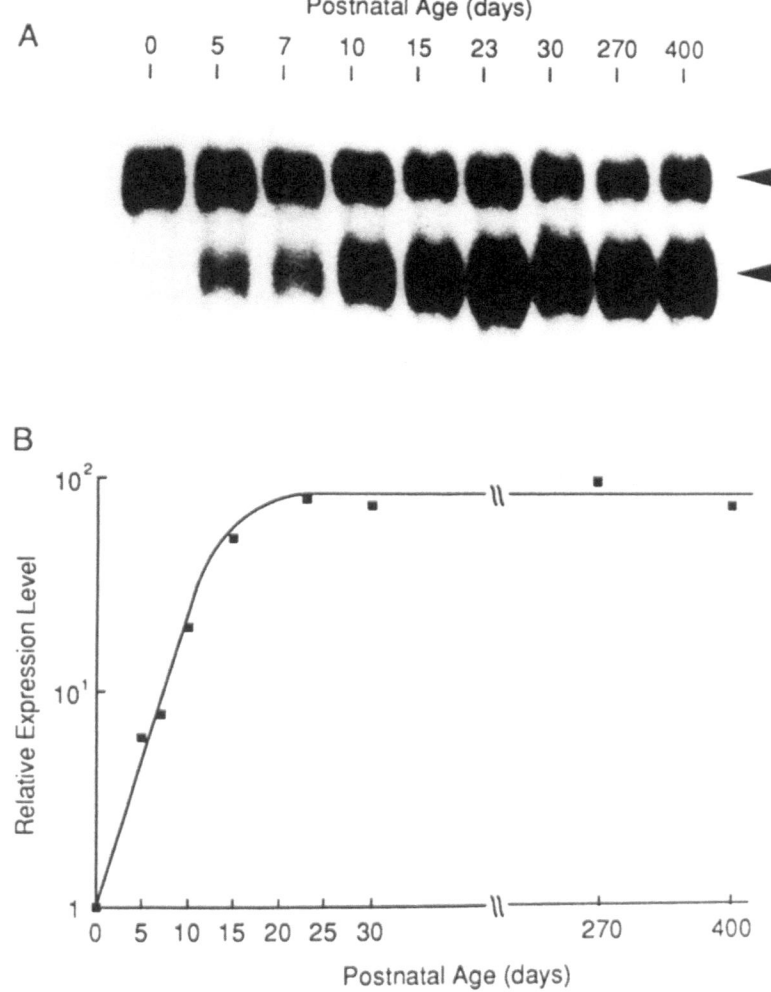

Figure 17. Developmental expression of RRG4 in rat retina. A. Representative northern blot analysis of RRG4 with 8 μg of total retinal RNAs obtained from rats of the indicated ages. Hybridization with RRG4 or actin probe is indicated by the lower and upper arrowheads, respectively. B. Densitometric analysis of the expression level of RRG4. The intensity of the hybridized RRG4 message at each postnatal age was standardized with that of actin, and the ratio compared to the value at postnatal day 0 was plotted in logarithmic scale.

bers inheriting the mutation, ranging from severe disease to apparent normal function (Fig. 19). This was similar to the phenotypic variability observed for peripherin/rds mutations (47, 48). Thus, a mutation in rom-1 alone may be able to cause retinopathy, or a mutation in a yet-to-be-uncovered second gene in addition to the rom-1 mutation may be the cause of the disease.

Recoverin appeared to be a good candidate gene for Leber's congenital amaurosis in view of the mapping of both to 17p (30, 50). The fact that circulating antibody against recoverin was found in cancer associated retinopathy (51) also suggested the possibility that recoverin may be involved in the pathogenesis of retinal degeneration. Screening of

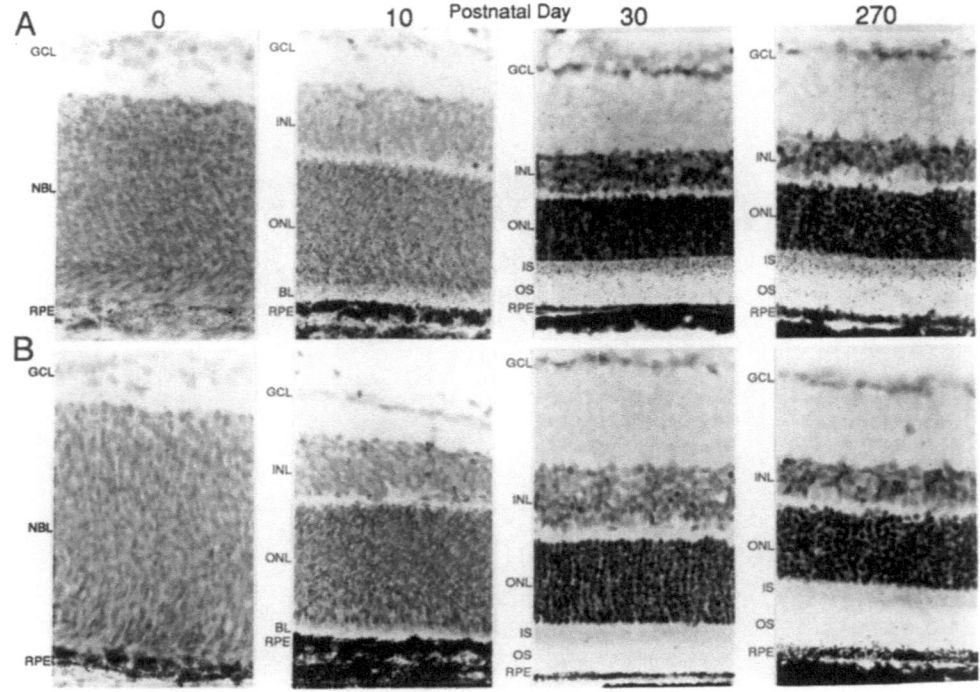

Figure 18. Developmental in situ hybridization of RRG4 in rat retina. Retinal sections from rats of the indicated ages were hybridized with antisense (A) or sense (B) riboprobe of RRG4 and subjected to liquid emulsion autoradiography and counter-staining with hematoxylin and eosin. GCL, ganglion cell layer; NBL, neuroblastic layer; RPE, retinal pigment epithelium; INL, inner nuclear layer; ONL, outer nuclear layer; BL, bacillary layer; IS, inner segment; OS, outer segment.

numerous cases of retinal disease, however, has not demonstrated a definitive causal relationship of this gene to disease yet (52). X-arrestin and HRG4 are beginning to be screened as candidate genes, the former especially for X-linked cone and cone-rod diseases. Considering the features of these new retinal genes and their protein products, it is very likely that retinal diseases will be identified that are caused by defects in these genes.

In conclusion, the strategy of isolating new retinal genes using a subtractive cloning approach in order to learn more about retinal biology and physiology and to obtain additional candidate genes for retinal degenerations has been fairly successful so far. Four new retinal genes that appear to be important for photoreceptor function have been isolated: rom-1, important for structural/functional integrity of the rod photoreceptor disk membrane; recoverin, most likely important for calcium-dependent modulation of phototransduction; X-arrestin, most likely important for cone phototransduction; and HRG4, important for a yet-to-be-described photoreceptor-specific function. One of these genes, rom-1, has already been shown to be related to human RP, and ρ ꞌogenicity of the other genes will most likely be demonstrated. Continued study of these isolated genes and additional genes from our subtracted library should yield additional important and useful information regarding retinal biology, physiology, and pathology.

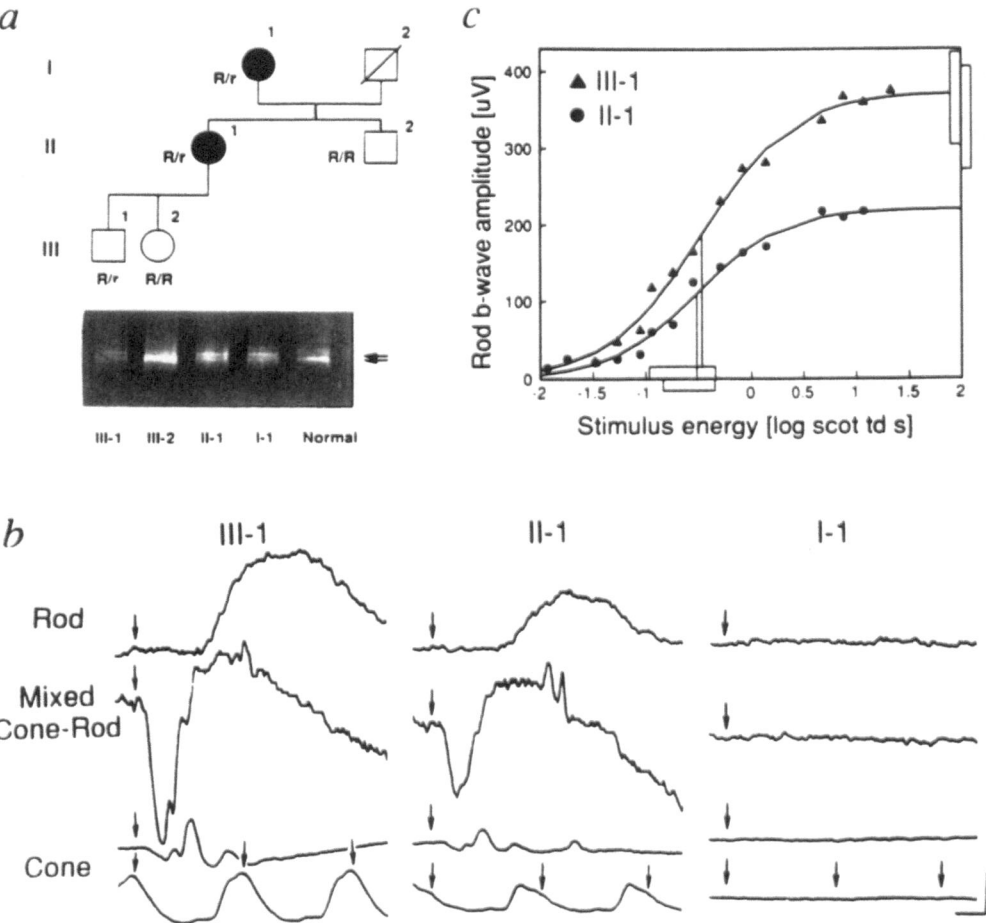

Figure 19. a, The pedigree of the family with RP and the DGGE analysis of exon 1(a). The presence of the hetero-duplex band (upper arrow), indicative of a heterozygous mutation, is shown for the proband (I-1), daughter (II-1), and grandson (III-1). The lower arrow points to the normal exon 1(a) band. The mutation, along with genotype of the son (II-2) was determined by direct sequencing of PCR-amplified exon 1(a) DNA. Circle, female; square, male; solid circle, affected female; diagonal line over symbol, deceased. R, normal rom-1 allele; r, mutant rom-1 allele. b, Standard rod, mixed cone-rod and cone (1Hz and 29 Hz) ERGs (47, 49) from the three family members with the heterozygous putative null mutation in rom-1. Results of III-1 are normal; II-1 shows a reduced rod b-wave amplitude and borderline mixed cone-rod ERG; cone ERGs fall within our normal limits. I-1 has no detect-able waveforms to any stimuli. Arrows indicate stimulus onset. Calibrations for the upper 3 traces of each individual are: 100 microvolts (uV) and 20 milliseconds (ms); for the lower trace: 100 uV and 10 ms. c, Rod ERG b-wave amplitude as a function of stimulus energy for II-1 and III-1. Solid curves are the Naka-Rushton function (49) fit to the patient data. For III-1, the amplitude at response saturation (V_{max}) and the intensity at $1/2 \, V_{max}$ (K) fall within the results of an age-related normal population (n=10; ages 13–28 years). For II-1, K is normal but V_{max} is reduced below the age-related normals (n=10; ages 40–55 years). Bars on vertical axis are mean normal V_{max}-2 s.d.; bars on horizontal axis are mean normal K+/-2 s.d. Bars inside the graph represent the younger normals and the outside bars are the older normals.

ACKNOWLEDGMENTS

This work was supported by Public Health Service Research Grant (EY10848), The Foundation Fighting Blindness, Baltimore, MD, and Research to Prevent Blindness, New York, NY.

REFERENCES

1. Dryja, T.P., McGee, T.L., Reichel, E., Hahn, L.B., Cowley, G.S., Yandell, D.W., Sandberg, M.A., and Berson, E.L., 1990, A point mutation of the rhodopsin gene in one form of retinitis pigmentosa, *Nature* 343:364–366.
2. McLaughlin, M.E., Sandberg, M.A., Berson, E.L., and Dryja, T.P., 1993, Recessive mutations in the gene encoding the β-subunit of rod phosphodiesterase in patients with retinitis pigmentosa, *Nature Genet.* 4:130–134.
3. Gal, A., Orth, U., Baehr, W., Schwinger, E., Rosenberg, T., 1994, Heterozygous missense mutation in the rod cGMP phosphodiesterase β-subunit gene in autosomal dominant stationary night blindness, *Nature Genet* 7:64–68.
4. Bernhard, H.F.W., Vogt, G., Pruett, R.C., Stohr, H., and Felbor, U., 1994, Mutations in the tissue inhibitor of metalloproteinases-3 (TIMP3) in patients with Sorsby's fundus dystrophy. *Nat. Genet.* 8:352–356.
5. Weil, D., Blanchard, S., Kaplan, J., Guilford, P., Gibson, F., Walsh, J., Mburu, P., Varela, A., Levilliers, J., Weston, M.C., Kelley, P.M., Kimberling, W.J., Wagenaar, M., Levi-Acobas, F., Larget-Piet, D., Munnich, A., Steel, K.P., Brown, S.D.M., Petit, C., 1995, Defective myosin VIIA gene responsible for Usher syndrome type 1B, *Nature* 374:60–61.
6. Fuchs, S., Nakazawa, M., Maw, M., Tamai, M., Oguchi, Y. and Gal, A., 1995, A homozygous 1-base pair deletion in the arrestin gene is a frequent cause of Oguchi disease in Japanese. *Nat. Genet.* 10: 360–362.
7. Murakami, A., Yajima, T., and Inana, G., 1992, Isolation of human retinal genes: recoverin cDNA and gene, *Biochem. Biophys. Res. Comm.* 187:234–244.
8. Southern, E.M., 1975, Detection of specific sequences among DNA fragments separated by gel electrophoresis, *J. Mol. Biol.* 98:503–517.
9. Sambrook, J., Fritsch, E.F., and Maniatis, T., 1989, *Molecular cloning, a Laboratory manual,* Cold Spring Harbor, Cold Spring Harbor, New York.
10. Chirgwin, J.M., Przybyla, A.E., MacDonald, R.J., and Rutter, W.J., 1979, Isolation of biologically active ribonucleic acid from sources enriched in ribonuclease, *Biochemistry* 18:5294–5299.
11. Lehrach, H., Diamond, D., Wozney, J.M., and Boedther, H., 1977, RNA molecular weight determinations by gel electrophoresis under denaturing conditions, a critical reexamination, *Biochemistry* 16:4743–51.
12. Inana, G., Totsuka, S., Dougherty, T., Redmond, M., Nagle, J., Shiono, T., Ohura, T., Kominami, E. and Katunuma, N., 1986, Molecular cloning of human ornithine aminotransferase mRNA, *Proc. Natl. Acad. Sci. USA* 83:1203–1207.
13. Feinberg, A. and Vogelstein, B., 1983, A technique for radiolabeling DNA restriction endonuclease fragments to high specific activity, *Anal. Biochem.* 132:6–13.
14. Sanger, F., Nicklen, S., and Coulson, A.R., 1977, DNA sequencing with chain-terminating inhibitors, *Proc. Natl. Acad. Sci. USA* 74:5463–67.
15. Lerea, C.L., Bunt-Milam, A.H., and Hurley, J.B., 1989, α Transducin is present in blue-, green-, and red-sensitive cone photoreceptors in the human retina, *Neuron* 3:367–376.
16. Kelley, M.W., Turner, J.K., and Reh, T.A., 1994, Retinoic acid promotes differentiation of photoreceptors in vitro, *Development* 120:2091–2102.
17. Donoso, L.A., Merryman, C.F., Edelberg, K.E., Naids, R., and Kalsow, C., 1985, S-antigen in the developing retina and pineal gland: a monoclonal antibody study, *Invest. Ophthalmol. Vis. Sci.* 26:561–567.
18. Rohlich, P., Adamus, G., McDowell, J.H., and Hargrave, P.A., 1989, Binding pattern of anti-rhodopsin monoclonal antibodies to photoreceptor cells: an immunocytochemical study, *Exp. Eye Res.* 49:999–1013.
19. Blin, N. and Stafford, D.W., 1976, Isolation of high-molecular-weight DNA, *Nucleic Acids Res.* 3:2303–2308.
20. Murakami, A., Yajima, T., Sakuma, H., McLaren, M.J., and Inana, G., 1993, X-arrestin: a new retinal arrestin mapping to the X chromosome, *FEBS Lett.* 334:203–209.
21. Murakami, A., Akaki, Y., Ara, F., Duval, E., and Inana, G., 1992, Isolation of new candidate genes for human retinal diseases: differential cloning approach, *Invest. Ophthalmol. Vis. Sci.* 33:498A.

22. Bowes, C., Li, T., Danciger, M., Baxter, L.C., Applebury, M.L., Farber, D.B., 1990, Retinal degeneration in the rd mouse is caused by a defect in the β subunit of rod cGMP-phosphodiesterase, *Nature* 347:677–680.

23. Pittler, S.J. and Baehr, W., 1991, Identification of a nonsense mutation in the rod photoreceptor cGMP phosphodiesterase β-subunit gene of the rd mouse, *Proc. Natl. Acad. Sci. USA* 88:8322–26.

24. Farrar, G.J., Kenna, P., Jordan, S.A., Kumar-Singh, R., Humphries, M.M., Sharp, E.M., Sheils, D.M., and Humphries, P., 1991, A three-base-pair deletion in the peripherin-RDS gene in one form of retinitis pigmentosa, *Nature* 354:478–480.

25. Kajiwara, K., Hahn, L.B., Mukai, S., Travis, G.H., Berson, E.L., 1991, Mutations in the human retinal degeneration slow gene in autosomal dominant retinitis pigmentosa, *Nature* 354:480–482.

26. Bascom, R.A., Manara, S., Collins, L., Molday, R.S., Kalnins, V.I., and McInnes, R.R., 1992, Cloning of the cDNA for a novel photoreceptor membrane protein (rom-1) identifies a disk rim protein family implicated in human retinopathies, *Neuron* 8:1171–1184.

27. Dizhoor, A.M., Ray, S., Kumar, S., Niemi, G., Spencer, M., Brolley, D., Walsh, K.A., Philipov, P.P., Hurley, J.B., and Stryer, L., 1991, Recoverin: a calcium sensitive activator of retinal rod guanylate cyclase, *Science* 251:915–918.

28. Yamagata, K., Goto, K., Kuo, C.-H., Kondo, H., and Miki, N., 1990, Visinin: a novel calcium binding protein expressed in retinal cone cells, *Neuron* 2:469–476.

29. Gorodovikova, E.N., Gimelbrant, A.A., Senin, I.I., and Philippov, P.P., 1994, Recoverin mediates the calcium effect upon rhodopsin phosphorylation and cGMP hydrolysis in bovine retina rod cells, *FEBS Lett* 349:187–90.

30. Camuzat, A., Dollfus, H., Rozet, J.M., Gerber, S., and others, 1995, A gene for Leber's congenital amaurosis maps to chromosome 17p, *Hum. Mol. Genet.* 4:1447–52.

31. Greenberg, J., Goliath, R., Beighton, P., and Ramesar, R., 1994, A new locus for autosomal dominant retinitis pigmentosa on the short arm of chromosome 17, *Hum. Mol. Genet.* 3:915–918.

32. Bardien, S., Ebenezer, N., Greenberg, J., Inglehearn, C.F., Bartmann, L., Goliath, R., Beighton, P., Ramesar, R., and Bhattacharya, S.S., 1995, An eighth locus for autosomal dominant retinitis pigmentosa is linked to chromosome 17q, *Hum. Mol. Genet.* 4:1459–1462.

33. Kuhn, H., Hall, S.W., and Wilden, U., 1984, Light-induced binding of 48 kDa protein to photoreceptor membrane is highly enhanced by phosphorylation of rhodopsin, *FEBS Lett* 176:473–478.

34. Lohse, M.J., Benovic, J.L., Codina, J., Caron, M.G., and Lefkowitz, R.J., 1990, β-arrestin: a protein that regulates β-adrenergic receptor function, *Science* 248:1547–1550.

35. Yamaki, K., Tsuda, M., and Shinohara, T., 1988, The sequence of human retinal S-antigen reveals similarities with α-transducin, *FEBS Lett* 234:39.

36. Wistow, G.J., Katial, A., Craft, C., and Shinohara, T., 1986, Sequence analysis of bovine retinal S-antigen, *FEBS Lett.* 196:23–28.

37. Sakuma, H., Inana, G., Murakami, A., and McLaren, M.J., 1996, Immunolocalization of X-arrestin in human cone photoreceptors, *FEBS Lett.* 382:105–110.

38. Nir, I. and Ransom, N., 1992, S-antigen in rods and cones of the primate retina: Different labeling patterns are revealed with antibodies directed against specific domains in the molecule, *J. Histochem. Cytochem.* 40:343–352.

39. Nork, T.M., Mangini, N.J., and Millechia, L.L., 1993, Rods and cones contain antigenically distinctive S-antigens, *Invest. Ophthalmol. Vis. Sci.* 34:2918–2925.

40. Higashide, T., Murakami, A., McLaren, M.J., and Inana, G., 1996, Cloning of the cDNA for a novel photoreceptor protein, *J. Biol. Chem.* 271:1797–1804.

41. Maduro, M. and Pilgrim, D., 1995, Identification and cloning of unc-119, a gene expressed in the Caenorhabditis elegans nervous system, *Genetics* 141:977–988.

42. Nichols, B.E., Sheffield, V.C., Vandenburgh, K., Drack, A.V., Kimura, A.E., and Stone, E.M., 1993, Butterfly-shaped pigment dystrophy of the fovea caused by a point mutation in codon 167 of the RDS gene, *Nature Genet.* 3:202–207.

43. Wells, J., Wroblewski, J., Keen, J., Inglehearn, C., Jubb, C., Eckstein, A., Jay, M., Arden, G., Bhattacharya, S., Fitzke, F., and Bird, A., 1993, Mutations in the human retinal degeneration slow (RDS) gene can cause either retinitis pigmentosa or macular dystrophy, *Nature Genet.* 3:213–218.

44. Kajiwara, K., Sandberg, M.A., Berson, E.L., and Dryja, T.P., 1993, A null mutation in the human peripherin/RDS gene in a family with autosomal dominant retinitis punctata albescens, *Nature Genet.* 3:208–212.

45. Kajiwara, K., Berson, E.L., and Dryja, T.P., 1994, Digenic retinitis pigmentosa due to mutations at the unlinked peripherin/rds and rom 1 loci, *Science* 264:1604–1608.

46. Sakuma, H., Inana, G., Murakami, A., Yajima, T., Weleber, R.G., Murphey, W.H., Gass, J.D.M., Hotta, Y., Hayakawa, M., Fujiki, K., Gao, Y.Q., Danciger, M., Farber, D., Cideciyan, A.V., and Jacobson, S.G., 1995,

A heterozygous putative null mutation in rom-1 without a mutation in peripherin/RDS in a family with retinitis pigmentosa, *Genomics* 27:384–386.

47. Kemp, C.M., Jacobson, S.G., Cideciyan, A.V., Kimura, A.E., Sheffield, V.C., and Stone, E.M., 1994, RDS gene mutations causing retinitis pigmentosa or macular degeneration lead to the same abnormality in photoreceptor function, *Invest. Ophthalmol. Vis. Sci.* 35:3154–3162.

48. Weleber, R.G., Carr, R.E., Murphey, W.H., Sheffield, V.C., and Stone, E.M., 1993, Phenotypic variation including retinitis pigmentosa, pattern dystrophy, and fundus flavimaculatus in a single family with a deletion of codon 153 or 154 of the peripherin/RDS gene, *Arch. Ophthalmol.* 111:1531–1542.

49. Jacobson, S.G., Kemp, C.M., Cideciyan, A.V., Macke, J.P., Sung, C.-H., and Nathans, J., 1994, Phenotypes of stop codon and splice site rhodopsin mutations causing retinitis pigmentosa, *Invest. Ophthalmol. Vis. Sci.* 35:2521–2534.

50. Wiechmann, A.F., Akots, G., Hammarback, J.A., Pettenati, M.J., and others, 1994, Genetic and physical mapping of human recoverin: a gene expressed in retinal photoreceptors, *Invest. Ophthalmol. Vis. Sci.* 35:325–31.

51. Polans, A.S., Buczylko, J., Crabb, J., and Palczewski, K., 1991, A photoreceptor calcium binding protein is recognized by autoantibodies obtained from patients with cancer-associated retinopathy, *J. Cell Biol.* 112:981–989.

52. Parminder, A.H., Murakami, A., Inana, G., Berson, E.L., and Dryja, T.P., Evaluation of the human gene encoding recoverin in patients with retinitis pigmentosa or an allied disease, *Invest. Ophthalmol. Vis. Sci.* in press.

25

STUDIES ON THE CONE CYCLIC GMP-PHOSPHODIESTERASE α' SUBUNIT GENE

Debora B. Farber,[1] Natik Piriev,[1] Yong Qing Gao,[1] Michael Danciger,[1] and Andrea Viczian[1,2]

[1]Jules Stein Eye Institute
[2]Interdepartmental Neuroscience Program
UCLA School of Medicine
Los Angeles, California
Loyola Marymount University
Los Angeles, California

INTRODUCTION

Studies on animal models of retinitis pigmentosa and allied retinal degenerations have provided very useful information about the molecular basis of these diseases. The genes identified as responsible for the primary cause of the disorders in animals also have been found to be involved in some types of human retinal degenerations. For example, mutations in a rod photoreceptor-specific gene encoding the β-subunit of cGMP-phosphodiesterase (β-PDE) (1–3) placed this gene as a strong candidate for the disease of *rd* mice. The β-PDE gene abnormality was confirmed to be the cause of the *rd* mouse disorder when the ill-fated photoreceptor cells were rescued in transgenic *rd* mice that had integrated the normal β-PDE gene into their genome (4). Later, another mutation in this gene was found to be responsible for the retinal degeneration of Irish setter dogs (5, 6). Hence, the rod β–PDE gene became a candidate for the cause of hereditary retinal degenerations in man. To date, 13 mutations of this gene have been described in families affected with autosomal recessive retinitis pigmentosa (RP) (7–11) and another mutation has been associated with autosomal dominant congenital stationary night blindness (12).

Rod and cone photoreceptors have different phosphodiesterases that hydrolyze cGMP. Each of these enzymes has two catalytic subunits; the α and β subunits of rods form a heterodimer (13) whereas the two identical subunits of cones constitute a homodimer (14, 15). Although these three PDE subunits function similarly, they are three distinct proteins that are encoded by different genes (16–18).

In addition to the mutations in human rod β-PDE mentioned above, abnormalities in the gene encoding rod α-PDE also have been found in patients affected with autosomal recessive RP (19). This led us to hypothesize that the cone α'-PDE gene could be responsible for some inherited human cone dystrophies. Only one animal model for selective cone degeneration has been described at this time in the literature, the *cd* dog, but the defective gene product causing the *cd* disease has not yet been identified (20, 21). Thus, cloning of the human α'-PDE gene, determination of the intronic sequences flanking each of its exons, and mapping it to its chromosomal locus were essential to finding any potential links between this gene and hereditary retinal disease.

We present here a summary of the studies that were carried out with this purpose in our laboratory. We also compare the deduced amino acid sequence of the proteins encoded by the α'-PDE gene with those of the α- and β-PDEs and of cGMP-PDEs present in other tissues. In addition, we analyze elements in the 5' flanking region that could be involved in transcriptional regulation of the human cone α'-PDE gene and comment on our preliminary results related to the screening for mutations in the α'-PDE gene in the DNA of patients affected with different cone dystrophies.

THE α'-PDE GENE

Primary Structure and Organization

The human α'-PDE gene is comprised of 22 exons (Fig. 1) and spans approximately 48 kb (22). Its sequence, as well as that for its cDNA have been deposited in GENBANK (accession numbers U20196–20212). All exon/intron splice sites are in agreement with the GT/AC rule (23) except for the donor recognition site in intron 9. The sequence for this 5' splice site is AG/**gc**aagt instead of the consensus sequence AG/**gt**^a_gagt. We determined the homology score for AG/**gc**aagt by calculating the frequency of each nucleotide at each position of the 8 highly conserved nucleotides(24). Homology scores for splice sites are based on the concept of weight matrix. For donor splice sites, this matrix has four rows (one for each possible nucleotide) and 8 columns (signal length). Despite the absence of **gt**, the intron 9 splice site has a score of 81.75 (22) which is within the range of possible scores for donor recognition sites. All other nucleotides match perfectly with the consensus sequence, compensating for the absence of **gt** and making possible the interaction of the 5' splice site with the U1 snRNA for the production of α'-PDE mRNA (Fig. 2). The α'-PDE gene is not the only one that has a non-canonical 5' splice site. Interestingly, the gene encoding human rod β-PDE also contains a non-consensus donor splice junction for exon 20: AG/**gc**gagt.

Similar to cone α'-PDE, rod α-PDE (25) and rod β-PDE (26) are encoded by 22 exons. The almost identical intron placement and amino acid homology of the putative functional domains in these three genes suggests that cone and rod PDEs have a close phylogenetic relationship and, therefore, may have a common origin. Knowledge about

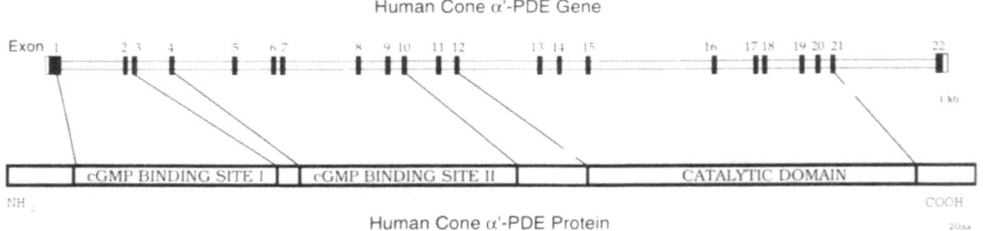

Figure 1. Physical map of the human cone α'-PDE gene and correlation between its exon-intron structure and putative functional domains of the protein it encodes. Exons and introns are shown as solid and shaded rectangles, respectively. The open boxes in exons 1 and 22 indicate the 5' and 3' UTRs, respectively. Reprinted with permission of Academic Press, Inc. from Piriev *et al.*, 1995.

Figure 2. A schematic representation of the interaction between the splice site for exon 9 and a U1 small nuclear ribonucleoprotein particle (snRNP) which is a complex of snRNA and a polypeptide called Sm (23). The splice site in the α'-PDE would be recognized by the U1 snRNP by sequence complementarity.

the gene organization of other PDEs isolated from different species and tissues will help us to better understand the evolution of these proteins.

Transcription Start Site and 5'-Flanking Region Analysis

The 5' flanking region of the human cone α'-PDE gene begins 138 base pairs from the ATG start codon as determined by primer extension and RNase protection assays (27). A putative initiator sequence is within the consensus area of −3 to +5 bp (28). However, this sequence has a C instead of the characteristic A, yet it is surrounded by pyrimidines on both sides (Fig. 3a).

The promoter of a gene is composed of basal elements that bind RNA polymerase II and initiate transcription. These include the initiator and sometimes a TATA box sequence that is usually at −25 or within 100 base pairs of the transcription start site (28). In the α'-PDE gene, the conventional TATA box is distinctly absent at −25 bp (Fig. 3a). However, the nucleotides of the 5' flanking region corresponding to −19 to −25 bp form a TATA-like box similar to those in the genes for human rod arrestin (29) and rod β-PDE (30) (Fig. 3b), suggesting that this sequence may act as a TATA element for photoreceptor TATA-less promoters. Interestingly, a canonical TATA box is found at −376 bp in the α'-PDE gene 5' flanking region. This element is too far from the start site of transcription to be considered functional. There is also an initiator sequence precisely at −354, 22 bp downstream of this TATA box. It is not clear what function, if any, these elements have, but one possibility is that they are part of another gene that begins just upstream of the α'-PDE gene.

Upstream promoter element-binding sites, such as CAAT boxes, enhance transcription initiation. Like TATA boxes, they are usually found within 100 base pairs of the transcription start site (28). The human cone α'-PDE promoter has no CAAT box in that area; however, we have identified two CAAT boxes further upstream at −286 and −296 bp. We have also found a PEA3 element at −454 bp, an AP-1-like site at −422 bp, a GC box at −181, a Ck-1 like element at +40 bp, and a Ck-2 site at +81 bp.

The binding of a PEA3 transcription factor to a PEA3 site is regulated by a number of oncogenes. Very often PEA3 acts synergistically on PEA3 and AP-1 sites to achieve maximal levels of transcription activation by phorbol esters and non-nuclear oncoproteins. Furthermore, the spacing between the PEA3 and AP-1 sites seems to be an important factor in the modulation of transcriptional activity (31, 32). An AP-1-like site is near the PEA3 sequence in the α'-PDE gene 5' flanking region and, possibly, together these two cis-acting elements could enhance transcription of this gene. We have previously shown that an AP-1 element is absolutely necessary for the transcriptional activation of the homologous photoreceptor-specific β-PDE gene (33).

a)

```
    -862   TCGAGGTCGA CCTGCAGGTC AACGGATCCT AGCAACAGAC AAGATATCTA

    -812   TCATCTAGTA GATTTAGTAG ACTGTATGTT GTAGCAAAAT CGCATGACCT

    -762   CTGGCTATTA CTATTGAAAA CAAAATCCTA ACTAAGATAC TTAGCAGGAA

    -712   AAACCCAGCC AGCAACTATG TAGGTTACCT TTTGTAATTA CAGTCTCTCG

    -662   ACAAGGGTTA TTATTCAATT CATGACTTGA TTAGGTTTGG ATCATTGGCA

    -612   AATATTTTGC TATGTTCACC TCTTTAGTAA TAAGCTTGAG GTTTACTCCT

    -562   TTAAAACACT CTGAGGTCA AGCTTTCTGA AGTGACAGGG ATTGAGAGGT

    -512   GAAGGTTCTA GAGAAGAGG AGGCCAAGCA ACCTTGGTGA TCAAAAGACT
                       PEA3                                   AP-1-like
    -462   TTTAAGAAAG GAAAGATTAA TATTCACCTT TATGTTGCAG TGATGCAAAC
                                                     TATA box
    -412   TGTTCCATAT CTCCCAGATG TTCATTACAG TTGCTATATA AACAAAAAAA

    -362   ATCCAAGCAC TTCCGAGGTC TAAATAGTTT TACTTCCTAT AGCTATCATA
                              CAAT boxes
    -312   CAAGAACGAT TCAAAGTGCC AATTGTTGCC AATGTCGTGT CATTCATGGA

    -262   TATGTCAGTG ATTCTTCCTT GCTGTGGACA GCAGCCGACA TTCAGTTATT
                                                     GC box
    -212   TTTCTGGAAT AGACACCTCA GCTTGCTACT TGCTGGGCTG ATCTAAGATA
                                                         PCE-1 like
    -162   CTCTGAAGAT TCGCTAGGGA CTTATGAGCT ACTAATGCTC AGGGATTTAG

    -112   TGTCCTGAAG AACTCTTTAA TCCTGTGCTT CCTGAATCC CACCCTAAGC
                                                 TATA box area
     -62   CAGGGCCTTC TGTCACATCC CAAGAGTTAC AGGCAGTTT GAAAGCTCTGC
                       *
     -12   TTTCTGCTTG ACctttggaa gtcctatgag ggaccattta cggtttcctc
               deg Ck-1                                    Ck-2
     +39   agtaatttcc accaggatga atttccttct catcactctg cctcaggtag

     +89   tgctctgaag gtcgtccttt ctgaacaaac gcagcaaagc aagccacacc

    +139   ATG
```

b)

```
    Hum cone α'-PDE CAGTTTGAAAGCTCT
     Hum rod β-PDE CAcTaaGAAAGactT
    Hum cone arres agcTTTcAtttagCT
```

Figure 3. Sequence of the 5' flanking region of the human cone α'-PDE gene. a) The translation initiation codon (ATG) is shown in bold. The translation initiation site is shown with an asterisk. The putative initiator sequence is underlined. All potential elements have been boxed and labeled. b) Comparison of the TATA box area from the 5' flanking region of human cone α'-PDE, rod β-PDE, and cone arrestin. The putative TATA-like box is in black.

GC boxes bind the ubiquitously expressed transcription factor Sp-1 that assists in forming the RNA polymerase II complexes (34). It has been shown that Sp-1 can effect transcription initiation at a distance of up to ~1.8 kb (35) and therefore, at −181, this element may be active in enhancing transcription of the α'-PDE gene.

Ck-2 ($^{5'}$TCAG^A_GTA$^{3'}$) and Ck-1 ($^{5'}$G^A_G G^A_GTT^T_C CA^T_C $^{3'}$) are cytokine-binding elements that are found in several genes (36). In the promoter of the granulocyte-colony stimulating factor gene, Ck-1 alone has been shown to activate transcription in fibroblasts (37) but in the IL-3 promoter, Ck-1 together with Ck-2 seem to repress transcriptional activation in the same cells (38). We are determining whether the cone α'-PDE gene is regulated by these two elements.

Response elements bind factors that are present at a certain time in a particular tissue. In photoreceptor-specific genes, PCE I (consensus sequence $^{5'}$CAATTAG$^{3'}$) has been identified as such an element (39). PCE I binds a retina-specific nuclear factor (Ret-1) which has been found in rat retinal extracts (40, 41). A degenerate form of this element ($^{5'}$GATTTAG$^{3'}$) located in the 5' flanking region of human cone α'-PDE may be involved in conferring tissue specificity to this gene.

We have begun studies intended to delineate the human cone α'-PDE promoter and its efficacy in driving expression of a reporter gene. Constructs containing different lengths of the α'-PDE 5' flanking region (the longest starting at −850 bp) were cloned upstream of the luciferase gene and transiently transfected into human Y-79 retinoblastoma cells in culture. Our preliminary results indicate that 850 bp of upstream sequence drive expression of the reporter gene at about half the levels found with an SV40 promoter. Transcriptional activity is reduced to basal levels when 740 bp are deleted from the −850 bp construct, suggesting that the minimal promoter may be constituted by the first 110 bp upstream of the transcription start site. We are currently completing the characterization of the promoter and using DNase I footprinting and gel shift assays to study the DNA-binding proteins that enhance or repress transcription. These studies may reveal novel, yet uncharacterized elements.

Mapping of the α'-PDE Gene

We have used fluorescence in situ hybridization to map the α'-PDE gene to human chromosome 10q24 (Fig. 4), in a region homologous to mouse chromosome 19 (42). None of the few currently known cone disease loci correspond to 10q24. However, by virtue of its map position, this gene is a potential candidate for any of the inherited heterogeneous cone diseases that may be localized to 10q24 by linkage analysis in the future.

COMPARISON OF THE DEDUCED AMINO ACID SEQUENCE OF CONE α'-PDE WITH THOSE OF OTHER cGMP-PDEs

Cone α'-PDE is a highly conserved protein. Human α'-PDE shares 89% and 83% amino acid identity with bovine and chicken α'-PDEs, respectively. Furthermore, it is homologous to human rod α-PDE (67% identity) and rod β-PDE (66% identity).

α'-PDE has five putative domains: the N terminus, two repeated segments which bind guanine nucleotides, the catalytic domain, and a C terminus containing a CAAX box that allows geranylgeranylation for membrane binding. We have compared these five putative domains of human cone α'-PDE to those of the enzymes from bovine and chicken cones and also to those of human rod α-PDE (43) and β-PDE (1). In addition, we have

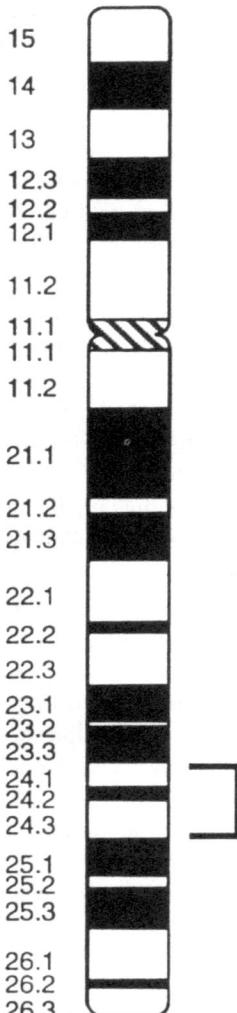

Figure 4. Localization of the human cone cGMP-PDE gene (PDE6C, previously known as PDEA2) to chromosome 10 at the q24 region.

analyzed the catalytic domain and the C-terminus region of the bovine heart cGMP-stimulated PDE (cGS-PDE (44)) and the cGMP-binding PDE (cGB-PDE (45)).

It has been established that the least conserved areas among photoreceptor PDEs are the N terminus and C-terminus (Figs. 5a and 5b). N-terminal amino acids ~15 to 35 encompassing the putative binding site for the inhibitory γ-PDE, may serve a function unique to cones and rods. Thus, it is not surprising to find low percent identity of this region of cone α'-PDE with cGB-PDE and cGS-PDE since these do not bind γ-PDE. The C-termini of all photoreceptor PDEs end with an isoprenylation site (the CAAX sequence). This is not the case for the cGB-PDE and cGS-PDE which are not membrane-associated proteins; however, cGS-PDE has a CAAX box located 3 amino acids from the end of the protein. At this location, the CAAX box is probably not allowing for isoprenylation of the enzyme (46), yet it is interesting that this area remained conserved but not functional.

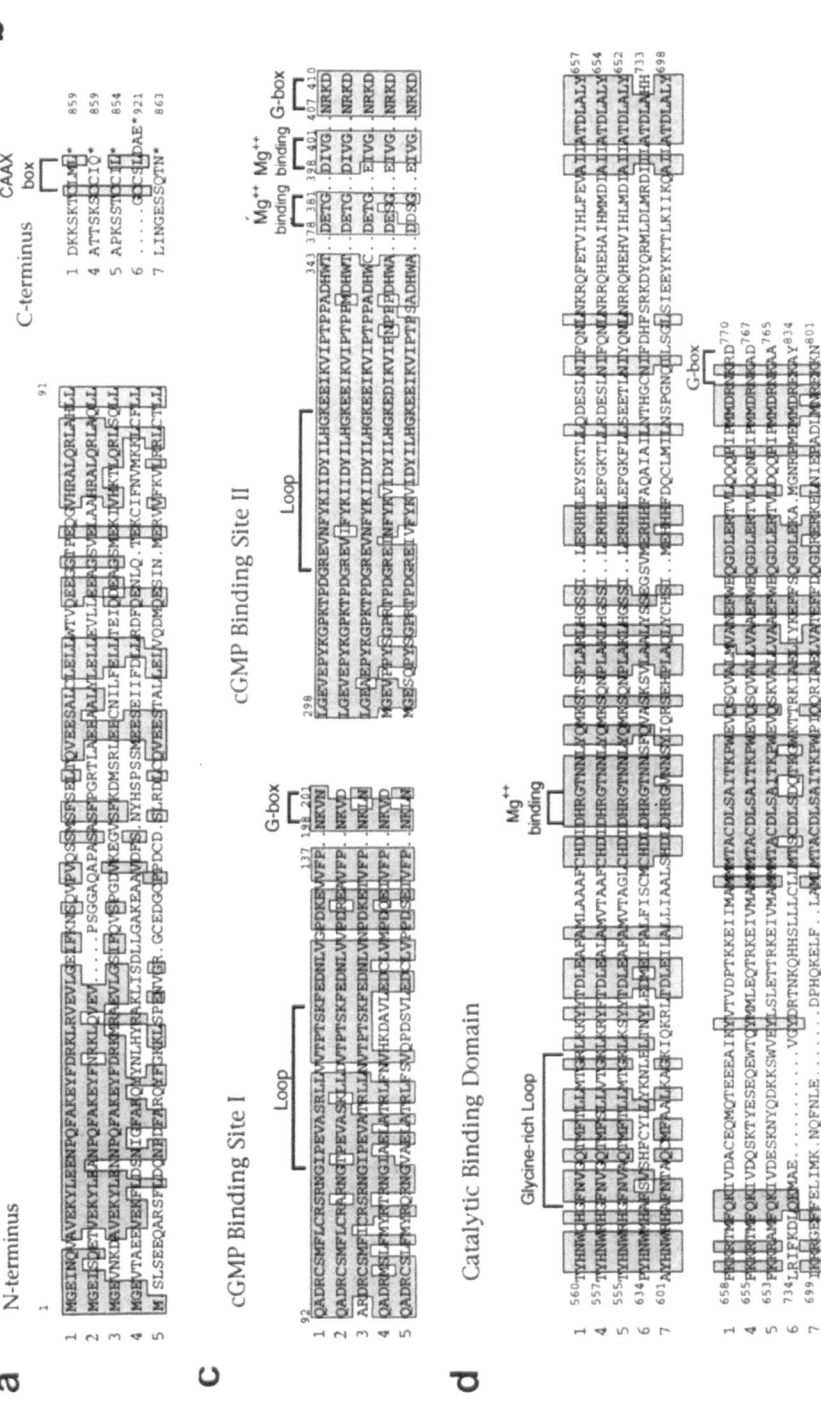

Figure 5. The five putative functional domains (a–d) of human cone α′-PDE (1) compared to bovine (2) and chicken (3) cone α′-PDEs, human rod α- (4) and β-PDEs (5), bovine cGS-PDE (6), and bovine lung cGB-PDE (7).

cGMP-binding domains have been shown to have a glycine-rich loop, a guanine-ring binding G-box (NKXD) and a Mg^{++}-binding sequence (DXXG) (47, 48). We found that the photoreceptor PDEs have all three of these characteristic peptide sequences conserved in the second non-catalytic cGMP-binding domain (Fig. 5c). The first non-catalytic GMP-binding domain has only the loop and the G-box. We have not shown in Fig. 5c the non-catalytic cGMP-binding domains of cGB-PDE and cGS-PDE since these are not conserved. However, we have examined the similarities between cone α'-PDE and the non-photoreceptor PDEs in their catalytic region (Fig. 5d). As expected, there is a 100% identity of the Mg^{++}-binding site among all PDEs but not of the glycine-rich ring and the G-box.

'Signature sequences' are defined by well conserved areas from multiple sequence alignments that reveal specific motifs in families of proteins. For example, signature sequences found by Bentley and Beavo (49) uniquely distinguish cyclic nucleotide PDEs from all others in the SwissProt Data base. We have identified two new amino acid strings that pull out only retinal PDEs when BLASTMAILed to a non-redundant database containing 128,089 sequences (35,506,170 letters): [442] MNKLENRKDIAQ [453] and [722] CDLSAITKPWEVQS [735]. The former was found in the cGMP-binding domain and the latter in the catalytic region.

SEARCH FOR MUTATIONS OF THE CONE α'-PDE GENE IN THE DNA OF PATIENTS AFFECTED WITH DIFFERENT CONE DYSTROPHIES

To date we have completed exon screening of the α'-PDE gene in more than 100 patients with various forms of autosomal recessive or simplex cone disease (including cone-rod degeneration, cone degeneration, fundus flavimaculatus, Bardet-Biedl syndrome and macular degeneration). To do this, we amplified by PCR each of the 22 exons of the α'-PDE gene (exon 1 in two amplicons) with flanking intronic primers and then carried out SSCPE and DGGE analyses on each one of the amplicons. Although we have found several sequence variants, none of them could be associated with disease. It is possible that mutations in the α'-PDE gene responsible for cone degeneration may be found with further screenings.

REFERENCES

1. Bowes, C., Li, T., Danciger, M., Baxter, L.C., Applebury, M.L., and Farber, D.B., 1990, Retinal degeneration in the *rd* mouse is caused by a defect in the β subunit of rod cGMP-phosphodiesterase, *Nature* 347: 677–80.
2. Pittler, S.J. and Baehr, W., 1991, Identification of a nonsense mutation in the rod photoreceptor cGMP phosphodiesterase beta-subunit gene of the *rd* mouse., *Proc. Natl. Acad. Sci. U S A* 88: 8322–6.
3. Bowes, C., Li, T., Frankel, W.N., Danciger, M., Coffin, J.M., Applebury, M.L., and Farber, D.B., 1993, Localization of a retroviral element within the *rd* gene coding for the beta subunit of cGMP phosphodiesterase, *Proc. Natl. Acad. Sci. U S A* 90: 2955–9.
4. Lem, J., Flannery, J.G., Li, T., Applebury, M.L., Farber, D.B., and Simon, M.I., 1992, Retinal degeneration is rescued in transgenic *rd* mice by expression of the cGMP phosphodiesterase beta subunit, *Proc. Natl. Acad. Sci. U S A* 89(10): 4422–6.
5. Farber, D.B., Danciger, J. and Aguirre, G., 1992, The beta subunit of cyclic GMP phosphodiesterase mRNA is deficient in canine rod-cone dysplasia 1, *Neuron* 9: 349–56.
6. Suber, M.L., Pittler, S.J., Qin, N., Wright, G.C., Holcombe, V., Lee, R.H., Craft, C.M., Lolley, R.N., Baehr, W., and Hurwitz, R.L., 1993, Irish setter dogs affected with rod/cone dysplasia contain a nonsense mutation in the rod cGMP phosphodiesterase beta-subunit gene., *Proc. Natl. Acad. Sci. U S A* 90: 3968–72.

7. McLaughlin, M.E., Sandberg, M.A., Berson, E.L., and Dryja, T.P., 1993, Recessive mutations in the gene encoding the β-subunit of rod phophodiesterase in patients with retinitis pigmentosa, *Nature Gen.* **4**: 130–133.

8. Danciger, M., Blaney, J., Gao, Y.Q., Zhao, D.Y., Heckenlively, J.H., Jacobson, S.G., and Farber, D.B., 1995, Mutations in the PDE6B gene in autosomal recessive retinitis pigmentosa, *Genomics* **30**: 1–7.

9. McLaughlin, M.E., Ehrhart, T.L., Berson, E.L., and Dryja, T.P., 1995, Mutation spectrum of the gene encoding the beta subunit of rod phophodiesterase among patients with autosomal recessive retinitis pigmentosa, *Proc. Natl. Acad. Sci. U S A* **92**: 3249–53.

10. Bayes, M., Giordano, M., Balcells, S., Grinberg, D., Vilageliu, L., Martinez, I., Ayuso, C., Benitez, J., Ramos-Arroyo, A.M., Chivelet, P., Solans, T., Valverde, D., Amselem, S., Goossens, M., Baiget, M., Gonzalez-Duarte, R., and Besmond, C., 1995, Homozygous tandem duplication within the gene encoding the beta-subunit of rod phosphodiesterase as a cause for autosomal recessive retinitis pigmentosa, *Hum. Mutat.* **5**(3): 228–34.

11. Valverde, D., Solans, T., Grinberg, D., Balcells, S., Vilageliu, L., Bayes, M., Chivelet, P., Besmond, C., Goossens, M., Gonzalez, D.R., and Baiget, M., 1996, A novel mutation in exon 17 of the beta-subunit of rod phosphodiesterase in two RP sisters of a consanguineous family, *Hum. Genet.* **97**(1): 35–8.

12. Gal, A., Orth, U., Baehr, W., Schwinger, E., and Rosenberg, T., 1994, Heterozygous missense mutation in the rod cGMP phosphodiesterase β-subunit gene in autosomal dominant stationary night blindness, *Nature Gen.* **7**: 64–7.

13. Baehr, W., Devlin, M.J., and Applebury, M.L., 1979, Isolation and characterization of cGMP phosphodiesterase from bovine rod outer segments, *J. Biol. Chem.* **254**: 11669–11677.

14. Hurwitz, R.L., Bunt-Milam, A.H., Chang, M.L., and Beavo, J.A., 1985, cGMP phosphodiesterase in rod and cone outer segments of the retina, *J. Biol. Chem.* **260**: 568–573.

15. Gillespie, P.G. and Beavo, J.A., 1988, Characterization of a bovine cone photoreceptor phosphodiesterase purified by cyclic GMP-sepharose chromatography, *J. Biol. Chem.* **263**: 8133–8141.

16. Ovchinnikov, Y.A., Gubanov, V.V., Khramtsov, N.V., Ischenso, K.A., Zagranichny, V.E., Muradov, K.G., Shuvaeva, T.M., and Lipkin, V.M., 1987, Cyclic GMP phosphodiesterase from bovine retina: Amino acid sequence of the α-subunit and nucleotide sequence of the corresponding cDNA, *FEBS Letters* **223**: 169–173.

17. Lipkin, V.M., Khramtsov, N.V., Vasilevskaya, I.A., Atabekova, N.V., Muradov, K.G., Gubanov, V.V., Li, T., Johnston, J.P., Volpp, K.J., and Applebury, M.L., 1990, β-Subunit of bovine rod photoreceptor cGMP phosphodiesterase, *J. Biol. Chem.* **265**: 12955–12959.

18. Li, T., Volpp, K. and Applebury, M.L., 1990, Bovine cone photoreceptor cGMP phosphodiesterase structure deduced from a cDNA clone, *Proc. Natl. Acad. Sci. USA*, **87**: 293–297.

19. Huang, S.H., Huang, X., Pittler, S.J., Oliveira, L., Berson, E.L., and Dryja, T.P., 1995, A mutation in the gene encoding the α-subunit of rod cGMP-phosphodiesterase (PDEA) in retinitis pigmentosa., *ARVO Abstract* **36**(4): S825.

20. Gropp, K., Szel, A., Huang, J., Acland, G., Farber, D., and Aguirre, G., 1996, Selective absence of cone outer segment beta(3)-transducin immunoreactivity in hereditary cone degeneration (CD), *Exp. Eye Res.* **63**(3): 285–296.

21. Akhmedov, N. and Farber, D.B., 1997, Structure and analysis of the transducin β3-subunit gene, a candidate for inherited cone degeneration (cd) in the dog, *Gene*, **in press**.

22. Piriev, N.I., Viczian, A.S., Ye, J., Kerner, B., Korenberg, J.R., and Farber, D.B., 1995, Gene structure and amino acid sequence of the human cone photoreceptor cGMP-phosphodiesterase α' subunit (PDEA2) and its chromosomal localization to 10q24, *Genomics* **28**(3): 429–35.

23. Sharp, P.A., 1987, Splicing of messenger RNA precursors, *Science* **235**: 766–71.

24. Shapiro, M.B. and Senapathy, P., 1987, RNA splice junctions of different classes of eukaryotes: Sequence statistics and functional implications in gene expression, *Nuc. Acids Res.* **15**(17): 7155–7174.

25. Pittler, S.J., personal communication.

26. Weber, B., Riess, O., Hutchinson, G., Collins, C., Lin, B., Kowbel, D., Andrew, S., Schappert, K., and Hayden, M.R., 1992, Genomic organization and complete sequence of the human gene encoding the β-subunit of the cGMP phophodiesterase and its localization to 4p16.3, *Nuc. Acids Res.* **19**: 6263–68.

27. Viczian, A.S., Piriev, N.I., and Farber, D.B., 1995, Isolation and characterization of a cDNA encoding the α' subunit of human cone cGMP-phosphodiesterase, *Gene*, **166**: 205–11.

28. Lewin, B., 1994, *Genes V*, Oxford University Press, Oxford.

29. Yamaki, K., Tsuda, M., Kikuchi, T., Chan, K.-H., Huang, K.-P., and Shinohara, T., 1990, Structural organization of the human S-antigen gene, *J. Biol. Chem.* **265**: 20757–762.

30. DiPolo, A., Bowes-Rickman, C.B., and Farber, D.B., 1996, Isolation and initial characterization of the 5' flanking region of the human and murine cyclic guanosine monophosphate-phosphodiesterase beta-subunit genes, *Invest. Ophthalmol. Vis. Sci.* **37**: 551–560.

31. Gutman, A. and Wasylyk, B., 1990, The collagenase gene promoter contains a TPA and oncogene-responsive unit encompassing the PEA3 and AP-1 binding sites, *Embo J.* **9**(7): 2241–6.

32. Wasylyk, B., Wasylyk, C., Flores, P., Begue, A., Leprince, D., and Stehelin, D., 1990, The c-ets proto-oncogenes encode transcription factors that cooperate with c-Fos and c-Jun for transcriptional activation, *Nature* **346**(6280): 191–3.

33. Farber, D.B., DiPolo, A., and Ogueta, S.B., 1996, Human β-PDE 5' region: *In vitro* and *in vivo* studies of potential elements involved in cell-specific transcription, *Exp. Eye Res. Supplement* **63**: 413.

34. Schmidt, M.C., Zhou, Q., and Berk, A.J., 1989, Sp1 activates transcription without enhancing DNA-binding activity of the TATA box factor, *Mol. Cell Biol.* **9**(8): 3299–307.

35. Courey, A.J., Holtzman, D.A., Jackson, S.P., and Tjian, R., 1989, Synergistic activation by the glutamine-rich domains of human transcription factor Sp1, *Cell* **59**(5): 827–36.

36. Shannon, M.F., Gamble, J.R., and Vadas, M.A., 1988, Nuclear proteins interacting with the promoter region of the human granulocyte/macrophage colony-stimulating factor gene, *Proc. Natl. Acad. Sci. U S A* **85**(3): 674–8.

37. Shannon, M.F., Coles, L.S., Fielke, R.K., Goodall, G.J., Lagnado, C.A., and Vadas, M.A., 1992, Three essential promoter elements mediate tumor necrosis factor and interleukin-1 activation of the granulocyte-colony stimulating factor gene, *Growth Factors* **7**: 181–193.

38. Ryan, G.R., Vadas, M.A., and Shannon, M.F., 1994, T-cell functional regions of the human IL-3 proximal promoter, *Mol. Reprod. Dev.* **39**(2): 200–7.

39. Kikuchi, T., Raju, K., Breitman, M.L., and Shinohara, T., 1993, The proximal promoter of the mouse arrestin gene directs gene expression in photoreceptor cells and contains an evolutionarily conserved retinal factor-binding site, *Molec. Cell. Biol.* **13**: 4400–4408.

40. Morabito, M.A., Yu, X., and Barnstable, C.J., 1991, Characterization of developmentally regulated and retina-specific nuclear protein binding to a site in the upstream region of the rat opsin gene, *J. Biol. Chem.* **266**: 9667–72.

41. Yu, X. and Barnstable, C.J., 1994–5, Characterization and regulation of the protein binding to a cis-acting element, RET 1, in the rat opsin promoter, *J. Molec. Neurosc.* **5**(4): 259–71.

42. Danciger, M., Kozak, C., Applebury, M.L., Polans, A., and Farber, D.B., 1993, Chromosomal mapping in the mouse of the genes for two proteins expressed in photoreceptors: Cone cGMP-PDE α' and recoverin, *Invest. Ophthalmol. Vis. Sci.* **34**: 1461.

43. Pittler, S.J., Baehr, W., Wasmuth, J.J., McConnell, D.G., Champagne, M.S., vanTuinen, P., Ledbetter, D., and Davis, R.L., 1990, Molecular characterization of human and bovine rod photoreceptor cGMP phosphodiesterase alpha-subunit and chromosomal localization of the human gene, *Genomics* **6**(2): 272–83.

44. Charbonneau, H., Beier, N., Walsh, K.A., and Beavo, J.A., 1986, Identification of a conserved domain among cyclic nucleotide phosphodiesterases from diverse species, *Proc. Natl. Acad. Sci. USA* **83**: 9308–9312.

45. McAllister-Lucas, L.M., Sonnenburg, W.K., Kadlecek, A., Seger, D., Trong, H.L., Colbran, J.L., Thomas, M.K., Walsh, K.A., Francis, S.H., Corbin, J.D., and Beavo, J.A., 1993, The structure of a bovine lung cGMP-binding, cGMP-specific phosphodiesterase deduced from a cDNA clone, *J. Biol. Chem.* **268**(30): 22863–73.

46. Clarke, S., 1992, Protein isoprenylation and methylation at carboxyl-terminal cysteine residues, *Annu. Rev. Biochem.* **61**(355): 355–86.

47. Jurnak, F., 1988, The three-dimensional structure of c-H-ras p21: Implications for oncogene and G protein studies, *TIBS* **13**(6): 195–198.

48. Holbrook, S. and Kim, S.H., 1989, Molecular model of the G protein alpha subunit based on the crystal structure of the HRAS protein, *Proc. Natl. Acad. Sci. USA* **86**: 1751–5.

49. Bentley, J.K. and Beavo, J.A., 1992, Regulation and function of cyclic nucleotides, *Curr. Opin. Cell Biol.* **4**: 233–240.

26

MUTATIONS IN *PDE6A*, THE GENE ENCODING THE α-SUBUNIT OF ROD PHOTORECEPTOR cGMP-SPECIFIC PHOSPHODIESTERASE, ARE RARE IN AUTOSOMAL RECESSIVE RETINITIS PIGMENTOSA

M. Meins,[1] A. Janecke,[1] C. Marschke,[1] M. J. Denton,[2]
G. Kumaramanickavel,[2] S. Pittler,[3] and A. Gal[1]

[1]Institut für Humangenetik
Universitäts-Krankenhaus Eppendorf
Butenfeld 42, D-22529, Hamburg, Germany
[2]Department of Biochemistry
University of Otago, Box 56, Dunedin, New Zealand
[3]Department of Biochemistry and Molecular Biology
Ophthalmology and the Center for Eye Research
College of Medicine
University of South Alabama
Mobile, Alabama 36688-0002

SUMMARY

Rod photoreceptor cGMP-specific phosphodiesterase (PDE) is a key enzyme in the phototransduction cascade. Mutations in the gene (*PDE6A*) for the PDE α-subunit have been implicated in autosomal recessive retinitis pigmentosa (RP) in a very small proportion (2/173) of patients from the USA. In order to determine what proportion of *PDE6A* mutations exists in patients with autosomal recessive RP from other populations, we examined 85 German and 9 Indian unrelated patients by single strand conformation polymorphism (SSCP) analysis and direct sequencing. SSCP band shifts were detected in 9 different *PDE6A* gene fragments amplified by polymerase chain reaction. Direct sequencing revealed 6 single base substitutions that are most likely disease unrelated polymorphisms. Three likely pathogenic exonic mutations (788G→T/Arg237Leu, 2041C→T/His655Tyr, and 2624G→T/Gly849Val) were found to be heterozygous whereas a second *PDE6A* mutation has not yet been identified in these patients. One patient from an Indian consanguineous family was homozygous for a 1-base pair deletion (2196dclA, exon 17). This mutation leads to a frame shift with premature stop codon and was shown to cosegregate with the disease phenotype. While we cannot rule out mutations in the coding region that escaped detection or those outside of the coding region examined in this study our results suggest that *PDE6A* mutations are also rare in the European population.

INTRODUCTION

Retinitis pigmentosa (RP) is a heterogeneous group of hereditary retinal dystrophies. To date, almost 20 different RP loci have been mapped in the human genome and seven disease-related genes have already been identified (for a recent compilation see ref. 1). Among these, there are several genes encoding proteins involved in the phototransduction cascade, e.g. rhodopsin, the α-subunit of rod cGMP-gated cation channel as well as the α- and β-subunits of rod photoreceptor cGMP-specific phosphodiesterase (PDE). Rod PDE is a heterotetrameric enzyme ($\alpha\beta\gamma_2$) that function is to reduce cytoplasmic levels of cGMP, a central event in phototransduction (for a recent review see 2). Catalytic PDE activity is associated with the α- and β-subunits. These two subunits have a rather similar primary structure and the structure of the corresponding genes is also largely conserved.

The human gene (*PDE6A*) encoding the α-subunit of PDE maps to the long arm of chromosome 5 (5q31.2-q34; ref. 3). *PDE6A* consists of 22 exons encoding a protein of 859 amino acids. Mammalian phosphodiesterase α- and β-subunits show 72% sequence homology at the protein level. There are two non-catalytic domains for cGMP binding and several regions for interaction with the γ-subunit in the N-terminal part. In the C-terminal part, both the α- and β-subunits possess a catalytic domain common to other cyclic nucleotide PDE (see ref. 4). A C-terminal CAAX motif found both in the α- and β-subunits was shown to be posttranslationally modified by addition of farnesyl and geranylgeranyl lipid groups, respectively, and carboxymethylation which are necessary for membrane attachment of PDE (5). Mutations in *PDE6A* have been implicated in autosomal recessive RP in a very small proportion (2/173) of patients from the USA. In order to investigate whether mutations in *PDE6A* might be responsible for autosomal recessive RP in patients from other populations, we examined 85 German and 9 Indian unrelated patients by single strand conformation polymorphism (SSCP) analysis and direct sequencing.

METHODS

Extraction of genomic DNA from blood leukocytes of the probands was performed by standard methods. Polymerase chain reaction (PCR) was carried out using 100 ng of genomic DNA. The oligonucleotide primers used were described by Huang et al. (6). Final concentration of the reagents in a 25 µl reaction volume was 10 pmol of each primer pair, 10 mM Tris-HCl (pH 8.3), 50 mM KCl, 1.5 mM $MgCl_2$, 200 µM of dATP, dCTP, dGTP, and dTTP, and 0.5 unit of Taq DNA polymerase (Gibco). PCR was performed on a Hybaid Omnigene cycler with 35 cycles of 94°C for 1 min, 55°C for 1 min, and 72°C for 1.5 min. For some fragments the amplification protocol was slightly modified (see Table 1).

Single strand conformation polymorphism (SSCP) analysis was performed essentially as described elsewhere (7) using two different gel conditions routinely [6% acrylamide (cross linking 2.6), and 10% glycerol in 1x TBE buffer; 8% acrylamide (cross linking 2.6) in 1 x TBE buffer without glycerol] and with some exons a third gel recipe [8% acrylamide (cross linking 1.3), 10% sucrose in 0.5x TBE buffer] was used to optimize mutation detection sensitivity. Samples were electrophoresed at 12–30 W for 4–16 hours

Table 1. PCR conditions used for *PDE6A* mutation screening by SSCP

PCR conditions	Exon no.																					
	1	2	3	4	5	6	7	8	9	10	11	12	13	14	15	16	17	18	19	20	21	22
35 cycles	x	x	x	x	x	x	x	x	x	x	x	x	x	x	x	x	x	x	x	x	x	x
Anneal. 1′ at 55°C or instead	x	x	x	x	x	56	x	54	x	54	57	x	x	57	54	56	x	56	56	x	x	x
Extension at 72°C for 90′′ or instead	x	x	x	x	x	10	x	30	x	30	30	x	x	30	30	20 +2	x	20 +2	20 +2	x	x	x

at room temperature or at 4°C. Gels were silver stained. Samples showing band shifts in SSCP analysis were analysed by direct sequencing using standard protocols (7).

RESULTS

We examined *PDE6A* in a total of 94 unrelated patients (85 patients from Germany and 9 patients from India) affected by autosomal recessive retinitis pigmentosa by SSCP analysis and direct sequencing. The amplification products for each of the 22 exons included the donor and acceptor splice sites and about 40 bp of flanking intronic sequences (the exeption being exon 9 in which case the reverse primer was located only 7 bp apart from the exon/intron boundary; Table 1).

Band shifts were detected in 9 different amplicons. Direct sequencing revealed 6 single base substitutions (most of them in more than one person, see Table 2) which are considered most likely disease unrelated polymorphisms. The four sequence changes listed in Table 3 have been found in all patients and unaffected controls analysed in our laboratory so far suggesting that, in each case, the *PDE6A* sequence originally reported (6) was a rare variant. Three likely pathogenic exonic mutations were each heterozygous in three separate

Table 2. Summary of most likely non-pathogenic sequence variants of the human *PDE6A* gene found in this study. The amino acid and nucleotide numbering follows that reported by Pittler et al. (3)

Exon	Nucleotide change	Codon	Restriction pattern change	Number of probands showing the alteration (n=94)
1	409 A/C	AGG → CGG(Arg111)	+BcnI	3 heterozygous, 1 homozygous
1	543 C/T	AAC → AAT(Asn155)	—	3 heterozygous, 1 homozygous
18	2278+21 A/C	Intron 18	—	4 heterozygous
20	2415 T/C	TTT → TTC(Phe779)	—	2 heterozygous
21	2478 C/T	GAC → GAT(Asp800)	+FokI	4 heterozygous
21	2502 G/A	GAG → GAA(Glu808)		1 heterozygous

Table 3. Summary of sequence variants of the human *PDE6A* gene found in all individuals analysed when compared to the sequence originally reported (3)

Exon	Nucleotide	Codon
3	TGG → GTG	Trp243Ser
7	1120A → G	Thr348Ala
8	TCA → GCG	Ser364Ala
22	2611AACCAGCCCAGGGGTGCA → AACCCCAGCCCAGGGGTGCA	Cd 845 ff (originally reported)

Table 4. Summary of the likely pathogenic sequence variants of the human *PDE6A* gene found in this study. The amino acid and nucleotide numbering follows that reported by Pittler et al. (3)

Exon	Nucleotide change	Mutation	Restriction pattern change	Number of probands showing the alteration (n=94)
3	788G →T	Arg237Leu	−MaeII	1 heterozygous
16	2041C → T	His655Tyr	−MslI	1 heterozygous
17	2196delA	1-bp deletion	+MspI	1 homozygous
22	2624G → T	Gly849Val	−	1 heterozygous

patients whereas a second *PDE6A* mutation has not yet been identified in these individuals (Table 4). None of these latter changes were seen on 60 alleles of unaffected controls (not shown) suggesting that they are not frequent polymorphisms (see also Discussion).

SSCP showed a unique band shift in exon 17 for an Indian RP patient from a consanguineous marriage (family PMK2, Figure 1). The index case has a history of night blindness from the age of three years. On ophthalmological examination at the age of 27 years, peripheral vision was clearly reduced. Although central vision was fairly preserved, visual acuity was low (RE 6/12, LE 6/18).

The proband was found to be homozygous for the mutation 2196delA (Figure 2). The deletion results in a frame shift predicting a premature stop of translation at codon 714 in the first half of exon 18 (Figure 3). The deletion creates a new MspI site. Amplification of exon 17 for all available family members and digestion of DNA samples with MspI showed that the gene alteration cosegregated with the disease phenotype. Both the index case and his affected sister were homozygous while the parents were heterozygous for the 1-base pair deletion (Figures 3 and 4).

DISCUSSION

Mutation screening of human *PDE6A* on 188 alleles of unrelated patients with autosomal recessive RP detected 4 most likely pathogenic alterations in the coding sequence of the gene. These results suggest that mutations in *PDE6A* are rarely the cause of autosomal recessive RP. Our data confirm the results of Huang et al. (6) who studied 167 and 173 unrelated patients with autosomal dominant and autosomal recessive RP, respec-

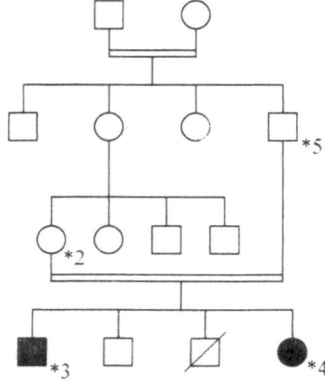

Figure 1. Pedigree of family PMK30. Family members available for genetic analysis are marked with *. Numbering in the pedigree corresponds to that of lanes in Figure 4.

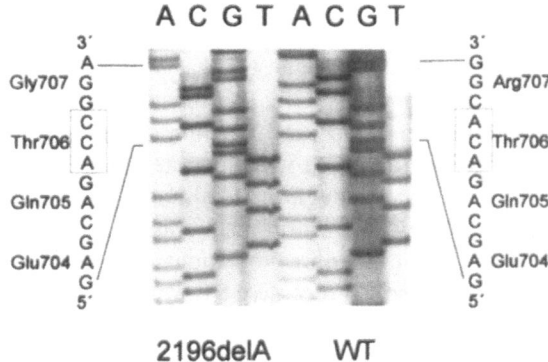

Figure 2. Partial DNA sequence of *PDE6A* exon 17. The proband (left) carries a homozygous deletion of nucleotide 2196 (numbering according to Pittler et al. (3)) causing a frame shift. On the right side the wild-type sequence is shown.

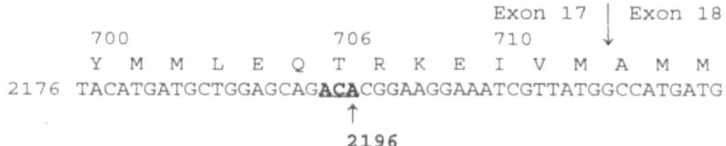

Figure 3. Predicted effect of the 1-bp deletion on protein translation. The frame shift leads to a premature stop of translation at codon 714.

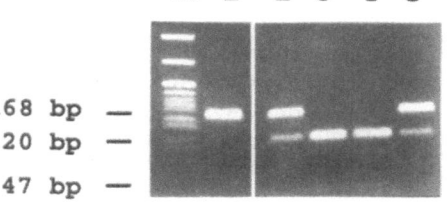

Figure 4. Detection of 2196delA by restriction digestion with MspI. The 1 base-pair deletion creates a new recognition site. Both parents (2 and 5) are heterozygotes, while the index case and his affected sister are homozygotes for the deletion. Lane 1=unaffected control, M=molecular weight marker pBR322, MspI digested.

tively, and found 3 point mutations in *PDE6A* in affected members of two families with autosomal recessive RP. Mutations in *PDE6A* thus seem to be responsible for approximately 1% of all autosomal recessive RP cases. Similarly, the frequency of mutations of other genes implicated in autosomal recessive RP identified to date is quite low, usually below 5% in each case.

A homozygous 1 bp-deletion (2196delA) was found to cosegregate with the disease phenotype in a consanguineous Indian family. It is very likely that this mutation is pathogenic by alterating the amino acid composition of *PDE6A* downstream from codon 707 due to frame shift and premature translation stop signal after residue 713. The mutant α-subunit thus lacks the posterior part of the putative catalytic domain ranging from amino acids 556 to 779. Moreover, the C-terminal CAAX motif required for membrane attachment of PDE is also missing. It was suggested that the CAAX motif of the α- and β-subunits is of primary importance for the activity and function of the enzyme (8). This is also consistent with the observation of a missense mutation changing the last residue (L854V) of the CAAX motif of the β-subunit which was suggested to be the cause of recessive RP in a compound heterozygote (9). Although the total number of clearly disease related mutations identified so far is very low, it is worth mentioning that the majority of them are nonsense mutations in *PDE6A*. Two of the three mutations identified by Huang et al. (6) in members of two families with autosomal recessive RP were stop mutations located early in the putative catalytic domain. It is very likely that these mutations, like the nonsense mutation we detected, result in a "functional null allele" as shown for nonsense mutations in other cases.

Three patients were each shown to carry a single heterozygous missense mutation predicting nonconservative amino acid changes (R237L, H655Y or G849V) of residues conserved throughout evolution whereas a second *PDE6A* mutation has not been detected. One (H655Y) of these mutations lies within the putative catalytic domain of the α-subunit. Histidine-655 is evolutionarily conserved between human and bovine α-subunits and is located in a domain highly conserved among many cyclic nucleotide phosphodiesterases from diverse species (3, 5). Arginine-237 is evolutionarily conserved between human and bovine α-PDE and is located in a region of the protein that is strictly conserved between both species. Glycine-849 is an evolutionarily conserved residue between human and bovine α-PDE, although it is situated in a less conserved part of the protein near the carboxyl-terminal end as arginine-848 and alanine-850 in human are replaced by glycine in the bovine sequence. Though each of the above mentioned substitutions might result in a decrease of enzyme activity, it is very unlikely that they cause retinal degeneration in heterozygous state alone as none of the parents of the three patients show(ed) signs of a retinal degeneration (data not presented). Although we have repeatedly screened all exons of the α-PDE gene of these three probands by SSCP and heteroduplex analyses using various experimental conditions to increase detection sensitivity no aberration was seen (data not shown). Clearly, we can not exclude that we have missed the second mutation in these DNA samples as neither SSCP nor HA, even if combined, reveals all mutations present in the amplicons. In addition, certain types of heterozygous mutations, e.g. exon-spanning deletions or inversions, escape detection if single-exon amplification is used. We also cannot rule out the presence of disease causing mutations in the promoter region or within the introns. One may also speculate that the missense mutations identified in our investigation are pathogenic only in combination with mutations in other genes (digenic or polygenic RP). This mechanism of allelic noncomplementation was suggested for heterozygous mutations in the peripherin/RDS and ROM1 genes (10). Although mutations in the genes encoding rhodopsin, the β-subunit of PDE, and the α-subunit of the cGMP-gated photoreceptor cation channel have already been excluded in the respective patients, further research is necessary to deal with this issue.

Light stimulus of a vertebrate rod cell results in closure of cGMP-gated cation channels leading to hyperpolarization of the external plasma membrane of the photoreceptor outer segment. Mutations in the genes encoding different proteins of the phototransduction cascade have been implicated in hereditary retinal diseases. Heterozygous mutations in the rhodopsin gene are responsible for approximately 25% of all dominant RP cases (see ref. 11) by a yet unknown cytotoxic or dominant negative effect. On the other hand, a minor proportion of mutant rhodopsin alleles was shown to be causative for recessive RP (12, 13) suggesting a loss of function as possible pathomechanism in those cases.

Mutations in other genes encoding proteins involved in the phototransduction cascade have only been reported in autosomal recessive retinal dystrophies. Mutations in the gene (*PDE6B*) for the β-subunit of rod cGMP-specific phosphodiesterase or in the gene (CNCG1) for the γ-subunit of the cGMP-gated cation channel may lead to recessive RP (6, 9, 14–21). Mutations in *PDE6A* have been reported in patients with autosomal recessive RP by Huang et al. (6) and in this paper. Mutations in the gene (*PDE6G*) encoding the γ-subunit of PDE have not been found in human retinal disease, but targeted disruption of the gene in knockout mice missing the γ-subunit of PDE results in a recessive retinal degeneration (22, 23).

The observations gathered to date seem to be in line with the hypothesis that recessive mutations in the above mentioned genes are pathogenic by loss of function. Yet, the exact pathomechanism by which mutations in human *PDE6A* lead to RP is not known. No animal model for retinal degeneration due to *PDE6A* mutations have been identifed so far. For *PDE6B*, encoding the other catalytic subunit of rod photoreceptor cGMP phosphodiesterase, the *rd* mouse and the Irish setter dog with rod-cone dysplasia have been described as animal models. In both cases, nonsense mutations eliminate more than half of the protein including the catalytic domain and the CAAX motif needed for membrane attachment (24, 25). Experiments on these animal models have shown that PDE cDNA and protein concentrations are significantly lower than in unaffected animals. Enzyme activity is minimal what results in elevated cGMP levels in rod outer segments (see ref. 2). It was shown that *rd* mice can be rescued from rapid retinal degeneration by integrating a β-PDE transgene. High expression of the transgene corresponded perfectly with preservation of rod cells (26). Gene therapy therefore might be especially helpful for patients suffering from autosomal recessive RP caused by nonsense mutations in the *PDE6A* or *PDE6B* genes.

ACKNOWLEDGMENTS

This study was financially supported by grants from the Deutsche Forschungsgemeinschaft (Ga 210/5-4), the FAUN-Stiftung (Nürnberg, Germany), The Foundation Fighting Blindness (USA), the New Zealand Health and Research Council, and the Australian Retinitis Pigmentosa Association.

REFERENCES

1. Inglehearn, C.F., and Hardcastle, A.J., 1996, Nomenclature for inherited diseases of the retina. *Am. J. Hum. Genet.* **58**: 433–435.
2. Farber, D.B., 1995, From mice to men: The cyclic GMP phosphodiesterase gene in vision and disease. The proctor lecture, *Invest. Ophthalmol. Vis. Sci.* **36**: 263–275.
3. Pittler, S.J., Baehr, W., Wasmuth, J.J., McConnell, D.G., Champagne, M.S., van Tuinen, P., Ledbetter, D., and Davis, R.L., 1990, Molecular characterization of human and bovine rod photoreceptor cGMP phosphodiesterase α-subunit and chromosomal localization of the human gene, *Genomics* **6**: 272–283.
4. Charbonneau, H., Beier, N., Walsh, K.A., and Beavo, J.A., 1986, Identification of a conserved domain among cyclic nucleotide phosphodiesterases from many species, *Proc. Natl. Acad. Sci. U.S.A.* **83**: 9308–9312.
5. Ong, O.C., Ota, I.M., Clarke, S., and Fung, B.K., 1989, The membrane binding domain of rod cGMP phosphodiesterase is posttranslationally modified by methyl esterification at a C-terminal cysteine, *Proc. Natl. Acad. Sci. U.S.A.* **86**: 9238–9242.
6. Huang, S.H., Pittler, S.J., Huang, X., Oliveira, L., Berson, E.L., and Dryja, T.P., 1995, Autosomal recessive retinitis pigmentosa caused by mutations in the α-subunit of rod cGMP phosphodiesterase, *Nature Genet.* **11**: 468–471.

7. Bunge, S., Fuchs, S., and Gal, A., 1996, Simple and nonisotopic methods to detect unknown gene mutations in nucleic acids, in *Methods in Molecular Genetics*, Volume 8 (K.W. Adolph, ed.), pp.26–39, Academic Press, Orlando.

8. Qin, N., and Baehr, W., 1994, Expression and mutagenesis of mouse rod photoreceptor cGMP phosphodiesterase in insect cells. *J. Biol. Chem.* **269**: 3265–3271.

9. Veske, A., Orth, U., Rüther, K., Zrenner, E., Rosenberg, T., Baehr, W., Gal, A., 1995, Mutations in the gene for the β-subunit of rod photoreceptor cGMP-specific phosphodiesterase in patients with retinal dystrophies and dysfunction, in: *Degenerative Diseases of the Retina* (R.E. Anderson, J. Hollyfield, and M.LaVail, eds.), pp. 313–322, Plenum Press, New York.

10. Kajiwara, K., Berson, E.L., and Dryja, T.P., 1994, Digenic retinitis pigmentosa due to mutations at the unlinked peripherin/RDS and ROM1 loci, *Science* **264**: 1604–1608.

11. Gal, A., Apfelstedt-Sylla, E., Janecke, A.R., Zrenner, E., 1997, Rhodopsin mutations in inherited retinal dystrophies and dysfunction, *Prog. Retin. Eye Res.* (N.N. Osborne and G.J. Chader, eds.), **16**: 51–79.

12. Kumaramanickavel, G., Maw, M., Denton, M.J., John, S., Srikumari, S.C.R., Orth, U., Oehlmann, R., and Gal, A., 1994, Missense rhodopsin mutation in a familiy with recessive RP, *Nature Genet.* **8**: 10–11.

13. Rosenfeld, P.J., Cowley, G.S., McGee, T.L., Sandberg, M.A., Berson, E.L., and Dryja, T.P., 1992, A null mutation in the rhodopsin gene causes rod photoreceptor dysfunction and autosomal recessive retinitis pigmentosa, *Nature Genet.* **1**: 209–213.

14. Bayes, M., Giordano, M., Balcells, S., Grinberg, D., Vilageliu, L., Martinez, I., Ayuso, C., Benitez, J., Ramos-Arroyo, M.A., Chivelet, P., Solans, T., Valverde, D., Anselem, S., Goossens, M., Baiget, M., Gonzales-Duarte, R., and Besmond, C., 1995, Homozygous tandem duplication within the gene encoding the β-subunit of rod phosphodiesterase as a cause for autosomal recessive retinitis pigmentosa, *Hum. Mutat.* **5**: 228–234.

15. Danciger, M., Blaney, J., Gao, Y.Q., Zhao, D.Y., Heckenlively, J.R., Jacobson, S.G., and Farber, D.B., 1995, Mutations in the PDEB6 gene in autosomal recessive retinitis pigmentosa, *Genomics* **30**: 1–7.

16. Danciger, M., Heilbron, V., Yong-Qing, G., Dan-Zhu, Z., Jacobson, S.G., and Farber, D.B., 1996, A homozygous PDEB6 mutation in a family with autosomal recessive retinitis pigmentosa, *Mol. Vis.* **2**: 10.

17. Dryja, T.P., Finn, J.T., Peng, Y-W., McGee, T.L., Berson, E.L., and Yau, K-W., 1995, Mutations in the gene encoding the α-subunit of the rod cGMP-gated channel in autosomal recessive retinitis pigmentosa, *Proc. Natl. Acad. Sci. U.S.A.* **92**: 10177–10181.

18. McLaughlin, M.E., Sandberg, M.A., Berson, E.L., and Dryja, T.P., 1993, Recessive mutations in the gene encoding the β-subunit of rod phosphodiesterase in patients with retinitis pigmentosa, *Nature Genet.* **4**: 130–134.

19. McLaughlin, M.E., Ehrhart, T.L., Berson, E.L., and Dryja, T.P., 1995, Mutation spectrum of the gene encoding the β subunit of rod phosphodiesterase among patients with autosomal recessive retinitis pigmentosa, *Proc. Natl. Acad. Sci. U.S.A.* **92**: 3249–3253.

20. Valverde, D., Baiget, M., Seminago, R., del Rio, E., Garcia-Sandoval, B., del Rio, T., Bayes, M., Balcells, S., Martinez, A., Grinberg, D., Ayuso, C., 1996, Identification of a novel R552Q mutation in exon 13 of the β-subunit of rod phosphodiesterase gene in a spanish family with autosomal recessive retinitis pigmentosa, *Hum. Mutat.* **8**: 393–394.

21. Valverde, D., Solans, T., Grinberg, D., Ballcells, S., Vilageliu, L., Bayes, M., Chivelet, P., Besmond, C., Goossens, M., Gonzales-Duarte, R., Baiget, M., 1996, A novel mutation in exon 17 of the beta-subunit of rod phosphodiesterase in two RP sisters of a consanguineous family, *Hum. Genet.* **97**: 35–38.

22. Tsang, S.H., Gouras, P., Yamashita, C.K., Kjeldbye, H., Fisher, J., Farber, D.B., Goff, S.P., 1996, Retinal degeneration in mice lacking the α subunit of the rod cGMP phosphodiesterase, *Science* **272**: 1026–1029.

23. Hahn, L.B., Berson, E.L., and Dryja, T.P., 1994, Evaluation of the gene encoding the gamma subunit of rod phosphodiesterase in retinitis pigmentosa, *Invest. Ophthalmol. Vis. Sci.* **35**: 1077–1082.

24. Pittler, S.J., and Baehr, W., 1991, Identification of a nonsense mutation in the rod photoreceptor cGMP phosphodiesterase β-subunit gene of the *rd* mouse, *Proc. Natl. Acad. Sci. U.S.A.* **88**: 8322–8326.

25. Suber, M.L., Pittler, S.J., Qin, N., Wright, G.C., Holcombe, V., Lee, R.H., Craft, C.M., Lolley, R.N., Baehr, W., Hurwitz, R.L., 1993, Irish setter dogs affected with rod/cone dysplasia contain a nonsense mutation in the rod cGMP β-subunit gene, *Proc. Natl. Acad. Sci. U.S.A.* **90**: 3968–3972.

26. Lem, J., Flannery, J.G., Li, T., Applebury, M.L., and Farber, D.B., 1992, Retinal degeneration is rescued in transgenic *rd* mice by the cGMP-phosphodiesterase β-subunit, *Proc. Natl. Acad. Sci. U.S.A.* **89**: 4422–4426.

27

MOLECULAR ANALYSIS OF THE HUMAN PEDF GENE, A CANDIDATE GENE FOR RETINAL DEGENERATION: LOCALIZATION TO 17p13.3

Joyce Tombran-Tink,[1] Gerald Chader,[1] and Robert Koenekoop[2]

[1]National Institutes of Health
National Eye Institute
Laboratory of Retinal, Cell and Molecular Biology
Bethesda, Maryland 20892
[2]McGill University
Ophthalmology, Montreal Children's Hospital
Montreal, Quebec, Canada

INTRODUCTION

Retinitis pigmentosa (RP) are diseases of the rod photoreceptor and represent a large group of hereditary retinal degenerations with a prevalence of 1 in 4000 births. RP can be inherited as an autosomal dominant, autosomal recessive, X-Linked, or digenic disease. Substantial genetic heterogeneity has been found, with 20 chromosomal loci mapped and mutations documented in eight genes, namely rhodopsin (1), the alpha and beta subunits of rod cyclic GMP phosphodiesterases (2, 3), and the alpha subunit of rod cGMP cation-gated channel protein of the rod phototransduction cascade (4). The peripherin/RDS and rod outer membrane protein-1 (ROM-1) genes (5, 6) encode proteins involved in maintaining photoreceptor outer segment structure. Defects in the myosin VIIa gene have been found in a syndromic form of RP with congenital deafness called Usher syndrome type 1 (7). Mutations in the RPGR (retinitis pigmentosa GTPase regulator) gene cause the most common form of X-linked RP (8). Mutations in these eight genes together probably account for more than 25–30% of cases of retinitis pigmentosa in North America, which leaves 70–75% of cases of RP unaccounted for genetically. Phenotypic variability is striking as well, with differing severity and onset of disease at a given age both within and between families, even among patients with the same gene defect.

Of the uncharacterized RP loci, the 17p13.3 locus may be the most common one remaining, illustrated by three recent, separate linkage studies of large families from three different continents (9–11). The gene for pigment epithelium-derived factor (PEDF) also maps to 17p13 (12) and is a strong candidate gene for retinal degenerative diseases assigned to this chromosomal location. PEDF is a 50 kD protein expressed in young adult retinal pigment epithelial cells. It is secreted in abundance by the RPE cells into the interphotoreceptor matrix (IPM) (13–14) and was first identified as a biological activity in medium conditioned by human fetal RPE cells (hfRPE-CM) (15–16). Phenotypic changes characteristic of neuronal differentiation are induced in human Y79 retinoblastoma cells after treament with either PEDF or medium condition ed by fetal human RPE cells cultures (hfRPE-CM). Extension of neurites, upregulation of neuronal marker molecules and promotion of neuron survival are among the effects of PEDF observed *in vitro*. Moreover, soluble components of the IPM, or PEDF purified from the IPM also contain this neurotrophic activity (13, 17–18). In subsequent experiments we show that the biological activity of PEDF, the secreted protein and RNA transcript decreases in an age related manner by cultured RPE cells (14, 19). Although the 1.5 kb PEDF cDNA share homology with members of the serine protease inhibitor (serpin) gene family PEDF may not function as an inhibitory serpin since little homology is found between this protein and the classical reactive domain of serpins (20).

PEDF maps to human chromosome 17p13 (12), a region to which several neurodegenerative disease loci have been assigned. Among these are the retinal degeneration loci progressive cone dystrophy (CORD5), central aerolar cone dystrophy (cacd), retinitis pigmentosa (RP-13) and Lebers congenital amaurosis (LCA). Taken together, the studies on PEDF raise the interesting possibility that the presence of this neurotrophic factor in the interphotoreceptor matrix may be essential in promoting differentiation and survival of retinal neurons. Thus, abberant forms or loss of PEDF in the retina may contribute to rod and/or cone photoreceptor degeneration, or impede proper differentiation of these and other retinal neurons. Therefore, for both genetic and physiological reasons, PEDF is an intriguing and important candidate gene for retinal disorders.

PEDF GENE ANALYSIS

A human cosmid and a lambda DASH II genomic libraries were screened using a radioactively labeled PEDF cDNA. Initial screenings resulted in two clones each containing approximately 7 kb inserts of the PEDF gene. By standard sequencing techniques the cosmid clone was shown to contain the entire promoter region and the lambda DASH II clone, most of the 3'end and flanking region of the gene (21). Several oligonucleotides were subsequently constructed based on the known human PEDF cDNA sequence and used in PCR amplification reactions containing human genomic DNA as a template. This resulted in 3 large PCR amplification products of 2 kb, 3.3 kb and 2.3 kb which, when subcloned and sequenced, were shown to contain most of the coding sequences of the PEDF gene. Two P1 clones were subsequently isolated and all gaps in the PEDF gene were completed by automated DNA fluorescent sequencing of selected PCR products from the P1 clones. All intron-exon boundaries were similarly confirmed using DNA from the P1 clone. We determined from the sequence analysis data that the human PEDF gene spans approximately 16 kb and contains 8 exons (Fig. 1), the junctions of which adhere to the AG/GT consensus rule. The intron/exon junctions and their flanking sequences are shown in Table 1. A CAAT box, sequences identical or similar to octomer, HNF-1, PEA3, c/EPB, TREp, an RAR motif and two large Alu repeats are present in the promoter region (Fig 2). The intron-exon arrangement and sequence homology now place PEDF in the ovalbumin/PAI-2 subgrouping of serpins (21).

Table 1. The sequences of the intron/exon boundaries and
the size of each intron and exon

No.	bp.	EXON	Donor splice-site	Acceptor splice-site	EXON	No.
			PROMOTER	..aaaggagta	GCTGTAATC	1
1	128	TATCCACAG	gtaaagtag..	4793 bp..ttcttgcag	GCCCCAGGA	2
2	92	CCGGAGGAG	gtcagtagg..	2863 bp..tcctgccag	GGCTCCCCA	3
3	199	TCTCGCTGG	gtgagtgct..	980 bp..ctctggcag	GAGCGGACG	4
4	156	TTGAGAAGA	gtgagtcgc..	688 bp..tcttctcag	AGCTGCGCA	5
5	204	ACTTCAAGG	gtgagcgcg..	2982 bp..tctttccag	GGCAGTGGG	6
6	143	AGCTGCAAG	gtctgtggg..	1339 bp..ttgtctcag	ATTGCCCAG	7
7	211	AGGAGATGA	gtatgtctg..	444 bp..tctctacag	AGCTGCAAT	8
8	377	TTTATCCCT	aacttctgt			

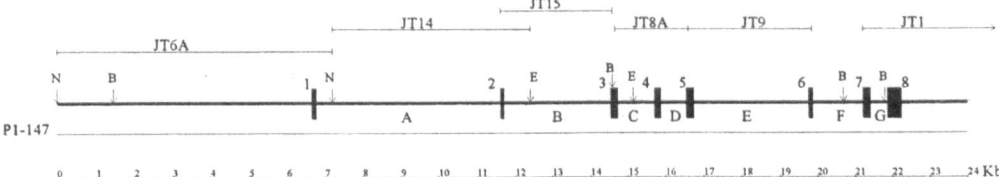

Figure 1. Scale map of the human PEDF gene: Exons 1–8 are numbered and shown as black boxes. BamH1 (B), EcoR1 (E) and Not1 (N) sites are shown with letters and arrows. The gene is approximately 16 kb in length. A rule bar with the relative size in kilobases (kb) is shown at the bottom Several overlapping clones from which the sequence were obtained are shown above the P1 clone.

```
-1050  tgggaggctgaggggggcgggatcacctgaggtcaggagtttgagacaag  -1001
-1000  cgtgaccaatgtggtgaaacctgtctctactaaaaatacaaaaattagc  -951
 -950  cgggcatgctcgtgcacacctatagtcccaactactcagcagggtgaggc  -901
 -900  aggagaacctcttgaacccgggaagcggaggttgcagtgagccgacattg  -851
 -850  cacccctgcactccagcctgggtgacagagtgagtctccactggaaaaaa  -801
 -800  aaaaaaaagaacagtgtgatacattgacctaagtttaagaacatgcaaa  -751
 -750  ctgatactatatatcacttagggacaaaaacttacatggtaaaagtaaaa  -701
                                              C/EBP
 -700  agaaatgtacgaaaataataaaaatcaaattcaagatggtggttatggtg  -651
 -650  acgggaaagaactgaggcggaaatataaggttgtcactatattgagaaat  -601
 -600  ttttctatctttttttctttttttcttttttttgagacggggtctcgctctg  -551
 -550  tcgcccaggatggagtgcagtggtgtgatctcagctcactgcaacctccg  -501
 -500  cctcccaggtttaagtgattctcctgcctcagactcccaagtagctggga  -451
 -450  ctacaggtgcgcgccaacacacctgggtaattttgtttgtattttttagta  -401
 -400  gagatggggtttcaccgtgttgactaggctggtctcgaactcctgacctc  -351
 -350  aggtgatcccccggcctcggtctcccaaagtgctgggataacaagcgtga  -301
 -300  gccactgcgcccagctttgtttgcattttttaggtagagatggggtttcacc  -251
                                              TREp/RAR
 -250  acgttggccaggctggtcttgaactcctgacctcaggtgatgcacctgcc  -201
 -200  tcagtctcccaaagtgctggattacaggcgttagcccctgcgcccggccc  -151
       PEA3              PEA3   PEA3      Oct
 -150  ctgaaggaaaatctaaaggaagagggaaggtgtgcaaatgtgtgcgcctta  -101
                                 HNF-1
 -100  ggcgtaatgatggtggtgcagcagtgggttaaacttaacacgagacagtg  -51
       CAAT Box
  -50  atgcaatcacagaatccaaattgagtgcaggtcgctttaagaaaggagta  -1
   +1  GCTGTAATCTGAAGCCTGCTGGACGCTGGATTAGAAGGCAGCAAAAAAAG  +50
       CTCTGTGCTGGCTGGAGCCCCCTCAGTGTGCAGGCTTAGAGGGACTAGGC
       TGGGTGTGGAGCTGCAGCGTATCCACAG
```

Figure 2. 5′-Flanking region of the PEDF gene. The first exon (capital letters) and the first 1050 bp of the 5′-flanking region are shown. Two Alu repetitive sequences are underlined. Possible binding sites for HNF-1, PEA3, Octomer (Oct) and c/EBP and a CAAT box (-43 to -47) are underlined and labeled. The TREp/RAR site is double underlined. The bold sequence (+1 through +20) in exon 1 was determined by 5′-RACE (21).

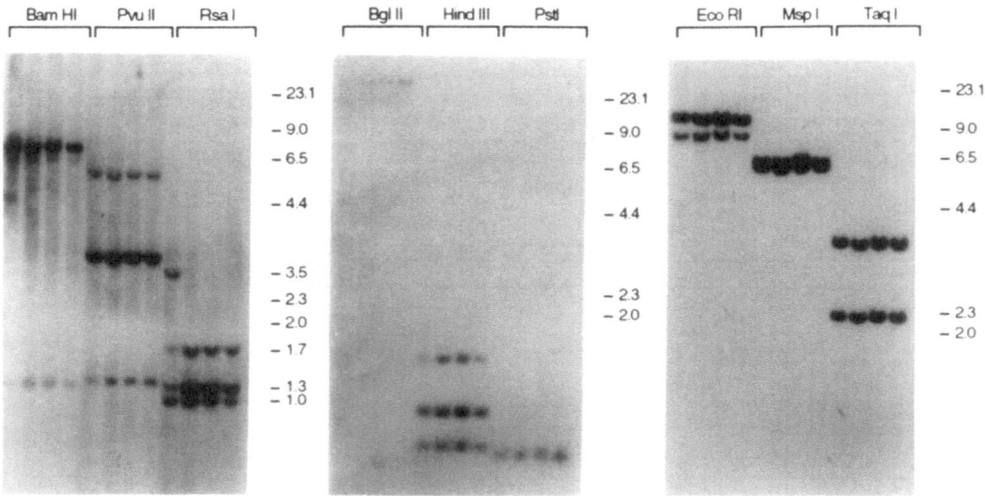

Figure 3. 10 ug of genomic DNA from 4 unrelated Caucasian individuals of CEPH pedigrees was digested with a battery of 9 restriction enzymes listed above each blot. Fragments were separated on 1% agarose gels, transfered to nylon membranes at 25 V overnight and baked for 2 hours. Blots were hybridized with an α - [^{32}P] dCTP ran-dom- primed 667 bp product amplified from human PEDF cDNA. Hybridization was performed for 1 hour at 68° C using Stratagene's QuikHyb solution. The blots were washed twice at room temperature with 2X SSC, 0.1% SDS (15 min each) and once at 68° C with 0.1X SSC, 0.1% SDS (30 min) and exposed at -70° C for 18 hrs. A 3.5 kb fragment is unique to the first individual and three constant fragments of 1.7 kb, 1.3 kb and 1.0 kb are observed in the lanes containing DNA digested with Rsa1.

LOCALIZATION OF THE PEDF GENE TO 17p13.3

DNA samples from 4 unrelated Caucasion individuals of CEPH pedigrees were di-gested with a battery of 9 restriction enzymes chosen as the most likely to reveal polymor-phisms. The fragments were separated on agarose gels, transblotted and hybridized using a radioactively labeled 667 bp PCR amplified product of the PEDF cDNA. Of the enzymes tested only the DNA sample digested with Rsa1 contained a 3.5 kb variant band in the first individual on the CEPH family blot (Fig 3). This variant was used in further analysis of the CEPH families.

Heritability Determination: Parental Screening

DNA samples from 16 individuals representing the parents of 8 separate families were digested with Rsa1 and analysed similarly by southern blot (Fig 4). Three constant frag-ments of 1.7 kb, 1.3 kb and 1.0 kb are observed in all the individuals. A fourth allele of ap-proximately 3.5 kb is seen in six of the individuals screened. An additional, strongly hybridizing fragment of approximately 2 kb is seen in the parental DNA sample of the indi-viduals in lane 10. Two sets of parents, 1,2 and 11,12, representing CEPH families 13291 and 1333 with family sizes of 13 and 14 respectively, were chosen for heritability studies.

Heritability Determination: Family Allelotyping

Chromosomal crossover maps for CEPH-family pedigrees have been previously documented. Figs 5 and 6 show meiotic breakpoint maps of human chromosome 17 for

1 2 3 4 5 6 7 8 9 10 11 12 13 14 15 16

— 3.5 kb

— 1.7 kb

— 1.3 kb
— 1.0 kb

Figure 4. Heritability determination: parental screening: DNA samples were digested with Rsa1 and subjected to southern blot analysis as described in Fig 3. Each consecutive pair of numbers represent the parents of a three-generation CEPH family. Several individuals carry the 3.5 kb variant. The positions of lambda/Hind 111 molecular weight markers are shown to the right.

each sibling in CEPH families 13291 and 1333. These allelotyping mapping panels and CEPH database Version 5 were used in Mendelian inheritance studies of the Rsa1 variant. In the family allelotyping analysis, DNA samples were obtained from 13 members of CEPH family 13291 and 14 members of CEPH family 1333. These were screened for the inheritance of the 3.5 kb variant using Rsa1 (Fig 5). The maternal grandmother, grandfather, mother and the siblings 3, 4, 6, 7, and 9 showed the strongly hybridizing 3.5 kb variant. This segregation pattern establishes that the siblings inherited the variant fragment from the maternal grandfather. Localization of the PEDF probe to the smallest subfragment inherited from the maternal grandfather is shown in individual 9 on family 13291 (Fig 5). This data confirms our previously reported localization of the gene to 17p and positions it between the genetic markers D17S30 and D17S67, a genetic interval of approximately 32.2 cM and 7.8 cM from the telomere of the short arm of chromosome 17.

In family 1333, PEDF hybridizes to the smallest subfragment in individual 10 inherited in this case from the maternal grandmother (Fig 6). Analysis of the meiotic crossover map for this family now positions the human PEDF gene between the markers D17S30 and D17S28 which is approximately 7.8 cM from the telomere to within 11.2 cM on the map. This region represents a genetic interval of approximately 3.4 cM between the two known loci which are found in the region 17p13.3. The studies, thus fine tune the mapping of PEDF to the 17p13.3 region.

Human Allelic Frequency

To determine if there are differences in the distribution of the Rsa1 variant within human subpopulations, Rsa1-digested DNA from a mixed-race population were examined by southern blot (Fig 7). The analysis showed 5/7 individuals in the Afro-American population, 7/7 individuals in the Oriental grouping and 5/6 in the Caucasian population carried the 3.5 kb Rsa1 fragment. A total of 21 unrelated Caucasion individuals (4 from the initial RFLP studies, 16 in the parental analysis and 7 in the population analysis; DNA from 6 individuals were repeated) studied contained the Rsa1 fragment and allele frequency analysis indicates a heterozygosity for PEDF of 0.34 and a Polymorphism information content (PIC) of 0.28.

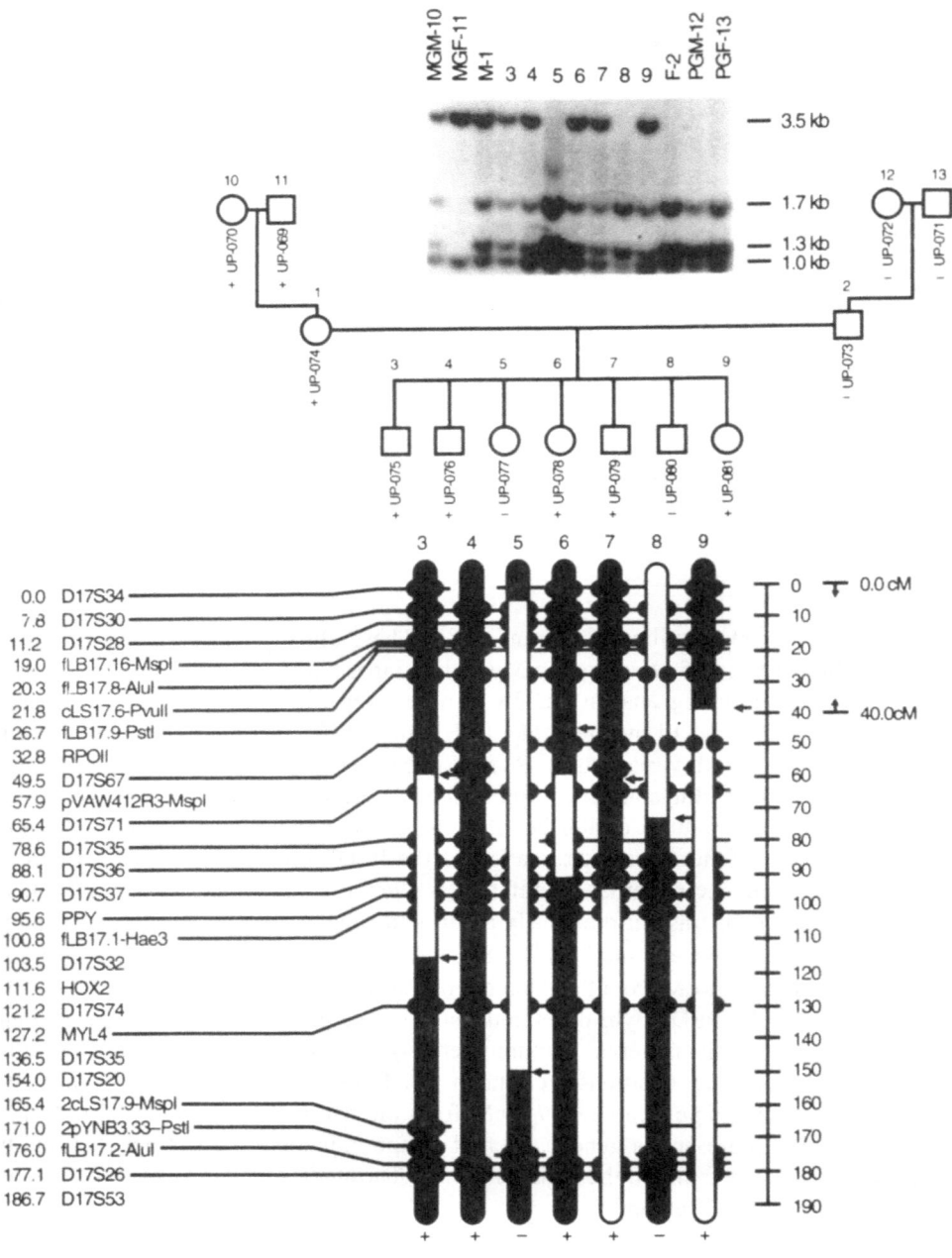

Figure 5. Family allelotyping: CEPH family 13291: DNA samples from CEPH family 13291 were digested with RsaI and processed by southern blot analysis (shown above). The pedigree for this family is depicted in the center. Variation is denoted by the presence (+) or absence (−) of the 3.5 kb fragment. A linear meiotic crossover map of chromosome 17 of sibs from this family is shown below. The chromosome fragments derived from the maternal grandfather are depicted in solid black; those from the maternal grandmother are in white. Arrows indicate reliable breakpoints. The data are taken from CEPH database version 5. Known loci and genetic map distances of each are listed on the left. The 3.5 kb fragment is inherited from the maternal grandfather. The data positions PEDF approximately 7.8 - 40 cM from the telomere of 17p.

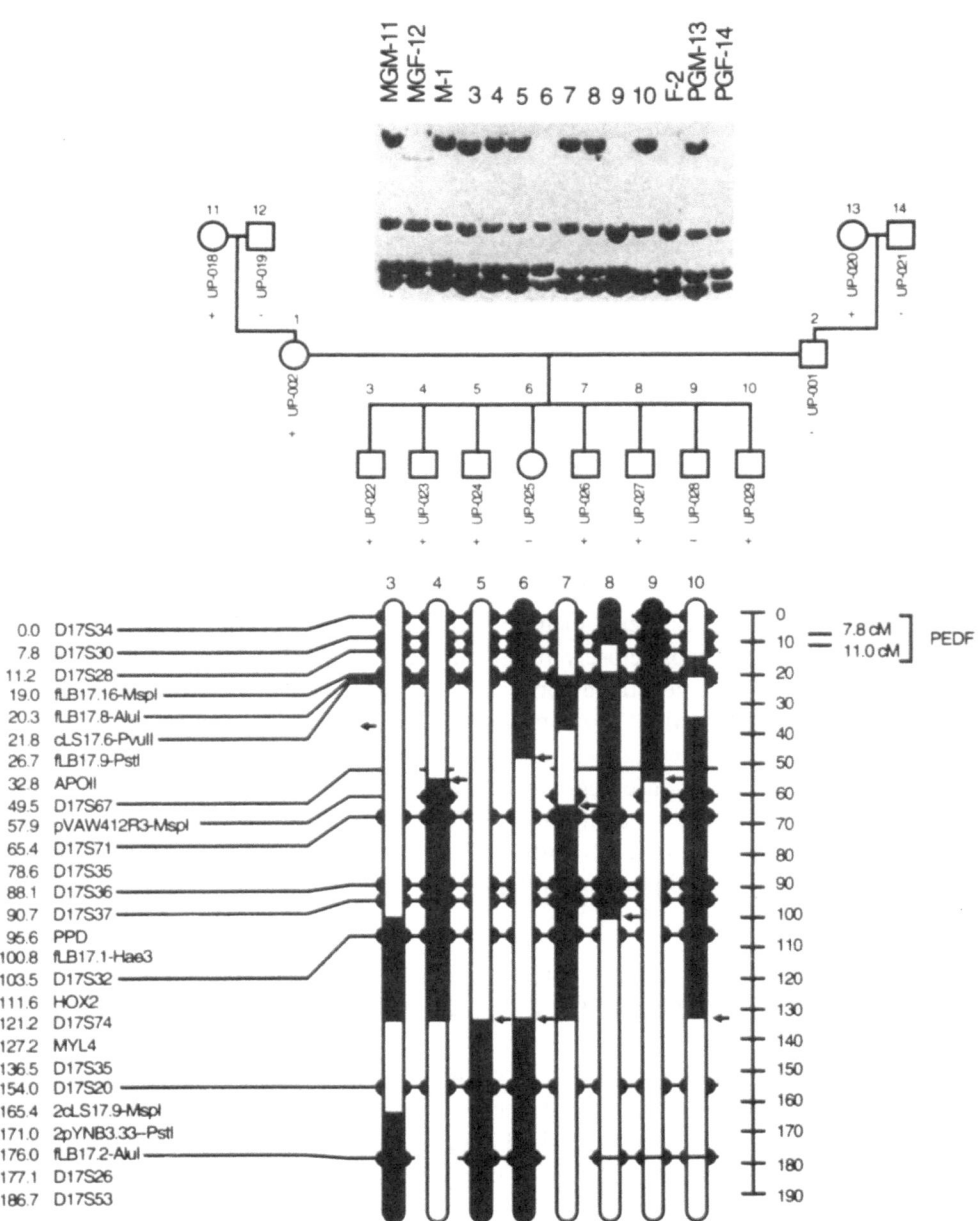

Figure 6. Family allelotyping: CEPH family 1333: DNA from this family was analyzed similar to that for family 13291. In this case, the variant fragment is inherited from the maternal grandmother. The data obtained positions PEDF between 7.8 cM and 11.2 cM from the telomere of 17p, a genetic interval of 3.4 cM. This locus is between the genetic markers D17S30 and D17S28 on 17p13.3.

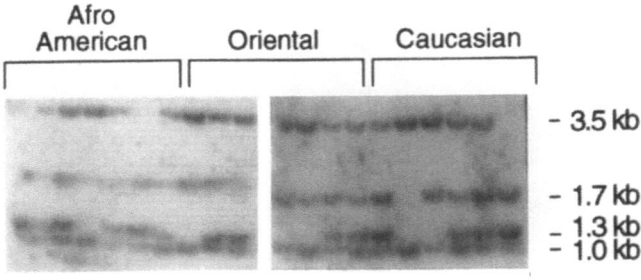

Figure 7. Allelic frequency: Southern blot analysis of Rsa1-digested DNA from a mixed-race population showing the presence of the 3.5 kb fragment in 5/7 Afro-Americans, 7/7 Orientals and 5/6 Caucasians. Heterozygosity = 0.32 and PIC = 0.28.

TIGHT LINKAGE OF PEDF TO AN AUTOSOMAL DOMINANT RETINITIS PIGMENTOSA LOCUS (RP 13)

An autosomal dominant retinitis pigmentosa (RP13) gene locus in a large South African family has recently been assigned to chromosome 17p13.3 (10,22). The gene for PEDF which also maps to this region was thus evaluated as a candidate disease causing gene for this family (23). A highly informative polymorphism in exon 4 of the PEDF gene, identified by SSCP analysis and direct sequencing of the PCR products (24, 23), facilitated the genotyping of this family with PEDF. Multipoint linkage analysis was assessed between the ADRP gene locus, several DNA markers at 17p13.3 including D17S1529 which is the most closely linked marker to the RP locus, and the candidate PEDF gene. Linkage data generated between the disease phenotype and the exon 4 polymorphism using the LINKMAP program produced a significant lod score of 7.1 with no recombinations thereby showing tight linkage between the PEDF gene and the RP13 locus. Although no definitive cosegregation between the disease locus and the exon 4 polymorphism of PEDF has been confirmed, PEDF is a viable candidate for the disease causing gene in this South African family. Other highly informative polymorphisms have been identified and will facilitate genotyping PEDF in these families.

POLYMORPHISMS IN THE PEDF GENE: EXCLUSION OF PEDF AS A CANDIDATE FOR LEBER'S CONGENITAL AMAUROSIS (LCA)

Leber's congenital amaurosis (LCA) is an autosomal recessive retinal disease characterized by severe visual impairment at or shortly after birth. Pathological studies suggest that LCA is a retinal aplasia with impaired development of the rod and cone photoreceptors (26). Recently a gene for LCA was mapped to the short arm of chromosome 17 (17p13) and genetic heterogeneity established by homozygosity mapping and linkage studies (27, 28). PEDF was thus screened in LCA patients for pathological mutations.

Patients with the LCA phenotype used in this study were collected following detailed histories, ocular examinations and electroretinography. Eighteen families (5 multiplex, 13 simplex) entered into this investigation. Inclusion criteria were: 1) Severe visual impairment at birth or in the first six months of life with nystagmus; 2) An extinguished or severely diminished photopic and scotopic ERG; 3) A normal or abnormal retinal exam; 4) Associated systemic diseases such as mental retardation, hydrocephalus and kidney disease. Exclusion criteria following serum assays or conjunctival biopsy included cerebro-hepatorenal syndrome of Zellweger (a peroxisomal disorder), abetalipoproteinemia,

Table 2. Four new polymorphisms with their respective exon,
created or destroyed enzyme and relative frequencies

EXON	MUTATION	ENZYME	FREQUENCY IN LEBERS PATIENTS	FREQUENCY IN NORMALS
3	Met72Thr	BssSI Created	9/17 (53%)	14/24 (58%)
4	Thr130Thr	BstEII Created	9/17 (53%)	5/13 (38%)
5	G→A Intronic (9bp after the exon)	BstUI Destroyed	4/17 (24%)	6/24 (25%)
7	Tyr321Tyr	MaeIII Created	8/17 (47%)	13/24 (54%)

infantile phytanic acid storage disease (Refsum disease), and neuronal ceroid lipofusci-
nosis (Batten disease). Patients were examined and diagnosed and the clinical phenotypes
previously reported (29, 30).

Eight segments of the PEDF gene were amplified by PCR using intronic primers
(24). By PCR-SSCP analysis four new intragenic basepair alterations were found but none
cosegregate with the disease phenotype. These were: a Met72Thr polymorphism in exon
3, a Thr130Thr polymorphism in exon 4, a G to A transition in intron 5 and a Tyr321Tyr
change in exon 7 (Table 2). These RFLP's were used in linkage analysis to exclude PEDF
as the causative gene in the families studied.

CONCLUSION

Because of its neurotrophic activity, its abundant presence in the IPM, its synthesis
and secretion specifically by the RPE cells and its map location, it is critical to define the
role of PEDF in the vertebrate retina. The identification of new intrageneic polymorphisms is warranted in future linkage analysis to assess the role of the PEDF gene in the
pathogenesis of several inherited retinal disorders.

REFERENCES

1. Dryja, T.P., McGee, T.L., Reichel, E., Hahn, L.B., Cowley, G.S., Yandell, D.W., 1990, A point mutation of the rhodopsin gene in one form of retinitis pigmentosa, Nature **43**:364–366.
2. Dryja, T.P., Finn, J.T., Peng, Y.W., McGee, T.L., Berson, E.L., and Yau, K.W., 1995, Mutations in the gene encoding the alpha subunit of the rod cGMP gated channel protein in Ar RP, Proc. Natl. Acad. Sci. USA **92**:10177–10181.
3. McLaughlin, M.E., Sandberg, M.A., Berson, E.L., 1993, Recessive mutations in the gene encoding the B-subunit of rod phosphodiesterase in patients with retinitis pigmentosa, Nature Genetics **4**:130–134.
4. Huang, S.H., Pittler, S.J., Huang, X., Oliveira, L., Berson, E.L., and Dryja, T.P., 1995, Ar RP caused by mutations in the alpha subunit of rod cGMP phosphodiesterase, Nature Genetics **11**:468–471.
5. Farrar, G.J., Kenna, P., Jordan, S.A., Kumar-Singh, R., Humphries, M.M., Sharp, E.M., 1991, A three-base-pair deletion in the peripherin-RDS gene in one form of retinitis pigmentosa, Nature **354**:478–480.
6. Sakuma, H., 1995, A heterozygous putative null mutation in ROM 1 without a mutation in peripherin/RDS in a family with RP, Genomics **27**:384–386.

7. Weil, D., Blanchard, S., Kaplan, J., Guilford, P., 1995, Defective myosin VIIa gene responsible for Usher syndrome type IB, Nature (London) **374**:60–61.
8. Meindl, A., Dry, K., Hermann, K., Manson, F., 1996, A gene (RPGR) with homology to the RCC1 guanine nucleotide exchange factor is mutated in X-linked retinitis pigmentosa (RP3), Nature Genetics **13**:35–42.
9. Tarttelin, E.E., Plant, C., Weissenback, J., Bird, A., Bhattacharya, S.S., Inglehearn, C.F., 1996, A new family linked to the RP13 locus for autosomal dominant retinitis pigmentosa on distal 17p, J. Med. Genet. **33**:518–520.
10. Goliath, R., Shugart, Y., Janssens, P., 1995, Fine localization of the locus for autosomal dominant retinitis pigmentosa on chromosome 17p, Am. J. Hum. Genet. **57**:962–965.
11. Kojis, T.L., Heinzmann, C., Flodman, P., 1996, Map refinement of locus RP13 to human chromosome 17p13.3 in a second family with autosomal dominant retinitis pigmentosa, Am. J. Hum. Genet. **58**:347–355.
12. Tombran-Tink, J., Pawar, H., Swaroop, A., and Chader, G., 1994, Localization of the gene for Pigment epithelium-derived factor to chromosome 17p13.1 and expression in cultured retinoblastoma cells, Genomics **19**:266–272.
13. Tombran-Tink, J., Li, A., Johnson, M.A., Johnson, L.V., and Chader, G.J., 1992, Neurotrophic activity of interphotoreceptor matrix on human Y79 retinoblastoma cells, J. Comp. Neurol. **317**:175–186.
14. Tombran-Tink, J., Shivaram, S., Chader, G.J., Johnson, L.V, and Bok, D., 1995, Expression, secretion and age-related downregulation of pigment epithelium-derived factor, a serpin with neurotrophic activity, J. Neurosci **15** (7):4992–5003.
15. Tombran-Tink, J., and Johnson, L.V., 1989, Neuronal differentiation of retinoblastoma cells induced by medium conditioned by human RPE cells, Invest. Ophthalmol. Vis. Sci. **30**:1700–1707.
16. Tombran-Tink, J., Chader, G.J., and Johnson, L.V, 1991, PEDF: A pigment epithelium-derived factor with potent neuronal differentiative activity, Exp. Eye Res. **53**:411–414.
17. Seigel, G.M., Tombran-Tink, J., Becerra, P.A., Chader, G.J., Diloreto, Jr., D.A,, Del Cerro, C., Lazar, E.S., and Del Cerro, M., 1994, Differentiation of Y79 retinoblastoma cells with Pigment epithelial derived factor and interphotoreceptor matrix wash:Effects on tumorigenicity, Growth Factors 10:289–297.
18. Taniwaki, T., Becerra, S.P., Chader, G.J., and Schwartz, J.P., 1995, Pigment epithelial-drived factor is a survival factor for cerebellar granule cells in culture, J. Neurochem. **64**:2509–2517.
19. Pignolo, R.J., Cristofalo, V.J., Rotenberg, M.O., 1993, Senescent WI-38 cells fail to express EPC-1, a gene induced in young cells upon entry in the G_o state, J. Biol. Chem. **268**:8949–8957.
20. Steele, F.R., Chader, G.J., Johnson, L.V., and Tombran-Tink, J, 1992, Pigment epithelium-derived factor:Neurotrophic activity and identification as a member of the serine protease inhibitor gene family, Proc. Natl. Acad. Sci. USA **90**:1526–1530.
21. Tombran-Tink, J., Mazaruk, K., Chung, D., Linker, T., Rodriguez, I., Chader, G., 1996, Organization, evolutionary conservation, expression and unusual Alu density of the human gene for pigment epithelium derived factor, Mol. Vis 2:11 http://www.emory.edu/molvis/v2/tink/>
22. Greenberg, J., Goliath, R.G., Beighton, P., and Ramesar, R., 1994, A new locus for autosomal dominant retinitis pigmentosa on the short arm of chromosome 17, Hum. Mol. Genet. **3**:915–918.
23. Goliath, R., Tombran-Tink, J., Chader, G., Ramesar, R., Greenberg, J., 1996, Genetic localization of the gene for pigment epithelium-derived factor (PEDF) on the short arm of chromosome 17 (17p13.3), Mol. Vis. **2**:9604.
24. Davidson, J.K., Tombran-Tink, J., Loyer, M., Traboulsi, E., Maumenee, I., Koenekoop, R.K., 1996, Pigment epithelium-derived factor as a candidate gene for Leber's congenital amaurosis, Invest. Ophthal. Vis. Sci. **37** (3):S991.
25. Sorsby, A., and Williams, C.E., 1969, Retinal aplasis as a clinical entity, Br. Med. J. I:293.
26. Camuzat, A., Dollfus, H., Rozet, J.M., et al., 1995, A gene for Leber's congenital amaurosis maps to chromosome 17p. Human Molecular Genetics 4(8): 1447–145.
27. Camuzat, A., Rozet, J., Dolffus, H., et al., 1996, Evidence of genetic heterogeneity of Leber's congenital amaurosis (LCA) and mapping of LCA1 to chromosme 17p13, Hum. Genet 97:798–801.
28. Schroeder, R., Mets, M.B., and Maumenee, I.H., 1987, Leber's congenital amaurosis. Retrospective review of 43 cases and a new fundus finding in two cases, Arch. Ophthalmol. 105:356–359.
29. Heher, K.L., Traboulsi, E.T., and Maumenee, I., 1992, The natural history of Leber's congenital amaurosis, Ophthal. 99:241–245.

28

SCREENING OF CANDIDATE GENES ON JAPANESE RETINAL DYSTROPHIES

Yoshihiro Hotta,[1] Keiko Fujiki,[1] Mutsuko Hayakawa,[1] Hitoshi Sakuma,[1] Hiroyuki Kawano,[1] Atsushi Kanai,[1] Akira Murakami,[2] Masaru Yoshii,[2] Kiyoshi Akeo,[2] Shigekuni Okisaka,[2] Masayuki Matsumoto,[3] Seiji Hayasaka,[3] Yasushi Isashiki,[4] Norio Ohba,[5] Takashi Shiono,[6] and Makoto Tamai[6]

[1]Department of Ophthalmology
Juntendo University School of Medicine
[2]Department of Ophthalmology
National Defense Medical College
[3]Department of Ophthalmology
Faculty of Medicine
Toyama Medical and Pharmaceutical University
[4]Center for Chronic Viral Diseases
Kagoshima University
Faculty of Medicine
[5]Department of Ophthalmology
Kagoshima University Faculty of Medicine
[6]Department of Ophthalmology
Tohoku University School of Medicine

ABSTRACT

Since the causative genes for retinal dystrophies were recently clarified and genetic defects were reported, we could screen the candidate genes in patients with retinal dystrophies. However, we have to check many candidate genes because most of the retinal dystrophies correlate with the various gene defects. Rhodopsin and perpherin/RDS genes of Japanese 70 retinitis pigmentosa and 80 atypical retinitis pigmentosa patients were screened using restriction enzyme, dot blot hybridization and polymerase chain reaction-single strand conformation polymorphism (PCR-SSCP). Three point mutations in the rhodopsin gene were detected. No peripherin/RDS gene defect was detected. We established and evaluated the SSCP condition using a gyrate atrophy family and four choroideremia families. The direct sequence analyses confirmed the accuracy of our PCR-SSCP analyses. Our investigation revealed that PCR-SSCP analysis was good procedure if the best condition for differentiating bands was determined. Further screening of the candidate genes using the improved PCR-SSCP procedure in Japanese retinal dytrophies is now on going.

Degenerative Retinal Diseases, edited by LaVail *et al.*
Plenum Press, New York, 1997

INTRODUCTION

In these five years, the causative genes have been dramatically isolated and genetic defects for the retinal degenerations have been detected. The mutation of the rhodopsin gene causes autosomal dominant retinitis pigmentosa (ADRP), autosomal recessive retinitis pigmentosa (ARRP), congenital stationary night blindness and retinitis punctata albescens. The mutation of the peripherin/RDS gene causes ADRP, cone-rod dystrophy, pattern dystrophy, macular dystrophy, fundus flavimaculatus and retinitis punctata albescens. The correlation between the genotype and phenotype of retinal dystrophies is more complicated than we expected. For example, gyrate atrophy and choroideremia are caused by a single gene defect (1–3). However, some of retinitis pigmentosa (RP) family are caused by both peripherin/RDS and Rom-1 gene defects (digenic RP)(4). Generally, the causative gene for a retinal dystrophy spans several genes. If we would like to check the genetic defect of the retinal dystrophies, we have to check many candidate genes. A simple and effective procedure to detect the genetic defect is required for the ophthalmologist. We report here the results of the screening of the rhodopsin and peripherin/RDS genes in Japanese patients with retinitis pigmentosa from 1992 to 1995 and the establishment and evaluation of the PCR-SSCP (polymerase chain reaction-single strand conformation polymorphism) quality using patients with gyrate atrophy or choroideremia.

MATERIALS AND METHODS

Screening of Japanese Patients with Retinal Dytrophies

The screening of the rhodopsin and peripherin/RDS genes in Japanese patients with 70 retinitis pigmentosa and 80 atypical retinitis pigmentosa was performed. DNAs were extracted from peripheral leukocytes. The entire exons were amplified by PCR. The PCR condition was described previously(5–7). We employed restriction enzyme digestion, dot blot hybridization and PCR-SSCP. We analyzed 11 sites of the rhodopsin gene using the restriction enzymes (Table 1)(6). The procedure for the dot blot hybridization was described previously(8). The SSCP analysis was performed using 8–10% polyacrylamide(PAG) gel with or without 10% glycerol. The gel was visualized by both silver staining and radioisotope.

Table 1. Restriction enzymes used for the analyzing the rhodopsin gene mutations

Codon	Mutation	Restriction enzyme
89	GGT → GAT	*Eco* T14I
106	GGG → TGG	*Apa* I
110	TGC → TAC	*Fok* I
178	TAC → TGC	*Afa* I
181	GAG → AAG	*Ava* I
182	GGC → AGC	*Hae* III
190	GAC → AAC	*Taq* I
267	CCC → CTC	*Ban* I
344	CAG → TAG	*Mva* I
345	GTG → ATG	*Mva* I
347	CCG → TCG, CTG, CGG	*Msp* I

Establishment and Evaluation of the SSCP Condition

We evaluated the accuracy of the SSCP analysis. A family of 2 patients with gyrate atrophy and 4 individual family with choroideremia were employed. DNAs were extracted from peripheral leukocytes. Exons from 3 to 11 of ornithine aminotransferase (OAT) gene (2) of gyrate atrophy patients and exons from 11 to 14 of choroideremia (CHM) gene (3) were amplified by PCR. SSCP analysis was performed in various condition and the entire amplified DNAs were directly sequenced.

RESULTS

Screening of Japanese Patients with Retinal Dystrophies

Three causative point mutations Pro347Leu (9,10), Thr17Met (8,11) and Asn15Ser (12,13) of the rhodopsin gene were detected in Japanese patients with ADRP. No causative peripherin/RDS gene defect was detected. Several polymorphisms (Table 2) in both genes were also observed(5,7) . A Pro347Leu mutation was detected in a family with ADRP and the mutation segregated with the disease. This mutation was found by the restriction enzyme digestion to detect the codon 347 mutation of the rhodopsin gene. Another mutation, Thr17Met was detected by the dot blot hybridization with allele specific oligonucleotides. This mutation was found by the dot blot hybridization to detect the Thr17Met mutation of the rhodopsin gene. Another rhodopsin mutation, Asn15Ser was found by the direct sequence by chance. Further investigation revealed the presence of a heterozygous mutation, Asn15Ser in five affected individuals in this pedigree. No point mutation was found in any patient analyzing 11 sites of the rhodopsin gene using the restriction enzymes in 30 unrelated Japanese patients with ADRP. Although we checked the codon 23 of the rhodopsin gene using both the restriction enzyme and the dot blot hybridizarion of allele specific oligonucleotides, the mutation of codon 23 which was frequently observed in American ADRP was not recognized in Japanese patients with ADRP.

Table 2. Base substitutions of rhodopsin and peripherin/RDS genes

Region	Nt / Codon	Base change	Amino acid	Consequence
Rhodopsin				
5' non coding	nt 269	A → G	-	Polymorphism
exon 1	codon 15	AAT → AGT	Asn → Ser	Point mutation
exon 1	codon 17	ACG → ATG	Thr → Met	Point mutation
intron 4	nt 5145	G → A	-	Polymorphism
exon 5	codon 347	CCG → CTG	Pro → Leu	Point mutation
3' non coding	nt 5321	C → A	-	Polymorphism
Peripherin/RDS				
exon 1	codon 98	CTG → TTG	-	Polymorphism
exon 1	codon 106	GTC → GTT	-	Polymorphism
exon 1	codon 189	TCC → TCT	-	Polymorphism
exon 1	codon 266	GGT → GGA	-	Polymorphism
exon 3	codon 303	AGC → AGT	-	Polymorphism
exon 3	codon 304	GAG → CAG	Glu → Gln	Polymorphism
exon 3	codon 338	GGC → GAC	Gly → Asp	Polymorphism

Nt, nucleotide position

Establishment and Evaluation of the SSCP Condition

We concluded that 8 or 9% PAG with glycerol gel running at 4°C visualized by silver staining is the best procedure for SSCP analysis. We determined the gel concentration by the length of the sample DNA size. The exon 13 of CHM gene of the choroideremia samples was amplified by PCR. Fig. 1 shows electrophoresis patterns where the PCR products were run at 4°C on either 8% PAG gel or 10% PAG gels containing 10% glycerol . Both gels are visualized by silver staining or using the ^{32}P-dCTP. Sharp band shift in lane M is observed in 8% gel, but not in 10% gel. Silver staining is enough quality for the SSCP analysis.

A band shift was observed in the exon 13 of CHM gene in a patient with choroideremia. The direct sequence showed a hemizygous A to CC mutation in nucleotide position 1608 of the CHM gene in the patient resulting in the absence or truncation of the predicted CHM protein. Fig. 2 shows the direct sequence analysis of exon 13 of the carrier who have both wild and mutant allele indicating the heterozygote of the wild and the mutant CHM genes. The clinical features of the patient have already been reported(14). In brief, the patient is a 31-year-old male, complained visual loss, diplopia and night blindness at age 12 years. His maternal grandfather and cousin are also affected suggesting X-linked inheritance. The single flash ERG assessment demonstrated an extinguished response. Fundus examination disclosed diffuse atrophy of the choroid and retina. During 18 years of follow up, visual field became highly constricted and fundus changes markedly worsened. He had no general findings until now except a pinealoma diagnosed at age 12 years, and V-P shunt was performed. His mother is a 57-year-old female and is asymptomatic. She is diagnosed as a carrier showing diffuse spotty area within the pigment epithelium at the mid-peripheral area of the ocular fundus. The other amplified DNAs showed no band shift.

A case with gyrate atrophy showed severe chorioretinal atrophy with typical gyrus style(15). Consanguineous marriage of parents of patient suggested an autosomal recessive trait. This patient was vitamin 6 non-responsive type. OAT assay of cultured skin fibroblast showed 1/10 activity of normal control. No band shift was observed in the SSCP analysis. No mutaion was found in the direct sequence procedure corresponding to the coding area.

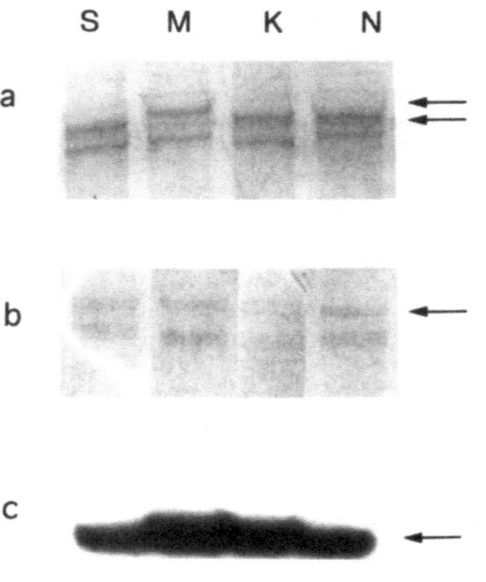

Figure 1. The exon 13 of the CHM gene of the choroideremia samples was amplified by PCR. Four DNA samples were run at room temperature using 10% glycerol using either 8% PAG gel (a,c) or 10% PAG gel (b). Both gels are visualized by silver staining (a,b) or using the $B#3#2(JP-dCTP(c). Sharp band shift in lane M is observed in the 8% gel, but not in the 10% gel. Sharp band shift in lane M is observed in silver staining, but not using the radioisotope.

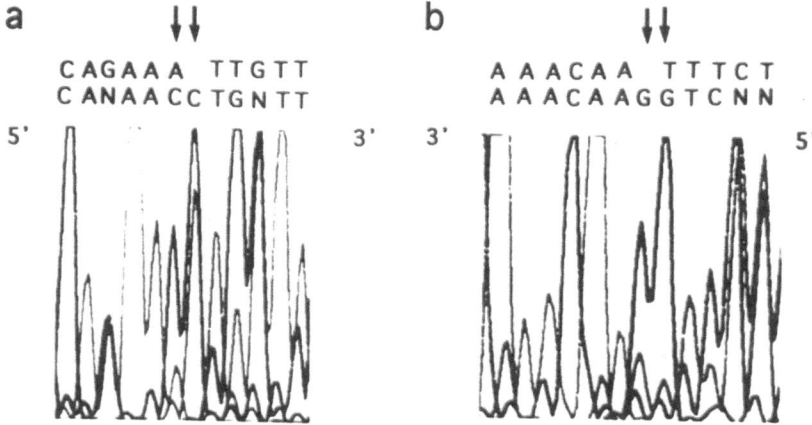

Figure 2. The direct sequence analysis of exon 13 of the carrier who have both wild and mutant allele indicating heterozygote of the wild and the mutant CHM genes. The sequence was performed using both sense (a) and antisense (b) primers.

DISCUSSION

Although the mutations cauing retinal dystrophies have recently clarified, the genetic defects are more complicated than we expected. It is difficult to detect the genetic defect in the clinical laboratory. If we would like to check the causative genes for ARRP, we have to check the rhodopsin, α- and β-*cGMP-PDE* gene. The screening procedure including SSCP, DGGE and TGGE is sometimes not reliable, and the possibility of the false negative result exists. Since we found a few mutations in Japanese retinal degenerations, we have to consider the false negative data in our screening. To tune up the quality of the SSCP analysis, we checked the condition of the SSCP. Retinal degenerations including RP and choroideremia were employed for analysis. The critical condition of the SSCP analysis included concentration of the PAG gel, addition of urea and/or glycerol, temperature for electrophoresis and gel staining. After we try the various condition for the SSCP, we concluded that 8 or 9% PAG with glycerol gel running at 4°C visualized by silver staining is the best procedure for SSCP analysis. We determined the gel concentration by the length of the sample DNA size.

The rhodopsin gene mutation which was detected in 30% of the ADRP pedigree in the USA(16) and 20% in Europe(17,18) was few in Japanese pedigree with ADRP. The high percentage in the USA is due to including high percentage of the Pro23His mutation(16,19). No Pro23His mutation was detected in Japanese patients with ADRP. Our screening results suggest that the frequency of the rhodopsin gene mutation in Japanese patients with ADRP is low. We could not detect the peripherin/RDS gene defect. Nakazawa et al reported several mutations of peripherin/RDS gene in patients with ADRP, cone-rod dystrophy and macular dystrophy(20–22). Our and their results suggest that the peripherin/RDS mutaion exist in Japanese atypical retinitis pigmentosa, however, frequency of the mutation may be low.

The genetic defect of the CHM gene causing choroideremia was reported mainly in Europe(3). Although we could not investigate the other family member, the hemizygous mutation was detected in a Japanese pateint and the heterozygous pattern in his mother who is the obligate carrier, suggesting that this mutation caused the disease. The mutation was found in a patient with different racial background from families reported previously confirming that the CHM gene defect causes choroideremia.

Since exons 1 and 2 are noncoding in OAT gene, we sequenced exons from 3 to 11 including exon-intron junction area in a gyrate atrophy patient. No mutation was found in this case. Since OAT level of the patient is one tenth of the normal level, the OAT gene mutation is strongly suspected in this case. Since most of the cases with gyrate atrophy have the genetic defect in exon or exon-intron junction area of the OAT gene(2), the case we experienced is exceptional case. We plan the further investigation including analysis of promotor and intron area of the OAT gene in this case. Furthermore, we plan to screen many DNA samples with Japanese patients with retinal dystrophies using the established SSCP procedure and direct sequencing.

ACKNOWLEDGMENTS

This study was supported by a grant for the study of chorioretinal degeneration from the Japanese Ministry of Health and Welfare, and a Grant-in-Aid for Scientific Research B01480420, B05807166 from the Ministry of Education of Japan.

REFERENCES

1. Hotta, Y., Kennaway, N.G., Weleber, R.G. and Inana, G., 1989, Inheritance of ornithine aminotransferase gene, mRNA, and enzyme defect in a family with gyrate atrophy of the choroid and retina, Am. J. Hum. Genet. 44: 353–357.
2. Valle, D. and Simell, O., 1995, The hyperornithinemias, in: The metabolic and molecular bases of inherited disease, Volume 1 (C.R.Scriver, A.L.Beaudet, W.S.Sly, and D.Valle, eds.), pp. 1147–1185, McGraw-Hill, Inc., New York.
3. van Bokhoven, H., Schwartz, M., Andreasson, S., van den Hurk, J.A.J.M., Bogerd, L., Jay, M., Ruther, K., Jay, B., Pawlowitzki, I.H., Sankila, E.-M., Wright, A., Ropers, H.-H., Rosenberg, T. and Cremers, F.P.M., 1994, Mutation spectrum in the CHM gene of Danish and Swedish choroideremia patients, Hum. Mol. Genet. 7:1047–1051.
4. Kajiwara, K., Berson, E.L. and Dryja, T.P., 1994, Digenic retinitis pigmentosa due to mutations at the unlinked peripherin/RDS and ROM1 loci. Science 264:1604–1608.
5. Fujiki, K., Kawano, H., Hotta, Y., Hayakawa, M., Nicilas, M.G., Takeda, M., Iwata, F., Ohta, N., Kanai, A., Hashimoto, T and Furuyama, J., 1995, Frequency of polymorphisms in the rhodopsin gene of Japanese retinitis pigmentosa and normal individuals, Jpn. J. Human Genet. 40:203–206.
6. Kawano, H., Hotta, Y., Fujiki, K., Takeda,M., Iwata,F., Sakuma,H., Hayakawa,M, Kanai, A., Shiono,T., Tamai,M., Hashimoto,T. and Furuyama, J., 1995, A study on the rhodopsin gene in Japanese retinitis pigmentosa, screening of mutation by restriction endonucreases and frequencies of DNA polymorphisms, J. Jpn. Ophthalmol. Soc. 99:1151–1157.
7. Fujiki, K., Hotta, Y., Hayakawa, M., Doi, R., Kohno, N., Takeda, M., Isashiki, Y., Ohba, N. and Kanai, A., 1996, Analysis of peripherin/RDS gene in Japanese retinal degenerations, Exp. Eye Res. 63:S.133 (suppl).
8. Fujiki, K., Hotta, Y., Hayakawa, M., Sakuma, H., Shiono, T., Noro, M., Sakuma, T., Tamai, M., Hikiji, K., Kawaguchi, R., Hoshi, A., Nakajima, A. and Kanai, A., 1992, Point mutations of rhodopsin gene found in Japanese families with autosomal dominant retinitis pigmentosa(ADRP), Jpn. J. Human Genet. 37:125–132.
9. Hotta, Y., Shiono, T., Hayakawa, M., Hashimoto, T., Kanai, J., Nakajima, A., Noro, M., Sakuma, T., Tamai, M., Fujiki, K., 1992, Malecular biological study of the rhodopsin gene in Japanese patients with autosomal dominant retinitis pigmentosa, Acta. Soc. Ophthalmol. Jpn. 96:237–242.
10. Shiono, T., Hotta, Y., Noro, M., Sakuma, T., Tamai, M., Hayakawa, M., Hashimoto, T., Fujiki, K., Kanai, A. and Nakajima, A., 1992, Clinical features of Japanese family with autosomal dominant retinitis pigmentosa caused by point mutaton in codon 347 of rhodopsin gene, Jpn. J. Ophhtalmol. 36:69–75.

11. Hayakawa,M., Hotta, Y., Imai,Y., Fujiki,K., Nakamura,A., Yanashima, K. and Kanai, A., 1993, Clinical features of autosomal dominant retinitis pigmentosa with rhodopsin gene codon 17 mutation and retinal neovascularization in a Japanese patient, Am. J. Ophthalmol.115:168–173.

12. Fujiki, K., Hotta, Y., Murakami A., Yoshii, M., Hayakawa, M., Ichikawa, T., Takeda, M., Akeo, K., Okisaka, S. and Kanai, A., 1995, Missense mutation of rhodopsin gene codon 15 found in Japanese autosomal dominant retinitis pigmentosa. Jpn. J. Human Genet. 40:271–277.

13. Yoshii, M., Murakami, A., Okisaka, S., Yanashima, K., Fujiki, K. and Hotta, Y., 1995, A case of sectorial retinitis pigmentosa associated with rhodopsin gene codon 15 mutation. Folia Ophthalmol. Jpn. 46: 560.

14. Moriwaki, K., Hayakawa, M. and Kishishita, H., 1989, Choroideremia: report of an affected male with a pinealoma. Folia Ophthalmol. Jpn. 40: 1106–1110.

15. Nakajima, A., Hayakawa, M., Kitagawa, T. and Sakiyama, T., 1982, A study of hyperornithinemia with gyrate atrophy, Report of Research Committee on Chorioretinal Degenerations, The Ministry of Health and Welfare of Japan, 1981, 106–110.

16. Vaithinathan, R., Berson, E.L. and Dryja, T.P., 1994, Further screening of the rhodopsn gene in patients with autosomal dominant retinitis pigmentosa, Genomics 21:461–463.

17. Inglehearn, C.F., Keen, T.J., Bashir, R., Jay, M., Fitzke, F., Bird, A.C., Crombie, A. and Bhattacharya, S., 1992, A completed screen for mutations of the rhodopsin gene in a panel of patients with autosomal dominant retinitis pigmentosa. Hum. Mol. Genet. 1:41–45.

18. Bunge, S., Wedemann, H., David, D., Terwilliger, D.J., van den Born, L.I., Aulehla-Schulz, C., Samanns, C., Horn, M., Ott, J., Schwinger, E., Schinzel, A., Denton, M.J. and Gal, A., 1993, Molecular analysis and genetic mapping of the rhodopsin gene in families with autosomal dominant retinitis pigmentosa, Genomics 17:230–233.

19. Dryja, T.P., Hahn, L.B., Cowley, G.S., McGee, T.L. and Berson, E.L., 1991, Mutation spectrum of the rhodpsin gene among patients with autosomal dominant retinitis pigmentosa, Proc. Natl. Acad. Sci. USA 88:9370–9374.

20. Nakazawa, M., Kikawa, E., Kamio, K., Chida, Y., Shiono, T. and Tamai, M., 1994, Ocular findings in patients with autosomal dominant retinitis pigmentosa and transversion mutaion in codon 244 (Asn244Lys) of the peripherin/RDS gene, Arch. Ophthalmol. 112:1567–1573.

21. Nakazawa, M., Wada, Y. and Tamai, M., 1995, Macular dystrophy associated with monogenic Arg172Trp mutation of the peripherin/RDS gene in a Japanese family, Retina 15:58–523.

22. Nakazawa, M., Kikawa, E., Chida, Y., Wada, Y., Shiono, T. and Tamai, M., 1996, Autosomal dominant cone-rod dystrophy associated with mutations in codon 244(Asn244His) and codon 184 (Try184Ser) of the peripherin/RDS gene, Arch. Ophthalmol. 114:72–78.

STRATEGIES FOR THE GENETIC ANALYSIS OF AUTOSOMAL RECESSIVE RETINITIS PIGMENTOSA IN SPANISH FAMILIES

Roser González-Duarte,[1] Mónica Bayés,[1] Amalia Martínez-Mir,[1] Diana Valverde,[2] Susana Balcells,[1] Montserrat Baiget,[2] Lluïsa Vilageliu,[1] and Daniel Grinberg[1]

[1]Departamento de Genética
Facultad de Biología
Universidad de Barcelona
Av. Diagonal 645, E-08071 Barcelona, Spain
[2]Unitat de Genética Molecular
Hospital de la Santa Creu i Sant Pau
Av. Sant Antoni Maria Claret 167, E-08025 Barcelona, Spain

INTRODUCTION

Retinitis pigmentosa (RP) is a group of inherited eye disorders that affect photoreceptor and pigment epithelial function. Prominent clinical features include night blindness and constriction of visual fields, generally leading to complete blindness (1). It affects about 1 in 4000 people (2).

Rod electroretinograms (ERG) are invariably reduced or absent in patients with RP. Funduscopically, the disease is characterised by the presence of clumps of pigment dispersed throughout the peripheral retina with a bone-spicule appearance (3). However, there is remarkable variability both in the age of onset and progression of symptoms and in the degree of peripheral and macular involvement (4).

The genetic heterogeneity of the disorder is even more marked. Several modes of inheritance have been described (1, 5), including autosomal dominant (ADRP), autosomal recessive (ARRP) and X-linked (XLRP), as well as isolated cases. ARRP is the most common mode of inheritance; in Spain it accounts for 39% of all cases (6).

There is further genetic heterogeneity within these categories. Recent advances in molecular genetics have led to the identification of several genes that are responsible for the phenotype (see 7 for a review). Mutations in the rhodopsin gene are found in 30% of patients with ADRP but in less than 1% of ARRP cases; mutations in the gene encoding peripherin/RDS are associated either with ADRP (3–5% of cases) or with a digenic peripherin/ROM1 form of the disease (3 cases described so far); mutations in the α and β subunits of rod phosphodiesterase (PDEA and PDEB, respectively) and in the α subunit of the cGMP gated channel (CNCG) have been identified in a few patients with ARRP. Lately, mutations that cosegregate with XLRP have been described in the RPGR (retinitis pigmentosa GTPase regulator) gene (8). In addition, nine as yet unidentified RP genes have been localised by linkage analysis. ADRP is linked genetically to loci on chromosomes 7p, 7q, 8cen, 17p, 17q and 19q; linkage studies in two large families indicate that their ARRP loci are located on 6p and 1q; finally, an X-linked RP gene has been mapped to the short arm of the X chromosome. To date, however, despite extensive studies, the molecular defect responsible for ARRP remains unidentified in most cases.

The identification of human disease genes can be accomplished by either functional or positional cloning. Unfortunately, because the primary biochemical alterations in RP are unknown, the *conventional* functional cloning approach is unattainable. Instead, the candidate gene approach has been fruitfully used (9). A number of proteins involved in retinal function have been described, and their genes cloned. These are, therefore, plausible candidates for RP.

Positional cloning is difficult in disorders showing complex genetics (10). In RP, the extensive genetic heterogeneity is the most serious drawback to linkage studies, but this can be overcome via the analysis of single, large pedigrees. For the autosomal recessive form of the disease, however, affected families are generally too small to yield significant linkage data.

Positional cloning in single, large pedigrees and candidate gene analysis have been combined giving rise to the positional-candidate approach. Rhodopsin, the most common RP gene so far, was identified in this way: in a large Irish pedigree strong linkage was reported (11) between ADRP and C17 (D3S47). This polymorphism maps to the long arm of chromosome 3, near the rhodopsin gene in 3q21-q24. Rhodopsin is the visual pigment of the rod cells and was thus a candidate gene for ADRP. Subsequent studies showed that mutations in this gene account for approximately 30% of all ADRP cases (see 12 for a review). However, molecular genetic studies of such an heterogeneous disorder are arduous: most RP genes identified account for only a few cases.

We collected DNA from members of 48 Spanish ARRP families. In order to elucidate the molecular basis of the disease we followed two approaches depending on family size. In small families the analysis of candidate genes is the only possibility, while in large families a genome-wide search for a new ARRP gene can also be pursued. Here we present the results from the study of eleven retina-specific genes: rhodopsin, PDEB, recoverin (RCV1), S antigen (SAG), γ subunit of rod phosphodiesterase (PDEG), peripherin/RDS, rod outer segment membrane protein 1 (ROM1), neural retinal leucine zipper (NRL), cellular retinaldehyde binding protein (CRALBP), interstitial retinol binding protein (IRBP) and guanylate cyclase activating protein (GCAP), as well as the ARRP loci at 1q and 6p. For two of these genes, RCV1 and SAG, a fine genetic mapping was initially performed to identify the closest polymorphisms. In two large pedigrees, a systematic genome search is currently being undertaken after discarding all known RP loci.

MATERIALS AND METHODS

Families

Forty-eight Spanish ARRP pedigrees (46 small + 2 large), comprising 31 consanguineous and 17 non consanguineous families were studied. A dignosis of ARRP was based on detailed family history and fundus examination, visual-field testing, electroretinography and measurement of visual acuity.

A total of 8 non-RP families including 104 meioses were used for the genetic mapping of RCV1 and SAG genes.

Informed consent was obtained for all individuals from whom blood was drawn. Genomic DNA was extracted from peripheral blood as described (13).

Analysis of DNA Polymorphisms

Intragenic polymorphisms used to analyse candidate genes and loci are shown in Table 1. In most cases PCR amplification products were resolved on 6% non-denaturing polyacrylamide gels and stained with ethidium bromide. The polymorphism in exon 16 of the S antigen gene was detected by hybridization with allele-specific oligonucleotides (ASOH). The nucleotide substitutions at the PDEG, peripherin/RDS and CRALBP genes were tested by digestion with the corresponding enzymes.

Polymorphic DNA markers known to harbour an RP locus are listed in Table 3. They were all from the MapPairs set (Research Genetics, Huntsville, AL) and were analysed according to the manufacturer's instructions.

Linkage Analysis

The reference genetic maps were those reported by Weissenbach et al. (14) and Gyapay et al. (15).

Two-point and multipoint linkage analyses were performed using the MLINK and the CMAP programs, respectively, from the LINKAGE package version 5.2 (16). To evaluate linkage between a candidate gene and the disease, lod scores were calculated at a recombination fraction (θ) of 0 in each family. Other values of θ were not considered since polymorphic markers were within or very close to the candidate gene. As accepted, lod scores below -2 were assumed to be evidence for excluding linkage. The detection of heterozygosity in patients from consanguineous marriages was also considered an exclusion criterion (17).

PCR-SSCP analyses

Pairs of oligonucleotide primers surrounding exonic regions of the candidate genes analysed were designed to give PCR products in the range of 200–300 bp. Most of the primer sequences and PCR conditions have been described elsewhere (18). SSCP analyses were performed as described previously (19). Each PCR amplified fragment was assayed under three conditions, combining acrylamide and glycerol concentrations and running temperatures.

DNA Sequencing

PCR products showing aberrant SSCP patterns were subsequently purified by Wizard™ PCR preps (Promega, Madison, WI) and cloned into pUC18 vector using the Sure-Clone™ Ligation Kit (Pharmacia Biotech., Uppsala, Sweeden). PCR amplification and SSCP analysis of the clones allowed selection of those carrying the mutation/polymorphism. In each case, three to five clones were sequenced by the dideoxy chain-termination method using the T7 sequencing™ Kit (Pharmacia). In addition, direct sequencing from the purified PCR products was also performed using the Sequenase™ version 2.0 DNA Sequencing Kit (United States Biochemical Co., Cleveland, OH).

RESULTS

Genetic Mapping of the Recoverin and S Antigen Genes

The chromosomal location of the RCV1 and SAG genes was not well defined when this study was initiated. The RCV1 gene had been assigned to chromosome 17 by somatic cell analysis (20), while the SAG gene had been mapped to 2q24-q37 by *in situ* hybridization (21). A dinucleotide polymorphim at the 3′ UTR of RCV1 had been described (22), but its informativeness was low (Het=0.46). For SAG, an internal triallelic polymorphism had been reported (Het=0.72) (23), which could be detected by ASOH.

We used these two intragenic polymorphisms to localise the RCV1 and SAG genes with respect to markers from high-density maps of chromosomes 17 and 2, respectively. Two-point and multipoint linkage analyses enabled us to identify the closest markers to these genes (Figure 1). RCV1 is tightly linked to the anonymous marker D17S786

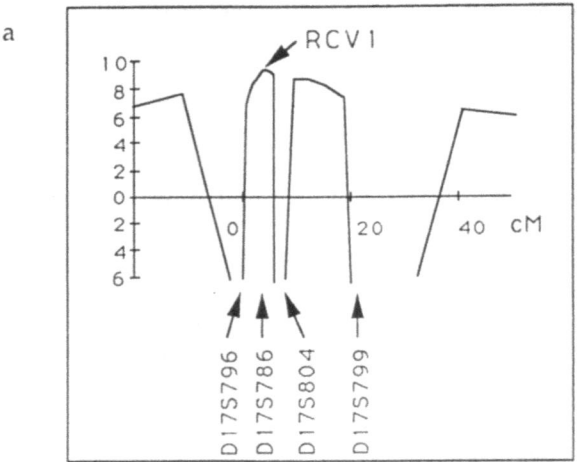

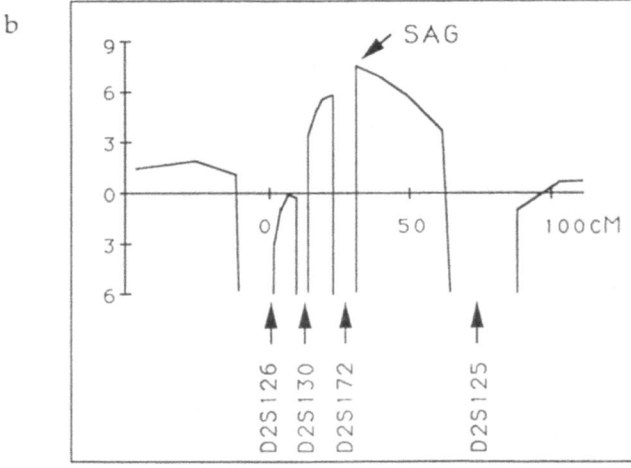

Figure 1. Multipoint analyses of RCV1 (a) and SAG (b) genes with microsatellite markers from 17p and 2q, respectively. Likelihood estimates are given in $\log_{10}$ on the *Y-axis*. Genetic distances (cM) are shown on the *X-axis*.

(Z_{max}=6.92 at θ=0.00) (19), while SAG appears to be located 4 cM distal to D2S172 (Z_{max}=9.25 at θ=0.038) (24). Both D17S786 and D2S172 are highly informative microsatellite markers (Het=0.77 and Het=0.92, respectively) and therefore their use could substantially improve the resolution of cosegregation analyses of these genes with several eye disorders, such as RP.

The Candidate Gene Approach: Analyses of 11 Genes and 2 Loci in 46 Small ARRP Families

Any gene encoding structural proteins or enzymes involved in the visual process could be considered a candidate for ARRP (9). In 46 Spanish ARRP families we evaluated the involvement in the disease of PDEB and rhodopsin, two known ARRP genes, as well as other candidates such as peripherin/RDS, ROM1, RCV1, SAG, PDEG, NRL, CRALBP, IRBP and GCAP. Most of these genes are expressed only in the retina and some are specific to the rod photoreceptor cells. Rhodopsin and PDEB are directly involved in the phototransduction pathway; RCV1, SAG, PDEG and GCAP are responsible for the restoration of the dark state; CRALBP and IRBP participate in vitamin A metabolism; finally, peripherin/RDS and ROM1 are essential for the structure of the disk membrane (see 25 for a review). The two ARRP loci at 1q and 6p were also analysed.

In our study we combined linkage, homozygosity and mutation analyses. As a starting point, cosegregation and homozygosity studies with polymorphisms (Table 1) at these loci were carried out. The results presented in Table 2 show that the nine genes and two loci analysed were excluded as the cause of the disease in an average of 70% of the cases.

Based on the results of linkage studies, mutation analysis of the candidate genes cosegregating with the disease in particular pedigrees was performed. Because of the absence of highly informative polymorphisms within the ROM1 gene, we screened the affected members of all the ARRP families for the presence of mutations. Taking into account the possibility of digenic RP due to mutations at the peripherin/RDS and ROM1

Table 1. Polymorphic markers within or very close to the candidate genes used for the linkage and homozygosity analyses in the ARRP families

Candidate genes	Polymorphisms analysed
Rhodopsin	Microsatellite in intron 1 D3S1238, at 3q21
PDEB	Microsatellite in the 3′ UTR region D4S432, at 4p16.3
Recoverin	Microsatellite in the 3′ UTR region D17S786, at 17p13
S antigen	ATT/GCT/GTT polymorphism in exon 16 D2S172, at 2q37
PDEG	A1447G polymorphism in exon 4 delC1683 in the 3′ UTR region
Peripherin/RDS	Microsatellite in the 3′ UTR region C558T in exon 1
NRL	D14S64, MYH7 and D14S54, at 14q11.1-q11.2
CRALBP	Intragenic RFLP (Southern)
IRBP	Microsatellite in the 3′ UTR region
RP-12	F13B and D1S158, at 1q31-q32
RP-14 (GCAP)	D6S291 and D6S439, at 6p21

Table 2. Results obtained by the candidate gene approach: linkage and mutation analyses. The percentages of families excluded by linkage and homozygosity analysis are shown, as well as the number of disease-causing mutations and polymorphisms found

Candidate genes	Exclusion by linkage analysis	Disease-causing mutations	Polymorphisms
Rhodopsin	85%	0	3
PDEB	80%	3	2
Recoverin	93%	0	0
S antigen	61%	0	7
PDEG	52%	0	2
Peripherin/RDS	56%	0	6
ROM1	–	0	1
NRL	92%	0	2
CRALBP	47%	0	0
IRBP	48%	0	0
RP-12	64%	–	–
RP-14 (GCAP)	80%	0	0

genes (26), we also performed direct SSCP analysis of the peripherin/RDS gene in all the non-consanguineous pedigrees. Finally, the GCAP gene, which has been mapped to the vicinity of RP-14 on 6p21 (27), was evaluated in the families that could not be excluded by linkage or homozygosity studies using RP-14 markers.

We found 26 variant bands by SSCP analysis (Table 2). DNA sequence and cosegregation analysis led to the identificaion of three disease-causing mutations in the PDEB gene, thus accounting for 6% of cases. All three mutations, 260 (71 bp dup) (28), Leu699Arg (29) and Arg552Gln (30), were present in homozygosis in the affected members of three consanguineous families (Figure 2). The 71 bp duplication in exon 1 causes a frameshift leading to a

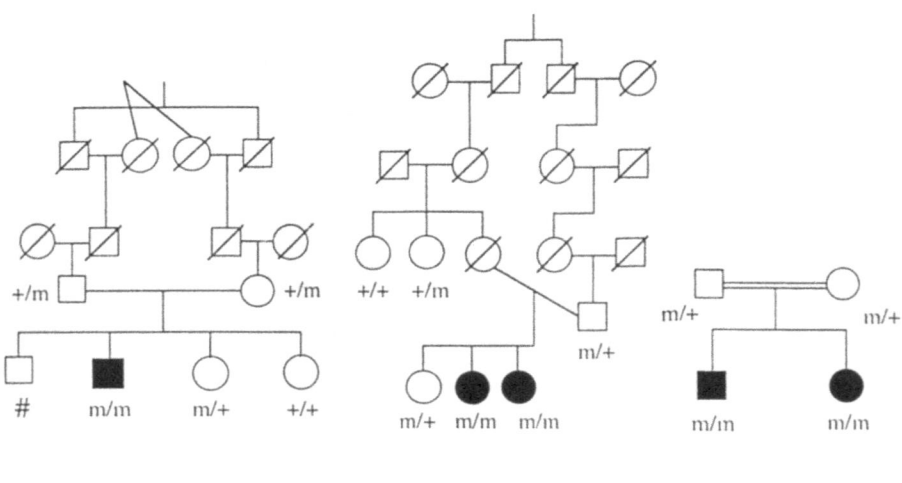

B-4: 260 (71bp dup)   B-27: Leu699Arg   M-9*: Arg552Gln

Figure 2. Pedigrees of the consanguineous ARRP families in which PDEB mutations cosegregating with the disease were found. Alleles below the pedigree symbols correspond to the mutated (m) or normal (+) chromosomes. * The degree of consanguinity in family M-9 is not known, but both parents come from the same small village.

premature stop codon at position 213. The predicted product is a truncated protein that lacks the catalytic and the cGMP binding domains. The two mutations, Leu699Arg and Arg552Gln, are located within or very close to the catalytic domain of the enzyme, which has been mapped between amino acids 555 and 790 (31). The absence of these mutations in 50 control individuals suggests that they are responsible for RP in these families.

The other abnormal SSCP patterns correspond to polymorphisms or rare variants that are not associated with the disease: some of them do not cosegregate with the disease, while others are silent substitutions (data not shown). Nevertheless, the high number of variant bands detected validates the PCR-SSCP tecnique used in this study.

Our results provide evidence that none of the candidate genes and loci analysed, with the sole exception of the PDEB gene, should be considered a major ARRP gene.

The Positional Cloning Approach: Genome-Wide Search in Two Large ARRP Families

Two large Spanish ARRP families were used in this study (Figure 3). Individuals from the first family, named M-33, suffer from an aggressive form of the disease, with an onset of night blindness in the first decade of life and choriocapillaris atrophy as a distinc-

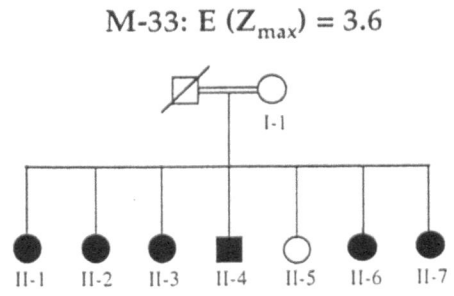

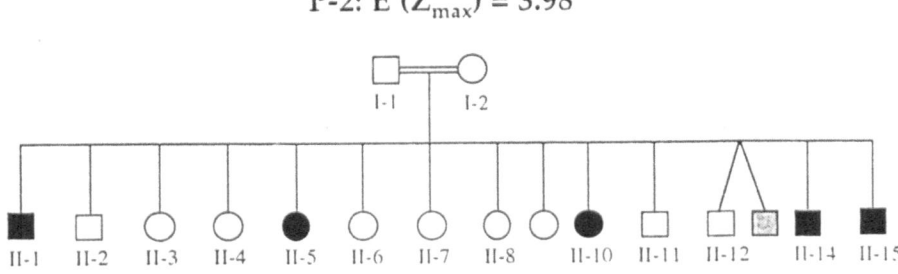

Figure 3. Spanish pedigrees M-33 and P-2 segregating ARRP. The expected maximum lod scores, $E(Z_{max})$, are shown.

tive feature. Despite the large number of affected individuals in this sibship (six out of seven), an autosomal recessive pattern of inheritance was assumed because both parents were unaffected at advanced ages (72 and 82 years, respectively) and because consanguinity was well-established (parents were second cousins).

In the second family, known as P-2, the disease has led to disturbances in both peripheral and central vision in the second decade of life. Intraretinal pigment deposits are evident although not in a bone-spicule configuration. The disease exhibits high intrafamilial variability in the age of onset of night blindness, in the preservation of visual acuity and in the ERG responses.

Both families are sufficiently large to show linkage independently. The expected maximum lod scores for pedigrees M-33 and P-2 are 3.6 and 3.98 respectively. Thus a systematic search of the genome for a new ARRP locus can be undertaken.

The first step of our linkage strategy was to examine polymorphisms at loci and genes that have been reported to be involved in the non-syndromic and syndromic forms of RP in these families. We also evaluated several retina-specific genes. Significantly negative lod scores ($Z<-2$) were achieved in all informative cases (Table 3), which indicates that a different locus is responsible for ARRP in these two pedigrees.

Table 3. Exclusion of RP genes and loci by linkage analysis in families P-2 and M-33. Values of θ corresponding to $Z=-2$ are shown

Loci analysed	Map location	Exclusion θ	
		Family P-2	Family M-33
Candidate genes			
PDEB	4p16.3	0.11	0.014
RHO	3q21-q24	0.09	0.035
RDS	6p12	0.03	0.017
ROM1	11q13	0.05	0.026
RCV1	17p13	0.04	0.075
SAG	2q37	0.06	0.11
NRL (D14S64)	14q11	0.09	0.055
PDEG	17q25	NI**	0.10
ADRP loci			
RP-1 (D8S165)	8q11	0.14	0.05
RP-9 (D7S435)	7p14	0.04	0.07
RP-10 D7S480)	7q31	0.12	0.05
RP-11 (D19S180)	19q13.4	0.10	0.015
RP-13 (D17S938)	17p13	0.09	0.035
RP-17 (D17S807)	17q22-q24	0.03	0.09
ARRP loci			
RP-12 (D1S158)	1q31-q32.1	0.06	0.085
RP-14 (D6S291)	6p	0.02	0.09
Syndromic RP loci *			
USH1A(D14S65)	14q32	0.01	—
USH1B (D11S527)	11q13.5	0.04	0.035
USH1C (D11S419)	11p15.2-p14	0.01	0.025
USH2A (D1S237)	1q41	0.008	0.07
USH3 (D3S1299)	3q21-q25	0.10	NI**
BBS1 (INT2)	11q13	0.08	0.13
BBS2 (D16S265)	16q21	0.03	0.025
BBS3 (D3S1752)	3p13-p12	0.001	NI**
BBS4 (D15S204)	15q22-q23	0.03	NI**

* USH: Usher syndrome; BBS: Bardet-Biedl syndrome
** NI: noninformative

In the absence of linkage to either of the candidate loci, a sequential study of the whole genome is currently being undertaken, using microsatellite markers of known chromosomal locations.

DISCUSSION

The candidate gene approach allowed the identification of six out of the seven RP genes known so far. The involvement of rhodopsin and peripherin/RDS in RP was, in fact, established by a combination of linkage mapping followed by a search for candidate genes in the interval defined. This is the positional-candidate approach.

Linkage studies have implicated several additional loci, but RPGR (8) is the only RP gene that has been isolated by positional cloning to date. So, even though the candidate gene approach has not been rewarding, positional cloning appears to be even harder, especially in the autosomic recessive form of the disease. Similar results have been obtained in other retinal diseases (see Table 4) (32).

The Candidate Gene Approach: So Many Plausible Candidates

A number of genes involved in either the phototransduction, restoration of the dark state, vitamin A metabolism or the structure of the disk membranes have been cloned. So, in RP, as in many complex diseases, such as coronary heart disease and psychiatric disorders, there are a large number of plausible candidates. The sequence of these genes is examined in DNA from patients in the search for mutations. This can be arduous when there are so many candidates, the genes are large and split into many exons, as is the case of the PDEB, PDEA and CNCG genes. In the latter two cases, the analysis of more than one thousand PCR fragments led to the identification of only four (33) and six (34) disease-causing mutations respectively.

Table 4. Human eye disease genes identified by positional cloning, candidate gene approach or a combination of both (see ref. 32 for a review)

Eye disorder*	Inheritance**	Gene/protein***	Strategy
Aniridia	AD	PAX6	Positional cloning
Retinitis pigmentosa	AD	Rhodopsin	Positional-candidate approach
	AD/DIG	Peripherin/RDS	Positional-candidate approach
	AD/DIG	ROM1	Candidate gene approach
	AR	Rhodopsin	Candidate gene approach
	AR	PDEB	Candidate gene approach
	AR	PDEA	Candidate gene approach
	AR	CNCG	Candidate gene approach
	XL	RPGR	Positional cloning
Usher syndrome 1B	AR	Myosin VIIA	Positional-candidate approach
CSNB	AD	Rhodopsin	Candidate gene approach
	AD	PDEB	Candidate gene approach
Macular dystrophy	AD	Peripherin/RDS	Candidate gene approach
SFD	AD	TIMP3	Positional-candidate approach
Oguchi disease	AR	S antigen	Positional-candidate approach
Choroideremia	XL	GGT	Positional cloning
Gyrate atrophy	AR	OAT	Candidate gene approach

  * CSNB: congenital stationary night blindness; SFD: Sorby's fundus dystrophy.
  ** AD: autosomal dominant; AR: autosomal recessive; XL: X-linked; DIG: digenic.
  *** GGT: geranyl-geranyl tranferase; OAT: ornithine amino-transferase.

We undertook a combined linkage and mutation analysis approach to test the reported genes. Linkage and homozygosity studies allowed us to rule them out in most of the families (Table 2), and thus mutation screening was performed in a few pedigrees. We found three disease-causing mutations in the PDEB gene but none in the rest of the genes studied. These results leave 45 out of the 48 ARRP families with the molecular defect underlying the disease still unidentified.

Among the high number of plausible RP candidate genes only very few are well supported by specific physiological evidence. When examining these genes in RP patients mutations can be found, but they often account for less than 1% of the cases. In the absence of functional studies, it is difficult to associate these mutations with the phenotype. If they are missense mutations, they may be rare, non-pathogenic variants that cosegregate with the disease by chance, because the families are small. It is even more risky to attest that an aminoacid substitution causes the disease when the mode of inheritance is deduced *a posteriori*, as described in the digenic form of RP attributed to simultaneous mutations in the peripherin/RDS and ROM1 genes (26). Furthermore, in one case of simplex RP the same ROM1 mutation has been described (35), but does not cosegregate with the disease in the family and a second mutation in the peripherin/RDS gene has not been found. In these cases conclusive evidence of pathogenicity remains to be established, awaiting further experiments such as *in vitro* functional tests or *in vivo* gene knock-outs.

The Positional Cloning Approach: So Few Large Families

Linkage studies in large RP pedigrees have defined nine additional RP loci on 7p, 7q, 8cen, 17p, 17q and 19q for ADRP; on 1q and 6p for ARRP; and on Xp for XLRP. Our preliminary linkage data in families M-33 and P-2 preclude at least one more recessive gene for RP, thus emphasizing the genetic heterogeneity in this group of retinopathies.

For disorders in which few meioses are available for linkage analysis, the resolution of the genetic mapping may not be better than 5 cM. It is not possible to narrow the interval and thus the number of positional candidate genes that need to be isolated and examined is large (assuming that there is a gene every 30 Kb, we would expect between 100 and 200 genes in a 5 cM interval). Any mutation found in such genes may cosegregate with the disease in the family and so it is difficult to establish which mutation is associated with the disease phenotype. Therefore, again, functional studies are required.

The case for RP-3 is instructive: the gene was genetically mapped to Xp21.1 in 1988 (36) but was cloned in 1996 (8). Fortunately, the availability of patients with chromosomal rearrangements in this region allowed the precise localization of the disease locus. Despite these clues it was not until 1995 (37, 38) that attempts at positional cloning identified a novel retinal-expressed gene, called SRPX or ETX1. However, no mutations were found in the great majority of XLRP patients. Finally, a new gene, named RPGR, has been isolated from the region, and preliminary results pinpoint it as being the *true* RP-3 gene (8). Thus the search for the gene for RP-3 has taken much longer than expected.

Positional cloning of genetically heterogeneous disorders is a painstaking task. Differential diagnosis is frequently unattainable so genetic subgroups cannot be established. In RP clear clinical differences have only been established in those patients presenting para-arteriolar preservation of the retinal pigment epithelium (PPRPE) as a distinctive feature (39). The locus responsible for PPRPE has been mapped to 1q in two large ARRP pedigrees (40, 41). A clear-cut differential diagnosis would greatly facilitate the positional cloning of RP genes.

Future Prospects

In the near future new RP genes may be identified following either the candidate gene approach or a positional cloning strategy. Only if the systematic study of tissues from patients leads to a consistent biochemical clue, will functional cloning also be plausible. Progress in transgenic technology may provide a more straightforward approach to the identification of those genes that can cause RP.

Knowledge of all RP genes may allow a better classification of this group of retinopathies and lead to the development of reliable diagnostic tests and the design of more effective therapies.

ACKNOWLEDGMENTS

The authors thank the families for their generous cooperation and Drs. C. Ayuso, M. Beneyto, J. Benítez, M.A. Ramos-Arroyo and I. Tejada who kindly contributed blood samples of some of their patients. Thanks also to Drs. P. Chivelet, T. Solans, B. García-Sandoval, B. Goldaracena and T. del Río for clinical assesment, and to R. Rycroft for revising the English.

This work was supported by Spanish CICYT (SAF93-0479-062-01 and SAF96-0329) and the "Federación de Asociaciones de Afectados de Retinosis Pigmentaria del Estado Español" (FAARPE).

REFERENCES

1. Heckenlively, J.R., Yoser, S.L., Friedman, L.H., and Oversier, J.J., 1988, Clinical findings and common symptoms in retinitis pigmentosa, *Am J Ophtalmol* 105: 504–511.
2. Berson, E.L., 1993, Retinitis-Pigmentosa: The Friedenwald Lecture, *Invest Ophthalmol Visual Sci* 34: 1659–1676.
3. Marmor, M.F., Aguirre, G., Arden, G.O., Berson, E., Birch, D.G., Boughman, J.A., Carr, R., Chatrian, G.E., Del Monte, M., Dowling, J., Enoch, J., Fishman, G.A., Fulton, A.B., Garcia, C.A., Gouras, P., Heckenlively, J., Hu, D.N., Lewis, R.A., Niemeyer, G., Parker, J.A., Perlman, I., Ripps, H., Sandberg, M.A., Siegel, I., Weleber, R.G., Wolf, M.L., Wu, L., and Young, R.S.L., 1983, Retinitis pigmentosa: A symposium on terminology and methods of examination, *Ophthalmology* 90: 126–131.
4. Kaplan, J., Bonneau, D., Frézal, J., Munnich, A., and Dufier, J.L., 1990, Clinical and genetic heterogeneity in retinitis pigmentosa, *Hum Genet* 85: 635–642.
5. Boughman, J.A., Conneally, P.M., and Nance, W.E., 1980, Population genetic studies of retinitis pigmentosa, *Am J Hum Genet* 32: 223–235.
6. Ayuso, C., Garcia-Sandoval, B., Najera, C., Valverde, D., Carballo, M., and Antiñolo, G., 1995, Retinitis pigmentosa in Spain, *Clin Genet* 48: 120–122.
7. Dryja, T.P., and Li, T., 1995, Molecular genetics of retinitis pigmentosa, *Hum Mol Genet* 4: 1739–1743.
8. Meindl, A., Dry, K., Herrmann, K., Manson, F., Ciccodicola, A., Edgar, A., Carvalho, M., Achatz, H., Hellebrand, H., Lennon, A., Migliaccio, C., Porter, K., Zrenner, E., Bird, A., Jay, M., Lorenz, B., Wittwer, B., Durso, M., Meitinger, T., and Wright, A., 1996, A gene (RPGR) with homology to the RCC1 guanine nucleotide exchange factor is mutated in X-linked retinitis pigmentosa (RP3), *Nat Genet* 13: 35–42.
9. Dryja, T.P., 1990, Deficiencies in sight with the candidate gene approach, *Nature* 347: 614.
10. Lander, E.S., 1988, Mapping complex genetic traits in humans, in: *Genome analysis: A practical approach*, Davies, K.E. ed, pp. 171–189, IRL, Oxford.
11. McWilliam, P., Farrar, G.J., Kenna, P., Bradley, D.G., Humphries, M.M., Sharp, E.M., McConnell, D.J., Lawler, M., Sheils, D., Ryan, C., Stevens, K., Daiger, S.P., and Humhpries, P., 1989, Autosomal dominant retinitis pigmentosa (ADRP): Localization of an ADRP gene to the long arm of chromosome 3, *Genomics* 5: 619–622.
12. Daiger, S.P., Sullivan, L.S., and Rodriguez, J.A., 1995, Correlation of phenotype with genotype in inherited retinal degeneration, *Behav Brain Sci* 18: 452–467.

13. Miller, S.A., Dyke, D.D., and Polesky, H.F., 1988, A simple salting out procedure for extracting DNA from human nucleated cells, *Nucl Acids Res* 16: 1215.

14. Weissenbach, J., Gyapay, G., Dib, C., Vignal, A., Morissette, J., Millasseau, P., Vaysseix, G., and Lathrop, M., 1992, A second-generation linkage map of the human genome, *Nature* 359: 794–801.

15. Gyapay, G., Morissatte, J., Vignal, A., Dib, C., Fizames, C., Millasseau, P., Marc, S., Bernardi, G., Lathrop, M., and Weissenbach, J., 1994, The 1993–94 Généthon human genetic linkage map, *Nature Genet* 7: 246–338.

16. Terwilliger, J.D., and Ott, J., 1994, *Handbook of human genetic linkage*, 1st ed., Johns Hopkins University Press, Baltimore and London.

17. Lander, E.S., and Botstein, D., 1987, Homozygosity mapping: a way to map human recessive traits with the DNA of inbred children, *Science* 236: 1567–1568.

18. Bayés, M., Martínez-Mir, A., Valverde, D., del Río, E., Vilageliu, Ll., Grinberg, D., Balcells, S., Ayuso, C., Baiget, M., and Gonzàlez-Duarte, R., 1996, Autosomal recessive retinitis pigmentosa in Spain: Evaluation of 4 genes and 2 loci involved in the disease, *Clin Genet* (in press).

19. Bayés, M., Valverde, D., Balcells, S., Grinberg, D., Vilageliu, Ll., Benítez, J., Ayuso, C., Beneyto, M., Baiget, M., and Gonzàlez-Duarte, R., 1995, Evidence against involvement of recoverin in autosomal recessive retinitis pigmentosa in 42 Spanish families, *Hum Genet* 96: 89–94.

20. Murakami, A., Yajima, T., and Inana, G., 1992, Isolation of human retinal genes: Recoverin cDNA and gene, *Biochem Biophys Res Commun* 187: 234–244.

21. Ngo, J.T., Klisak, I., Sparkes, R.S., Mohandas, T., Yamaki, K., Shinohara, T., and Bateman, J.B., 1990, Assignment of the S-antigen gene (SAG) to human chromosome 2q24-q37, *Genomics* 7: 84–87.

22. Dollfus, H., Rozet, J.M., Musarella, M.A., Kaplan, J., and Munnich, A., 1993, Dinucleotide repeat polymorphism at the human recoverin RCVI gene locus on chromosome 17p, *Hum Mol Genet* 2: 1081.

23. Sheffield, V.C., Beck, J.S., Nichols, B., Cousineau, A., Lidral, A.C., and Stone, E.M., 1992, Detection of multiallele polymorphisms within gene sequences by GC-clamped denaturing gradient gel electrophoresis, *Am J Hum Genet* 50: 567–575.

24. Valverde, D., Bayés, M., Martínez, I., Grinberg, D., Vilageliu, Ll., Balcells, S., Gonzàlez-Duarte, R., and Baiget, M., 1994, Genetic fine localization of the arrestin (S-antigen) gene 4 cM distal from D2S172, *Hum Genet* 94: 193–194.

25. Palczewski, K., 1994, Is vertebrate phototransduction solved? New insights into the molecular mechanism of phototransduction, *Invest Ophthalmol Visual Sci* 35: 3577–3581.

26. Kajiwara, K., Berson, E.L., and Dryja, T., 1994, Digenic retinitis pigmentosa due to mutations at the unlinked peripherin/RDS and ROM1 loci, *Science* 264: 1604–1608.

27. Subbaraya, I., Ruiz, C.C., Helekar, B.S., Zhao, X.Y., Gorczyca, W.A., Pettenati, M.J., Rao, P.N., Palczewski, K., and Baehr, W., 1994, Molecular characterization of human and mouse photoreceptor guanylate cyclase-activating protein (GCAP) and chromosomal localization of the human gene, *J Biol Chem* 269: 31080–31089.

28. Bayés, M., Giordano, M., Balcells, S., Grinberg, D., Vilageliu, Ll., Martínez, I., Ayuso, C., Benítez, J., Ramos-Arroyo, M.A., Chivelet, P., Solans, T., Valverde, D., Amselem, S., Goossens, M., Baiget, M., Gonzàlez-Duarte, R., and Besmond, C., 1995, Homozygous tandem duplication within the gene encoding the beta subunit of rod phosphodiesterase as a cause for autosomal recessive retinitis pigmentosa, *Hum Mutat* 5: 228–234.

29. Valverde, D., Solans, T., Grinberg, D., Balcells, S., Vilageliu, Ll., Bayés, M., Chivelet, P., Besmond, C., Goossens, M., Gonzàlez-Duarte, R., and Baiget, M., 1996, A novel mutation in exon 17 of the beta subunit of rod phosphodiesterase in two RP sisters of a consanguineous marriage, *Hum Genet* 97: 35–38.

30. Valverde, D., Baiget, M., Seminago, R., del Río, E., García-Sandoval, B., del Río, T., Bayés, M., Balcells, S., Martínez-Mir, A., Grinberg, D., and Ayuso, C., 1996, Identification of a novel Arg552Gln mutation in exon 13 of the beta subunit of rod phosphodiesterase gene in a Spanish family with autosomal recessive retinitis pigmentosa, *Hum Mutat* (in press).

31. Chabonneau, H., Prusti, R.K., LeTrong, H., Sonnenburg, W.K., Mullaney, P.J., Walsh, K.A., and Beavo, J.A., 1990, Identification of a noncatalytic cGMP-binding domain conserved in both the cGMP-stimulated and photoreceptor cyclic nucleotide phosphodiesterases, *Proc Natl Acad Sci USA* 87: 288–292.

32. Rosenfeld, P.J., McKusick, V.A., Amberger, J.S., and Dryja, T.P., 1994, Recent advances in the gene map of inherited eye disorders: Primary hereditary diseases of the retina, choroid, and vitreous, *J Med Genet* 31: 903–915.

33. Huang, S.H., Pittler, S.J., Huang, X.H., Oliveira, L., Berson, E.L., and Dryja, T.P., 1995, Autosomal recessive retinitis pigmentosa caused by mutations in the alpha subunit of rod cGMP phosphodiesterase, *Nature Genet* 11: 468–471.

34. Dryja, T.P., Finn, J.T., Peng, Y.W., McGee, T.L., Berson, E.L., and Yau, K.W., 1995, Mutations in the gene encoding the alpha subunit of the rod cGMP-gated channel in autosomal recessive retinitis pigmentosa, *Proc Natl Acad Sci USA* 92: 10177–10181.

35. Sakuma, H., Inana, G., Murakami, A., Yajima, T., Weleber, R.G., Murphey, W.H., Gass, J.D.M., Hotta, Y., Hayakawa, M., Fujiki, K., Gao, Y.Q., Danciger, M., Farber, D.B., Cideciyan, A.V., and Jacobson, S.G., 1995, A heterozygous putative null mutation in ROM-1 without a mutation in peripherin/RDS in a family with retinitis pigmentosa, *Genomics* 27: 384–386.

36. Mussarella, M.A., Burghes, A., Anson-Cartwright, L., Mahtani, M.M., Argonza, R., and Tsui, L.C., 1988, Localization of the gene for X-linked recessive type of retinitis pigmentosa (XLRP) to Xp21 by linkage analysis, *Am J Hum Genet* 43: 484–494.

37. Meindl, A., Carvalho, M., Herrmann, K., Lorenz, B., Achatz, H., Apfelstedtsylla, E., Wittwer, B., Ross, M., and Meitinger, T., 1995, A gene (SRPX) encoding a sushi-repeat-containing protein is deleted in patients with X-linked retinitis pigmentosa, *Hum Mol Genet* 4: 2339–2346.

38. Dry, K.L., Aldred, M.A., Edgar, A.J., Brown, J., Manson, F., Ho, M.F., Prosser, J., Hardwick, L.J., Lennon, A.A., Thomson, K., Vankeuren, M., Kurnit, D.M., Bird, A.C., Jay, M., Monaco, A.P., and Wright, A.F., 1995, Identification of a novel gene, ETX1, from Xp21.1, a candidate gene for X-linked retinitis pigmentosa (RP3), *Hum Mol Genet* 4: 2347–2353.

39. Porta, A., Pierrottet, C., Aschero, M., and Orzalesi, N. 1992, Preserved para-arteriolar retinal pigment epithelium retinitis pigmentosa, *Am J Ophtalmol* 113: 161–164.

40. van Soest, S., Vandenborn, L.I., Gal, A., Farrar, G.J., Bleekerwagemakers, L.M., Westerveld, A., Humphries, P., Sandkuijl, L.A., and Bergen, A.A.B., 1994, Assignment of a gene for autosomal recessive retinitis pigmentosa (RP12) to chromosome 1q31-q32.1 in an inbred and genetically heterogeneous disease population, *Genomics* 22: 499–504.

41. Leutelt, J., Oehlmann, R., Younus, F., Vandenborn, L.I., Weber, J.L., Denton, M.J., Mehdi, S.Q., and Gal, A., 1995, Autosomal recessive retinitis pigmentosa locus maps on chromosome 1q in a large consanguineous family from Pakistan, *Clin Genet* 47: 122–124.

30

PROGRESS IN POSITIONAL CLONING OF RP10 (7q31.3), RP1 (8q11-q21), AND VMD1 (8q24)

Stephen P. Daiger,[1,2] Rachel E. McGuire,[1,3] Lori S. Sullivan,[1] Melanie M. Sohocki,[1] Susan H. Blanton,[4] Peter Humphries,[5] Eric D. Green,[3] Helen Mintz-Hittner,[2] and John R. Heckenlively[6]

[1]Human Genetics Center
School of Public Health
[2]Department of Ophthalmology and Visual Science
The University of Texas Health Science Center
Houston, Texas
[3]Genome Technology Branch
National Center for Human Genome Research
National Institutes of Health
Bethesda, Maryland
[4]Department of Pediatrics
University of Virginia
Charlotte, Virginia
[5]Ocular Genetics Unit, Department of Genetics
Trinity College
Dublin, Ireland
[6]Jules Stein Eye Institute
University of California

INTRODUCTION

The goal of our research is to determine the genes and mutations causing autosomal dominant retinitis pigmentosa (adRP) and related diseases. As is now common knowledge, this deceptively-simple goal is confounded by the exceptional heterogeneity of retinitis pigmentosa and other forms of retinal degeneration. This heterogeneity includes allelic heterogeneity, i.e., different mutations in the same gene causing different clinical phenotypes or modes of inheritance; genetic heterogeneity, i.e., different genes causing similar diseases; and clinical heterogeneity, i.e., the same gene—even the same allele—causing disimilar diseases in different individuals, even within the same family.

Degenerative Retinal Diseases, edited by LaVail *et al.*
Plenum Press, New York, 1997

This heterogeneity is particularly striking for adRP. For example, there are now at least 10 district loci known to cause this disease[*](1). Of these, two, rhodopsin and peripherin/RDS, are known to cause approximately 30% and 5% of adRP cases, respectively. Mutations in ROM1 may cause an additional 1% of cases. None of the remaining mapped genes, though, have been identified, and the fraction each contributes to the total is unknown. Further, it is highly likely that additional adRP genes will be identified eventually.

In view of this heterogeneity we are using three interrelated approaches to determine the genes and mutations causing adRP, first, mutation screening in individual patients for mutations in rhodopsin or peripherin/RDS; second, linkage testing in adRP families without known mutations; and, third, positional cloning of two mapped adRP genes, the RP10 locus on 7q31.3 (2,3) and the RP1 locus on 8q11-q21 (4). The "related diseases" mentioned in our goal include other forms of retinal degeneration which may be caused by mutations in rhodopsin or peripherin/RDS, and families which appear to have adRP but whose disease locus, instead, maps to the X chromosome (5). In addition, we are engaged in positional cloning of a dominant disease locus previously mapped close to the RP1 locus, atypical vitelliform macular degeneration (VMD1) (6).

This paper summarizes our progress in each of these areas, with particular emphasis on positional cloning. All of these efforts are dependent on the critical contribution of our many clinical and research collaborators.

ACCESS TO FAMILIES AND PATIENTS WITH adRP OR RELATED DISEASES

Our research projects are conducted under the auspices of the Laboratory for Molecular Diagnosis of Inherited Eye Diseases, a joint program of the School of Public Health and the Hermann Eye Center, UT-Houston. The Diagnostic Laboratory serves as a referral point for contacts with patients, families and clinicians. All research projects involving human subjects are approved by the Committee for Protection of Human Subjects, UT-Houston.

Patients and families are ascertained through our clinical collaborators or through referrals from retina specialists. In addition, we are participants in the Southwest Eye Registry (SER), Dallas, Texas, a facility for registering patients with inherited eye diseases, principally retinopathies, and the retinal specialists who serve these patents. Dr. David Birch, Retina Foundation of the Southwest, Dallas, is Director. At present, the SER, which serves Texas and bordering states, has registered 75 participating clinicians and incorporated two large patient databases.

We have enrolled approximately 220 families in our studies, each with one or more members affected with retinal degeneration. Of these, approximately 75% have an apparent diagnosis of adRP, that is, dominant transmission across 3 or more generations. The remainder are either isolated cases or other forms of retinopathy. We estimate that the ethnic origin of these families is 80% Caucasian, 10% Hispanic, 5% African American, and 5% "other".

All individuals with retinitis pigmentosa or related diseases are screened for mutations in rhodopsin and peripherin/RDS. Our present protocol uses SSCP with MDE[TM] gels (FMC, Inc.) under two conditions, ambient temperature and 4^0 C. The 5 exons of rhodopsin are screened with 10 primer pairs and the 3 exons of peripherin/RDS are screened with 8 primer pairs (Table 1) (7,8; and unpublished). In our experience, this protocol will detect approximately 90% of mutations in the coding sequence or within intron-exon junctions.

---

* Tables of cloned and/or mapped genes causing retinal degeneration can be found at RetNet (Retinal Information Network), http://utsph.sph.uth.tmc.edu/www/utsph/RetNet/home.htm.

Table 1. SSCP primers

Primer	Location	Length	Fragment size
Rhodopsin Exon 1			
OPS1A-F	273–292	20	161 bp
OPS1A-R	414–433	20	
OPS1B-F	389–408	20	168 bp
OPS1B-R	537–556	20	
OPS1C-F	516–535	20	167 bp
OPS1C-R	665–682	18	
Rhodopsin Exon 2			
CPRHO2A	2418–2437	20	133 bp
OPS2A-R	2531–2550	20	
RHO2B-F	2496–2576	20	171 bp
RHO2B-R	2646–2667	22	
Rhodopsin Exon 3			
OPS3A-F	3789–3808	20	170 bp
OPS3A-R	3939–3958	20	
OPS3B-F	3908–3927	20	140 bp
CPRHO3B	4028–4047	20	
Rhodopsin Exon 4			
OPS4A-F	4072–4088	20	189 bp
CPRHO4B1	4241–4260	20	
CPRHO4A1	4207–4236	20	158 bp
OPS4B-R	4348–4364	17	
Rhodopsin Exon 5			
OPS5-F	5141–5160	20	164 bp
OPS5-R	5285–5304	20	
Peripherin/RDS Exon 1			
EXON1AF	200–219	20	195 bp
PER1A-R	375–394	20	
PER1B-F	320–349	20	218 bp
PER1B-R	518–537	20	
PER1C-F	497–516	20	196 bp
PER1C-R	673–692	20	
PER1D-F	649–668	20	200 bp
EXON1BR	+8 to +27	20	
Peripherin/RDS Exon 2			
EXON2F	−32 to +13	20	192 bp
PER2A-R	962–981	20	
PER2B-F	941–960	20	160 bp
EXON2R	+13 to +32	20	
Peripherin/RDS Exon 3			
EXON3-F	−32 to +13	20	150 bp
PER3A-R	1169–1188	20	
PER3B-F	1150–1169	20	180 bp
EXON3-R	1310–1329	20	

Table 2 lists the families tested to date and the mutations observed. Of the patients and families tested, 41 have mutations in rhodopsin and 7 have mutations in peripherin/RDS; the disease loci in 5 additional families map to other sites.

Rhodopsin and peripherin/RDS mutations account for 20% and 3% of the total, respectively, or approximately 25% and 4% of the adRP cases. These values are consistent with the experience of other laboratories. The most common rhodopsin mutation is the Pro23His variant (41% of rhodopsin families); mutations in codons 135 and 347 account

Table 2. Summary of rhodopsin and peripherin/RDS
testing in families with adRP

A. Sample population	
Total tested	211
Family evidence of AdRP	166
Rhodopsin mutation	41
Peripherin/RDS mutation	7
Other mapped loci	5

B. Rhodopsin mutations

Mutation	N Families	Mutation	N Families
Pro23His	17	Glu181Lys	2
Leu46Arg	1	Asp190Asn	1
Gly106Trp	1	Met207Arg	1
Lys110Phe	1	His211Arg	1
Arg135Leu	2	Ser270Arg	1
Arg135Pro	1	Leu328Pro	1
Arg135Trp	2	Pro347Leu	2
Pro171Glu	1	Pro347Thr	1
Pro171Ser	1	delta 332 stop	1
Pro180Ala	1	3′ intron 4 splice	2

C. Peripherin/RDS mutations

Mutation	N Families	Mutation	N Families
Arg13Trp	1	Gly266Asp	2
Leu45Phe	1	delta206-209	1
Pro216Ser	1	5′ intron 2 splice	1

D. Other mapped loci

Locus	N Families
RP1	1
RP2/3	1
RP10	1
RP13	1
RP15	1

for an additional 12% and 7% respectively. The rest are one-of-a-kind. Likewise, all of the peripherin/RDS mutations, except Gly260Asp (seen in 2 "unrelated" families), are unique.

LINKAGE TESTING

AdRP families without rhodopsin or peripherin/RDS mutations and with at least 7 informative meioses (based on the maximum possible lod score with highly informative markers) are tested for linkage to known "adRP" sites. At present we are testing 2 intragenic markers per site at 14 chromosomal locations (Table 3). These include the 10 known adRP sites (RP18–1cen, rhodopsin-3q21, RDS-6p21, RP9-7p, RP10-7q, RP1-8q, ROM1-11q, RP13-17p, RP17-17q and RP11-19q), 3 recessive sites with known genes (PDEB-4p, CNCG-4p and PDEA-5q) and 1 X-linked site (RP15-Xp) (1). Inclusion of the X-linked site is because some X1RP families may appear to be dominant due to clinical expression in females (5).

Table 3. Linked microsatellite markers for retinitis pigmentosa screening

Locus or gene	Location	Marker	Marker type	Heterozygosity
RP18	1cen	D1S498	PCR; dinucleotide	0.82
		D1S534	PCR; tetranucleotide	0.93
rhodopsin	3q21.3-q24	RHO	PCR; dinucleotide	0.33
		D3S1589	PCR; dinucleotide	0.68
PDE6B	4p16.3	PDEB	PCR; dinucleotide	0.75
		D4S43	PCR; dinucleotide	0.76
CNCG	4p14-q13	D4S1635	PCR; dinucleotide	0.70
		D4S1627	PCR; dinucleotide	0.81
PDE6A	5q31.2-q34	CSF1R	PCR; dinucleotide	0.86
		D5S413	PCR; dinucleotide	0.70
peripherin/RDS	6p21.2-cen	RDS	PCR; $(A)_n$	0.62
		D6S271	PCR; dinucleotide	0.87
RP9	7p	D7S460	PCR; tetranucleotide	0.95
		D7S435	PCR; dinucleotide	0.69
		D7S795	PCR; tetranucleotide	0.66
RP10	7q31.3	D7S514	PCR; dinucleotide	0.72
		D7S686	PCR; dinucleotide	0.76
RP1	8q11-q21	D8S593	PCR; tetranucleotide	0.93
		D8S591	PCR; tetranucleotide	0.84
ROM1	11q13	D11S480	PCR; dinucleotide	0.77
		D11S956	PCR; tetranucleotide	0.88
RP13	17p	D17S1529	PCR; dinucleotide	0.81
		D17S1528	PCR; dinucleotide	0.81
RP17	17q	D17S808	PCR; dinucleotide	0.68
		D17S787	PCR; dinucleotide	0.82
RP11	19q13	D19S180	PCR; dinucleotide	0.75
		D18S572	PCR; dinucleotide	0.81
RP15	Xp22.13-p22.15	DXS989	PCR; dinucleotide	0.82
		DXS1048	PCR; dinucleotide	0.68

To date we have mapped several adRP families to known loci but, as we have come to expect given the exceptional genetic heterogeneity of adRP, at least one family tested fails to map to any of these sites (unpublished). Therefore there is at least one more adRP locus, and possibly several.

POSITIONAL CLONING OF RP10

Jordan et al. mapped the disease locus in a Spanish family with adRP to 7q, thus defining a new locus, RP10 (2). In linkage testing against known adRP loci, we mapped the disease locus in a second, unrelated American family to the same site. The maximum, combined two-point lod score for both families is 13.1 at 0% recombination to D7S514, which maps to 7q31 (3).

More recently, the disease locus in a third adRP family, also of Spanish origin, has been mapped to the RP10 region (9). However, the lod score for this family, corrected for reduced penetrance, is less than 2.0. Also, testing in our laboratory appears to exclude the disease locus from the RP10 critical region (10; and unpublished).

A fourth adRP family, of Scottish orgin, has also been mapped to this site (11).

To identify the RP10 gene, in collaboration with Dr. Peter Humphries and Dr. Eric Green, we are following what has become the "traditional" approach to positional cloning: linkage mapping and haplotype analysis to determine the minimum critical region; physical characterization of the critical region; and identification and testing of potential candidate genes mapping within the region. We expect that the disease-causing gene is expression in the adult retina, though not, necessarily, exclusively in the retina. This rule has been true for all retinitis pigmentosa genes cloned to date but, admittedly, nature may have some unpleasant surprises in store.

Linkage mapping and haplotype analysis of the Spanish and American RP10 families place the disease locus on 7q31.3 between the flanking markers D75686 and D75530, a distance of 5cM (Figure 1). This distance is spanned by a series of YAC contigs, with 5-fold redundancy, containing 45 mapped STS's (Figure 2) (12). We are using this YAC and STS resource to identify candidate genes within the RP10 critical region.

To date more than 20 ESTs have been mapped within the critical region (10,12). Of these, 8 are expressed in the adult human retina (Table 4). We have confirmed retinal expression by PCR amplification of EST primers in 3 independently-derived retinal cDNA libraries (13,14; and Stratagene, Inc.) Of the 8 ESTs, 2 (ACHE and GNGT1) have been eliminated because fine-structure linkage mapping and/or physical mapping place them outside of the RP10 critical region. A third gene, BCP, is an unlikely candidate since af-

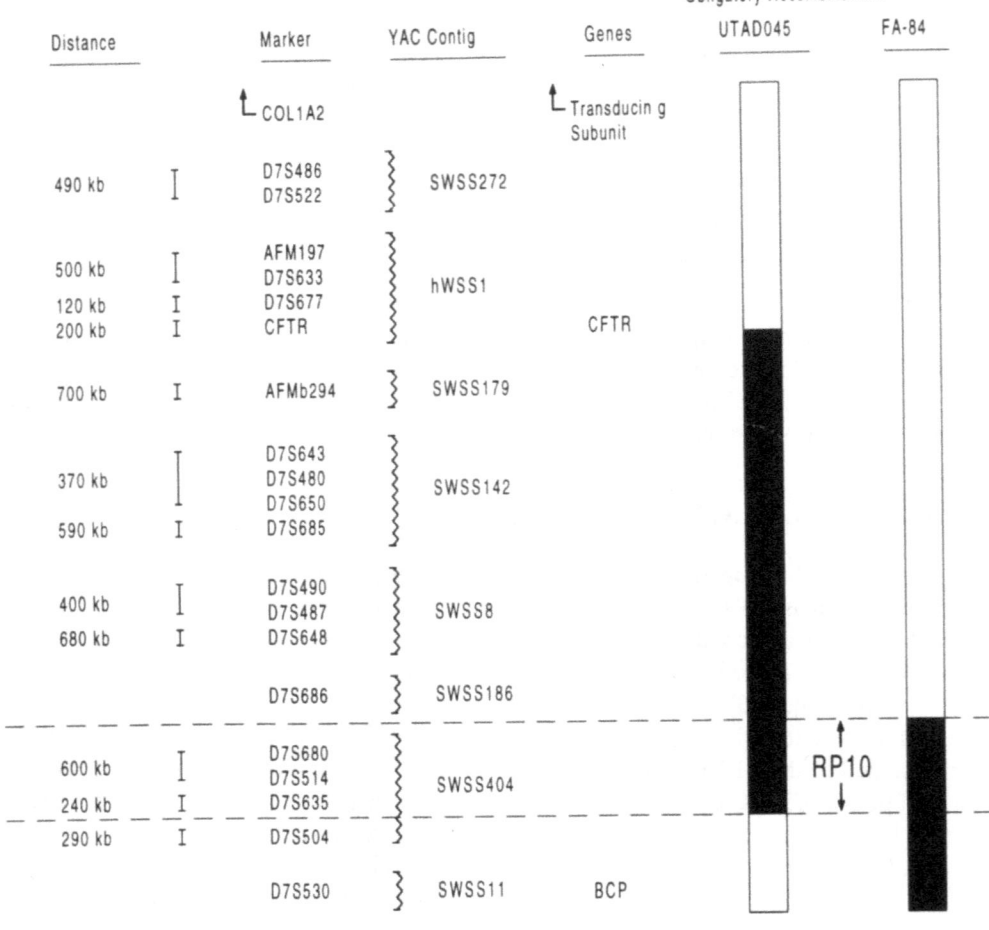

Figure 1. Minimum region of overlap for RP10.

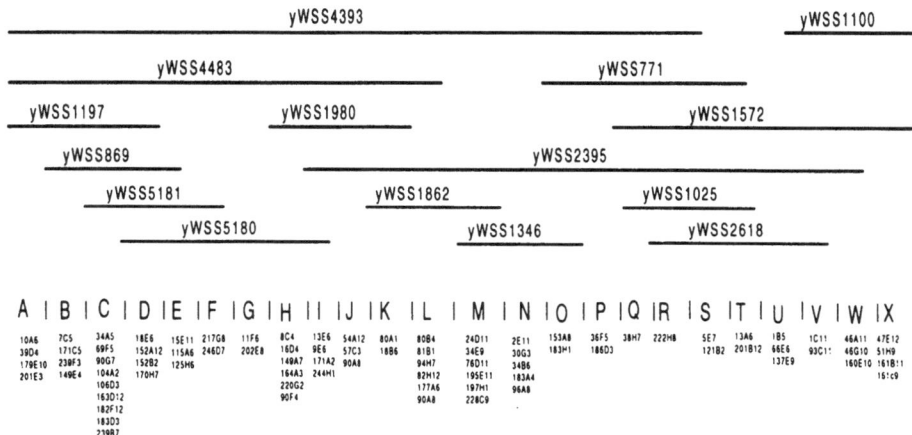

Figure 2. YAC contig and binned cosmids for RP10.

fected members of both families have normal color vision. However, to convincingly eliminate this candidate we sequenced the BCP coding sequence in both families. No disease-causing mutations were found.

Of the remaining 5 EST's, 2 are particularly interesting, GRM8 and ARF5. GRM8 is a glutamate receptor expressed in the retina and olfactory bulbs (15). Dr. Humphries is currently testing this candidate. ARF5 is an ADP ribosylation factor expressed in the retina and elsewhere (16). It is an intriguing candidate because ARF proteins are involved in vesicular transport. Thus, ARF5 may participate in disc transport and shedding in the photoreceptor.

To further characterize ARF5, we obtained cDNAs of homologous ARF's, screened a retinal-specific cDNA library for the ARF5 primary transcript and determined the full-length sequence (17). Using this information to design oligonucleotide to screen a chromosome 7 cosmid library, we determined the genomic sequence and intron/exon structure of human ARF5 (Figure 3). This information, in turn, was used to design primers for genomic sequencing of RP10 patients.

Sequencing of affected members of the American RP10 family failed to detect potential diseases-causing mutations. Testing in the Spanish family is underway. Thus, ARF5 remains a potential, but unlikely, candidate.

Our experience with ARF5 demonstrates the technical difficulty of testing candidate EST's in affected individuals. Genomic sequencing requires, at least, knowledge of the full-length cDNA sequence and intronic sequences around intron-exon junctions. In turn, actual genomic sequencing must be very reliable to detect the expected heterozygous nu-

Table 4. Summary of candidate genes for RP10

Symbol	Name or characteristic	Current status
1. ACHE	Acetylcholinesterase	Excluded by mapping
2. ARF5	ADP-ribosylation factor 5	Excluded by sequencing
3. BCP	Blue cone pigment	Excluded by sequencing
4. EST385792	Retina-expressed EST	Testing...
5. EST387484	Retina-expressed EST	Testing...
6. GNGT1	Transducin gamma subunit	Excluded by mapping/linkage
7. GRM8	Metabotropic glutamate receptor	Testing...
8. TGR-A002N36	Retina-expressed EST	Testing...

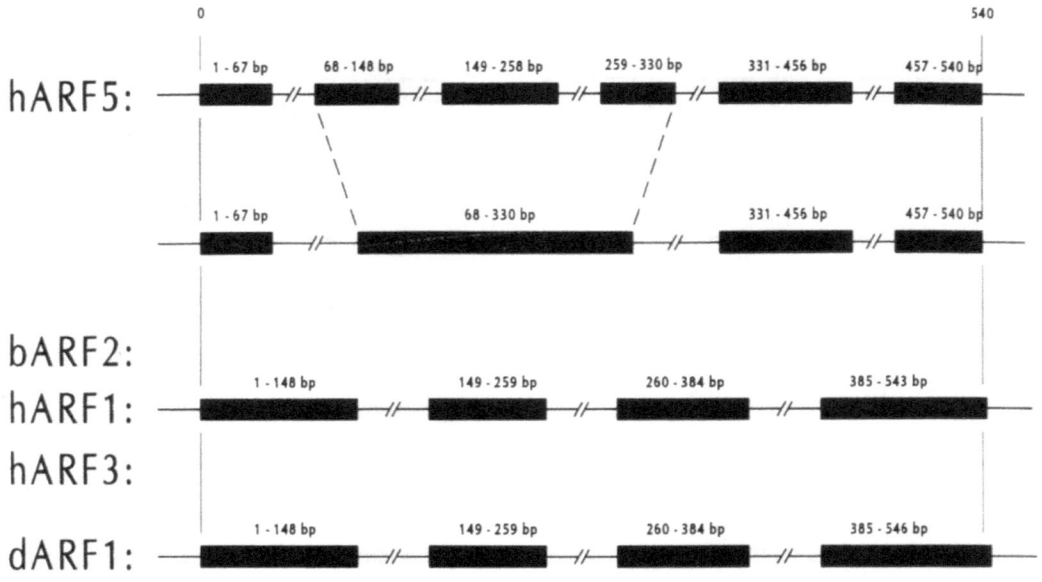

Figure 3. ARF5 genomic structure across species.

cleotide substitution in patients with dominant diseases. In addition, there are likely to be many more potential candidate genes within the 5 cM RP10 region.

There are several resources for dealing with these concerns. First, the Unigene database currently lists over 4,000 EST clusters with one or more members expressed in the retina (18). In collaboration with the South African National Bioinformatics Institute (19) we are identifying retinal-related EST clusters which map to 7q31 (and to other chromosomal sites of interest). Fortunately, many of the original cDNAs are available from the IMAGE Consortium through Research Genetics, Inc. (20). We are currently using IMAGE cDNA's to determine the sequence of the remaining 3 candidate ESTs in Table 2.

Finally, the RP10 critical region on 7q has been targeted for large-scale sequencing by the National Center for Human Genome Research, NIH. Thus, within a few years full-length genomic sequences will be known. Nonetheless, the remaining problem—reliable genomic sequencing in affected individuals—is still daunting.

POSITIONAL CLONING OF RP1

The primary resource for the study of RP1 is a large, nine-generation family known as UCLA-RP01 in which the disease locus was first mapped. UCLA-RP01 has over 150 living, affected members, all of whom can trace their disease to an ancestor living in the early 19th century. The family has classical type 2 adRP with relatively late onset of night blindness, usually by the second decade of life, and slow progression (21). Like other families with type 2 adRP, there is extensive clinical heterogeneity.

We have focused linkage analysis in a subset of the UCLA-RP01 family with 195 individuals, 101 of whom are affected. In 1991, linkage testing in this subset placed the RP1 locus on chromosome 8q, near the centromere (4). Since then, we have tested a total of 34 chromosome 8 markers in the family. We have also collaborated in establishing new markers for chromosome 8 (22–26) and in improving the global linkage map of the chromosome (27;28).

Linkage mapping in UCLA-RP01 places the RP1 locus on 8q11-q21 with a maximum two-point lod score, to D85531, of 16.9 (unpublished). A single, extended haplotype on 8q, including seven microsatellite markers spaning 6cM, tracks with the disease in 95 affected individuals in two large branches of the family. Although multipoint linkage analysis in the family is prohibitively difficult (in addition to the large number of alleles and genotypes, the family has two inbreeding loops), the numbers informative meioses without recombination implies a multipoint lod score of over 28.

Though there is no doubt that the RP1 locus in UCLA-RP01 maps to 8q11-q21, we were surprised to find that the disease locus in a third, large branch of the family, with clinically-similar adRP, was excluded from 8q (unpublished). In fact, this branch of the family has an Arg135Leu mutation rhodopsin. This is a useful reminder that genetic heterogeneity can exist within a single large family, even though the coincidence seems unlikely.

Recently, the disease locus in 3 additional adRP families has been mapped to 8q. Two of these families are too small to contribute to linkage localization of RP1 (S Bhattacharya, personal communication; E Stone, personal communication). However, the third, a large Australian family, shows a maximum two-point lod score of 5.8 to the RP1 region, and helps refine the linkage localization (29).

Combining linkage data from UCLA-RP01 and the Australian family places the disease locus between D8S601 and D8S285, a 4cM distance within 8q11-q21 (Figure 4). In collaboration with investigators from the University of Houston, Texas, we have developed a YAC contig, with 2-fold redundancy, spanning this region (Figure 5). The mini-

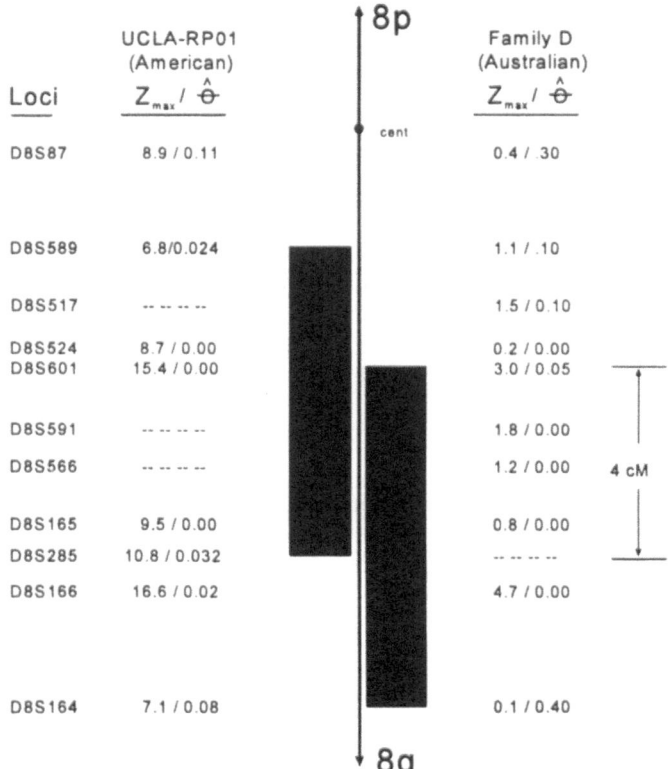

Figure 4. Minimum region of overlap of RP1.

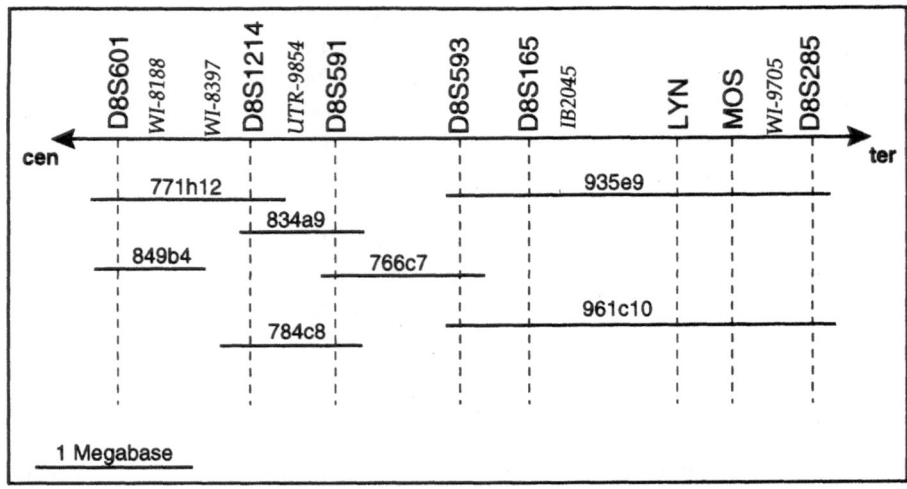

YAC contig spanning RP1 region (8q11-q13)

Figure 5. YAC contig spanning the RP1 region.

mum tiling path through this contig incorporates 4 YACs, with an approximate total insert size of 5mb. The minimal contig can be placed within the much-larger Whitehead STS map of chromosome 8 thus identifying mapped ESTs in this region (30).

Following a strategy similar to our project for cloning RP10, several potential candidate genes have been mapped to the RP1 critical region. Table 5 lists the EST clusters mapped to this region and summarizes their current testing status of each. As for RP10, potential candidate genes for RP1 can be identified from EST databases, but considerable, additional research is required to characterize and eliminate—or confirm—each candidate.

CHARACTERIZATION OF THE VMD1 LOCUS

The disease locus for an autosomal dominant disorder characterized as "atypical vitelliform macular dystrophy" (VMD1) (31) was linked to the GPT locus in the early 1980's (6). The disease is found in a single large Louisiana family. GPT (glutamate pyruvate transaminase, EC 2.6.1.2) is a solvable enzyme, found in serum, with two major, polymorphic isozymes, GPT-1 and GPT-2. Linkage of VMD1 to GPT was established by serologic testing.

Table 5. Summary of candidate genes for RP1

Symbol	Name or characteristic	Current status
1. WI8188	DR-4 beta chain-like EST	Not retinal expressed
2. WI8397	Anonymous EST	Not retinal expressed
3. UTR9854	Transcription elongation factor	Testing...
4. IB2045	Retinal expressed cDNA	No open reading frame
5. LYN	Lymphocyte-associated	Not a likely candidate
6. Mos	C-mos transforming	Not a likely candidate
7. WI9705	405 ribosomal protein (520)	Not a likely candidate
8. WI2392	Anonymous EST	Testing...
9. WI8821	Cytochrome P450 superfamily member	Not a likely candidate

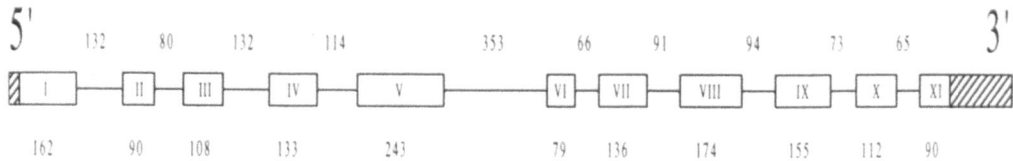

Figure 6. Human GPT genomic structure.

Because the GPT locus was poorly characterized, mapping to either chromosome 8q or 16, and because of the need to develop DNA-based markers to refine the localization of VMD1, we cloned and mapped the GPT gene (31). The human GPT cDNA was isolated from a liver cDNA library and the genomic sequence was isolated from a chromosome 8-specific cosmid library. The cDNA and genomic sequences were determined from these isolated clones (Figure 6). Mapping of GPT to 8q was established by several methods. First, fluorescence in situ hybridization mapped the human cDNA to 8q24.3. Second, chromosome 8-specific cosmids were shown to contain GPT. Third, rat gpt cDNA was mapped to the syntenic region of rat chromosome 7. Finally, PCR primers specific to human GPT amplify sequences within a "half-YAC" of chromosome 8, that is, a YAC containing the 8q telomere. This last finding places GPT within 19 kbp of the terminus of 8q.

To refine the localization of VMD1, we determined the DNA basis of the GPT serologic polymorphism and retested the VMD1 family. The GPT polymorphism (found in all human populations) is caused by a nucleotide substitution in codon 14, which is detectable as a restriction site polymorphism using NlaIII (31). In addition, a polymorphic microsatellite marker, D85421, was found in a GPT-containing cosmid from 8q.

Applying these DNA markers to the VMD1 family, including several newly-ascertained members, led to a disappointment: the maximum lod score to the GPT locus dropped below 2.0 and the contiguous marker, D85421, excluded linkage at a distance of 10 cM (32; and unpublished). Based on this evidence, we believed that VMD1 does not map to the terminus of 8q. Presently, we are testing linkage of the VMD1 locus to other mapped forms of autosomal dominant macular dystrophy. We can exclude the peripherin/RDS locus and the Best macular dystrophy locus (VMD2) on 11q, thus the actual chromosomal site of VMD1 is still undetermined.

SUMMARY AND CONCLUSIONS

We believe that the best approach for dealing with the heterogeneity in inherited retinal dystrophies is to focus on a well-defined (though complex) entity, such as autosomal dominant retinitis pigmentosa, and to apply several strategies for identifying the underlying genes and mutations. Because of rapid progress in mapping and sequencing human cDNAs, it is likely that most genes causing inherited retinal degeneration will be determined within the next few years. Then the truly interesting and important effort to design rational therapies can begin in earnest.

ACKNOWLEDGMENTS

This research was supported by grants from the Foundation Fighting Blindness, the George Gund Foundation, the Farish Fund, the MD Anderson Foundation, NIH Grants EY07142 and EY05235 and NIH Training Grant EY07024.

REFERENCES

1. SP Daiger, LA Sullivan, JA Rodriguez. Correlation of phenotype with genotype in inherited retinal degeneration. Behavioral Brain Sci. 18:491–506, 1995.

2. SA Jordan, GJ Farrar, P Kenna, MM Humphries, DM Sheils, R Kumar-Singh, EM Sharp, N Soriano, C Ayuso, J Benitez, P Humphries. Localization of an autosomal dominant retinitis pigmentosa gene to chromosome 7q. Nat. Genet. 4:54–58, 1993.

3. RE McGuire, AM Gannon, LA Sadler-Sullivan, JA Rodriguez, SP Daiger. Evidence for a major gene (RP10) for autosomal dominant retinitis pigmentosa on chromosome 7q: linkage mapping in a second, unrelated family. Hum. Genet., 95:71–74, 1995.

4. SH Blanton, JR Heckenlively, AW Cottingham, J Friedman, LA Sadler, M Wagner, LH Friedman, SP Daiger. Linkage mapping of autosomal dominant retinitis pigmentosa (RP1) to the pericentric region of human chromosome 8. Genomics, 11:857–873, 1991.

5. RE McGuire, LS Sullivan, SH Blanton, ME Church, JR Heckenlively, SP Daiger. X-linked dominant cone-rod degeneration: linkage mapping of a new locus for retinitis pigmentosa (RP15) to Xp22.13-p22.11. Am. J. Hum. Genet., 57:87–94, 1995.

6. RE Ferrell, HM Hittner, JH Antoszyk. Linkage of atypical vitelliform macular dystrophy (VMD-1) to the soulable glutamate pyruvate transaminase (GPT) locus. Amer. J. Hum.

7. JA Rodriguez, CA Herrera, DG Birch, SP Daiger. A leucine to arginine amino acid substitution at codon 46 of rhodopsin is responsible for a severe form of autosomal dominant retinitis pigmentosa. Hum. Mutation, 2:205–213, 1993.

8. LS Sullivan, SR Guilford, DG Birch, SP Daiger. A novel splice site mutation in the gene for peripherin/RDS causing dominant retinal degeneration. Invest. Ophthalmol. Vis. Sci., 37:1145, 1996.

9. JM Milldÿòüä n, F Martínez, C Vuilela, M Beneyto, F Prieto, C Ndÿòüä jero. An autosomal dominant retinitis pigmenmtosa family with close linkage to D7S480 on 7q. Hum. Genet., 96:216,218. 195.

10. RE McGuire. The Story of RP10: A Positional Cloning Approach to the Identification of an Autosomal Dominant Retinitis Pigmentosa Gene. PhD Thesis, The University of Texas Health Science Center, Houston, November 1996.

11. Z Mohamed, C Bell, HM Hammer, CA Converse, L Esakowitz, NE Haites. Linkage of a medium sized Scottish autosomal dominant retinitis pigmentosa family to chromsome 7q. J. Med. Gene.t 33:714–715, 1996.

12. RE McGuire, SA Jordan, VV Braden, GG Bouffard, P Humphries, ED Green, SP Daiger. Mapping the RP10 locus for autosomal dominant retinitis pigmentosa on 7q: refined genetic positioning and localization within a well-defined YAC contig. Genome Res. 6:255–266, 1996.

13. J Nathans, DS Hogness. Isolation and nucleotide sequence of the gene encoding human rhodopsin. Sci., 232:203–210, 1984.

14. A Swaroop, J Xu, N Agarwal, SM Weissman. A simple and efficient cDNA library subtraction procedure: isolation of human retina-specific cDNA clones. Nuc Acids Res, 19:1954, 1991.

15. SW Scherer, RM Duvoisin, R Kuhn, HQ Heng, E Belloni, L-C Tsui. Localization of two metabotropic glutamate receptor genes, GRM3 and GRM8, to human chromosome 7q. Genomics, 31:230–233, 1996.

16. J Moss, M Vaughan. Structure and functon of ARF proteins: activators of cholera toxin and critical components of intracellular vesicular transport proteins. J Biol. Chem. 21:12327–12330, 1995.

17. RE McGuire, SP Daiger, ED Green. Localization and characterization of the human ADP-ribosylation factor 5 (ARF5) gene. Submitted, Genomics, 1996.

18. Unigene, http://www.ncbi.nlm.nih.gov/Schuler/UniGene/index.html, 1996.

19. SANBI, http://techno.uwc.ac.za/~winhide/mission.html

20. GG Lennon, C Auffray, M Polymeropoulos, MB Soares. The I.M.A.G.E. Consortium: an integrated molecular analysis of genomes and their expression". Genomics 33, 151–152, 1996.

21. JR Heckenlively, JT Pearlman, RS Sparkes, AM Spence, D Zedalis, L Field, M Sparkes, M Crist, S Tiddman. Possible assignment of as dominant retintis pigmentosa gene to chromsome 1. Ophthalmopl. res. 14:46–53, 1982.

22. AW Cottingham, SH Blanton, D Retief, L Warnich, 7SP Daiger. A new StyI RFLP and haplotypes with the HindIII RFLP at D8S5 (TI11) locus at 8p23-q11. Nuc. Acids Res., 17:4790, 1991.

23. AW Cottingham, LA Sadler, SH Blanton, MJ Wagner, DE Wells, SP Daiger. A tight linkage cluster including two new RFLPs, D8S96 and D8S108, on 8q11-q13. Nuc. Acids Res., 20:1426, 1992.

24. J Gu, LA Sadler, SP Daiger, D Wells, M Wagner. Dinucleotide repeat polymorphism at the CRH gene. Hum. Mol. Genet., 2:85, 1993.

25. LA Sadler, SH Blanton, SP Daiger. Dinucleotide repeat polymorphism of the tissue plasminogen activator gene (PLAT). Nuc. Acids Res., 19:6058, 1991.

26. LA Sullivan, J Parrish, MJ Wagner, SH Blanton, SP Daiger. Tetranucleotide repeat polymorphism (D8S582) for human EST00680. Hum. Mol. Genet., 3:386, 1994.

27. J Tomfohrde, S Wood, J Schertzer, MJ Wagner, DE Wells, J Parrish, LA Sadler, SH Blanton, SP Daiger, Z Wang, PJ Wilkie, JL Weber. Human chromosome 8 linkage map based on short tandem repeat polymorphisms: effect of genotyping errors. Genomics, 14:144–152, 1992.

28. T Steinbrueck, C Read, SP Daiger, LA Sadler, JL Weber, S Wood, H Donis-Keller. Chromosome 8. Science, 258:71-ff, 1992.

29. S-Y Xu, M Denton, LS Sullivan, SP Daiger, A Gal. Genetic mapping of RP1 on 8q11-q21 in an Australian family with autosomal dominant retinitis pigmentosa reduces the critical region to 4 cM between D8S601 and D8S285. Submitted, Hum. Genet., 1996. Genet. 35:78–84, 1983.

30. Whitehead, http://www_genome.wi.mit.edu/cgi-bin/contig/phys_map, 1996.

31. MM Sohocki, LS Sullivan, WR Harrison, EJ Sodergren, FFB Elder, G Weinstock, S Tanase, SP Daiger. Human glutamate pyruvate transaminase (GPT): localization to 8q24.3, cDNA and genomic sequences, and polymorphic sites. In press, Genomics, 1997.

32. RJ Leach, SS Banga, K Ben-Othame, S Chughtai, R Clarke, SP Daiger, MM Sohocki, et al. Report of the Third International Workshop on Human Chromosome 8 Mapping 1996. In press, Cytogenet. Cell Genet., 1997.

31

GROWTH FACTORS IN THE RETINA

Pigment Epithelium-Derived Factor (PEDF) Now Fine Mapped to 17p13.3 and Tightly Linked to the RP13 Locus

Jacquie Greenberg,[1] Rene Goliath,[1] Joyce Tombran-Tink,[2] Gerald Chader,[2] and Rajkumar Ramesar[1]

[1]Department of Human Genetics
MRC Research Unit of Medical Genetics
University of Cape Town, Medical School
Observatory, Cape Town, 7925, South Africa
[2]Laboratory of Retinal Cell and Molecular Biology
National Eye Institute, NIH

INTRODUCTION

PEDF Locus

Pigment epithelium-derived factor (PEDF), a unique 50 kD neurotrophic protein secreted by fetal Retinal Pigment Epithelial (RPE) cells, has been shown to have neurotrophic affects on human Y-79 retinoblastoma cells in vitro. The gene for PEDF is highly expressed by both human fetal and young adult RPE cells and is down-regulated in senescent but not quiescent RPE cells. The secretion of PEDF and synthesis of its mRNA transcript, which decreases dramatically and specifically in an age-related manner, makes PEDF a very good candidate for any form of late onset retinal or macular degeneration. The cDNA of the human PEDF gene has been mapped to the short arm of chromosome 17 (17p13.1-ter)[1].

RP13 Locus

An autosomal dominant Retinitis Pigmentosa gene locus (RP13) has been assigned to the same chromosomal region as the PEDF locus in a large South African (SA) family[2]. The physical mapping of the PEDF gene locus has now been refined with respect to previously mapped microsatellite markers in this region. Evidence is provided for a more distal localization of the PEDF gene (17p13.3) with respect to initial findings (17p13.1-ter). As the PEDF gene is a prime candidate for retinal degeneration, it is possible that it is the disease-causing gene in this SA family where ADRP is linked to the RP13 locus.

Degenerative Retinal Diseases, edited by LaVail *et al.*
Plenum Press, New York, 1997

METHODS

SSCP Analysis

Using single stranded conformational polymorphism (SSCP) analysis a polymorphism in exon 4 of the the PEDF gene was used to genotype the RP13-linked family at this locus.

The intronic primers used to amplify exon 4 of the PEDF gene are:

4F 5'-TGAGTATAGTGTCTGTGTTCTGGGA-3'
4R 5'-AAGACCCCCCCAGCCTGCAGCATGG-3'

PCR reactions were performed using 200ng genomic DNA, 200uM of each dATP, dGTP, dCTP and dTTP, 100ng of each primer, 50mM KCl, 10mM Tris HCl pH 8.4, 1.5mM $MgCl_2$ and 1 unit of Taq DNA polymerase in a total volume of 25ul.

Cycling conditions were one cycle at 95°C denaturation for 3 minutes, followed by 30 cycles of denaturation at 95°C for 30 seconds, annealing at 62°C for 30 seconds and elongation at 72°C for 40 seconds, followed by one extended elongation cycle at 72°C for 5 minutes.

The PCR products were diluted 1:1 in loading buffer (7M urea, 50% formamide, 20mM Tris-HCl and 10mM EDTA) and used for the SSCP analysis. After being denatured, the samples were loaded directly onto a 0.5x MDE (Hydrolink) gel containing 5% glycerol and 0.5 x TBE. Gels were electrophoresed at 6 watts constant power overnight and silver-stained for visualisation of the bands.

Linkage Analysis

LINKMAP was used to performed a four-point linkage analysis where PEDF was moved up to and in between the 3 fixed loci D17S849- D17S1529-D17S831. The estimated theta value between these new markers is 0.03 (Genethon, unpublished data). A theta value of 0.5 was used as the starting point away from D17S849 and then 4 equal increment values were used to move in on the fixed marker loci.

RESULTS

The exon 4 SSCP pattern facilitated segregation analysis in the family (Figure 1).

Direct sequencing of the PCR products revealed a T to C and a C to G substitution in codons 130 and 132, resulting in a neutral and a missense mutation, respectively.

Two-point linkage analysis between the disease locus and this polymorphism produced a maximum lod score of 7.1 at a theta of 0.00. Subsequent multipoint linkage analysis allowed us to map the PEDF gene locus with respect to the tightly linked markers on the short arm of chromosome 17 that were used in the fine mapping study[3].

The pooled results are presented in Figure 2.

The maximum estimate for the location of PEDF was close to D17S1529 with a maximum lod score of 6.71. Two point analyses between PEDF and D17S1529, D17S1528 and D17S831 produced maximum lod scores of 6.53, 6.26 and 5.36 respectively, each at a recombination fraction of zero.

Regions of the promoter sequence of PEDF (unpublished data) that were used for screening did not exhibit any altered mobility shift on SSCP analysis to indicate a conformation change in the single DNA strands.

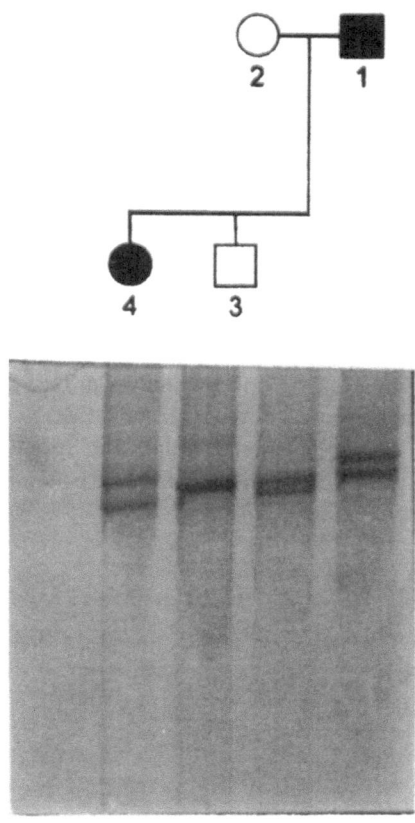

Figure 1. Mendelian pattern of inheritance of the PEFD exon 4 SSCP polymorphism.

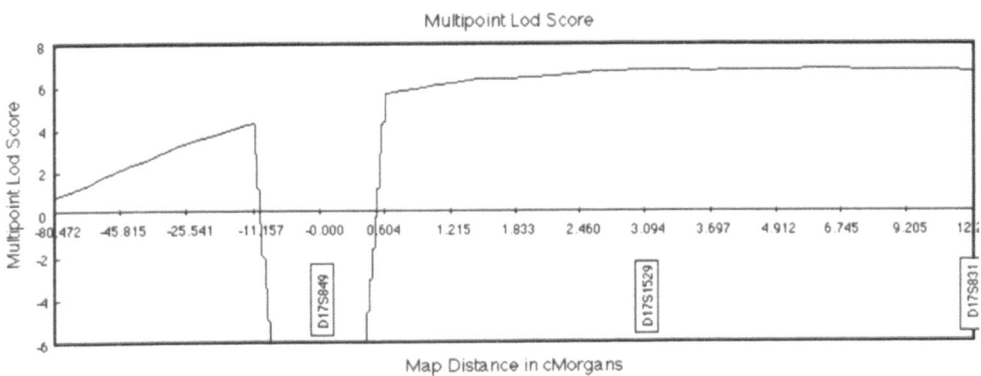

Figure 2. Multipoint map illustrating the multipoint results. As published in : Goliath R, J Tombran-Tink, IR Ro-driquez, G Chader, R Ramesar and J Greenberg. (1996). The gene for PEDF, a retinal growth factor, is a prime candidate for retinitis pigmentosa and is tightly linked to the RP13 locus on chromosome 17p13.3. http://www.cc.emory.edu/MOLECULAR_VISION/v2/goliath/.

CONCLUSIONS

The results indicate a more distal localization of the PEDF gene locus on the short arm of chromosome 17 with respect to previously fine mapped microsatellite markers[3].

Taken together, these data provide evidence for the localization of the PEDF gene to the 17p13.3 chromosomal region, very close to the RP13 locus. This result confirms the in situ sublocalization of PEDF to 17p13.1-pter[1] and places the gene on a more distal and possibly more confined physical location, very close to D17S1529 on 17p13.3[4].

The SSCP analysis of all 8 exons as well as some of the promoter sequence of PEDF showed no evidence of a mutation in the RP13-linked family. The investigation of the remaining promoter sequence is underway in an endeavour to verify whether PEDF is the disease-causing gene in this RP family.

The informative internal polymorphism in exon 4 of the PEDF gene illustrated here could facilitate further candidate-gene linkage studies in families with inherited retinal degeneration. For both genetic and physiological reasons, PEDF is an intriguing and important candidate gene for RP and other degenerative diseases. Intensive investigations are being undertaken to establish whether it is the disease-causing gene in this SA ADRP family.

ACKNOWLEDGMENTS

This research was supported by grants from the Retinal Preservation Foundation of South Africa, the South African Medical Research Council, the UCT staff research fund, the Mauerberger Foundation, and grants from NIH-EY07961.

REFERENCES

1. Tombran-Tink, J., Pawar, H., Swaroop, A., Rodriquez, I., and Chader, C. (1994). Localization of the gene for Pigment Epithelium-Derived Factor (PEDF) to Chromosome 17p13.1 and expression in cultured human retinoblastoma cells. *Genomics*. **19**: 266–272.
2. Greenberg, J., Goliath, R.G., Beighton, P., and Ramesar, R. (1994). A new locus for autosomal dominant retinitis pigmentosa on the short arm of chromosome 17. *Hum. Mol. Genet.* **3**: 915–918
3. Goliath, R., Shugart, Y.Y., Janssens, P., Weissenbach, J., Beighton, P., Ramesar, R., and Greenberg, J. (1995). Fine localization of the locus for autosomal dominant retinitis pigmentosa on chromosome 17p. *Am. J. Hum. Genet.* **57**: 962–965.
4. Goliath R, J Tombran-Tink, IR Rodriquez, G Chader, R Ramesar and J Greenberg. (1996). The gene for PEDF, a retinal growth factor is a prime candidate for retinitis pigmentosa and is tightly linked to the RP13 locus on chromosome 17p13.3. http://www.cc.emory.edu/MOLECULAR_VISION/v2/goliath/

32

GENETIC AND PHYSICAL LOCALISATION OF THE GENE CAUSING CONE-ROD DYSTROPHY (*CORD2*)

James Bellingham,[1] Sujeewa D. Wijesuriya,[1] Kevin Evans,[2] Alan Fryer,[3] Greg Lennon,[4] and Cheryl Y. Gregory[1]

[1]Department of Molecular Genetics
Institute of Ophthalmology
University College London
11-43 Bath Street, London EC1V 9EL, United Kingdom
[2]Moorfields Eye Hospital
City Road, London EC1V 2PD, United Kingdom
[3]Royal Liverpool University Hospital
Liverpool, L69 3BX, United Kingdom
[4]Human Genome Center
Lawrence Livermore National Laboratory
Livermore, California 94551

INTRODUCTION

Choroidoretinal dystrophies are incurable and essentially untreatable, representing the most common cause of genetic visual loss in childhood (1). They are a clinically and genetically heterogeneous group of disorders. Dystrophies which primarily affect rod function such as retinitis pigmentosa have been extensively studied and a number of genes have been implicated in the disease pathogenesis (2). In contrast dystrophies which primarily affect cone photoreceptors have been less well studied and examples of this group of diseases includes cone dystrophies and cone-rod dystrophies. Cone dystrophies are characterised by photophobia, loss of visual acuity and colour vision defects associated with reduced cone photoreceptor ERG responses. Abnormal pigmentation with atrophy is often seen at the macula. Cone-rod dystrophies are distinct from cone dystrophies in that abnormalities of cone dysfunction is seen with progressive peripheral retinal disease. Diminished visual acuity and loss of colour discrimination is followed by nyctalopia, progressive peripheral visual field deficit and decreasing rod photoreceptor ERG amplitudes from an early age. Advancing chorioretinal atrophy of the central and peripheral retina is characteristic (3). Autosomal dominant, recessive and X-linked patterns of inheritance for cone dystrophies and cone-rod dystrophy (CRD) have been described and studies have implicated a number of loci for the disease-causing genes. Loci associated with cone dystro-

Degenerative Retinal Diseases, edited by LaVail *et al.*
Plenum Press, New York, 1997

phies include a balanced translocation on chromosome 6q which was reported in a patient with mental retardation and cone dystrophy (4), an X-linked cone dystrophy mapping to Xp21.1-p11.3 (5) and two independent studies have used linkage analysis to identify an autosomal dominant cone dystrophy locus on chromosome 17p (6–7). Loci implicated in CRD include two case reports that have suggested localisation of CRD genes on chromosome 18q (8) and 17q (9), an autosomal dominant form of CRD mapping to chromosome 19q (10) and a transverse mutation in the peripherin/RDS gene (Asn244His) has been found in one Japanese CRD family (11).

We are interested in finding new genes causing primarily loss of central vision. This work will be relevant to studies aimed at identifying susceptibility loci for age-related macular degeneration, the most common cause of blindness in the elderly population (12). In this article we discuss the advances we have made in studying the chromosome 19q CRD locus (*CORD2*; MIM 120970). Recently, progress towards construction of an integrated physical map spanning the entire chromosome 19 has been reported (13). It consists of YACs, BACs, PACs, P1 and cosmid clones covering approximately 60 megabases of DNA. Although this map provided a framework for construction of a physical contig across the *CORD2* region, there were a number of uncloned regions and a paucity of informative polymorphic markers in the *CORD2* region. Thus, in this study we describe the ascertainment of a new branch of the original family that has now genetically refined the *CORD2* locus, isolation of new polymorphic markers and studies towards the construction of a complete cloned DNA contig across the *CORD2* critical region.

PATIENTS AND METHODS

Family Ascertainment and Genotyping

A new 6 generation branch of the original chromosome 19q CRD family (10) was ascertained which consisted of 8 affected patients, 3 unaffected individuals and 4 spouses (Fig. 1, Branch B). Patients were considered to be affected if they had a history of progressive loss of visual acuity, colour vision defects and abnormal visual fields within the first decade of life (3). All persons assigned normal had no visual problems and were over the age of 20. Genomic DNA was extracted from peripheral blood lymphocytes by standard procedures. The DNA from each individual in the study was radioactively genotyped with 6 microsatellite markers spanning the *CORD2* locus including and between flanking markers D19S219 and D19S246 (14). Additionally, the flanking recombinants from the original family were also genotyped for new marker loci not previously available (Fig. 1, Branch B).

Isolation of New Polymorphic Markers

Cosmids mapping to the region were digested with either Sau3AI or HaeIII and transferred to Hybond+ nylon filters by alkali Southern blotting. Filters were hybridized overnight with a $(GT)_{13}$ probe in 20 ml Churches buffer (0.125M NaPi, 1mM EDTA, 7% SDS) at 65°C and then washed under stringent conditions (0.1x SSC, 0.1% SDS, 65°C). Those cosmid fragments containing complementary (CA)n repeat sequences were subcloned into either pUC18 or pBluescript vectors and the flanking DNA sequence to each repeat was obtained by fluorescent dye termination sequencing on an automated ABI DNA Sequencer. Primers were designed from the obtained sequence to amplify the new microsatellite markers by PCR, which were then tested in the normal population for the numbers of alleles and heterozygosity (15).

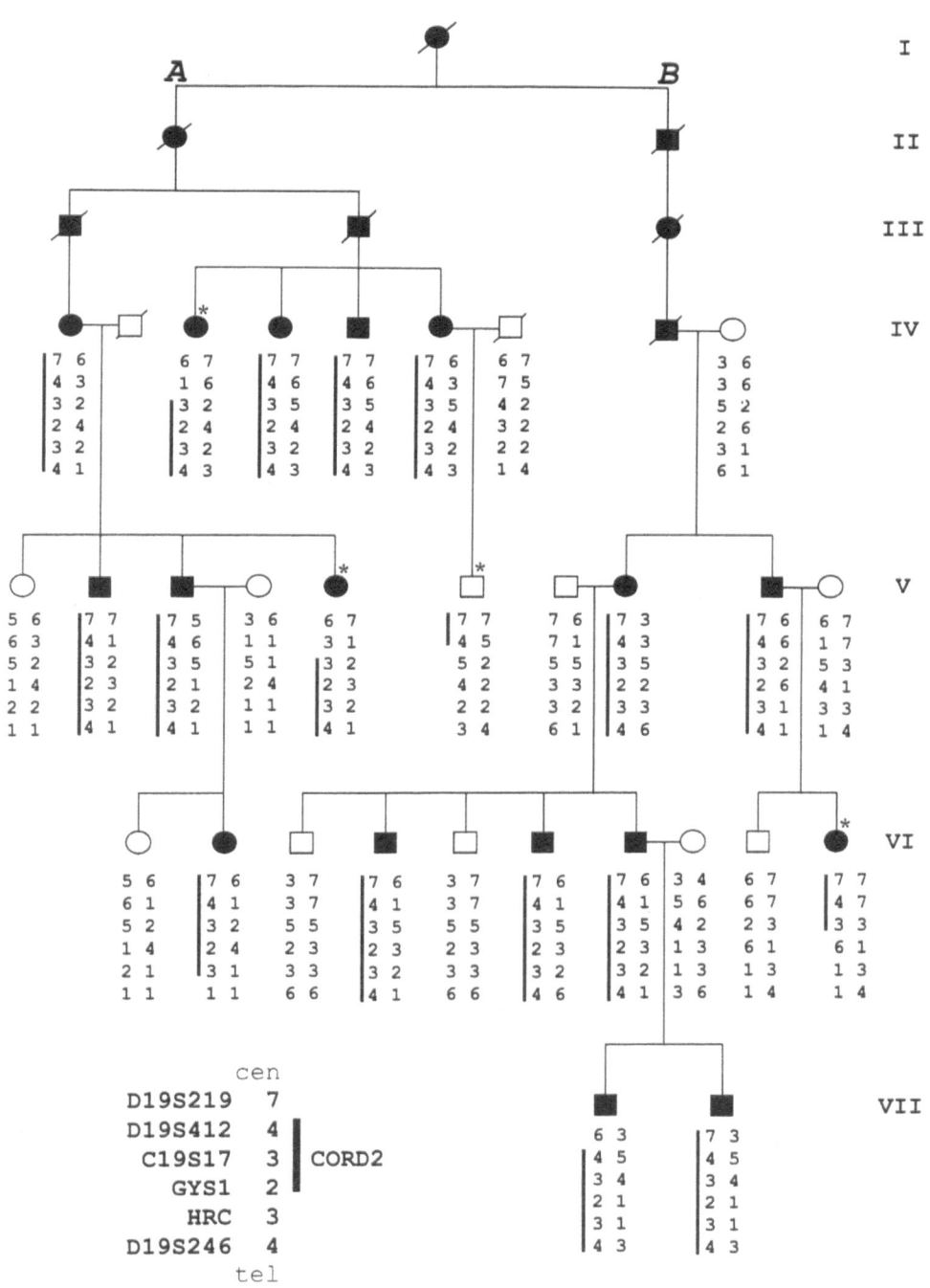

Figure 1. Haplotype analysis of 6 polymorphic markers in the CRD family. The haplotype linked to the disease is represented by a thick vertical line to the left of the haplotype data. Branch A is an abbreviated pedigree showing the original flanking recombinant individuals and branch B is the newly ascertained branch of the pedigree. Individuals IV-3, V-5 and V-6 refine the disease locus distal to marker locus D19S412. Individual VI-10 refines the disease locus proximal to GYS1. * denotes critical recombinant individuals.

Isolation of YAC Clones

The ICI and ICRF YAC libraries available from the UK Human Genome Mapping Project Resource Centre were screened by PCR amplification of YAC DNA hierarchical pools with microsatellite markers, sequence tagged sites (STSs), expressed sequence tags (ESTs) and YAC end-clones corresponding to the regions of the framework map which were poorly represented by cloned DNA. YAC end-clones were obtained either by Alu-vector arm PCR (16) or walking PCR (17) from yeast DNA. STSs were developed from the YAC end-clones and used to confirm the extent of overlap of the new YACs with CEPH YACs already available towards completing the physical contig. YAC sizes were determined by pulse-field gel electrophoresis (18).

RESULTS

Genetic Localisation of *CORD2*

The *CORD2* interval was most recently mapped to a 5 cM region (14) between marker loci D19S219 (3 recombinant individuals) and D19S246 (1 recombinant individual). Haplotype analysis (Fig. 1) with new marker loci on the 3 proximal flanking recombinant individuals in branch A (IV-3, V-5 and V-6) showed that they were all recombinant with marker locus D19S412, placing the disease gene telomeric to D19S412. Individual VI-10 from branch B was recombinant with marker loci D19S246, HRC and GYS1 placing the disease gene centromeric to GYS1. A new microsatellite marker that we identified (GDB Accession No. G29026) was fully informative and showed no recombination with disease. The refined genetic distance between D19S412 and GYS1 defining where the disease gene must lie is approximately 2 cM.

Characterisation of New Marker Loci

Nine cosmids in the region were found to be positive on hybridisation with a $(GT)_{13}$ probe. The cosmids were digested and subcloned into pUC18 or pBluescript vectors and the flanking sequences of 11 (CA)n repeats were characterised. Only three repeats were found to be polymorphic when tested in the general population. Table 1 shows the polymorphic nature of each of the three repeats, the primers used to amplify it and GDB Accession numbers. These new markers were tested in the CRD mapping panel, however, they did not refine the disease locus any further. Location of each new marker is shown in the physical map of the *CORD2* region (Fig. 2).

Physical Map of *CORD2* Critical Region

The progress towards a complete physical contig of the *CORD2* critical region is shown in Fig. 2. The estimated physical distance by high resolution FISH mapping is 1.6 megabases (13). We have still to close the gap between YAC 2DF7 and marker D19S902. In this gap there are 4 cosmid islands which have yet to be anchored to YACs. Two of these islands have been firmly mapped to this location by screening the Genebridge radiation hybrid panel (19) with STSs developed from the cosmids within the islands. Additionally, the region between YAC 4X119F5 and 4X117A10 containing three cosmid islands remains to be bridged. We have also isolated a number of floating YACs that are associated with specific genes that have been FISH mapped to 19q13.3, which have yet to

Table 1. Characteristics of new polymorphic markers and STSs

STS ID	GDB accession number	Type of STS	Size of repeat (bp)	Allele frequency		Primer sequences 5´-3´
SW114	G29025	CA repeat	153	1	0.17	F - CATTATCCCTGTCACTCAGC
				2	0.12	R - CAGACTTCCTTCTGTCTGC
				3	0.22	
C19S17	G29026	CA repeat	131	1	0.01	F - TCATGAATTAATCCCAGGAG
				2	0.40	R - CTGTATCTTGGATAAAGTGG
				3	0.10	
				4	0.01	
				5	0.48	
C19-12H73	G29027	CA repeat	137	1	0.24	F - CTATGTACAGCTGAGGTGAG
				2	0.54	R - TGCAGACACTTCCAGAGATG
				3	0.22	
A7LA	D19S160	YAC end STS	157		—	F - GGAATGTTTGGATCTGGG
						R - TAAACAAGTCAGCTTACC
F3RA	D19S1158	YAC end STS	185		—	F - ATTTACCTGCACCTTTGTC
						R - AGGGTAAATCTCAAGACTG
G8RA	Submitted	YAC end STS	129		—	F - AATCCTATGGGGCCAAGAT
						R - TGAGATTCCAGGGCTCGAAG
A10RA	Submitted	YAC end STS	229		—	F - CCTTCTCACCCATATACACAG
						R - TTATGTGCAGTGATCGGGTG
F4RA	D19S1161	YAC end STS	170		—	F - TCAATTTTCTTCCCATTGTC
						R - CCTAAAAGAAGTAACTGTTC
21H31	D19S1157	cosmid 14960 STS	170		—	F - GCAGGGTGTGGAAAGACTC
						R - CACTCTCAAACACAGGCAGA
LIG 1	L22709	cosmid 20031 STS	300		—	F - GGACGAGGACAGAGAAGCC
						R - GTGGTCGTAGCTGTCCTGC
16H48	D19S1156	cosmid 15217 STS	240		—	F - CACTAGCTGGGCATGGTAG
						R - GTGGTCGTAGCTGTCCTGC
F5RA	D9S1159	YAC end STS	214		—	F - GGAATTCTACCAGATGTG
						R - TGTATATACCGTAGAATGGC

anchored to the contig. These include YACs for a solute carrier (20), nucleobindin (21), neurotrophin-4 (22) and human ubiquitous receptor (23). Within the *CORD2* region are 8 genes, 2 ESTs, 3 polymorphic markers and 13 STSs which have been localised to cosmids contigs and YACs positioned on the map. Nine new STSs that we have developed in this study are listed in Table 1.

DISCUSSION

We have used a number of complementary approaches towards finding the gene causing cone-rod dystrophy on chromosome 19q13.3. Ascertainment of a new branch of the family has allowed us to refine the disease locus to a genetic distance of approximately 2 cM between D19S412 and GYS1. We have isolated a number of YACs and cosmids towards forming a complete physical contig of the disease region which is estimated to be 1.6 megabases of DNA. On completion this will then allow us to search systematically for the *CORD2* gene by techniques such as cDNA selection and exon trapping. Chromosome 19 is the most CpG island rich chromosome (24) suggesting there are many genes on this small chromosome. Most house-keeping genes and about 40% of tissue-specific

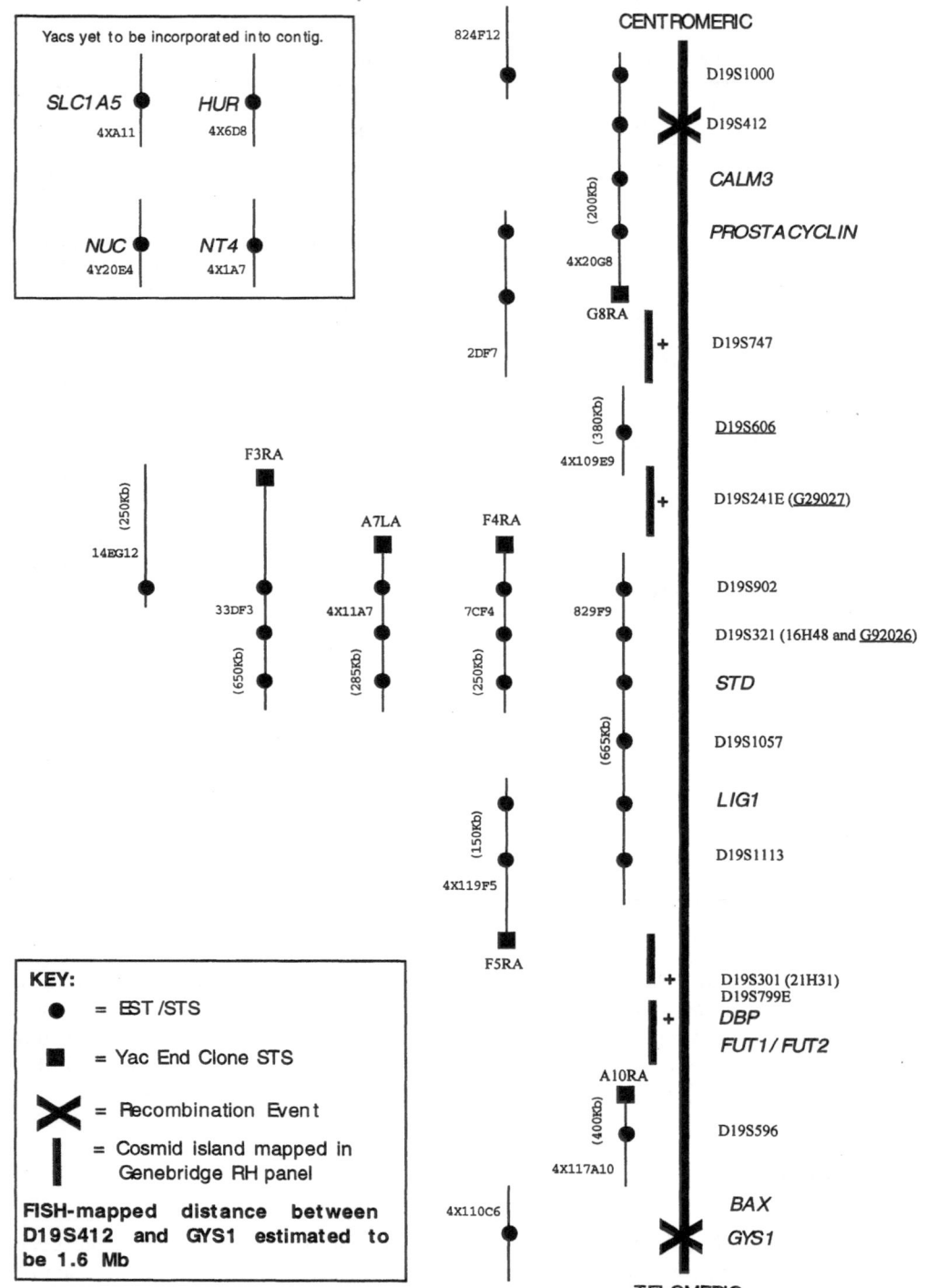

Figure 2. Integrated physical map of the *CORD2* critical region. The thick vertical line to the right represents the chromosome with genes, STSs, ESTs and polymorphic markers listed in order from centromere to telomere. YAC clones are represented by thin vertical lines with the YAC name and size annotated. The presence of an STS or EST in a YAC (^{2}) is indicated at the appropriate position. STSs developed from YAC-end termini (μ) are labelled as LA (left arm) or RA (right arm). The location of the critical recombination events are shown by an X. Polymorphic markers are underlined. Cosmid islands mapped to the critical region by STS mapping through the Genebridge radiation panel is depicted as a vertical grey bar. A plus sign (+) next to a cosmid island indicates the presence of the STS to the right.

genes are known to have CpG islands and thus we may have to analyse many genes in the region before finding the *CORD2* gene.

One of the genes already known to map to the critical region is CALM3 (25) which could be a candidate for CRD. Calmodulin is known to be present in the inner layers of the retina by immunohistochemical studies (26), although it is not known whether this represents CALM3 localisation since two other calmodulin genes CALM1 (27) and CALM2 (28) are also known to exist. Currently we are screening the CALM3 gene in the CRD family for mutations in order to determine whether it is the disease-causing gene. The two ESTs in the region (D19S241E and D19S799E) are derived from brain cDNA libraries and since the retina is part of the CNS they could be candidates for CRD. Thus, we are characterising the cDNAs for each of these ESTs. Two other disease phenotypes mapping to the region are DNA ligase 1 deficiency and Bombay phenotype, however neither of these conditions have any eye associations which might be allelic with CRD.

In summary, by combining genetic and physical mapping we have significantly reduced the size of the chromosomal region where the *CORD2* gene must lie. Using the YAC and cosmid clones described here, work is now underway to create a transcriptional map of this region. This information will hasten the cloning of the gene responsible for CRD and may provide candidates for other diseases mapping to the region. Additionally, the identification of this gene may be an important step in the search for effective treatments of this degenerative disorder and will provide an insight into the normal biochemical and molecular functioning of the retina.

ACKNOWLEDGMENTS

This work was support by The Wellcome Trust Grants 043825/Z/95/A (CYG) and 038650/Z/93/Z/1.5E (SDW). The authors acknowledge the support from the UK Human Genome Mapping Project Resource Centre during this research.

REFERENCES

1. Elston, J., 1992, Epidemiology of visual handicap in childhood, in: *Pediatric Ophthalmology*, (D. Taylor, ed), pp. 1–3, Blackwell Scientific Press, London.
2. Dryja, T.P., and Li, T., 1995, Molecular genetics of retinitis pigmentosa, *Hum. Mol. Genet.* **4**: 1739–1743.
3. Evans, K., Duvall-Young, J., Fitzke, F.W., Arden, G.B., and Bird, A.C., 1995, Chromosome 19q cone-rod retinal dystrophy. Ocular phenotype, *Arch. Ophthalmol.* **113**: 195–201.
4. Tranebjaerg, L., Sjö, O., and Warburg, M., 1986, Retinal cone dysfunction and mental retardation associated with a *de novo* balanced translocation 1:6(q44:27), *Ophthal. Paediat. Genet.* **7**: 167–173
5. Hong, H.K., Ferrel, R.E., and Gorin, M.B., 1994, Clinical diversity and chromosomal localisation of X-linked cone dystrophy (COD1), *Am. J. Hum. Genet.* **55**: 1173–1181.
6. Balciuniene, J., Johansson, K., Sandgren, O., Wachtmeister, L., Holmgren, G., and Forsman, K., 1995, A gene for autosomal dominant progressive cone dystrophy (CORD5) maps to chromosome 17p12-p13, *Genomics* **30**: 281–286.
7. Small, K., Syrquin, M., Mullen, L., and Gehrs, K.,1996, Mapping of autosomal dominant cone degeneration to chromosome 17p, *Am. J. Ophthalmol.* **121**: 13–18.
8. Warburg, M., Sjö, O., and Fledelius, H.C., 1991, Deletion mapping of a retinal cone-rod dystrophy: Assignment to 18q211, *Am. J. Med. Genet.* **39**: 288–293.
9. Kylstra, J.A., and Aylsworth, A.S., 1993, Cone-rod retinal dystrophy in a patient with neurofibromatosis type 1, *Can. J. Ophthalmol.* **28**: 79–80.
10. Evans, K., Fryer, A., Inglehearn, C., Duvall-Young, J., Whittaker, J.L., Gregory, C.Y., Butler, R., Ebenezer, N., Hunt, D.M., and Bhattacharya, S.S., 1994, Genetic linkage of cone-rod retinal dystrophy to chromosome 19q and evidence for segregation distortion, *Nat. Genet.* **6**: 210–213.

11. Nakazawa, M., Kikawa, E., Chida, Y., and Tamai, M., 1994, Asn244His mutation of the peripherin/RDS gene causing autosomal dominant cone-rod degeneration, *Hum. Mol. Genet.* **3**: 1195–1196.

12. Evans, J., and Wormald, R., 1996, Is the incidence of registrable age-related macular degeneration increasing?, *Br. J. Ophthalmol.* **80**: 9–14.

13. Ashworth, L.K., Batzer, M.A., Brandriff, B., Branscomb, E., de Jong, P., Garcia, E., Garnes, J.A., Gordon, L.A., Lamerdin, J.E., Lennon, G., Mohrenweiser, H., Olsen, A.S., Slezak, T., and Carrano, A.V., 1995, An integrated metric physical map of human chromosome 19, *Nat. Genet.* **11**: 422–427.

14. Gregory, C.Y., Evans, K., Whittaker J.L., Fryer, A., Weissenbach, J., and Bhattacharya, S.S., 1994, Refinement of the cone-rod retinal dystrophy locus on chromosome 19q, *Am. J. Hum. Genet.* **55**: 1061–1063.

15. Lindsay, S., Curtis, A.R.J., Roustan, P., Kamakari, S., Thiselton, D.L., Stephenson, A., and Bhattacharya, S.S., 1993, Isolation and characterisation of three microsatellite markers in the proximal long arm of the human X chromosome, *Genomics* **17**: 208–210.

16. Nelson, D.L., Ledbetter, S.A., Corobo, L., Victoria, M.F., Ramirez-Solis, R., Webster, T.D., Ledbetter, D.H., and Caskey, C.T., 1989, *Alu* polymerase chain reaction: A method for rapid isolation of human-specific sequences from complex DNA sources, *Proc. Natl. Acad. Sci. USA* **86**: 6686–6690.

17. Screaton, G.R., Bangham, C.R.M., and Bell, J.I., 1993, Direct sequencing of single primer PCR products: a rapid method to achieve short chromosomal walks, *Nucl. Acids Res.* **21**: 2263–2264.

18. Smith, C.L., Klco, S.R., and Cantor, C.R., 1990, Pulse-field gel electrophoresis and the technology of large DNA molecules, in: *Genome Analysis-A Practical Approach*, (K.E. Davies, ed), pp. 41–72, IRL Press, Oxford.

19. Gyapay, G., Schmitt, K., Fizames, C., Jones, H., Vega-Czarny, N., Spillett, D., Muselet, D., Prud'Homme, J-F., Dib, C., Auffray, C., Morissette, j., Weissenbach, J., and Goodfellow, P.N., 1996, A radiation hybrid map of the human genome, *Hum. Mol. Genet.* **5**: 339–346.

20. Jones, E.M.C., Menzel, S., Espinosa, R., Le Beau, M.M., Bell, G.I., and Takeda, J., 1994, Localisation of the gene encoding a neutral amino acid transporter-like protein to human chromosome band 19q13.3 and characterisation of a simple sequence repeat DNA polymorphism, *Genomics* **23**: 490–491.

21. Miura, K., Hirai, M., Kanai, Y., and Kurosawa, Y., 1996, Organisation of the human gene for nucleobindin (*NUC*) and its chromosomal assignment to 19q13.3-q13.4, *Genomics* **34**: 181–186.

22. Ip, N.Y., Ibanez, C.F., Nye, S.H., McClain, J., Jones, P.F., Gies, D.R., Belluscio, L., Le Beau, M.M., Espinosa, R., Squnito, S.P., Persson, H., and Yancopoulos, G.D., 1992, Mammalian neurotrophin-4: Structure, chromosomal localisation, tissue distribution and receptor specificity, *Proc. Natl. Acad. Sci. USA* **89**: 3060–3064.

23. Song, C., Kokontis, J.M., Hiipakka, R.A., and Liao, S., 1994, Ubiquitous receptor: A receptor that modulates gene activation by retinoic acid and thyroid hormone receptors, *Proc. Natl. Acad. Sci. USA* **91**: 10809–10813.

24. Larsen, F.G., Gundersen, R., Lopez, R., and Prydz, H., 1992, CpG islands as gene markers in the human genome, *Genomics* **13**: 1905–1107.

25. Berchtold, M.W., Egli, R., Rhyner, J.A., Hameister, H., and Strehler, E.E., 1993, Localisation of the human bona fide calmodulin genes CALM1, CALM2, and CALM3 to chromosomes 14q24-q31, 2p21.1-p21.3 and 19q13.2-q13.3, *Genomics*, **16**: 461–465.

26. Pochet, R., Pasteels, B., Seto-Ohshima, A., Bastianelli, E., Kitajima, S., and van Eldik, L.J., 1991, Calmodulin and calbindin localisation in retina from six vertebrate species, *J. Comp. Neurol.* **314**: 750–762.

27. Rhyner, J.A., Ottiger, M., Wicki, R., Greenwood, T.M., and Strehler, E.E., 1994, Structure of the human CALM1 calmodulin gene and identification of two *CALM1*-related pseudogenes CALM1P1 and CALM1P2, *Eur. J. Biochem.* **225**: 71–82.

28. SenGupta, B., Friedberg, F., and Detera-Wadleigh, S.D., 1987, Molecular analysis of human and rat calmodulin complementary DNA clones, *J. Biol. Chem.* **34**: 16663–16670.

33

USHER SYNDROME TYPE 1C

Localization to Chromosome 11p14 and Construction of a YAC Contig

Radha Ayyagari,[1] Anren Li,[2] Ann Nestorowicz,[3] Yan Li,[2] Richard J. H. Smith,[4] M. Alan Permutt,[3] and J. Fielding Hejtmancik[2]

[1]Kellogg Eye Center
University of Michigan
Ann Arbor, Michigan
[2]National Eye Institute
NIH
Bethesda, Maryland
[3]Division of Metabolism and Endocrinology
Washington Univesity School of Medicine
St. Louis, Missouri
[4]Department of Otolaryngology-Head and Neck Surgery
University of Iowa
Iowa City, Iowa

INTRODUCTION

The Usher Syndromes are a group of autosomal recessive disorders characterized by congenital sensorineural deafness and progressive pigementory retinopathy. Several types of Usher syndromes (Ush) have been distinguished by age of onset, rate of progression and severity of clinical symptoms.

Usher syndrome type 1 is the most severe form with a profound congenital hearing loss, vestibular dysfunction and early onset of retinitis pigmentosa. In Usher syndrome type 2 (Ush 2), deafness is moderate and the onset of retinitis pigmentosa tends to be delayed to the second or third decade with normal vestibular function. In Usher syndrome type 3 (Ush 3), both the sensorineural hearing loss and retinitis pigmentosa are progressive over time (1). At least five different loci causing Ush 1 (2–5) and one locus each for Ush 2 (6) and Ush 3 (7)have been identified so far indicating the complexity of genetic heterogeneity in various forms of Usher syndromes.

Usher syndrome type 1A identified in a population from southern France has been localized to chromosome 14q (2). The gene involved in causing Usher syndrome type 1B

Degenerative Retinal Diseases, edited by LaVail *et al.*
Plenum Press, New York, 1997

on chromosome 11q has been identified as myosin VIIA (8). Usher syndrome type 1C, which occurs in the French Acadian population of Louisiana has been localized to a 5 cM interval on the P arm of chromosome 11(9,10). Loci for Usher syndrome type 1D (4) and USH1E (5) have been mapped to 10p and 21q respectively.

In our earlier report we have presented the heterogeneity in Usher syndrome and localization of Usher syndrome type 1B and type 1C to the q and p arm of chromosome 11 respectively (11). Mutations in myosin VIIA were found to be associated with Usher syndrome type 1B (8). Here we present the precise localization of Usher syndrome type 1C to 11p14 and construction of a YAC contig encompassing the Usher syndrome type 1C interval.

METHODS

Clinical Studies

Seven Acadian families with a history of Usher syndrome and individual affected patients with a history of Usher syndrome in their families were studied in detail (3). Affected individuals and their family members underwent a detailed physical examination including a battery of tests suggested for the clinical diagnosis of Ush (1) including the Bruininks-Osteretsky subtests of balance function, an otologic examination and ice-water calorics using Frenzel's glasses to prevent fixation and to facilitate observation.

Isolation of Human Genomic DNA and Microsatellite Marker Analysis

Human genomic DNA was prepared from peripheral blood lymphocytes or extracted from cultured lymphoblastoid cell pellets as described earlier (12). Amplification of microsatellite repeats was performed by PCR using 60ng of human genomic DNA as described earlier (13).

Linkage Analysis

The LINKAGE program package version 5.1 was used for linkage analysis (14). USH1C was assumed to be fully a penetrant autosomal recessive trait with an affected allele frequency of 0.001. Marker allele frequencies were taken from the Genome Data base at the Welch Library, Johns Hopkins University.

Yeast Artificial Chromosome (YAC) Library Screening

YAC clones were obtained by screening total human genomic YAC libraries constructed at the Center for Genetics in Medicine (CGM: Washington University School of Medicine, St.Louis) (15) and the CEPH (16). YAC libraries were screened with the microsatellite markers D11S861, D11S419, D11S902, D11S921, D11S1310, D11S899 and the Sequence Tagged Site (STS) 966e8(L) using a PCR based library pooling strategy as described by Green and Olsen (17). Primer sequences for the six microsatellite markers were obtained from the Genome Data Base. Oligonucleotide sequences for the STS 966e8(L) are shown in Table 1. Additional YACs positive for the six microsatellite markers were identified from the CEPH YAC database (18). YACs 14FH2, 25ED3 and 39CC12, previously shown to contain the STSs MYOD1, TPH, SAA2, SAA3, LDH or TFIIH (19) were obtained from the ICI YAC library (20).

Isolation of YAC DNA

Total Yeast genomic/YAC DNA was isolated as previously described (21). Briefly, yeast cells grown in AHC medium were treated with lyticase and then with β-mercapi-toethanol and Sodium Dodecyl Sulfate. DNA from lysed cells was isolated by Phenol-Chloroform extraction and ethanol precipitation.

Sequence Analyses of YAC-Insert Termini and STS Development

Sequence tagged sites were developed from DNA sequences obtained from YAC insert termini and inter-*Alu* PCR products. YAC-end fragments were generated by a ligation-mediated PCR method as described by Kere et al (22). For generation of STSs from inter-*Alu* PCR products, YAC DNA was amplified by PCR with 5'- and 3'- *Alu* PCR primers (23). Unique inter-*Alu* PCR products were sequenced.

Sequences obtained from YAC insert termini and inter-*Alu* PCR products were used to develop STS using the computer program PRIMER and are listed in Table 3. Sequenced tagged sites are designated as the YAC number followed by the origin of the insert DNA relative to the left (L) or right (R) YAC vector arms.

STS-Content Screening of YACs

With the exception of STSs developed from YAC insert sequences and listed in Table 3, primer sequences and PCR amplification conditions for STSs were obtained from the Genome Data Base. PCR reactions were carried out in a 10 μl reaction volume containing 20ng total yeast DNA.

RESULTS

Localization of Microsatellite Markers

Markers D11S1397, D11S902, D11S921, D11S1890, D11S1888 and D11S1310 were localized to the 'g' segment of chromosome 11p, which spans the upper half of the 11p14 region. D11S1861 and D11S419 were localized to the 'f' segment which covers 11p15.1 (Fig.1) as described by Glasser et al. (24).

Linkage Analysis

Lod scores from the two-point analysis of chromosome 11p microsatellite markers with USH1C in the French-Acadian families are shown in Table 1. All the markers in this region show positive lod scores at zero recombination fraction except D11S861 and D11S928 identifying these as telomeric and centromeric flanking markers respectively for the Usher syndrome type 1C locus.

Haplotype Analysis in Acadian Families

Haplotypes of chromosomes with USH1C and normal alleles were determined in order to localize the USH1C gene to a small interval. Haplotypes were constructed for markers in the USH1C region in the following order, as determined by linkage analysis

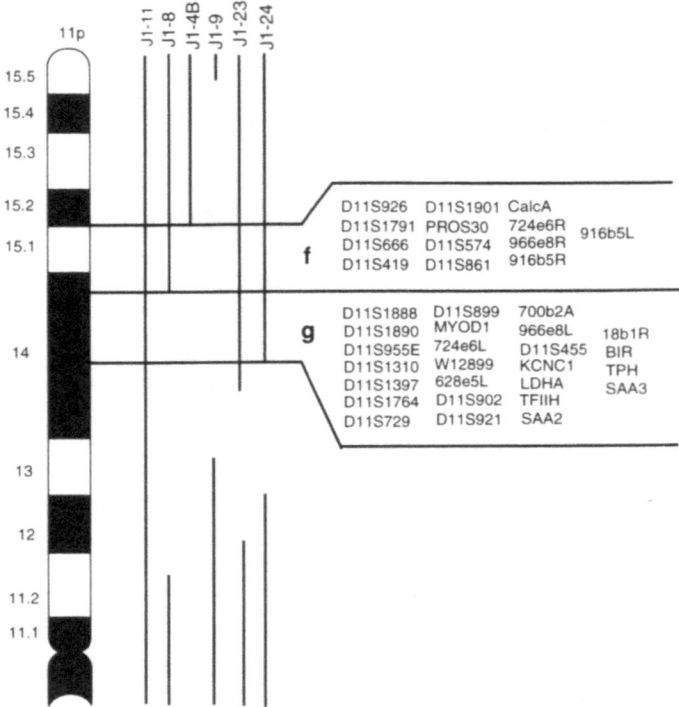

Figure 1. Regional localization of STSs used for USH1C mapping and YAC contig construction. Somatic cell lines and intervals shown are a subset of those defined by Glaser et al. (1998), with only intervals f and g shown. Bold vertical lines represent portions of chromosome 11p retained in each hybrid. Absence of such lines indicates deleted regions within chromosome 11p. Regions of 11q retained in these cell lines are not indicated.

and physical mapping: tel-D11S861-D11S419-D11S1397-D11S902-D11S921-D11S1890-D11S1888-D11S1310-cen (Fig 2). Haplotypes from 70 chromosomes with the Ush1C mutation and 13 non-Ush1C chromosomes from the French Acadian population were determined. Individuals belonging to the Acadian USH1C families used in the original localization of the USH1C gene to chromosome 11 (3) were used for haplotype analysis.

Table 1. Two-point lod scores of 11p microsatellite markers versus USH1C in the French Acadian families. Marker order corresponds to Figure 2

| Marker | lod scores at theta = | | | | | | | | | | |
	0	0.01	0.02	0.03	0.04	0.05	0.1	0.2	0.3	Z max	Theta
D11S861	$-\infty$	3.22	3.39	3.43	3.41	3.37	2.99	1.97	0.88	3.42	0.03
D11S419	2.95	2.83	2.77	2.71	2.65	2.58	2.25	1.54	0.82	2.95	0
D11S1397	1.26	1.22	1.18	1.14	1.11	1.11	0.87	0.51	0.21	1.26	0
D11S902	6.44	6.26	6.09	5.92	5.75	5.57	4.71	3.02	1.51	6.44	0
D11S921	3.31	3.24	3.16	3.08	3.01	2.92	2.51	1.67	0.86	3.31	0
D11S1890	2.26	2.21	2.13	2.07	2.01	1.95	1.64	1.06	0.54	2.26	0
D11S1888	6.41	6.23	6.05	5.88	5.71	5.53	4.66	2.97	1.47	6.41	0
D11S1310	3.84	3.74	3.63	3.53	3.43	3.32	2.81	1.79	0.88	3.84	0
D11S899	5.46	5.31	5.16	5.01	4.85	4.71	3.95	2.51	1.23	5.46	0
D11S928	$-\infty$	1.58	1.81	1.91	1.97	1.99	1.89	1.34	0.71	1.99	0.05

Table 2. Association of haplotypes with the USH1C mutation in
the French-Acadian population of Louisiana

Haplotype	D11S861	D11S419	D11S1397	D11S902	D11S921	D11S1890	D11S1888	D11S1310	USH	Normal
A	3	2	1	4	3	2	4	4	55	0
B	4	2	1	4	3	2	4	4	1	0
C	3	2	3	4	3	2	4	4	7	0
D	3	2	1	4	3	2	4	2	4	0
E	3	2	1	4	3	2	1	3	2	0
F	3	2	1	4	3	2	4	1	1	0
	2	2	3	3	3	1	1	3	0	1
	1	2	3	2	2	4	2	4	0	1
	4	2	1	2	3	2	2	3	0	1
	2	1	1	2	1	5	1	4	0	1
	5	2	1	3	3	1	3	4	0	1
	1	1	1	2	2	2	1	1	0	1
	1	1	3	2	3	2	2	1	0	1
	4	1	1	3	1	2	1	2	0	1
	1	2	1	2	1	3	1	2	0	1
	1	1	1	2	3	2	2	5	0	1
	1	1	1	1	3	3	3	3	0	1
	2	2	3	3	1	2	1	3	0	1
	6	1	2	2	3	2	2	2	0	1
Total									70	13

Additional individuals not in families 1–7 were analyzed only if they were affected and then were assumed to have two chromosomes carrying the USH1C allele, consistent with autosomal recessive inheritance of Usher syndrome type 1C.

Seventy eight percent of the affected chromosomes show haplotype A (Table 2). One affected chromosome with haplotype B differs from haplotype A by having allele 4 instead of allele 3 at the D11S861 locus. Haplotype C which is observed in seven affected chromosomes differs from haplotype A by carrying allele 3 instead of allele 1 at the D11S1397 locus. Four affected chromosomes with haplotype D differ from haplotype A by carrying allele 2 instead of allele 4 at the marker D11S1310. Haplotype E observed in 2 affected chromosomes differs from haplotype A by.having alleles 1 and 3 instead of al-leles 4 and 4 at D11S1888 and D11S1310 loci respectively. One affected chromosome shows haplotype F which differs from haplotype A by having allele 1 instead of allele 4 at the D11S1310 locus. None of the 13 chromosomes carrying normal alleles shows haplo-type similar to A-F which is seen on affected chromosomes.

Isolation of YAC Clones

The CEPH libraries were initially screened with six microsatellite markers located within or flanking the proximal or distal boundaries of the USH1C critical region. A total of 47 YAC clones were identified as positive for at least one of the six STSs used for screening. A subset of these YACs was selected for further construction of the contig on the basis of size and apparent lack of large deletions. Potential overlap between these YACs was examined by PCR analysis of all clones for the presence or absence of each of the six markers used for library screening.

STS Development

Since the density of microsatellite markers in the D11S861 and D11S899 interval was initially insufficient to establish map closure, new STSs were developed from YAC end-fragment sequences of 6 clones and from inter-*Alu* PCR product sequences to 1) determine overlap between clones, 2) screen YAC libraries for additional clones and 3) investigate whether selected YACs were chimeric. The chromosomal location of each STS was determined by PCR analysis of a rodent/human somatic cell hybrid panel (NIGMS). With the exceptions of the STSs 185b1(L) and 628e5(R), all STSs were specific for chromosome 11 (Table 3). All 10 chromosome 11 specific STSs as well as adjacent microsatellite markers were regionally assigned to the appropriate location on chromosome 11 by PCR analysis of the J1 somatic cell hybrid panel (Fig 1). These results are consistent with the previously described genetic and cytogenetic locations (19,25) of adjacent genetic markers. The STS 966e8(L) was used to rescreen both the CEPH 'mega'-YAC (16) and Washington University YAC (15) libraries, resulting in the identification of 4 additional clones that were incorporated into the contig.

YAC STS Content Analysis and Contig Construction

To establish the presence of overlapping regions, each YAC was screened for the presence of the 10 novel chromosome 11 STSs. During construction of the contig an addi-

Table 3. Primer sequences and chromosomal assignments
for STSs developed from YAC insert sequences

YAC	STS	Primer Sequence (5' - 3')	Annealing Temp (° C)	Size (bp)	Chromosomal Location
185b1	185b1 (L)	FOR: TGACTGGGTTTCTTCACTTTG REV: GGATGGATAGTTATCCAGAAAG	60	177	1,12
	185b1 (R)	FOR: GTGCAGACATTGCTAATTGTTC REV: CCTGGCCTAGTTTTATCTCATG	60	125	11
966e8	966e8 (L)	FOR: AGAAATGTCTGGAACATACCAG REV: CAGATCTCTCCTCAGCCTTG	58	150	11
	966e8 (R)	FOR: CTTTACCATGAGAACAGTATGG REV: CCTGTGCACAAAAGAGAAAAC	60	174	11
916b5	916b5 (L)	FOR: TAAAGTGTTGGACACATAGCTG REV: GTCATTGTAGCTACATGTACCC	60	125	11
	916b5 (R)	FOR: ATTTGTATTTCCTGGCATGC REV: TGGGGTCCAAAGATAGTTAAC	55	307	11
724e6	724e6 (L)	FOR: AGGGAGGCTATAGTAAGCAAAC REV: TCAGCTAGTAGGAGTTGAAATTG	55	195	11
	724e6 (R)	FOR: TATCAGCAGTATTTAGCAGTG REV: CTAAGGAGGTAATGCTTTGT	58	91	11
628e5	628e5 (L)	FOR: GATTCCAGGATAGCATGACAG REV: GCCCTTTTATACCAGTGAAAG	60	169	11
	628e5 (R)	FOR: TTATTGTGCTACAGATGTTGGC REV: GCTTACCCAAATAGATTGCC	58	107	10
700b2	700b2 A	FOR: TCCTCTACCCAATATGCTCC REV: CAAGGAGATAGTGCCAACAGC	52	80	11

tional 23 STSs, including 6 microsatellite markers (D11S1901, D11S1791, D11S1397, D11S1890, D11S1888, D11S1764) were assigned by radiation hybrid mapping to the D11S926-D11S899 interval (19,25). To further establish the degree of overlap between individual YACs and also to define their STS content, all YACs were screened by PCR analysis for the presence or absence of a total of 40 STSs, including the six STSs used for initial screening of YAC libraries. Results of the STS content screening are summarized in Figure 2. The sizes of selected YACs were determined by pulsed-field gel electrophoresis (PFGE).

DISCUSSION

Usher Syndrome type 1C was initially localized to a 5.5 cM interval on the p arm of chromosome 11 between the markers D11S861 and D11S899 (9) using linkage analysis. Since that time, the additional microsatellite markers D11S1397, D11S1310, D11S1890 and D11S1888 have been localized to the region between D11S861 and D11S899. Two point linkage analysis of all these markers gave significant positive lod scores at zero recombination fraction (Table 1) confirming the localization of these markers close to Ush 1C gene. The precise order of the new markers with respect to the existing markers was not known.

Analysis of these makers on a chromosome 11 somatic cell hybrid panel has localized D11S1397, D11S1890, D11S1888 and D11S1310 to the "g" segment to which other markers D11S902, D11S921 and D11S899 that show tight linkage with Ush 1C are also localized (Fig 1). Based on radiation hybrid mapping (19) and linkage mapping, these markers are ordered as follows: 11pter-D11S926-D11S861-D11S419-D11S902-D11S921-D11S1310-D11S899-11pcen. Markers D11S1890 and D11S1888 were localized to the region between D11S902 and D11S921 based on previous radiation hybrid mapping (25).

A YAC contig was constructed extending from D11S861 to D11S899 encompassing the USH1C region. This physical map comprises 28 YACs and 38 STSs spanning an estimated distance of about 4cM. Continuity of the contig was established by STS - content analysis of YACs, including STSs developed from YAC end-fragments and inter-*Alu* PCR products. Physical overlap was also establied by comparison of inter-*Alu* or L1 PCR fingerprints (data not shown). The contig is ordered on the genetic map by seven previously mapped microsatellite markers [11p ter-D11S926-D11S861-D11S419-(D11S902-D11S921)-D11S1310-D11S899-11p Cen] (19). With the exceptions of the (PROS30, D11S861)-916b5(R), (D11S1764, SAA3)-(D11S455, TFIIH, SAA2) and 700b2A-D11S899 intervals, the density of YACs comprising the contig is at least three fold. All YACs comprising the minimal path (770c6, 808b11, 916b5, 966e8, 693b10, 700b2) were localized to chromosome 11p14–15.1 by fluorescent in situ hybridization (FISH) (data not shown), consistent with the regional assignment of STSs developed from the insert-termini of YACs 916b5 and 966e8 to intervals of f or g on 11p (Fig 1).

A total of 38 STSs were positioned on the physical map. Providing an estimated average resolution of about 1 STS/100 kb. The order of STSs depicted in Fig.2 is derived from the STSs content of each YAC and analysis of chromosome 11p somatic cell hybrid panel. The order of adjacent STSs could not be establised definitively, specifically for those STSs within the groups (PROS30, D11S861), (D11S574, D11S1901, D11S1791), (WI-2899, D11S729), (D11S902, 185b1(R), SUR, BIR, D11S921, D11S1890), (MYOD1, KCNC1), (SAA3, D11S1764), and (D11S455, TFIIH, SAA2). In addition the relative placement of STSs in some regions is complicated by the apparent loss of STSs from several YACs. As not all YACs comprising the contig were examined for chimeric end-sequences, and the presence of small internal chimerism, and/or small internal deletions with in YACs would not have been detected in these studies, although these sequences have to be limited to the areas distal to the terminal STS.

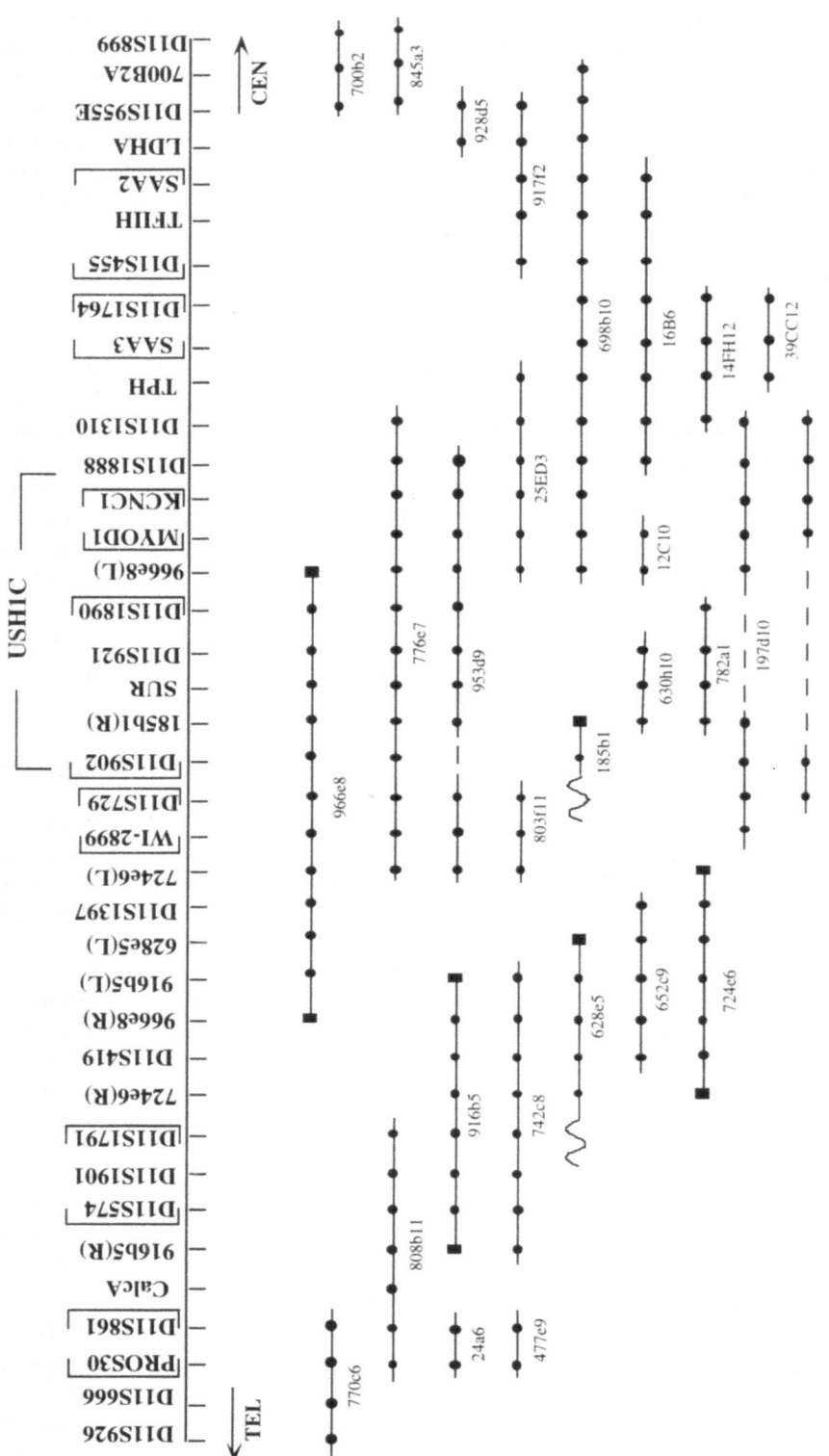

Figure 2. YAC contig encompassing the USH1C locus. YAC clones are represented by horizontal lines. (●) YAC DNA was amplified by the STS listed directly above. STS developed from sequences for YAC insert-termini or inter-*Alu* PCR products are represented by ■ and □, respectively. Dashed lines indicate regions of YACs that were negative by PCR analysis for the STS listed above. Uneven lines denote chimeric YAC end-sequences. STSs are placed equidistant to each other as the physical distances between adjacent STSs has not been determined. Adjacent STSs placed within brackets are those for which the relative orders could not be established definitively from YAC STS-content analyses.

In general, the physical map described here is in agreement with genetic, radiation hybrid and other physical maps of the region (19,25,26). For the STSs D11S574, D11S1901 and D11S1791, which could not be ordered relative to D11S419 by Radiation hybrid mapping, the data described here support a location distal to D11S419. The relative order of STSs at the proximal end of the contig, 11pter-(MYOD1, KCNC1)- D11S1888-D11S1310-TPH-(SAA3, D11S1764)-(D11S455, TFIIH, SAA2)-LDHA-D11S955E-11pcen is consistent with the integrated map of Fantes et al.(19) and the physical map of Sellar et al. (27), but the order described here differs in the placement of these STSs relative to the microsatellite markers D11S902 and D11S921. Fantes et al., (19) have assigned this group of STSs to the interval between D11S902 and D11S921. However, members of this group of STSs were not detected in YACs (966e8, 185b1, 630h10, 782a1) that are partly encompassed within or span the D11S902 and D11S921 interval. Fruther more , YACs 25ed3 and 693b10 are negative for D11S902 and D11S921 but contain the proximal end of 966e8 (966e8L) as well as the STSs MYOD1, KCNC1, D11S1888, D11S1310 and TPH. Centromeric placement of the above group of STSs is also supported by results obtained with YACs 16b6, 14fh12 and 39cc12 which are positive for all or some of the STSs D11S1888, D11S1310, TPH, SAA3, D11S1764, D11S455, TFIIH, SAA2 and LDH but are negative for D11S902 and D11S921. Together, these data support a location for D11S902 and D11S921 distal to this group of STSs.

Haplotype analysis of 70 USH1C chromosomes carrying the disease allele for USH1C and 13 chromosomes carrying normal allele from the French-Acadian population was carried out using the order of the markers derived from the physical map: Tel-D11S861-D11S419-D11S1397-D11S902-D11S921-D11S1890-D11S1888-D11S1310-D11S899-Cen (Fig 2). Seventy percent of the affected chromosomes possess a common haplotype for markers between D11S1397 to D11S1888. However, the occurrence of variant haplotypes in some chromosomes with the USH1C allele restrict the USH1C to approximately 1cM interval bounded by D11S1397 and D11S1888. These results also localize the USH1C locus to the upper one third of 11p14 based on the analysis of these flanking markers on the J1 hybrid panel (Fig 1).

ACKNOWLEDGMENTS

We thank Drs. Settara Chandrasekharappa and Craig Chinault for helpful advice and for providing YAC clones.

REFERENCES

1. Smith, R.J.,Berlin, C.I.,Hejtmancik, J.F.,Keats, B.J.,Kimberling, W.J.,Lewis, R.A.,Moller, C.G.,Pelias, M.Z.,Tranebjaerg, L. 1994, Clinical diagnosis of the Usher syndromes. Usher Syndrome Consortium. *Am. J. Med. Genet.* 50:32–38.
2. Kaplan, J.,Gerber, S.,Bonneau, D.,Rozet, J.M.,Delrieu, O.,Briard, M.L.,Dollfus, H.,Ghazi, I.,Dufier, J.L.,Frezal, J. 1992, A gene for Usher syndrome type I (USH1A) maps to chromosome 14q. *Genomics* 14:979–987.
3. Smith, R.J.,Lee, E.C.,Kimberling, W.J.,Daiger, S.P.,Pelias, M.Z.,Keats, B.J.,Jay, M.,Bird, A.,Reardon, W.,Guest, M. 1992, Localization of two genes for Usher syndrome type I to chromosome 11. *Genomics* 14:995–1002.
4. Wayne, S.,Der Kaloustian, V.M.,Schloss, M.,Polomeno, R.,Scott, D.A.,Hejtmancik, J.F.,Sheffield, V.C.,Smith, R.J.H. 1996, Localization of the Usher Syndrome type 1D gene (Ush1D) to chromosome 10 P. *Human Molecular Genetics* 5:1689
5. Chaib, H.,Kaplan, J.,Gerber, S.,Vincent, C.,Ayadi, H.,Slim, R.,Munnich, A.,Weissenbach, J.,Petit, C. 1997, A newly identified locus for Usher Syndrome type 1 USH1E, maps to chromosome 21q21. *Human Molecular Genetics* 6:27–32.

6. Kimberling, W.J.,Weston, M.D.,Moller, C.,van Aarem, A.,Cremers, C.W.,Sumegi, J.,Ing, P.S.,Connolly, C.,Martini, A.,Milani, M. 1995, Gene mapping of Usher syndrome type IIa: localization of the gene to a 2.1-cM segment on chromosome 1q41. *Am. J. Hum. Genet.* 56:216–223.

7. Sankila, E.M.,Pakarinen, L.,Kaariainen, H.,Aittomaki, K.,Karjalainen, S.,Sistonen, P.,de la Chapelle, A. 1995, Assignment of an Usher syndrome type III (USH3) gene to chromosome 3q. *Hum. Mol. Genet.* 4:93–98.

8. Weil, D.,Blanchard, S.,Kaplan, J.,Guilford, P.,Gibson, F.,Walsh, J.,Mburu, P.,Varela, A.,Levilliers, J.,Weston, M.D. 1995, Defective myosin VIIA gene responsible for Usher syndrome type 1B. *Nature* 374:60–61.

9. Ayyagari, R.,Smith, R.J.H.,Polymeropoulos, M.,Daiger, S.P.,Pelias, M.Z.,Wozencraft, L.,Kaiser-Kupfer, M.,Hejtmancik, J.F. 1995, Localisation of the Usher Syndrome Type 1 Gene in the French Acadian Population of Louisiana to Chromosome 11p14–15.1 by Linkage and Haplotype Analysis. *Ind. J. Hum. Genet* 1:93–103.

10. Keats, B.J.,Nouri, N.,Pelias, M.Z.,Deininger, P.L.,Litt, M. 1994, Tightly linked flanking microsatellite markers for the Usher syndrome type I locus on the short arm of chromosome 11. *Am. J. Hum. Genet.* 54:681–686.

11. Ayyagari, R.,Smith, R.J.H.,Lee, E.C.,Kimberling, B.J.,Jay, M.,Bird, A.,Hejtmancik, J.F. *Heterogeneity of Usher syndrome type 1.* In: *Retinal Degeneration*; Hollyfield, J. G.; Anderson, R. E.; La Vail, M. M. Eds. Plenum Press, New York: 1993,pp 127–133.

12. Padma, T.,Ayyagari, R.,Murty, J.S.,Basti, S.,Fletcher, T.,Rao, G.N.,Kaiser-Kupfer, M.,Hejtmancik, J.F. 1995, Autosomal dominant zonular cataract with sutural opacities localized to chromosome 17q11–12. *Am. J. Hum. Genet.* 57:840–845.

13. Ayyagari, R.,Li, Y.,Smith, R.J.H.,Pelias, M.Z.,Hejtmancik, J.F. 1995, Fine Mapping of the Usher Syndrome Type 1C to Chromosome 11p14 and Idnetification of Flanking Markers by Haplotype Analysis. *Mol. Vis. (http://www. emory. edu/MOLECULAR_VISION/v1/ayyagari.* 1:

14. Lathrop, G.M.,Lalouel, J.M. 1984, Easy calculations of lod scores and genetic risks on small computers. *Am. J. Hum. Genet.* 36:460–465.

15. Anand, R.,Riley, J.H.,Butler, R.,Smith, J.C.,Markham, A.F. 1990, A 3.5 genome equivalent multi access YAC library: construction, characterisation, screening and storage. *Nucleic. Acids. Res.* 18:1951–1956.

16. Albertsen, H.M.,Abderrahim, H.,Cann, H.M.,Dausset, J.,Le Paslier, D.,Cohen, D. 1990, Construction and characterization of a yeast artificial chromosome library containing seven haploid human genome equivalents. *Proc. Natl. Acad. Sci. U. S. A.* 87:4256–4260.

17. Green, E.D.,Olson, M.V. 1990, Systematic screening of yeast artificial-chromosome libraries by use of the polymerase chain reaction. *Proc. Natl. Acad. Sci. U. S. A.* 87:1213–1217.

18. Chumakov, I.M.,Rigault, P.,Le Gall, I.,Bellanne-Chantelot, C.,Billault, A.,Guillou, S.,Soularue, P.,Guasconi, G.,Poullier, E.,Gros, I. 1995, A YAC contig map of the human genome. *Nature* 377:175–297.

19. Fantes, J.A.,Oghene, K.,Boyle, S.,Danes, S.,Fletcher, J.M.,Bruford, E.A.,Williamson, K.,Seawright, A.,Schedl, A.,Hanson, I. 1995, A high-resolution integrated physical, cytogenetic, and genetic map of human chromosome 11: distal p13 to proximal p15.1. *Genomics* 25:447–461.

20. Brownstein, B.H.,Silverman, G.A.,Little, R.D.,Burke, D.T.,Korsmeyer, S.J.,Schlessinger, D.,Olson, M.V. 1989, Isolation of single-copy human genes from a library of yeast artificial chromosome clones. *Science* 244:1348–1351.

21. Chandrasekharappa, S.C.,Marchuk, D.A.,Collins, F.S. *Analysis of Yeast Artificial Chromosome Clones.* In: *Methods in Molecular Biology*; Burmeister, M.; Ulanovsky, I Eds. The Humana Press Inc., Totowa, NJ. 1992,pp 235–257.

22. Kere, J.,Nagaraja, R.,Mumm, S.,Ciccodicola, A.,D'Urso, M.,Schlessinger, D. 1992, Mapping human chromosomes by walking with sequence-tagged sites from end fragments of yeast artificial chromosome inserts. *Genomics* 14:241–248.

23. Tagle, D.A.,Collins, F.S. 1992, An optimized Alu-PCR primer pair for human-specific amplification of YACs and somatic cell hybrids. *Hum. Mol. Genet.* 1:121–122.

24. Glaser, T.,Housman, D.,Lewis, W.H.,Gerhard, D.,Jones, C. 1989, A fine-structure deletion map of human chromosome 11p: analysis of J1 series hybrids. *Somat. Cell Mol. Genet.* 15:477–501.

25. James, M.R.,Richard, C.W., 3rd,Schott, J.J.,Yousry, C.,Clark, K.,Bell, J.,Terwilliger, J.D.,Hazan, J.,Dubay, C.,Vignal, A. 1994, A radiation hybrid map of 506 STS markers spanning human chromosome 11. *Nat. Genet.* 8:70–76.

26. Qin, S.,Nowak, N.J.,Zhang, J.,Sait, S.N.,Mayers, P.G.,Higgins, M.J.,Cheng, Y.,Li, L.,Munroe, D.J.,Gerhard, D.S.,Weber, B.H.,Bric, E.,Housman, D.E.,Evans, G.A.,Shows, T.B. 1996, A high-resolution physical map of human chromosome 11. *Proc. Natl. Acad. Sci. U. S. A.* 93:3149–3154.

27. Sellar, G.C.,Jordan, S.A.,Bickmore, W.A.,Fantes, J.A.,van Heyningen, V.,Whitehead, A.S. 1994, The human serum amyloid A protein (SAA) superfamily gene cluster: mapping to chromosome 11p15.1 by physical and genetic linkage analysis. *Genomics* 19:221–227.

34

OGUCHI DISEASE, RETINITIS PIGMENTOSA, AND THE PHOTOTRANSDUCTION PATHWAY

Marion A. Maw and Michael J. Denton

Biochemistry Department
University of Otago
PO Box 56, Dunedin, New Zealand

ABSTRACT

Oguchi disease is a form of retinal dysfunction in which congenital night blindness is accompanied by a light-dependent discoloration of the fundus, extremely slow dark adaptation and an abnormal rod electroretinogram. This autosomal recessive condition is generally considered to be non-progressive however Oguchi disease and progressive retinal pigmentary degeneration or congenital stationary night blindness has occurred in the same families and even in the same patients. Oguchi disease may therefore represent a disorder involving features of both congenital stationary night blindness and retinitis pigmentosa. Here we summarise molecular genetic analyses which indicate that mutations in the arrestin gene are responsible for at least some cases of Oguchi disease in the Indian and Japanese populations. We speculate that the light-dependent fundus discoloration characteristic of Oguchi disease may reflect hyper-activity of the phototransduction pathway resulting from failure of arrestin to bind to and quench light-activated rhodopsin.

INTRODUCTION

Oguchi disease is a rare autosomal recessive form of congenital night blindness involving 3 characteristic clinical features:

1. Slow dark adaptation. Following exposure to bright light, normal cone and rod photoreceptors require respectively ten and thirty minutes of darkness to regain maximal sensitivity to light. In Oguchi disease cone dark adaptation proceeds normally but rod adaptation is slow to occur; it may take several hours of darkness for the rod branch to appear and for thresholds to descend to the normal level. By contrast the rate of rhodopsin regeneration as measured by fundus reflectometry appears normal (1–3).

2. Light-dependent fundus discoloration. When in a light environment, the fundus of patients with Oguchi disease has a golden discoloration. Normal fundus colora-

tion is achieved when the patient remains in darkness for a prolonged period. Upon re-exposure to light the golden fundus discoloration returns within about 20–30 minutes. This symptom is termed Mizuo-Nakamura phenomenon and has also been reported in X-linked retinoschisis and cone dystrophy patients (4–6). A partial explanation for the phenomenon was provided by de Jong *et al.* (1991) who observed that a similar discoloration can be produced by introducing potassium-containing solutions into the inner retina of cats and monkeys. Given that light-induced neuronal activity is associated with increased extracellular K+ and that the Muller cells normally transport K+ away from the retina, de Jong *et al.* suggested that a high extracellular K+ concentration (and golden fundus) might be created by excess K+ formation or decreased uptake of K+ by Muller cells (4).

3. Abnormal rod electroretinogram. After 1 hr of dark adaptation, dim blue light flashes fail to elicit a rod b-wave in Oguchi patients whereas white light produces a cornea-negative reponse. Following prolonged dark adaptation some patients achieve normal rod b-wave amplitude and implicit time but only in response to one or two flashes of light which suggests that in these patients the test flash is intense enough to cause the rod system to adapt (7). Because of the abnormal rod b-wave it has been suggested that in Oguchi patients the rod-derived signal passes through the retina without appropriate electrical response by Muller cells (3).

When considered collectively the clinical features of Oguchi disease provide clues to the molecular pathology of the disorder. The abnormal electroretinogram and fundus discoloration suggest that excess production of extracellular K+ may occur. Over-activity of the phototransduction pathway might perhaps produce the putative excess K+ and it is well-established that molecular genetic defects in phototransduction genes can cause both congenital stationary night blindness and retinitis pigmentosa (8–11). Finally that the fundus discoloration is light-dependent and dark adaptation is extremely slow suggests the function affected in Oguchi disease may normally be involved in inhibiting rather than stimulating phototransduction pathway activity.

Here we summarise molecular genetic evidence that recessive arrestin gene mutations are responsible for some cases of Oguchi disease (12,13, manuscript in preparation). The 48kDa arrestin protein together with rhodopsin kinase is responsible for quenching light-activated rhodopsin and has been implicated in dark adaptation (14). We speculate that these mutations represent functional null alleles and that the absence of arrestin results in a prolonged lifetime for photo-activated rhodopsin and in abnormal adaptation.

METHODS

This study of Oguchi disease is part of a larger project which aims to identify genes responsible for autosomal recessive forms of inherited retinal disorders by studying pedigrees from the Indian subcontinent which each contain multiple affected individuals. Such pedigrees are capable of providing some independent evidence of linkage, an attribute which is invaluable when studying genetically heterogeneous conditions. Pedigrees containing multiple individuals affected by autosomal recessive conditions such as Oguchi disease appear to be more common in India than in many western countries. There are two main reasons for this phenomenon. Firstly large sibships are common throughout the country and secondly in some regions up to 30% of marriages are consanguineous (15). This combination of factors tends to generate pedigrees in which multiple individuals are homozygous by descent for a particular recessive condition. Furthermore because India has historically had a large and diverse population nation-wide founder effects are unlikely. Hence in the case of Oguchi disease unrelated Indian pedigrees are each likely to involve homozygosity by descent for a different Oguchi allele.

Oguchi disease is also relatively common in Japan (1,2). That country has historically had a relatively small and isolated population and a strong founder effect may explain the relatively high prevalence of Oguchi disease ie apparently unrelated individuals affected by Oguchi disease may be homozygous for an allele originally carried by one specific ancestor.

Here we summarise linkage and mutational screening studies of the arrestin gene defects in some Indian and Japanese cases of Oguchi disease.

RESULTS

The first molecular genetic evidence that the arrestin gene is involved in Oguchi disease was provided by segregation analysis of a consanguineous Indian pedigree in which the parents were third cousins and three of four children were affected by Oguchi disease. The arrestin gene had previously been mapped to chromosome 2q and the three affected siblings were found to be homozygous for markers which mapped to that region. Multipoint analysis of the data enabled full utilisation of the meioses in the inbreeding loop and gave a maximum lodscore of 2.65 for this moderate-sized pedigree (12).

Single strand conformation polymorphism screening of the arrestin gene in the Indian pedigree and in six apparently unrelated Japanese Oguchi patients was subsequently undertaken (13). This gene consists of 16 exons of which exons 2–16 code for the 405 amino acid arrestin protein. The mutational screening was conducted using primers designed from previously published arrestin intron/exon boundary sequences. Because these primers consisted partially of exonic sequence, the entire coding sequence was not screened. This approach failed to identify an arrestin gene mutation in the previously linked Indian pedigree but did detect a bandshift in exon 11 in five of the six Japanese patients. DNA sequencing subsequently revealed a homozygous deletion of nucleotide 1147 in codon 309 of the arrestin gene, predicting a shift in reading frame and premature termination of translation after incorporation of an additional 11 amino acids unrelated to arrestin. This mutation was presumed to result in a functional null allele.

More recently we have screened several small Indian Oguchi pedigrees using the same primers (unpublished data). In one of those pedigrees segregation of an arrestin gene polymorphism indicated two affected siblings had inherited different alleles at arrestin. This finding suggested a different gene is likely responsible for Oguchi disease in that pedigree and indicated that Oguchi disease involves locus heterogeneity. In a second pedigree two siblings had congenital night blindness and a light-dependent fundus discoloration; in addition one sibling had unilateral signs of a more severe retinal dysfunction including visual acuity less than N36, macular degeneration and mottled pigment epithelium. These siblings were both found to be homozygous for a point mutation encoding a codon 193 Arg-to-ter mutation in the arrestin gene (manuscript in preparation). This alteration represents an even more severe truncation than the codon 309 frameshift allele previously detected in the Japanese patients.

DISCUSSION

Retinitis pigmentosa and allied retinal disorders are genetically and clinically heterogeneous conditions (8). Although the precise molecular genetic defects responsible for most cases remain to be identified, it now seems likely that many different molecular pathological pathways contribute to these disorders. Previous studies together with the current research have demonstrated that molecular genetic defects in phototransduction pathway genes are responsible for some cases of congenital stationary night blindness, Oguchi disease and retinitis pigmentosa. A close relationship between these disorders is supported by the occurrence of Oguchi disease and retinal pigmentary degeneration or congenital stationary night blindness without fundus coloration in the same families and even in the same patients (1,16,17).

Here we consider some of the phototransduction gene mutations which have been characterised to date (Table 1) and attempt to discern some general relationships between types of mutation and clinical outcome.

1. Constitutive mild phototransduction activity. The congenital stationary night blindness mutations identified to date are dominant and may possibly result in constitutive mild activity of the phototransduction pathway. The basal activity of the rod photoreceptors in such patients might resemble the response appropriate for a low light environment ie perhaps constitutive light adaptation occurs (9, 18–20).

2. Light-dependent phototransduction over-activity: We speculate that recessive arrestin mutations in Oguchi disease patients may be associated with a prolonged lifetime for photo-activated rhodopsin and failure of light adaptation; the result being an excessive response to light thereby producing the characteristic fundus discoloration.

3. Constitutive downstream phototransduction under-activity: Some autosomal recessive mutations in phototransduction pathway genes are associated with autosomal recessive retinal degeneration and elevated cGMP in animal models and with retinitis pigmentosa in humans(21–25). These defects may perhaps be associated with uncoupling of the phototransduction pathway and disrupted adaptation such that components upstream of the defective protein may be active whereas downstream components may be inactive.

4. Light-independent phototransduction over-activity: A K296E mutation in rhodopsin which causes autosomal dominant retinitis pigmentosa encodes an opsin without a chromophore binding site. This mutant protein constitutively activates transducin *in vitro* and so it was proposed that the mutation acts via over-activity of the phototransduction pathway. This mechanism now seems unlikely as a recent study of transgenic mice found that *in vivo* the mutant opsin was phosphorylated and stably bound to arrestin (26).

Several patients diagnosed with retinitis pigmentosa were shown to be compound heterozygotes for mutations affecting the gene encoding the alpha subunit of the rod cGMP-gated channel (27). In an assay system the mutant alleles were associated with decreased ion channel activity, a phenomenon which might be predicted to mimic phototransduction overactivity (during phototransduction photo-isomerisation of rhodopsin triggers a nerve signal by activating a cascade that closes cation-specific channels in the rod photoreceptor plasma membrane).

If this mechanism were present we would expect such patients to exhibit light-independent fundus discoloration at least in the early stages of the disease. Such fundus discoloration is however noticeable in Oguchi patients because the coloration changes with light exposure rather than because the coloration is distinctively different from that considered normal.

Table 1. Examples of phototransduction pathway gene mutations reported in congenital stationary night blindness, Oguchi disease and retinitis pigmentosa

Gene	Inheritance	Comment	Reference
Congenital stationary night blindness			
Rhodopsin	dominant	A292E, G90D	9,18, 20
Transducin alpha	dominant	G38D	19
Phosphodiesterase beta	dominant	H258N	10,11
Congenital night blindness with fundus discoloration (Oguchi)			
Arrestin	recessive	1147delA, R193ter	13, this study
Retinal degeneration			
Rhodopsin	dominant	various	8
	dominant	K296E	26
	recessive	E249ter	24
	recessive	E150K	25
Phosphodiesterase alpha	recessive	various	22
Phosphodiesterase beta	recessive	various	21
Phosphodiesterase gamma	recessive	mouse knockout	23
Ion channel alpha	recessive	various	27

In summary, some phototransduction pathway gene defects may involve relatively specific effects upon the activity of the phototransduction pathway. Other studies indicate mutations in phototransduction pathway genes can also have effects upon diverse cellular functions such as maturation of proteins in the endoplasm reticulum and formation and shedding of the outer segment discs (8, 28). Further investigations are required to determine which molecular pathological events are critical in determining the nature and severity of associated clinical symptoms.

ACKNOWLEDGMENTS

We wish to thank Australian, British, New Zealand and United States RP societies and the New Zealand Health Research Council and Lottery Grants Board for providing funds to support this project. Collaborations with the Sankara Nethralaya in Madras and Prof Andreas Gal's group in Germany and the families participating in this study of Oguchi disease are gratefully acknowledged.

REFERENCES

1. Carr, R.E., and Gouras, P., 1965, Oguchi's disease, *Arch. Ophthal.* **73**: 646–656.
2. Carr, R.E., and Ripps, H., 1967, Rhodopsin kinetics and rod adaptation in Oguchi's disease, *Invest. Ophthal.* **6**: 426–436.
3. Sharp, D.M., Arden, G.B., Kemp, C.M., Hogg, C.R., and Bird, A.C., 1990, Mechanisms and sites of loss of scotopic sensitivity: a clinical analysis of congenital stationary night blindness, *Clin. Vision Sci.* **5**: 217–230.
4. deJong, P.T.V., Zrenner, E., van Meel, G.J., Keunen, J.E.E., and van Norren, D., 1991, Mizuo phenomenon in X-linked retinoschisis: pathogenesis of the Mizuo phenomenon. *Arch Opthalmol* **109**: 1104–1108.

5. Heckenlively, J.R., and Weleber, R.G., 1986, X-linked recessive cone dystrophy with tapetal-like sheen: a newly recognised entity with Mizuo-Nakamura phenomenon, *Arch. Ophthal.* **104**: 1322–1328.

6. Noble, K.G., Margolis, S., and Carr, R.E., 1989, The golden tapetal sheen reflex in retinal disease, *Am. J. Ophthal.* **107**: 211–217.

7. Berson E.L., 1994, Retinitis pigmentosa and allied diseases, in: *Principles and practice of ophthalmology: clinical practice*, Volume 2 (D.M. Albert, and F.A. Jakobiec, eds), p1228, W.B. Saunders Company, Philadelphia.

8. Dryja, T.P., and Li, T., 1995, Molecular genetics of retinitis pigmentosa, *Hum Mol. Genet.* **4**: 1739–1743.

9. Rao, V.R., Cohen, G.B., and Oprian, D.D., 1994, Rhodopsin mutation G90D and a molecular mechanism for congenital night blindness, *Nature* **367**: 639–42.

10. Gal, A., Orth, U., Baehr, W., Schwinger, E., and Rosenberg, T., 1994, Heterozygous missense mutation in the rod cGMP phosphodiesterase β-subunit gene in autosomal dominant stationary night blindness, *Nature Genet.* **7**: 64–67.

11. Correction: Gal A, Orth U, Baehr W, Schwinger E, and Rosenberg T. Heterozygous missense mutation in the rod cGMP phosphodiesterase β-subunit gene in autosomal dominant stationary night blindness. *Nature Genet.* 1994; **7**: 551.

12. Maw, M.A., John, S., Jablonka, S., Muller, B., Kumaramanickavel, G., Oehlmann, R., Denton, M.J., and Gal, A., 1995, Oguchi disease: suggestion of linkage to markers on chromosome 2q, *J. Med. Genet.* **32**: 396–398.

13. Fuchs, S., Nakazawa, M., Maw, M., Tamai, M., Oguchi, Y., and Gal, A., 1995, A homozygous 1-base pair deletion in the arrestin gene is a frequent cause of Oguchi disease in Japanese, *Nature Genet.* **10**: 360–362.

14. Palczewski, K., Rispoli, G., and Detwiler, P.B., 1992, The influence of arrestin (48K protein) and rhodopsin kinase on visual transduction, *Neuron* **8**:117–126.

15. Bittles, A.H., Mason, W.M., Greene, J. and Rao, A.N., 1991, Reproductive behaviour and health in consanguineous marriages, *Science* **252**: 789–794.

16. Yamanaka, M., 1969, Histologic study of Oguchi's disease: its relationship to pigmentary degeneration of the retina, *Am. J. Ophthal.* **68**: 19–26.

17. Remler, B., Papst, N., and Bopp, M., 1988, New findings in Oguchi disease, *Klinische Monatsblatter fur Augenheilkunde* **192**: 239–43.

18. Dryja, T.P., Berson, E.L, Rao, V.R., and Oprian, D.D., 1993, Heterozygous missense mutation in the rhodopsin gene as a cause of congenital stationary night blindness, *Nature Genet.* **4**: 280–283.

19. Dryja, T.P., Hahn, L.B., Reboul, T., and Arnaud, B., 1996, Missense mutation in the gene encoding the α subunit of rod transducin in the Nougaret form of congenital stationary night blindness, *Nature Genet.* **13**: 358–360.

20. Sieving, P.A., Richards, J.E., Naarendorp, F., Bingham, E.L., Scott, K., and Alpen, M., 1995, Dark-light: model for nightblindness from the human rhodopsin Gly-90 -> Asp mutation, *Proc. Natl. Acad. Sci.* **92**: 880–884.

21. McLaughlin, M.E., Erhart, T.L., Berson, E.L., and Dryja, T.P., 1995, Mutation spectrum of the gene encoding the β subunit of rod phosphodiesterase among patients with autosomal recessive retinitis pigmentosa, *Proc. Natl. Acad. Sci.* **92**: 3249–3523.

22. Huang, S.H., Pittler, S.J., Huang, X., Oliveira, L., Berson, E.L., and Dryja, T.P., 1995, Autosomal recessive retinitis pigmentosa caused by mutations in the α subunit of rod cGMP phosphodiesterase, *Nature Genet.* **11**: 468–471.

23. Tsang, S.H., Gouras, P., Yamshita, C.K., Kjeldbye, H., Fisher, J., Farber, D.B., and Goff, S.P., 1996, Retinal degeneration in mice lacking the γ subunit of the rod cGMP phosphodiesterase, *Science* **272**: 1026–1029.

24. Rosenfeld, P.J., Cowley, G.S., McGee, T.L., Sandberg, M.A., Berson, E.L., and Dryja, T.P., 1992, A null mutation within the rhodopsin gene causes rod photoreceptor dysfunction and autosomal recessive retinitis pigmentosa, *Nature Genet* **1**: 209–213.

25. Kumaramanickavel, G., Maw, M., Denton, M.J., John, S., Srikumari, C.R.S., Orth, U., Oehlmann, R., and Gal, A., 1994, Missense rhodopsin mutation in a family with recessive RP, *Nature Genet* **8**: 10–11.

26. Li, T., Franson, W.K., Gordon, J.W., Berson, E.L., and Dryja, T.P., 1995, Constitutive activation of phototransduction by K296E opsin is not a cause of photoreceptor degeneration, *Proc Natl Acad Sci U.S.A.* **92**: 3551–3555.

27. Dryja, T.P., Finn, J.T., Peng, Y.-W., McGee, T.L., Berson, E.L., and Yau, K.-W., 1995, Mutations in the gene encoding the α subunit of the rod cGMP-gated channel in autosomal recessive retinitis pigmentosa, *Proc. Natl. Acad. Sci* **92**: 10177–10181.

28. Berson, E.L., 1996, Retinitis pigmentosa: unfolding its mystery, *Proc. Natl. Acad. Sci.* **93**: 4526–4528.

35

A PATIENT WITH PROGRESSIVE RETINAL DEGENERATION ASSOCIATED WITH HOMOZYGOUS 1147delA MUTATION IN THE ARRESTIN GENE

Y. Wada, M. Nakazawa, and M. Tamai

Department of Ophthalmology
Tohoku University School of Medicine
1-1 Seiryou-machi, Aoba-ku, Sendai 980-77, Japan

INTRODUCTION

Oguchi disease is known as a type of autosomal recessive congenital stationary night blindness with a golden yellow-reflex of the fundus(1–3). Characteristic clinical features are Mizuo-Nakamura phenomenon(4) and extremely retarded dark adaptation of rod photoreceptors. The cone function usually appears normal.

Recently, linkage analysis on an Indian family showed that the locus of Oguchi disease was on chromosomal 2q(5), in which arrestin gene is located(6). After the report, mutation screening was performed with Japanese patients with Oguchi disease to search for a mutation in the arrestin gene and 5 unrelated Japanese patients from 6 independent families have the same mutation in the arrestin gene, designated 1147delA, which is an identical 1-base pair (bp) deletion at codon 309(7).

After that, we reported phenotypic characteristics of Japanese patients with Oguchi disease associated with1147delA mutation(8). Our patients showed variable expressivity as for visual acuity, fundus appearance, and 30 Hz-flicker ERG.

To search genotype-phenotype correlation with the 1147delA mutation, we performed further clinical and molecular genetic investigation on a family with Oguchi disease.

PATIENTS AND METHODS

We screened 4 siblings of a family. One of these cases had been diagnosed as Oguchi disease and identified to have the1147delA mutation in the arrestin gene. So we screened genomic DNA samples of the other 3 siblings to search for the 1147delA mutation in exon 11 by polymerase chain reaction (PCR) followed by non- radioisotopic single

Degenerative Retinal Diseases, edited by LaVail *et al.*
Plenum Press, New York, 1997

strand conformation polymorphism (SSCP) analysis(9) with modification(10). The DNA fragment that showed abnormal mobility on SSCP was sequenced to identify the mutation. For sequence analysis, the DNA fragment was cloned into a plasmid vector (PGEM; Promega, USA) and subsequently sequenced with an automated DNA sequencer (A.L.F. DNA sequencer, Pharmacia, Sweeden)

Ophthalmic examination included corrected visual acuity, slit-lamp examination, fundus examination, electroretinography (ERG), dark adaptometry and fluorescein angiography. The ERG testing was perfomed under controlled conditions, as reported previously(11,12). For standard recordings, ERGs were obtained with the use of a dim blue flash in a regular dark adapted condition (30 minutes) to record rod-isolated responses, a standard flash (20J) under a regular dark-adapted state to record maximum responses of rods and cones, and a single flash or a 30 Hz flicker stimulus of red light under light-adapted conditions to isolate cone responses. As described above, one of the four siblings was a proband who had been diagnosed to have Oguchi disease associated with the 1147delA mutation in the arrestin gene. In addition to the proband, the subsequent ophthlmologic analyses disclosed that her older brother had progressive retinal degeneration. Therefore, we further concentrated to analyze ophthalmic and molecular genetic studies on this patient.

Case Report

A 58-year-old man had initially noticed nightblindness during his early teens. His parents were first cousins and his younger sister was diagnosed as Oguchi disease associ-

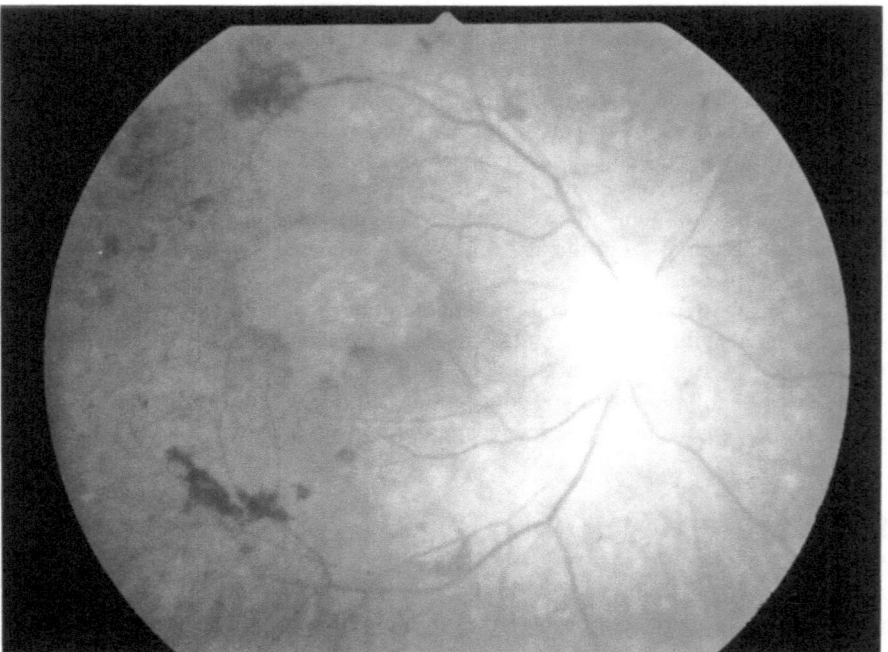

Figure 1. Fundus photograph of the case shows the pale disc, attenuated retinal vessels, retinal pigment epithelial atrophy with apparent pigmentation. Sharply demarcated chorioretinal atrophy in the macula. Retinal degeneration occurred mainly from the posterior pole to mid peripheral retina.

ated with 1147delA mutation. He has realized the progress of restricted visual field for these ten years. He had first visited our clinic in July 1996, at the age of 58-year-old. His corrected visual acuity was 0.6 O.D. and 0.3 O.S. Slit lamp examination disclosed normal appearances of cornea, anterior chamber, and iris bilaterally. Lenses had posterior subcapsular cataract in both eyes. Ophthalmoscopic examination showed the pale disc, attenuated retinal vessels, retinal pigment epithelial atrophy with apparent pigmentation and sharply demarcated chorioretinal atrophy in the macula. Retinal degeneration occurred mainly from the posterior pole to mid- peripheral retina. The far peripheral retina appeared to be relatively good color, although this area showed mild degeneration. Fluorescein angiography showed diffuse hyperfluorescent areas mainly in the pericentral retina corresponding to the atrophy in the retinal pigment epithelial layer. Hypofluorescent areas corresponded to chorioretinal atrophy along the vascular arcade. On ERG testing, scotopic ERG and photopic ERG showed non-recordable amplitudes, but both dark-adapted maximal rod-cone ERG and 30Hz-flicker ERG disclosed severely reduced amplitudes although they were still recordable.

RESULTS

The DNA fragments of the proband with Oguchi disease(lane2) and the present case with progressive retinal degeneration (lane1) showed abnormal band shifts, indicating homozygous mutation of the arrestin gene.analysis. Nucleotide sequencing analysis demonstrated an identical homozygous 1-bp deletion in codon 309 (1147delA) in both of affected patients.

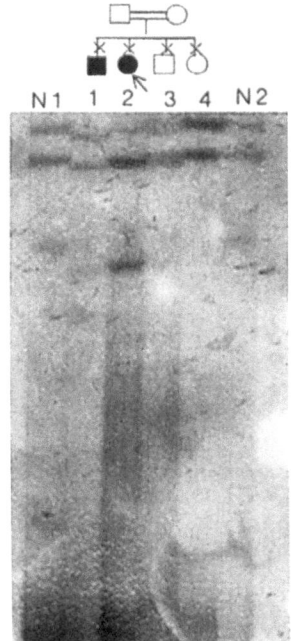

Figure 2. Results of non-radioisotopic SSCP. N1, N2 show normal controls. Lane 2 indicates the proband with Oguchi disease; lane 1 shows the present case with progressive retinal degeneration; lane 3 and lane 4 show clinically non-affected siblings.

DISCUSSION

Oguchi disease has been known as one form of the congenital stationary nightblindness that does not progress to damage visual acuity and visual field. Recently, mutation screening revealed that Japanese patients with Oguchi disease frequently had the 1147delA mutation in the arrestin gene. For the 1147delA mutation we reported that slight variability had been observed in the distribution of the typical fundus discoloration, in visual acuity, and in amplitude of 30 Hz-flicker ERG(8). Some patients showed decreased visual acuity, and in amplitude of 30 Hz-flicker ERG. Some patients showed decreased visual acuity, and reduced of 30 Hz-flicker ERG. In the present study, from the clinical features, our patient with 1147delA mutation could be diagnosed with progressive retinal degeneration. And the younger sister who had been diagnosed as Oguchi disease had the same mutation.

Now it is well known that mutations in the peripherin/RDS gene produce rod or cone predominant retinal degeneration, depending on the kinds and locations of mutations. It has been reported that 1147delA mutation frequently causes Oguchi disease(7,8). However in this study, our patient with 1147delA mutation indicated progressive retinal degeneration. This results suggests that 1147delA mutation in the arrestin gene can cause not only Oguchi disease but also progressive retinal degeneration. For our patients we will have to perform mutation screening on other candidate genes, such as peripherin/RDS, rhodopsin, ROM1, and bPDE genes, because it is possible that another mutation may cause retinal degeneration. To clarify these results, further molecular genetic analysis is needed on patients with retinitis pigmentosa with a particular focus on the arrestin gene.

REFERENCES

1. Oguchi C., 1907, Über eine Abart von Hemeralopie, *Acta Soc Ophthalmol Jpn.* 11: 123–134.
2. Carr RE, Gouras P., 1965, Oguchi disease, *Arch Ophthalmol.* 73: 646–656.
3. Carr RE, Ripps H., 1991, Rhodopsin kinetics and rod adaptation in Oguchi disease, *Invest Ophthalmol.* 6: 426–436.
4. Mizuo G., 1913, A new discovery in dark adaptation in Oguchi disease, *Acta Soc Ophthalmol Jpn.* 17: 1148–1150.
5. Maw MA, John S, Jablonka S, et al., 1995, Oguchi disease: suggestion of linkage to markers on choromosome 2q, *J Med Genet.* 32: 396–398.
6. Valverde D, Bayes M, Martinez L, et al., 1994, Genetic fine localization of the arrestin (S-antigen) gene 4cM distal from D2S172, *Hum Genet.* 94: 193–194.
7. Fuchs S, Nakazawa M, Tamai M, Oguchi, Oguchi Y, GalA., 1995, A homozygous 1-base pair deletion in the arrestin gene is a frequent cause of Oguchi disease in Japan, *Nature genet.* 10: 360–362.
8. Nakazawa M, Wada Y, Fuchs S, Gal A, Tamai M., Oguchi disease: Phenotypic Characteristics of Patients with the Frequent 1147delA Mutation in the Arrestin Gene, *Retina. (in press)*
9. Nakazawa M, Kikawa E, Chida Y, Shiono T, Tamai M., 1993, Nonradioactive single strand conformation polymorphism (PCR-SSCP); a simplified method applied to a molecular genetic screening of retinitis pigmentosa, In: Hollyfield JG, Lavil MM, Anderson RE eds. *Retinal Degeneration: Clinical and Laboratory Application.* New York, NY: Plenum Publishing Corp; 181–188.
10. Kikawa E, Nakazawa M, Chida Y, Shiono T, Tamai M., 1994, A novel mutation (Asn244Lys) in the peripherin/RDS gene causing autosomal dominant retinitis pigmentosa associated with bull's-eye maculopathy detected by nonradioisotopic SSCP, *Genomics.* 20: 137–139.
11. Nakazawa M, Kikawa E, Kamio K, Chida Y, Shiono T, Tamai M., 1994, Ocular findings in patients with autosomal dominant retinitis pigmentosa and transversion mutation in codon 244 (Asn244Lys) of the peripherin/RDS gene, *Arch Ophthalmol.* 112: 1567–1573.
12. Nakazawa M, Kikawa E, Chida Y, Wada Y, Shiono T, Tamai M., 1996, Autosomal dominant cone-rod dystrophy associated with mutations in codon 244 (Asn244His) and codon 184 (Tyr184Ser) of the peripherin/RDS gene, *Arch Ophthalmol.* 114: 72–78.

bFGF TRANSFECTED IRIS PIGMENT EPITHELIAL CELLS RESCUE PHOTORECEPTOR CELL DEGENERATION IN RCS RATS

M. Tamai, K. Yamada, N. Takeda, H. Tomita, T. Abe, S. Kojima, and S.-I. Ishiguro

Department of Ophthalmology
Tohoku University School of Medicine
Sendai 980-77, Japan

SUMMARY

We could insert rat bFGF-cDNA into a high-expression vector, pCXN2 and transfected it into cultured rat iris pigment epithelial cells (IPE). They showed high level of expression of mRNA of bFGF in vitro. These gene-modified iris PE were transplanted into the subretinal space of dystrophic RCS rat and could protect photoreceptors from their early death. In the future, this gene regulation technique could be applied for modifying DNA of iris or retinal PE and obtaining suitable characteristics for certain therapeutic purposes. Then, they could be transplanted in the subretinal space and prolong the survival period of photoreceptor cells or rescue from apoptosis.

INTRODUCTION

The vision is obtained by the function of the neural retina but it is maintained by the various functions of the retinal pigment epithelium (RPE)(1). In the last 10 years, the development of intraocular surgical procedure has made it possible to manage the subretinal neovascular membrane of age-related macular degeneration(ARMD)(2–4). After surgical excision of such neovascular membrane, it has become clear that RPE is severely damaged and the recovery of visual function is very limited in most cases. That means the surgical excision is not the final treatment but just the first step and we must care the dystroied or excisied RPE with some method.

Rescue effects of neurotrophic or growth factors have been reported in the central nervous system (5) and also in the retina (6). These results have suggested that one of these cytokines may be useful for getting some survival effects of photoreceptors. Also that means it might be possible to use them as one of modalities of treatment of ARMD. Some cytokines with such growth effects, especially basic fibroblast growth factor (bFGF) and its receptor have been shown to be produced in RPE and function as autocrine or paracrine fashion (7,8). So if RPE, whether it is auto- or allograft, could be transplanted in the subretinal region, they may secrete growth factors and it gives some important effects to photoreceptors in paracrine fashion and could rescue and keep photoreceptor cells alive and give visual function. If the secretary function of transplanted cells is larger, the rescue effects may be stronger.

Recently some institutes have been reported to try RPE transplantation for these cases (9–11). They have used RPE obtained from eye bank eye or fetus eye but in such cases, we must always be careful for tissue rejection. If we avoid such immunological reaction against allograft, we must use autograft, that means, RPE of patient themselves. But it is not so easy for getting them. If the Iris pigment epithelium (IPE) could work as a substitute for RPE, it must be very helpful for promoting transplantation, because it is easy for ophthalmologists to obtain the IPE by simple technique "peripheral iridectomy."

MATERIALS AND METHODS

This study was done in adherence with the ARVO Statement for the Use of Animals in Ophthalmic and Vision Research.

Animals and Cell Culture

After removing the cornea of 5 adult Long Evans rats under the anesthesia with xylazine hydrochloride (3.3mg/kg) and ketamine hydrochloride (66mg/kg) mixture intramuscularly, irises were excised. These irises were incubated with trypsin (0.25%)/ethylenediaminetetraacetic acid (EDTA)(0.53mM)(Gibco)solution in calcium and magnesium-free Dulbecco's phosphate buffered saline (DPBS-) for one hour. By performing scraping under visualization of the dissecting microscope, it was centrifuged at 1000 rpm for 5 min and IPE cell pellet was resuspended and incubated with F12 medium with 20% fetal bovine serum (FBS). The media were changed every 3 days.

Expression Vector

We used pCXN2 vector for permanent expression of bFGF. This vector, developed by Niwa et al. (12), allows efficient selection for transfectants that express foreign genes at high levels and was kindly supplied by Prof. Miyazaki, Institute of Aged Medicine, Tohoku University. The vector was digested by a restriction enzyme, EcoRl and was ligated with rat bFGFcDNA. The rat bFGFcDNA was reported by Kurokawa et al. (13) and kindly supplied by Takeda Phamaceutical Co.Ltd (Tsukuba,Japan).

Transfection

pCXN2-bFGF or pCXN2 vector was diluted in F12 medium and mixed with LipofectAMINE reagent (Life Technologies, Inc.,Gathersburg, MD) and incubated for 30 min in room temperature. The rat IPE was seeded in the 10 cm plate and cultured until obtaining 80% confluency, then washed 2 times with FBS free F12 medium and replaced with the DNA-liposome complex for 5 hours. The equal volume of F12 medium with 40%FBS was added and changed it after 24 hours. Seventy two hours after the transfection, the medium was replaced with 20% FBS -F12 medium containing 400ug/mi, G418 and the selection was started. Usually, it took 1–2 weeks for selection

Measurement of bFGF mRNA by Semi-Quantitative RT-PCR

mRNA was extracted from 3000 cells of pCXN2-bFGF or pCXN2 transfected IPE each, and single strand cDNA was synthesized by reverse transcriptase with random hexamer, then PCR was performed with bFGF and b-actin specific primers.

For bFGF

5'-GGCCACTTCAAGGACCCCAAG-3' and 5'-TCAGCTCTTAGCAGACA-3'

For b-actin

5'-CTACAATGAGCTGCGTGTGG-3' and 5'-CGGTGAGGATCTTCATGAGG-3'

The primer sets amplified 397 base pair (bp) and 313bp, respectively.

Transplantation of Iris Pigment Epithelium

Vector, bFGF-transfected IPE or IPE themselves (1–3 x 10^4 cells per eyeball) was injected into the subretinal space of the upper part of the eye of young RCS dystrophic rats. The transplantation was performed on the postnatal day 21 and the eyeball was obtained 4 weeks later, the postnatal day 49 or 50. Each eyeball was histologically examined by hematoxylene and eosin staining and observed the extent of rescued photoreceptor nuclei.

RESULTS

Gene Expression of bFGF

The expression of beta-actin and bFGF were shown in Figure 1. The expression of bFGF in bFGF-transfected IPE was much stronger than that in vector-transfected IPE, while the expression of beta-actin was the same level in vector or BFGF transformed IPE.

Effects of Transplantation

The effects of transplantation of bFGF-transfected IPE or vector transformed IPE are shown in Figure 2. It is clear that the rescue effect is very strong in the site of transplantation and outer nuclear layer in the other side to the optic disc was very thin. The effect of vector-transfected IPE was also observed and photoreceptors were rescued.

DISCUSSION

As the souce of autologous pigment cells, iris is ideal by the following two reasons. The first, it would be rather easy to obtain from the same patient by local anesthesia and peripheral iridectomy within short period. That means patients were operated in the outpatient clinic several days before the intraocular surgery. Secondly, the amount of obtained PE cells are usually enough for culture and management for transfection.

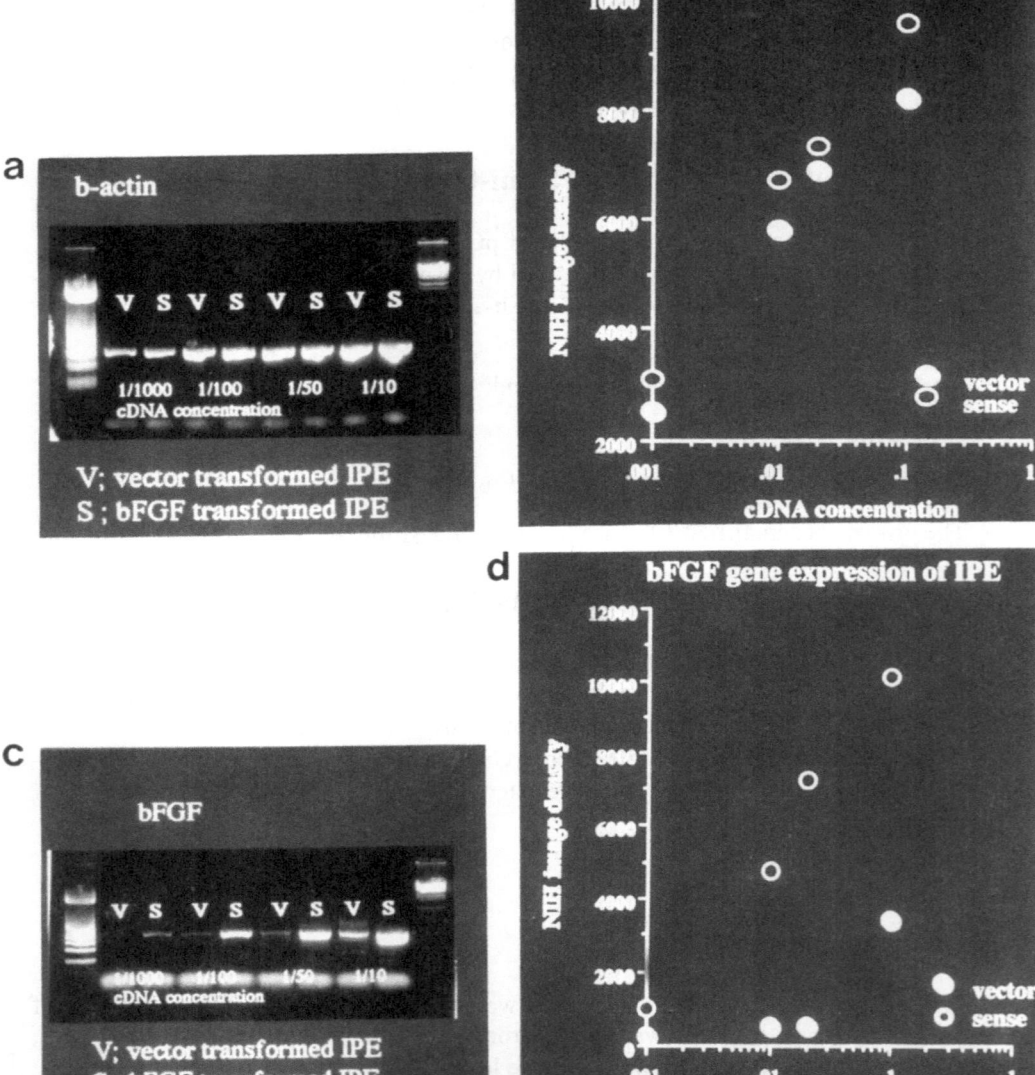

Figure 1. Beta-actin (a,b) and bFGF (c,d) gene expression of the transfected IPE cells. The amount of expression was measured by the NIH image. Beta-actin was the same amount both in vector-(a) and bFGF-transfected IPE(b) but the expresion of bFGF in the latter(d) was much stronger than the former(c).

The IPE is of course different from the RPE even if the origin is the same, so some characteristics of gene expression are different. For example, the former does not express the cellular retinaldehide binding protein (14). But if we use these cells as a kind of biological delivery tool for cytokines or other bioactive materials, they may play their role well.

In recent years, various techniques have been developed for efficient introduction of DNA into cultured cells, and have allowed expression vector for the studies of gene regulation and protein biosynthesis. We used high-expression transfectants with a novel eukaryotic vector, pCXN2, invented by Niwa et al. (12) and introduced rat bFGFcDNA for

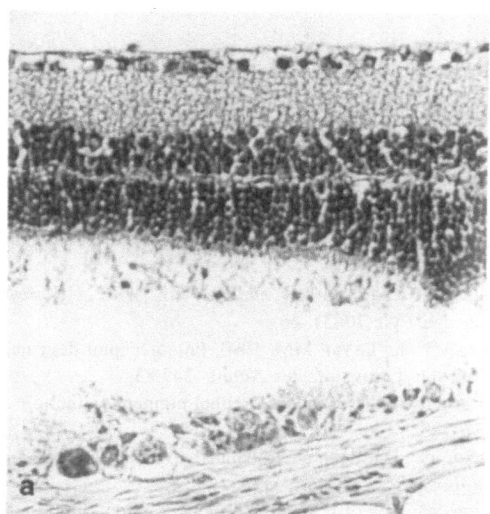

Figure 2. The effect of transplantation of bFGF-transfected IPE in the RCS dystrophtic rat. a. Photoreceptors were very well serviced and both inner and some outer segments like structures were also preserved light microscopically. It is remarkable that in these rescued area, almost no debris of photoreceptor outer segment was found. b. Inferior part of the eye ball, opposite side of the optic nerve showed severe photoreceptor cell death with thick debris of outer segment materials and only one layer of photoreceptors were remained.

stable expression and production of large quantities of bFGF. We could show the expression of high level of mRNA successfully.

The transplantation of these transfected IPE could show active rescue effect of photoreceptor cell death in RCS dystrophic rat. In these in vivo experiments, we could not show distinctive difference in rescue effects in bFGF transfected and vector transfected IPE. If wild IPE were placed in different condition in vivo, it could express more active mRNA. It must be a future work to evaluate more accurately in these conditions.

In the developing animal, cell death occurs in the form of apoptosis and these phenomenon is explained by the competition for limiting amounts of survival signals. If they have enough amount of survival factors, cell death could be decreased. Li et al (15) showed survived photorecetor cells contained much bFGF in the clinical samples of the retinal dystrophy. It is true that retinal dystrophy can be manifested by mutation of phototransduction molecules in photoreceptor outer segment. But it is also true that photoreceptor cell death occurs by the activation of a suicide program in the cells, so called apoptotic mechanism. If bFGF could be delivered to the site where complete disappearance or limitation of trophic factors in photoreceptors, it could help for decreasing their death. Transplantation of IPE with strong productive characteristics of some trophic factors by genetic technologies may be very effective for surviving photoreceptors.

ACKNOWLEDGMENTS

This work was supported by Grant-in-Aid of the Ministry of Education (06404061, 07307016) and by Grant-in-Aid of thejapan Medical Association for M.T.

REFERENCES

1. Bok D,1993, The retinal pigment epithelium: a versatile partner in vision. *J Cell Sci Supp.* 17:189–195,1993.

2. LambertHM, Capone A Jr, Aaberg TM, Sternberg P Jr, Mandell BA and Lopez PE.,1992, Surgical excision of subfoveal neovascular membranes in age-related macular degeneration. *Am. J Ophthalmol.,*113: 257–262.

3. Thomas MA, Grand MG, Williams DF, Lee CM, Pesin S R and Lowe MA, 1992, Surgical management of subfoveal choroidal neovascularization.*Ophthamol.,*99:952–968.

4. Berger AS and Kaplan HJ, 1992, Clinical ecperience with the surgical removal of subfoveal neovascular membranes. *Ophthalmol.,*99: 969–976.

5. Barres BA, Hart IK, Cloes HSR, Burne JF, Voyvodic JT, Richardson WD, and Raff MC., 1992, Cell death and control of cell survival in the oligodendrocyte lineage. *Cell,* 70: 31–46.

6. Faktrovich EG, Steinberg RH, Yasumura D, Matthes MT and LaVail MM, 1990, Photoreceptor degeneration in inherited retinal dystrophy delayed by basic fibroblast growth factor. *Nature.* 347:83.

7. Piftac C, Jones M and Reh TA, 1991,Basic fibroblast growth factor induced retinal pigment epithelium to generate neural retina in vitro. *Development,* 113:577–588.

8. Amin R, Puklin JE and Frank RN, 1994, Growth factor localization in choroidal neovascular membranes of age-related macular degeneration. *Invest . Ophthalmol. Vis. Sci.,* 35:3178–3188.

9. Algvere PV, Berglin L, Gouras P, and Sheng Y. Transplantation of fetal retinal pigment epithelium in age-related macular degeneration with subfoveal neovascularization. *Graefes Arch. Clin. Exp. Ophthalmol.,* 232:707–716,1994

10. Algvere PV, Berglin L, Gouras P and Sheng Y. Human fetal RPE transplants in age related macular degeneration (ARMD). *Invest. Ophthalmol. Vis.Sci.,* 37:S96,ARVO 1996.

11. Das TP, del Cerro M, Lazar ES, Jalali S, DiLoreto DA, Little CW et al. Transplantation of neural retina in patients with retinitis pigmentosa. *Invest. Ophthalmol.Vis. Sci.,* 37: S96,ARVO 1996.

12. Niwa H, Yamamura K-I and Miyazaki J-I., 1991, Feeicientselection for high- expression transfectants with a novel eukaryotic vector, *Gene,* 108:193–200.

13. Kurokawa T, Seno M and Igaarashi K.,1988, Nucleotide sequence of rat basic fibroblast growth factor cDNA. *Nucleic Acids Res.,* 16: 5201.

14. Tamai M.,1996, Retinal pigment epithelial cell transplantation: Perspective, *Nippon Ganka Gakkai Zasshi,* 100 (12): in press, 1996.

15. Li Z-Y,Chang J and Milam AH, 1996, Localization of FGF-2 in normal and RP human retinas. *Invest. Ophthalmol. Vis. Sci.,* 37: S789,1996.

37

TRANSPLANTATION OF NEONATAL NEURAL RETINA IN PHOTORECEPTOR DEGENERATION OF CATS

Kristina Narfström,[1] Lena Ivert,[1,2] Peter Naeser,[2] and Peter Gouras[3]

[1]Department of Medicine and Surgery
Faculty of Veterinary Medicine
Swedish University of Agricultural Sciences
Uppsala
[2]Departments of Ophthalmology and Medical Cell Biology
Uppsala University
Sweden
[3]Edward S. Harkness Eye Institute
Columbia University
New York

INTRODUCTION

Photoreceptor transplantation offers the possibility of restoring vision to a blind retina. Several laboratories have explored the feasability of this approach using allografts in normal animals such as rabbits (1) rats (2) and mice (3) as well as in rats, (4, 5) mice (3), and even in human retinas (6) with degeneration of the photoreceptors.

The Abyssinian cat has been well characterized as an animal model for human Retinitis Pigmentosa (RP) using clinical and laboratory techniques (7–10). In this strain of cats the retina at first develops normally, but then at the age of 1.5 to 2 years, displays slow degeneration. After 2–4 years the retina is completely atrophic. The rods are primarily affected by the defect, but there is a successive involvement also of the cone system. The disorder is inherited as a simple autosomal recessive trait.

The cat eye is large enough to use examination procedures and closed microsurgical techniques that are used in the study and treatment of human retinal disease processes. Because of the recent attempts to employ transplantation techniques in the treatment of such disease, we have been exploring the feasability of fetal neural retinal transplantation into dystrophic mutant cats. A preliminary report of morphologic and functional studies in conjunction with transplantation of neonatal allografts into the cat retina has previously been published (11).

Degenerative Retinal Diseases, edited by LaVail *et al.*
Plenum Press, New York, 1997

METHODOLOGY

Preparation of Microaggregates

Neural retinal tissue was obtained from 3–5 day old domestic short-haired kittens that were unrelated to the dystrophic strain of cats used as recipients. At this early age the kittens have not yet developed outer segments (12). The eyes were removed after euthanizing the kittens with an intraperitoneal injection of pentobarbital sodium. The eye-balls were hemisected at the ora serrata (pars plana region) and the eye-cups were submerged in Hank's solution saturated with 95% oxygen at pH 7.4. The procedure for preparing the microaggregates has been previously described in detail by Gouras et al., 1992 and 1994 (13, 3). In short the neural retina was peeled away from the pigment epithelial cell layer and placed in minimal essential medium (MEM) saturated with 95% O_2 and 5% CO_2. The neural retinas were then gently dissected into microaggregates containing relatively undifferentiated photoreceptors as well as other retinal neurons. Only small aggregates were used for transplantation. The microaggregate suspension was concentrated by sedimentation and sucked into sterilized glass micropipettes with a 0.15–0.2 mm tip diameter, to be used during the transplantation surgery. At least 25–50 microaggregates (Fig.1a and b) were drawn into the pipette in two aliquots separated by an air bubble.

Recipient Animals

Eleven affected dystrophic cats, homozygous for hereditary rod cone degeneration were used as recipients. The cats were between 1–3 years old and at a stage of early dis-

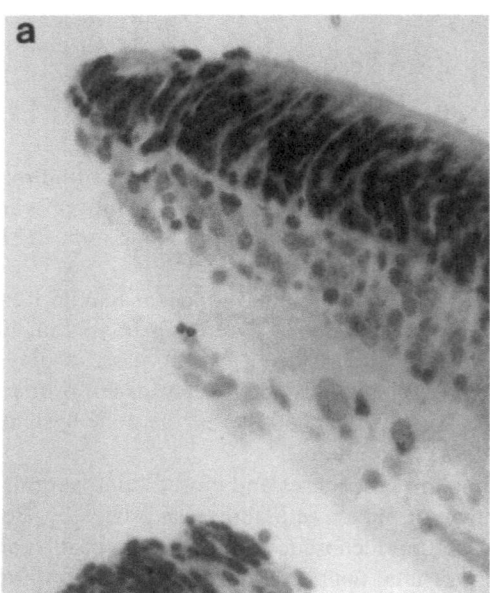 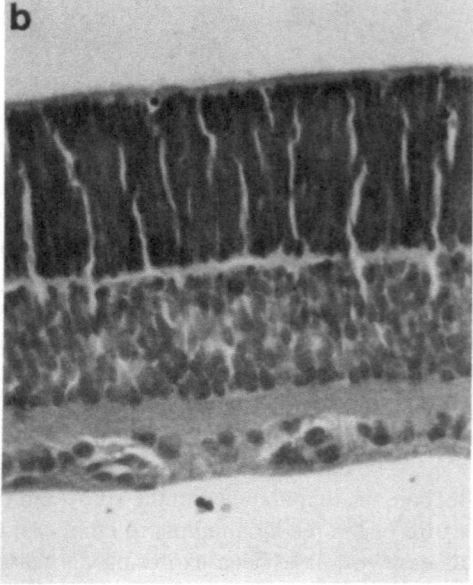

Figure 1. The appearance of two retinal microaggregates embedded in either epon and stained with tolouidine blue (a) or embedded in paraffin and stained with hematoxylin & eosin (H&E) (b). The different retinal layers are easily seen in each section. The retinal pigment epithelium is separated from the neural retina before preparation of the microaggregates, therefore not seen in the sections. Note that the neonatal receptors are undeveloped. (Original magnification: x 304).

ease (8). Usually transplantation was performed in the left eye. Preceeding the transplantation the cats were studied using indirect ophthalmoscopy, electroretinography and fundus photography. Postoperatively the recipient animals were given a single intramuscular injection of prednisolone (0.5 mg/kg) and topical cycloplegics (Isopto-Atropin 0.5% eye drops, Alcon Laboratories, Inc., U.S.A.) and antibiotics (Chloromycetin eye drops, Park-Davies, Park-Davies Scandinavia AB, Sweden) during approximately one week. The cats were followed daily, during the first week and later weekly to monthly until the cats were sacrificed at varying post-operative times (see below).

Microsurgery

Preoperatively the cats were sedated with xylazine, then anesthetized using sodium pentobarbithal intravenously and intubated. General anesthesia was maintained using isoflurane. The temporal lateral canthus area was prepared for surgery. The cats were put on their right side with the left eye under the operating microscope. After a cantothomy the sclera was uncovered and perforated six to seven mm from the limbus over the pars plana area. A glass micropipette was introduced into the eye under direct visualization, and directed towards the central fundus. After gentle perforation of the retina, the injection was performed in the central tapetal area, nasal to the optic disc, usually at 2 locations (one aliquot at each).

Approximately 5–10 µl of solution containing the microaggre–gates were injected into the subretinal space. During the procedure a clearly visible small elevated bleb (about 2 disc diameters) was obtained. Usually there was no bleeding in the area except for a distinct mark at the site of the retinotomy. The sclerotomy was closed with two simple interrupted 9/0 polyglactin sutures. The canto–thomy and conjunctiva were adapted and sutured.

Histology

The cats were sacrificed by an intravenous or intraperitoneal injection of an overdose of pentobarbital sodium at 1 hour, 1 and 2 weeks, 1, 2 (2 cats), 2.5, 3, 4, 12, and 24 months postoperatively. After euthanasia, both eyes were enucleated within 2 minutes and the posterior eye cups immersed into and kept in cold fixative (2% glutaraldehyde in 0.1 M sodium cacodylate buffer) for at least 24 hours. The areas of transplantation as well as specific representative areas (approximately 3 x 4 mm large pieces) from central temporal and nasal, and peripheral tapetal and non-tapetal areas were obtained, postfixed, dehydrated and embedded in epon or paraffin. The pieces were thereafter sectioned and examined by light-microscopy. In this publication the findings in two transplanted cats at post-operative times 1 and 10 weeks will be presented using a light-microscope (Axiophot, Zeiss, Germany). All experiments conformed to the ARVO statement for the Use of Animals in Ophthalmic and Vision Research.

RESULTS

Clinical Findings

In most cases there were no postoperative signs of inflammation in the transplanted eyes. One recipient cat showed signs of low-grade uveitis during the first and second post-

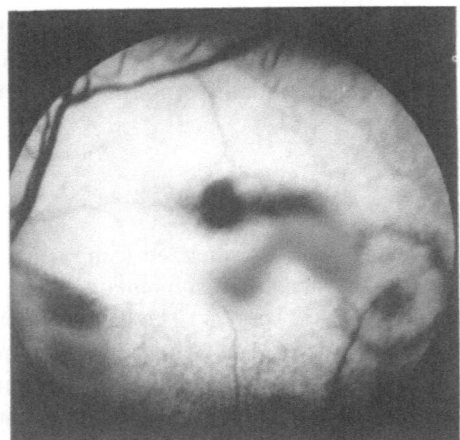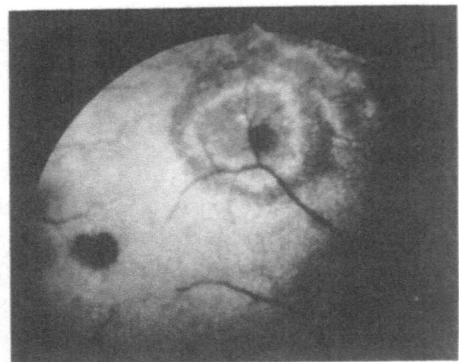

Figure 2. a. The fundus appearance 1 week post-transplantation in cat Athena. Two retinotomies were performed whereof microaggregates were injected subretinally into one of them. A well demarcated circle is seen indicating the extent of neuro-retinal elevation during surgery. Microaggregates were inadvertly deposited intravitreally (instead of through the second retinotomy). b. Fundus of same cat (Athena) at 10 weeks post-transplantation Note the bulls eye appearance at the transplantation site.

operative days. The fundi could easily be inspected in all animals. Fig. 2a shows the fundus of a dystrophic cat 1 week post-transplantation. A well demarcated circle is seen indicating the extent of the neuro-retinal elevation during surgery. Microaggregates were inadvertly deposited intravitreally, instead of through the second retinotomy in this cat. Fig. 2b shows the same cat 10 weeks after surgery, at euthanasia. Note the bulls-eye appearance of the transplantation site. There are no inflammatory signs in the fundus or in the vitreous at this post-surgical time.

Light-Microscopy

By light-microscopy and serial sectioning the retinotomies with adjacent transplanted tissue could be easily found. At one week after transplantation the host neural retina had flattened out over the transplanted cells overlying the host retinal pigment epithelium. The photoreceptor layer and inner retina of the host appeared healthy. There were no signs of rejection or inflammation in the transplant area. Figure 3 is a low magnification light-micrograph 1 week after surgery. There is a discrete boundary between the transplanted tissue and the host retina. In the transplant several rosettes are seen with still immature looking outer nuclear layer cells surrounding a lumen, possibly with inner segment material, seen more clearly in Fig. 4. There are no signs of outer segment formation. The outer retinal layers of the transplanted tissues appear to be more dispersed. There are no signs of inflammation or invasion by macrophages.

Ten weeks after surgery the transplant is separated from the host in part by a cleft (Fig. 5), whereas in some areas the boundary is more discrete. At this post-surgical time most of the host photoreceptors have degenerated and only a sparse number of outer nuclear layer cells are seen overlying the transplant. There is no inflammatory reaction or

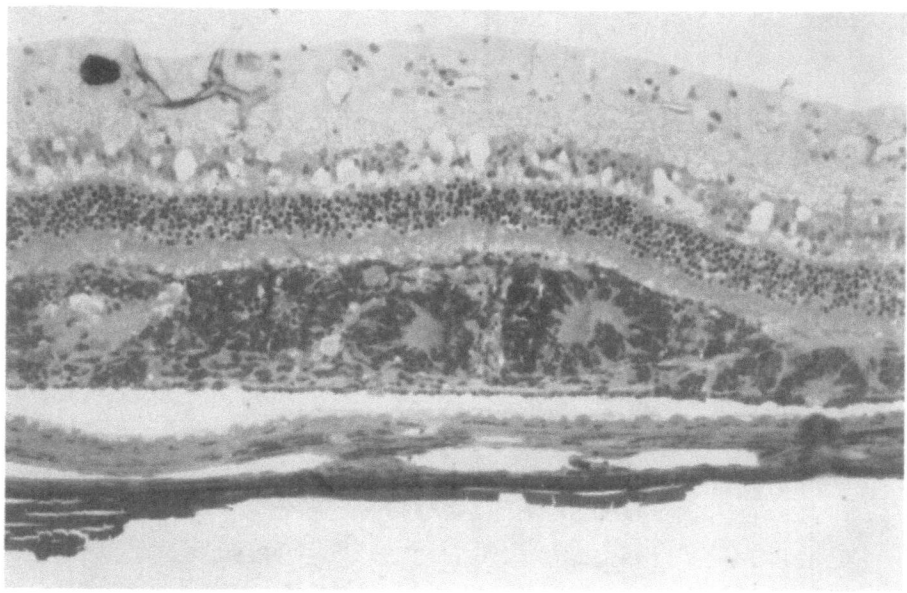

Figure 3. Light-micrograph at 1 week post-transplantation in cat Daim. The host outer nuclear and inner retinal layers are clearly visible. There is a discrete boundary between transplanted tissue and host retina. In the former several rosettes are seen. There are no signs of inflammation in the transplanted area. The section is embedded in epon and toloidine blue stained (x 304).

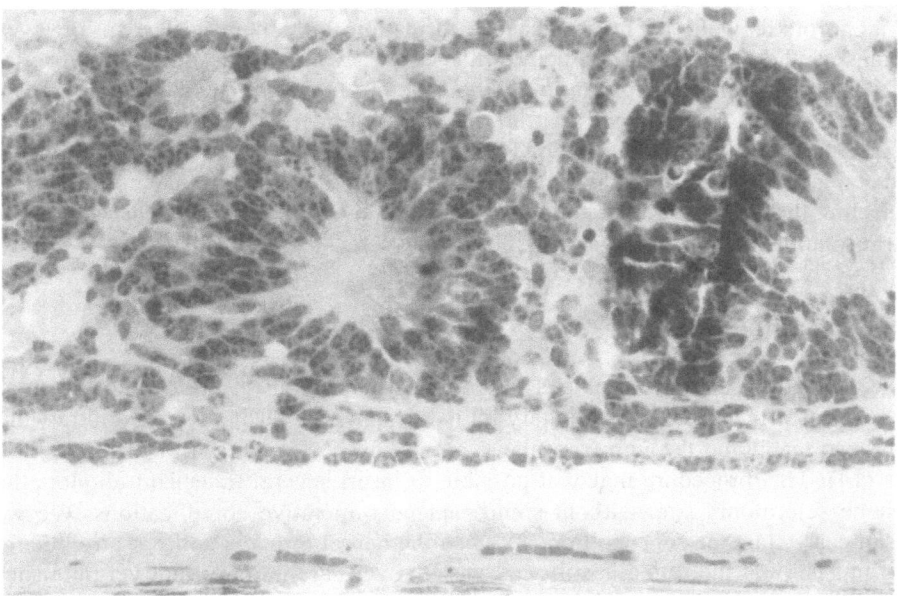

Figure 4. High magnification of central region of transplant seen in Fig. 3. Rosettes of outer retinal layers are formed in which the inner layers do not seem to participate. There is no evidence of photoreceptor outer segment formation yet and no macrophages are observed in or near the rosettes (x 960).

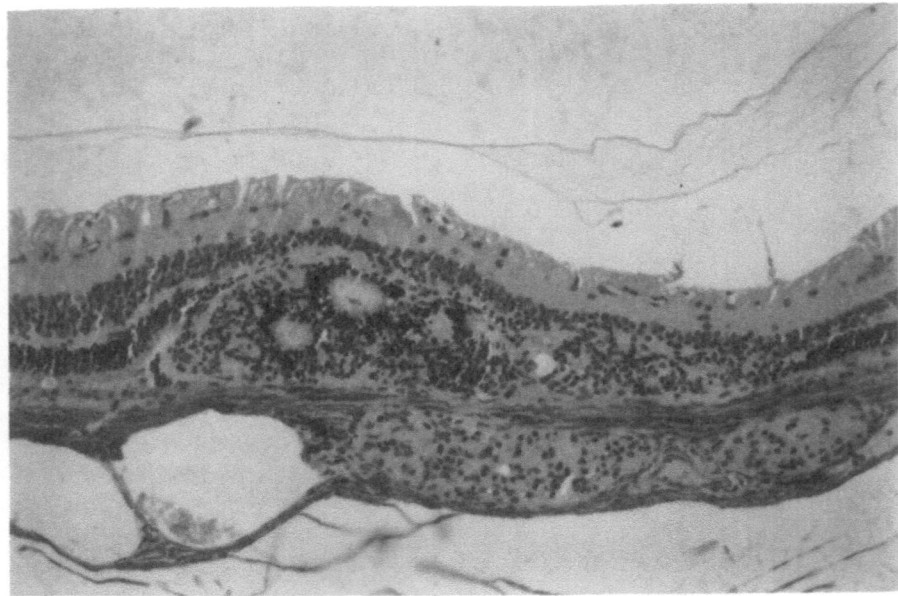

Figure 5. Light micrograph at 10 weeks post-transplantation time in cat (Athena). The transplant is separated from the host in part by a cleft, whereas in some areas the boundary is more discrete, theoretically facilitating cell contacts between transplant and host. At this post-surgical time most of the host photoreceptors have degenerated over the transplant area. No inflammatory reaction is obvious (some transplant has been inadvertly placed in the choroid as well, due to perforation of Bruch´s membrane). The section is embedded in paraffin and H&E stained (x 304).

other indication of a rejection process. Due to perforation of Bruchs membrane during surgery in the recipient illustrated in Fig. 5, some transplanted tissue was by accident placed in the choroid as well. There is no rosette formation in the choroidal area, however, and the transplanted cells appear as a homogeneous mass without signs of adverse reactions.

Ten weeks after transplantation photoreceptor outer segments have developed in the rosettes. There are macrophages within the central lumen of the rosettes (Figs. 6 and 7). Neovascularization is also observed. This neovascularization is seen in the boundaries of the transplant and host tissue as depicted in figs. 7 and 8a. Also in the choroid, seemingly penetrating the retinal pigment epithelial cell layer, there are small vessels at the border of the transplant and the host retina (Fig. 8b).

DISCUSSION

Our results show that it is possible to perform several injections into the subretinal space of cats affected with a hereditary retinal degeneration using a closed intraocular microsurgery technique through pars plana, an approach that is used in mice (13) and humans (14). This procedure makes it possible to insert several transplant aliquots through the same sclerotomy site, without significant post-operative complications. We worked with an early stage of degeneration (8) when the neural retina is still not atrophic, gliosis is not present and the retina is still well vascularized. Preliminary (not yet published) results by the first author, have shown that subretinal injection using the presently described method is difficult to perform in cats at advanced stages of retinal degeneration. Because of lack of normally functioning retinal vessels it might also be more difficult to obtain long-term viable transplants in retinas that have undergone generalized atrophy.

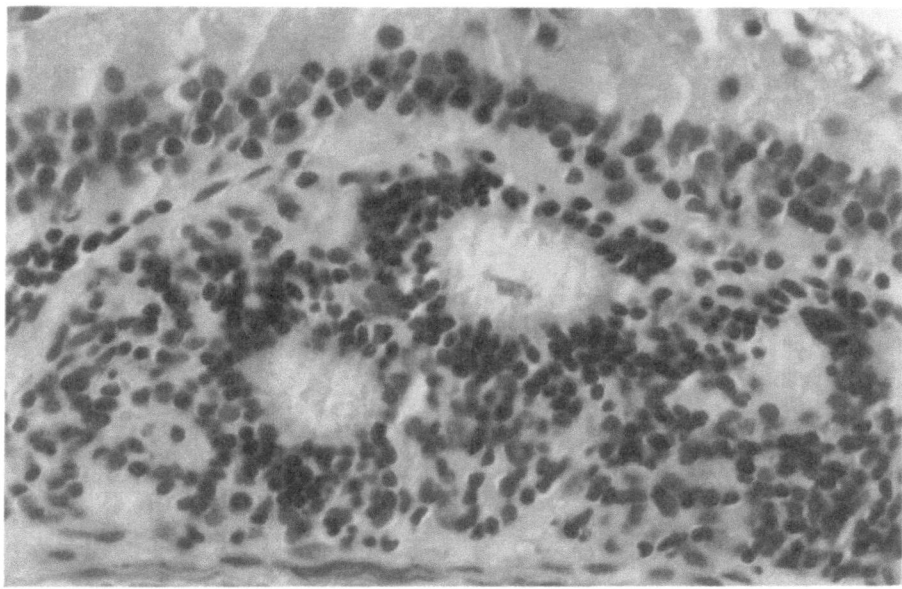

Figure 6. High magnification of central area of transplant seen in Fig. 5. Photoreceptor outer segments are now obvious in the center of the rosettes. A macrophage is seen in the center of one rosette (x 960).

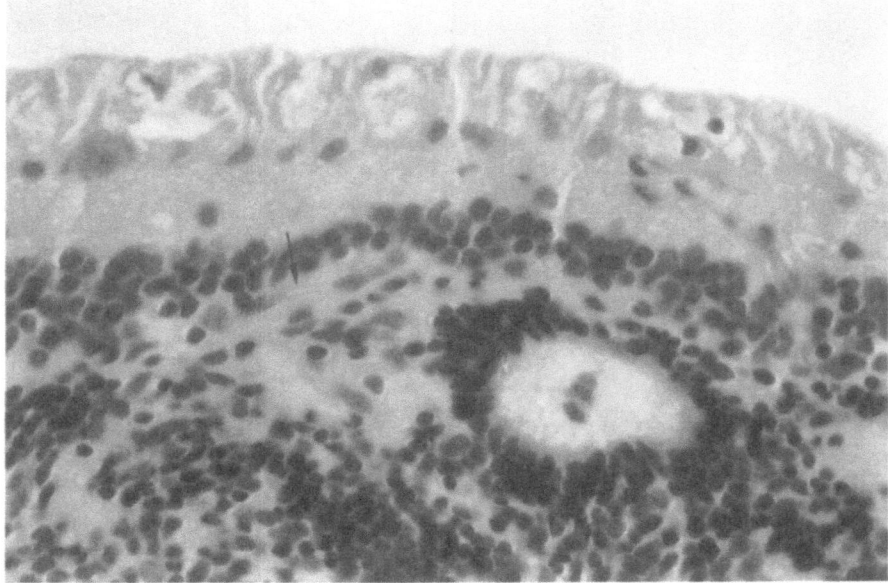

Figure 7. The same transplant shown in figures 5 & 6. Three macrophages are seen in the rosette. Also there are signs of neovascularization (arrow) in the boundary of the transplant and the host (x 960).

Resembling much of the research of others, we find that fetal neural retinal transplants survive in the subretinal space and may develop outer segments observable by light-microscopy. We examined the suggestions of Gouras et al., 1994 (3) that the use of relatively small microaggregates can, when properly oriented, grow on to the host retinal pigment epithelium. In only rare occasions we have confirmed this hypothesis (11). It ap-

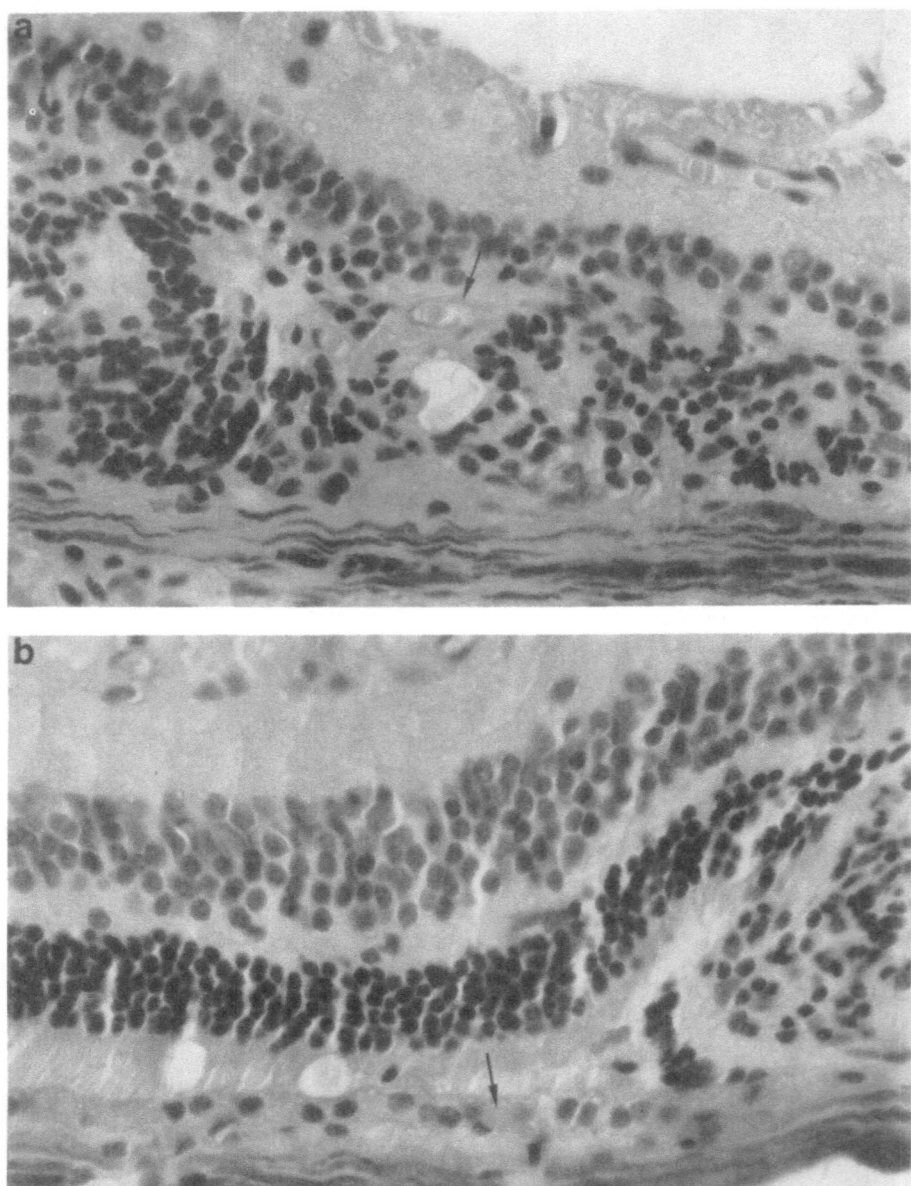

Figure 8. Light micrograph at high magnification of the boundaries of the transplant area seen in Fig. 5. a. Neovascularization can be observed, subretinally in the peripheral area of the transplant (arrow). b. Some neovascularization is also seen in the choroid and appears to penetrate the RPE (arrow) (x 960).

pears, however, that we used somewhat larger microaggregates in our experiments than they did. Our retinal transplants formed rosettes, just as has been described previously by several investigators (15, 1, 16). Other methods may be necessary to prevent rosette formation and facilitate correct orientation in developing neonatal photoreceptors.

Our method also has the disadvantage of including other neural and glial cells in addition to photoreceptors. By light-microscopy it was not possible to distinguish other retinal cell types in these transplants. It was apparent, however, that other cell types were present in the transplants. In this regard it was interesting to note that there was clear lamination of external plexiform layer separating the outer from the inner nuclear layers in the microaggregates (Fig. 1a & b). Such a distinction could not be detected in the transplants. It appears that the photoreceptors tend to form rosettes whereas the more inner retinal layers are dispersed around the rosette, often intimately associated with host retinal tissue. This suggests that connections between transplant and host are possible, as suggested by others (17).

We did not use any markers to identify our transplanted cells. Nevertheless we feel confident that the tissue we have identified within the subretinal space are fetal neural retinal transplants because of the following. At one week after transplantation the outer nuclear layer is clearly evident adjacent to the transplanted tissue in which a distinct and additional photoreceptor layer is evident. It seems impossible to have two very distinct outer nuclear layers be created in the same retina over such a large area by surgical trauma alone. In addition there were no outer segment material apparent in the transplants at 1 week after surgery whereas photoreceptor outer segments were present in the host retina. Furthermore retinal blood vessels could not be identified in the transplant at 1 week and were only observed at the edge of the transplant at 10 weeks after surgery. At this time the host outer nuclear layer is degenerating making the distinction between host retina and transplant difficult. Other factors provide evidence of the identity of the transplant, however. There is often a distinct space that separates much of the transplant from the host tissue and vessel ingrowth is virtually absent within the core of the transplant.

Blood vessel growth into transplants has not been reported before, to our knowledge. It was observed at the edge of the transplant at 10 weeks after surgery. There was also some neovascularization at the level of the host retinal pigment epithelial cells at this time. We believe that the neovascularization may be more of a complication from surgery than a response from the foreign tissue. In fact there was no evidence of any inflammation or rejection of these transplants even when some microaggregates entered the choroid. This confirms the reports of Bergström et al., 1994, (18) that embryonic retinal tissue can survive in the choroid without rejection.

We invariably found macrophages within the core of rosettes at 10 weeks but not at 1 week after surgery. This suggests that the development of outer segments may be responsible for attracting macrophages into the rosettes. This, of course, places them at the most strategic position to encounter outer segment material on all sides. The factor that lures these cells almost precisely in the geometric center of a rosette is most intriguing. Macrophages are commonly seen around degenerating photoreceptor outer segments in various retinal degenerative diseases such as the RCS rat (19) and the Abyssinian cat (10, 20). It is intersting that they are also seen in photoreceptor transplants that are contacting the host retinal pigment epithelium in mice (3) as well as in the center of rosettes shown in the present report.

ACKNOWLEDGMENTS

This work was supported by the Swedish Medical Research Council grant no. 19X-09938 and The Foundation Fighting Blindness (USA). Financial support from the former Laboratory for Comparative Pathology is also greatly acknowledged.

REFERENCES

1. Bergström, A., Ehinger, B., Wilke, K., Zucker, C.L., Adolph, A.R., Aramant, R. and Seiler, M., 1992, Transplantation of embryonic retina to the subretinal space in rabbits, Exp. Eye Res., 55:29–37.

2. Juliusson, B., Bergström, A., van Veen, T. and Ehinger, B., 1993, Cellular organization in retinal transplants using cell suspensions or fragments of embryonic retinal tissue, Cell Transplantation, 2:411–418.
3. Gouras, P., Du, J., Kjeldbye, H., Yamamoto, S. and Zack, D.J., 1994, Long-term photoreceptor transplants in dystrophic and normal mouse retina, Invest. Ophthalmol. Vis. Sci. 35:3145–3153.
4. Silverman, M.S., Hughes, S.E., Valentino, T.L. and Liu, Y., 1991, Photoreceptor transplantation to dystrophic retina, in: Retinal Degenerations, (R.E. Anderson, J.G. Hollyfield, M.M. LaVail, eds.) pp.321–335, CRC Press, Boston.
5. Du, J., Gouras, P., Kjeldbye, H., Kwun, R. and Lopez, R., 1992, Monitoring photoreceptor transplants with nuclear and cytoplasmic markers, Exp. Neurology, 115: 79–86.
6. Das, T.P., del Cerro, M., Lazar, E.S., Jalali, S., DiLoreto, D.A., Little, C.W., Streedharan, A., del Cerro, C. and Rao, G.N.,1996, Transplantation of neural retina in patients with retinitis pigmentosa, Invest. Ophthalmol. Vis. Sci. 37:S96.
7. Narfström, K., 1983, Hereditary progressive retinal atrophy in the Abyssinian cat. J. Heredity, 74:273–276.
8. Narfström, K., 1985, Progressive retinal atrophy in the Abyssinian cat: Clinical characteristics, Invest. Ophthalmol. Vis. Sci. 26:193–200.
9. Narfström, K., Nilsson, S.E. and Anderson, B.-E., 1985, Progressive retinal atrophy in the Abyssinian cat. Studies of the DC-recorded electroretinogram and the standing potential of the eye, Br. J. Ophthalmol. 69:618–623.
10. Narfström, K. and Nilsson, S.E., 1986, Progressive retinal atrophy in the Abyssinian cat: Electron microscopy. Invest. Ophthalmol. Vis. Sci. 27:1569–1576.
11. Narfström, K., Ivert, L., Li, W. and Gouras, P., 1995, Function of photoreceptor transplants in the hereditary rod-cone degeneration of the Abyssinian cat, Invest. Ophthalmol. Vis. Sci. 36:S253.
12. Tucker, G.S., Hamasaki, D.I., Labbie, A. and Muroff, J, 1979, Anatomic and physiologic development of the photoreceptor of the kitten, Exp. Brain Res. 37:459–474.
13. Gouras, P., Du, J., Kjeldbye, H., Yamamoto, S. and Zack, D. 1992, Reconstruction of degenerate rd mouse retina by transplantation of transgenic photoreceptors, Invest. Ophthalmol. Vis. Sci. 33:2579–2586.
14. Algvere, P., Berglin, L., Gouras, P. and Sheng, Y., 1994, Transplantation of fetal retinal pigment epithelium in age-related macular degeneration with subfoveal neovascularization, Graefe's Arch. Clin. Exp. Ophthalmol. 232:707–716.
15. del Cerro, M., Kordower, J., Lazar, E., Olschowka, J.A., Grover, D. and del Cerro, C., 1991, Development of rod and cone photoreceptors in intraretinal grafts, (R.E. Anderson, J.G. Hollyfield, M.M. LaVail, eds.), pp. 303–311, CRC Press, Boston.
16. Aramant. R., Seiler, M., Ehinger, B., Bergström, A., Adolph, A.R., Gustavii, B. and Brundin, P., 1991, Transplanting embryonic retina to the retina of adult animals, in: Retinal degenerations (R.E. Anderson, J.G. Hollyfield, M.M. LaVail, eds.), pp. 275–288, CRC Press, Boston.
17. Zucker, C.L., Ehinger, B., Seiler, M., Aramant, R.B., Adolph, A.R., 1994, Ultrastructural circuitry in retinal cell transplants to rat retina, J. of neurotransplantation & plasticity, 5:17–29.
18. Bergström, A., 1994, Embryonic rabbit retinal transplants survive and differentiate in the choroid, Exp. Eye Res., 59:281–289.
19. Essner, E., Gorrin, G., 1979, An Electron microscopic study of macrophages in rats with inherited retinal atrophy, Invest. Ophthalmol. Vis. Sci., 18:11–25.
20. Narfström, K., Nilsson, S. E., 1989, Morphological findings during retinal development and maturation in hereditary rod-cone degeneration in Abyssinian cats. Exp. Eye Res., 49:611–628.

38

MECHANICAL ASPECTS OF RETINAL PIGMENT EPITHELIAL TRANSPLANTATION

Devjani Lahiri-Munir,[1] Lichun Lu,[1] Charles A. Garcia,[1] Antonios Mikos,[2] and Emily Aguilar[1]

[1]Department of Ophthalmology and Visual Science
University of Texas Health Science Center
Houston, Texas
[2]Institute of Biosciences and Bioengineering
Rice University
Houston, Texas

INTRODUCTION

Retinal pigment epithelium (RPE) dysfunctionis implicated in various blinding diseases including Age-related Macular degeneration (ARMD) and Stargardt's disease. RPE transplantation is currently under investigation as a potential therapy for degenerative and/or hereditary retinal diseases. As RPE cells are polar with specific apical and basal characteristics, proper orientation of the grafted cells is expected to have an impact on the success of the transplant and rapid re-establishment of structural integrity of the host retina. Reasonable success has been achieved by randomly injecting RPE cell suspensions in the rescue of degenerating photoreceptors in Royal College of Surgeons rats. These studies have shown the transplanted cells to be functional evaluated by outer segment phagocytosis and photoreceptor transduction (1–5). But in these studies, no consideration was given to the fact that these dissociated cells settled randomly in the host retina often wandering in the subretinal space. Morever, the structural and functional polarity of RPE has also been ignored. As a possible treatment for conditions such as in ARMD, where degeneration of the retina is localized, use of correctly oriented, targeted RPE implants is the ideal method by which diseased retina can be restored to health. There have been other reports describing the use of substrates for RPE transplants (6–11), but none of the substrates were consistently proven effective. Some of the disadvantages of previous substrates include thickness of the substrate, poor degradability, rigidity causing retinal damage and poor permeability. Biodegradable polymers have been used as temporary scaffolds for hepatocyte (12–15)(Cima), fibroblast (Saltzman), chondrocyte (freed) and osteoblast (Ishaug) cultures. In this study, biodegradable polymer sheets as a matrix for RPE grafts have been employed. Polymers such as polyglycolic and polylactic acid are selected due to properties that include biodegradability, maleability and permeability. We have evaluated the polymer matrices as substrates for RPE cell culture, the technique for implanting such polymer/RPE cell complexes in the subretinal space and their subsequent inteactions with the adjacent retina and Bruch's membrane.

MATERIALS AND METHODS

Manufacture of Polymer Substrates

Two poly (-hydroxy esters) were used in this study: average molecular weight (MW=44700±700) 50:50 PLGA, and 75:25 PLGA (MW=74600±700).Each of the polymers were formed into thin films using a solvent casting technique (16)(Thompson). Briefly, 50 mg of raw polymer was dissolved in 2ml of chloroform. The solution was then cast onto circular, glass, cover slips (diameter 12mm) placed on a leveled table. Sufficient polymer solution was used to completely cover the glass. The samples were left under the hood for 8 hours to let the solvent evaporate and were then vacuum dried. The polymer film samples were stored in a nitrogen atmosphere, over anhydrous $CaSO_4$ (Drierite) at −5° until use. The polymer films were sterilized under UV light for 90 min prior to cell culture.

RPE Cell Culture

For transplants, human fetal eyes (10–16 weeks gestational) were obtained following therapeutic abortion. RPE was dissected and tranferred to 35 mm tissue culture dishes in medium 199 supplemented with 20% fetal bovine serum and other additives (17). Polymer film substrates were placed in 24-well plates and 32×104 cells/ml medium were plated on top of each substrate and on tissue culture wells for controls.

Surgical Implantation

In this study xenografts of fetal human RPE cultured on the polymer substrates were used. Young Sprague-Dawley rats were anesthetized using 50 mg/Kg body weight of Nembutal intraperitoneally. Xylocaine was injected into the retrobulbar space of the right eye. Pupils were dilated with 2.5% neosynephrine and 0.5% tropicamide. A modified version of the original closed-eye technique (1) was used. The eye was gently proptosed and a posterior incision was made through the sclera, choroid and peripheral retina. A knife was inserted at an angle, through the incision, avoiding the lens, to make a small retinotomy 1–2 DD inferior to the disc. The knife was then withdrawn and the RPE/polymer sheet was directed towards the site of the retinotomy and placed under the retina in the subretinal space in correct orientation. The eyes were monitored ophthalmoscopically. The transplanted animals were sacrificed at 1 week and 2 weeks and eyes were enucleated. Rat eyes were fixed in 2.5% glutaraldehyde and 2% paraformaldehyde in 0.1M phosphate for light microscopy Fixed eyes were rinsed and dehydrated in a series of alcohol grades and embedded in Epon. Semithin sections were stained with toluidine blue.

RESULTS

Polymer Substrates

Films of poly(lactic-co-glycolic acid) (PLGA) 75:25 with molecular weight 74600±700 and 50:50 with molecular weight of 44700±700 were manufactured. The films had a thickness of less than 20 m and had reproducible surface morphologies and mechanical properties. These polymer sheets are biocompatible and approved by the Food and Drug Administration. They degrade by hydrolysis into lactic and glycolic acids, which can be removed from the body by normal metabolic pathways.

RPE Cell Culture

Surgical Implantation. The surgical method employed proved to be highly effective. Ophthalmoscopic examination revealed within 24 hrs of implantation, the retina as reattached. No trauma or clinical signs of inflammation were noted at anytime. The small retinotomy at the transplant site healed effectively and no retinal detachment was noted. The peripheral incision was small enough that it healed without suturing. The lens and vitreous remained undamaged by the procedure.

Light Microscopy. In Vitro. Human fetal cells were found to cover large areas of each polymer substrate after 2 days in culture. Cells contained melanin granules and each substrate demonstrated a combination of tightly packed cells with areas of loosely arranged cells. The tightly packed cell groups assumed a more typical cuboidal cobblestone morphology of RPE cells. Fetal RPE cells were positive for anticytokeratin 8 and 18 indicating epithelial characteristics (17).

In Vivo. Similar to in vitro, the polymer film had a globular apearance and appeared transparent in the subretinal space. By 1 week post-transplantation, the polymer substrate had broken up into smaller pieces of varying dimensions (Figure 1). The smallest pieces were in the form of small droplets and had an appearance of polystyrene beads. The degradation of the polymers followed a similar pattern as observed in vitro. Due to the hydrophilic nature of the copolymer it expanded initially and then broke up into smaller pieces

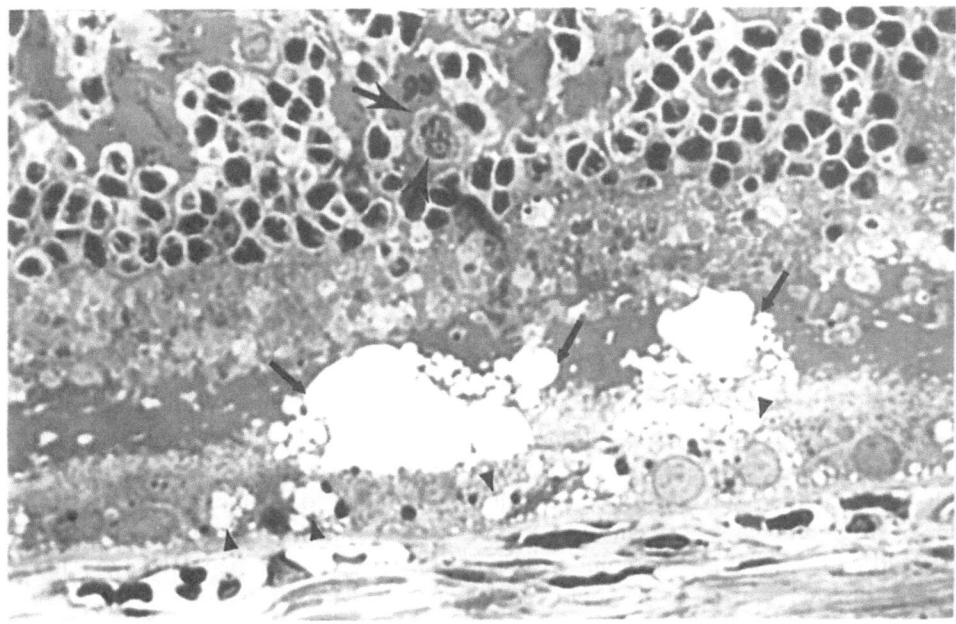

Figure 1.

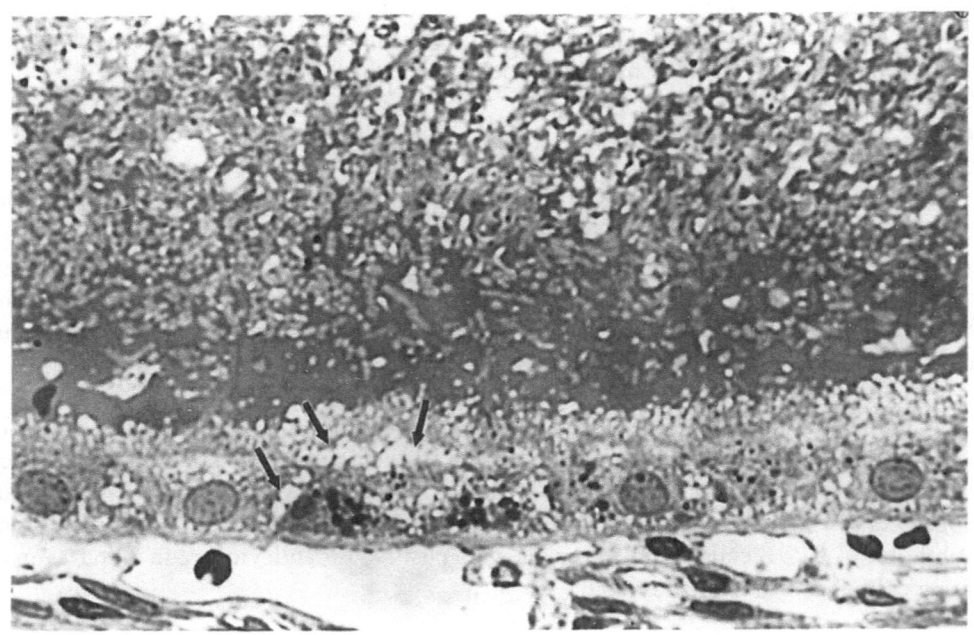

Figure 2.

leaving the donor cells behind (Figure 2). The donor RPE cells were left in intact sheets were left apposing the host cells. In some areas they replaced the host cells and had established contact with the Bruch's membrane (Figure 3), and in some occasions with the photoreceptor outer segments (Figure 3). The donor cell sheets appeared structurally intact with a few apical melanin granules and engulfed photoreceptor outer segments. These cells also contained a large number polymer degradation products in the form of small plastic beads scattered throughout the cell. These products were also seen in the interphotoreceptor space. By two weeks, most of the polymer was degraded (Figure 3). Also, the donor cell sheets appeared well established on the Bruch's membrane. The only traces remaining were in the RPE cells and occasionally in the interphotoreceptor space. Also, at two weeks there were indications of a slight cell mediated immune response evidenced by some mononuclear and giant cells. In two cases, which had some retinal hemorrhage, the inflammatory condition was highly evident (Figure 3).

DISCUSSION

Using a solvent casting technique, thin biodegradable polymer fims of PLGA were manufactured with reproducible surface morphologies and mechnical properties (15, 16). By this technique, it was possible to vary the thickness of the polymer films. The thickness of the manufactured polymers were extremely conforming to the morphology and dimensions of the subretinal space. These films revealed smooth surfaces without microporosity and could be manufactured with uniform thickness. We have also demonstrated that these polymer films serve as more than optimal substrates for RPE cell culture (16). Cell attachment and proliferation studies confirmed that these polymers to be as competent general tissue culture substrates (16). The present study confirms our predictions that RPE cells can be

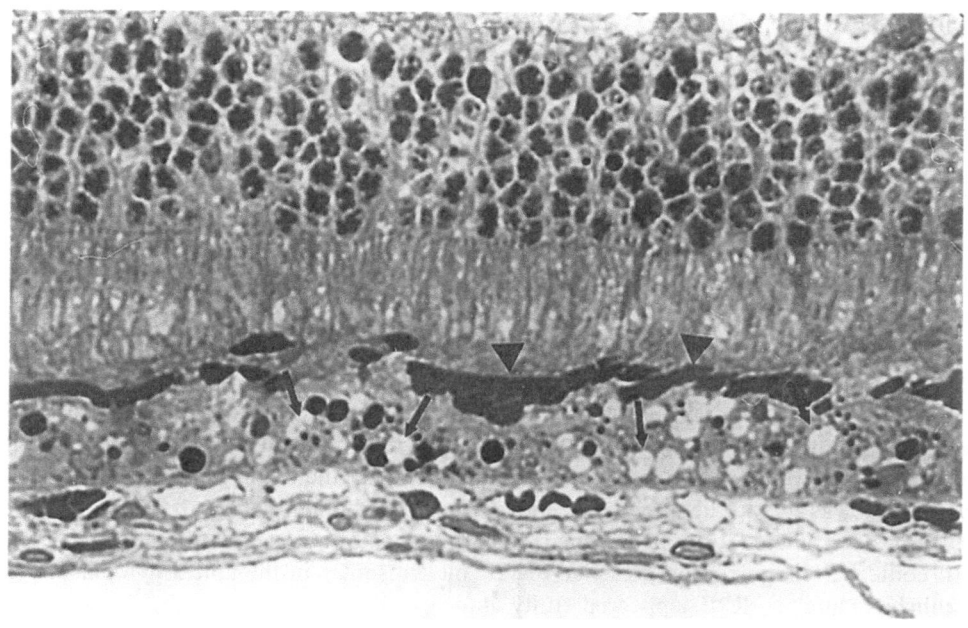

Figure 3.

transplanted as an intact monolayer, in correct orientation and the polymer film will gradually degrade leaving the donor cells in apposition to host retina.

Our primary goals for this study were, to establish a surgical procedure for transplanting RPE cells grown on a polymer substrate and to determine the time required for the polymer to degrade leaving the transplanted RPE cells in contact with the host retina. We employed a modified form of the original closed eye technique (1), entering the globe through an incision that was away from the transplant site and introducing the RPE/polymer sheet into the subretinal space through a small retinotomy. There are several advantages to the particular surgical procedure used. Firstly, the incision site is well away from the transplant site preventing choroidal blood components to enter the site and thus cause inflammation. Secondly, the placement of the graft is highly precise and targeted in comparison to random injection. Although, in two cases a retinal hemorrhage developed due to the retinotomy being too close to a retinal blood vessel, complete reattachment was observed following retinotomy and transplantation, removing the need for a endodiathermy to seal the retinotomy site. Patch or en bloc transfer of RPE has been successfully attempted by other investigators (17, 18) in rabbits. These investigators used a bleb detachment technique to transplant RPE monolayers in rabbit. While the bleb-transplant procedure is easily carried out in a larger animal such as the rabbit as has been used by us previously (19, 20), it is very difficult to perform in the rodent model. It is necessary to carry out RPE transplantation studies in the rodent particularly the RCS rat, in order to further substantiate the treatment for retinal degeneration. Therefore, the targeted transplant procedures have to conform to this model. Also, the injection of the RPE patches is highly random in the bleb-detachment technique and due to the presence of balanced salt solution in the bleb, transplanted RPE cells float freely in the subretinal space until the bleb is absorbed. This leads to considerable aggregation and multilayering of the transplanted RPE (18). Presence of free RPE cells in humans, also increases the chances of proliferative vitreoretinopathy. Our results demonstrate the presence of donor RPE patches limited to the retinotomy site proving the efficacy of the targeted technique. In addition, the patches remained intact and attached to the Bruch's membrane en bloc, replacing host RPE, although no attempt was made to remove the host RPE. Thus, targeted transplants on a degradable substrate seems much more promising than randomly injected cells in therapy for retinal degeneration.

It is apparent from this initial in vivo study with the polymer substrate that these co-polymers (PLGA) gets degraded very quickly in the subretinal space. Most of the polymer had been degraded by one week and all of it was gone by 2 weeks. It appears that the degradation proceeded at a faster pace than we anticipated, since our first chosen time point was a week. Although we have determined the end point (2 weeks) of the degradation process, the onset and the initial stages are still to be assessed. Our future studies will address these stages by observing early time points starting from a few hours to a week, post-transplantation. It is known that these copolymers are biocompatible and non-toxic to the eye (21), however, it is too early to speculate whether the slight immune response observed here was due to the polymer or the cross-species human xenograft. The polarity of the graft also remains to be determined, as earlier time points need to be studied. Donor RPE could be recognized by presence of few melanosomes in the apical cytoplasm. The established human RPE appeared fully functional evidenced by the presence of phagosomes containing host photoreceptor outer segments and others containing degradation products of the polymer membrane. Both at one and two week time points, donor cells were observed on the Bruch's membrane with intermittent single or patches of host RPE. In some cases, human RPE cells had settled on host RPE, forming apparent junctional complexes with the latter. Therefore, in future it may be necessary to employ a host RPE debridement method (23) to ensure the establishment of the entire sheet of donor RPE. However, these methods often lead to irreparable detachment, so gentler methods may have to be invetigated for human patients.

In summary, we have successfully formulated a polymer substrate for transplantation of RPE cells in the subretinal space, and have established a successful surgical procedure for such transplantation. We have proven this substrate to be highly effective for culture of RPE cells (15, 16). Results from our initial in vivo study indicate that the polymer film completely degrades by two weeks, leaving intact donor RPE sheets attached to the Bruch's membrane and in contact with the host photoreceptors. We observed mild inflammatory events in response to the graft but the cause and mechanism of such response remains to be determined. Therefore, such a substrate can now be effectively used for transplantation studies in animals with retinal degeneration.

ACKNOWLEDGMENTS

The authors thank Jeanie Park and Mary Keener for her valuable technical support and The Foundation Fighting Blindness and the Knights Templar Eye Foundation for funds to support this project.

REFERENCES

1. Lopez, R., Gouras, P., Brittis, M., Kjeldbye, H., 1987, Transplantation of cultured rabbit retinal epithelium to rabbit retina using a closed-eye method, Invest Ophthalmol Vis Sci 28; 1131–1137.

2. Yamamoto, S., Du, J., Gouras, P., Kjeldbye, H., 1993, Retinal pigment epithelial transplants and retinal function in RCS rats, Invest Ophthalmol Vis Sci 34; 3068–3075.
3. Li, L., Turner, J.E., Inherited retinal dystrphy in the RCS rat: prevention of photoreceptor degeneration by pigment epithelial cell transplantation, Exp Eye Res 47; 911–917.
4. Sheedlo, H.L., Li, L., and Turner, J.E., 1989, Functional and structural characteristics of photoreceptor cells rescued in RPE-cell grafted retinas of RCS dystrophic rats, Exp Eye Res 48; 841–854.
5. Seaton, A.D., Sheedlo, H.J., and Turner, J.E., 1994, A primary role for RPE transplants in the inhibition and regression of neovascularization in the RCS rat, Invest Ophthalmol Vis Sci 35; 162–169.
6. Liu, Y., Silverman, M.S., Berger, A.S., Kaplan, H.J., 1992, Transplantation of confluent sheets of adult human RPE, ARVO Abstracts, Invest Ophthalmol Vis Sci 33; 1128.
7. Moritera, T., Peyman, G.A., Rahimy, M.H., Lu, Q., Wafapoor, H., Gebhardt, B.M., 1993, Transplants of monolayer retinal pigment epithelium grown on biodegradable membrane in rabbits, ARVO Abstracts, Invest Ophthalmol Vis Sci 34; 1093.
8. Fang, S.R., Kaplan, H.J., Del Priore, L.V., Liu, Y., Wang, X., Hornbeck, R., Landgraf, M., Mason, G., Silverman, M.S., 1993, Development of a surgical procedure and instrument for transplantation of extended gelatin sheets to the subretinal space, ARVO Abstracts, Invest Ophthalmol Vis Sci 34; 1093.
9. Bhatt, N.S., Oliver, P.D., Fenech, T., Hessberg, T.P., Diamond, J.G., Miceli, M.V., Kratz, K.E., and Newsome, D.A., 1993, transplantation of Human retinal pigment epithelium into rabbits, ARVO Abstracts, Invest Ophthalmol Vis Sci 34; 1093.
10. Bhatt, N.S., Newsome, D.A., French, T., 1994, Experimental transplantation of human retinal pigment epithelial cells on collagen substrates, Am J Ophthalmol 117; 214–221.
11. Cima, L.G., Ingber, D.E., Vacanti, J.P., and Langer, R., 1991, Hepatocyte culture on biodegradable polymeric substrates, Biotechnol Bioeng 38; 145–158.
12. Saltzmann, W.M., Parsons-Wingerter, P., Leong, K.W., and Lin, S., 1991, Fibroblast and Hepatocyte behavior on synthetic polymer surfaces, J Biomed Mater Res 25; 741–759.
13. Freed, L.E., Marquis, J.C., Nohria, A., Emmanual, J., Mikos, A.G., amd Lnger, R., 1993, Neocartilage formation in vitro and in vivo using cells cultured on synthetic biodegradable polymers, J Biomaterial Res 27; 11–23.
14. Ishaug, S.L., Yaszemski, M.J., Bizios, R., and Mikos, A.G., 1994, Osteoblast function on synthetic biodegradable polymers, J Biomed Mater Res 28; 1445–1453.
15. Thompson, R.S, Giordano, G.G., Collier J.H., Ishaug, S.L., Mikos, A.G., Lahiri-Munir, D., and Garcia, C.A., 1996, Manufacture and characterization of poly(-hydroxyester) thin films as temporary substrates for retinal pigment epithelium cells, J Biomed Mater Res 17; 321–327.
16. Giordano, G.G., Thompson. R.C., Ishaug, S.L., Mikos, A.G., Cumber, S., Garcia, C.A., and Lahiri-Munir, D., 1997, Retinal pigment Epithelial cell culture on synthetic biodegradable polymers, J Biomed Mater Res 34; 87–34.
17. Gouras, P., Sheng, Y., Cao, H., Kjeldbye, H., Lopez, R., Weisman, G., 1994, Patch transplantation of fetal human RPE, ARVO Abstracts, Invest Ophthalmol Vis Sci 35; 1335.
18. Sheng, Y., Gouras, P., Cao, H., Berglin, L., Kjeldbye, H., Lopez, R., amd Rosskothen, H., 1995, Patch transplants of human fetal retinal pigment epithelium in rabbit and monkey retina, Invest Ophthalmol Vis Sci 36; 381–390.
19. Lahiri-Munir, D., 1995, Retinal pigment epithelial transplantation. R. G. Landes and Co. Austin, TX and Springer. Heidelberg, Germany, pgs 1–141.
20. Lahiri-Munir, D., B.X. Zhang, T.Li, M. Keener, P. Chevez-Barrios, S. Cumber, S.M. Stepkowski and C.A. Garcia. 1995, Immunological response in xenogeneic and allogeneic RPE transplantation. Exp Eye Res. (submitted).
21. Giordano, G.G., Arroyo, M.H., and Refojo, M.F., 1993, Sustained delivery of retinoic acid from microspheres of biodegradable polymer in PVR, Invest Ophthalmol Vis Sci 34; 2743–2751.
22. B. Parolini, I.K. Sugino, E. Gordon, M.A. Zarbin. 1995, A new method to debride RPE cells frm Bruch's membrane in vitro. Invest Ophthalmol Vis Sci 36; 1141.

39

MIDKINE FROM VARIOUS SOURCES IN CONSTANT LIGHT-INDUCED PHOTORECEPTOR DEGENERATION OF THE RAT

Kazuhiko Unoki,[1] Hisako Muramatsu,[2] Norio Kaneda,[2] Shinya Ikematsu,[2] Fumiyuki Uehara,[1] Norio Ohba,[1] and Takashi Muramatsu[2]

[1]Department of Ophthalmology
Kagoshima University Faculty of Medicine
Sakuragaoka 8-35-1, Kagoshima-shi 890, Japan
[2]Department of Biochemistry
Nagoya University School of Medicine, Syowa-ku
Tsurumai 65, Nagoya-shi 466, Japan
Nagoya-shi, Japan

INTRODUCTION

Midkine (MK) is a heparin binding growth factor that is expressed temporally during the early stages of retinoic acid-induced differentiation of embryonal carcinoma cells[1,2]. MK protein produced by L-cell which had been transformed stably with a mouse MK expression vector, enhances the survival and neurite outgrowth of cultured embryonic neurons and promote mitosis in some fibroblast cell lines[3–5]. In constant light induced retinal degeneration, we demonstrated that the intravitreal injection of MK from L-cell show significant rescue effects on photoreceptor damage by both histological and electrophysiological methods[6,7]. These results indicate that MK has neurotrophic activity or survival promoting activity in retina of the rat.

Although it is difficult to prepare large amounts of MK from L cells to be injected intravitreal space, Kaneda et al. produced a recombinant baculo virus clone which expresses a large amount of the mouse MK in the culture medium than L-cell8. In this report, we evaluated the survival-promoting activity of each midkine produced by L-cell system, baculo virus system in the constant light-induced photoreceptor degeneration of the rat. Also, rat pleiotrophin (PTN, known as heparin-binding growth-associated molecule) was examined in light damaged model of the rat.

Degenerative Retinal Diseases, edited by LaVail *et al.*
Plenum Press, New York, 1997

FACTORS

We tested the three preparations namely two MK and a related substance PTN. MK from L-cell system was purified from the culture medium of L-cells transfected with a mouse MK expression vector, as described in details elsewhere[3]. MK from baculo virus system was purified to homogeneity by heparin-Sephrose column chromatography from the culture medium of Trichoplusia ni High Five cells infected with the recombinant virus, which transfected with the mouse MK-2 cDNA[8]. The concentration each MK from various sources was 1 g/l. Rat recombinant Pleiotrophin (Heparin-binding growth-associated molecule) (0.5g/5l: Beckton & Dickinson Labware, Bedford, MA) was commercially available. The control vehicle was phosphate buffered saline (PBS).

HISTOLOGY OF LIGHT DAMAGED RETINA AND FACTOR INJECTED RETINA

In this constant light-induced retinal degeneration experiment, we used Sprague-Dawley albino rats at 2–3 months of age. Two days before constant light exposure, rats were anesthetized and then injected each factors into the vitreous of one eye, as described elsewhere[9]. The other eye of each rat was injected with the same volume of PBS as a control. Two days later, the rats were placed into constant light at an illuminance of 130–150 foot-candles for 1 week. After the constant light exposure, the histological procedure was performed as reported before[9].

Degeneration of photoreceptor cells after 1 week of constant light in uninjected rats or those injected with PBS was most severe in the posterior to equatorial region of the superior hemisphere. The ONL was reduced in thickness from the normal 9–10 rows of photoreceptor nuclei (Fig. 1A) to 13 rows (Fig.1B). Only a few fragments of photoreceptor inner and outer segments were saved in this most damaged region. The RPE did not show any damage. In other parts of the light damaged-retina, the inner and outer segments of photoreceptors were damaged to a lesser degree and the ONL was thicker. The peripheral region of the inferior hemisphere had the least damage from the constant light exposure.

Intravitreal injection with MK from L-cell and MK from baculo virus significantly rescued the photoreceptors and ONL (Fig.1C and D). The photoreceptors had inner and outer segments and sometimes appeared even normal, although they were a little shortened and disorganized. The ONL had 7–8 rows of nucleus, although pyknotic nuclei were scattered throughout the layer. However, the eyes injected with PTN did not show any rescue effect compared with PBS control eye.

DEGREE OF RESCUE EFFECT FROM FACTORS

We measured the thickness of the outer nuclear layer(ONL), to quantify the light induced retinal degeneration, as an index of photoreceptor loss[9–12]. A mean ONL thickness was obtained from 48 measurements in two hemispheres of almost the entire a single section. The thickness of the ONL in the uninjected and PBS-injected eyes was about 25% of that seen in normal cyclic light (Fig.2). The thickness of the eye injected with MK from

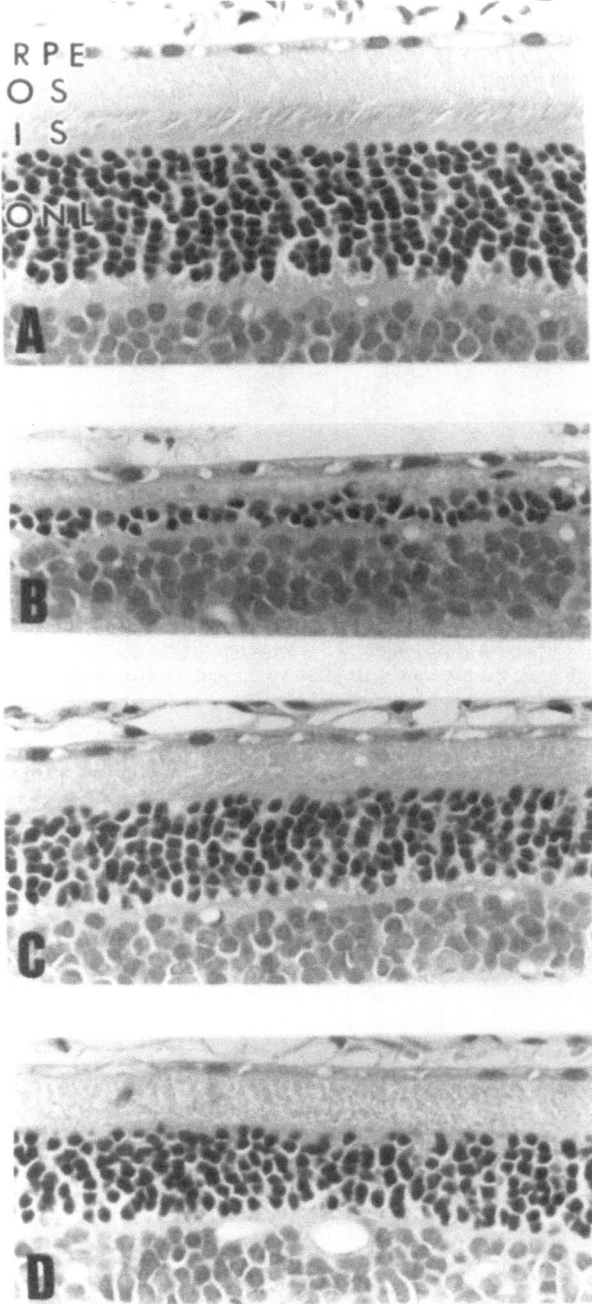

Figure 1. Light micrographs of the posterior retinas of Sprague-Dawley rats. (A) Normal retina of the rat in cyclic light. RPE: retinal pigment epithelium; OS: outer segment of photoreceptor; IS: inner segments of photoreceptor; ONL: outer nuclear layer. The photoreceptor outer and inner segments are distinct, and the ONL has 9–10 rows of photoreceptor nuclei. (B) The retina exposed to constant light for 7 days. The outer and inner segments of photoreceptors are almost lost in this section. The ONL is decreased in thickness, composed of only 2–3 rows of nucleus, significantly decreased from the normal ONL. (C) The retina rescued from light damage by an intravitreal injection of MK from L-cell. The inner segments of photoreceptors are present and almost normal, while the outer segments of the photoreceptors are shorter and partially destroyed. The ONL has 7–8 rows of nucleus, although pyknotic nuclei are seen. (D) The retina rescued by MK from baculo virus. The outer and inner segments as well as the ONL show features similar to those injected with MK from L-cell. (Hematoxylin-eosin; bar = 20m).

two sources showed a considerable rescue in constant light damaged retina, (P<0.0001). The rescue activity of PTN was not seen.

To analyzed the rescue effects of each agent, we assigned a relative score compared with the control eye described[9]. MK from L-cell showed a little higher score than MK from baculo virus (Fig. 3). The eyes injected with PTN had a low score.

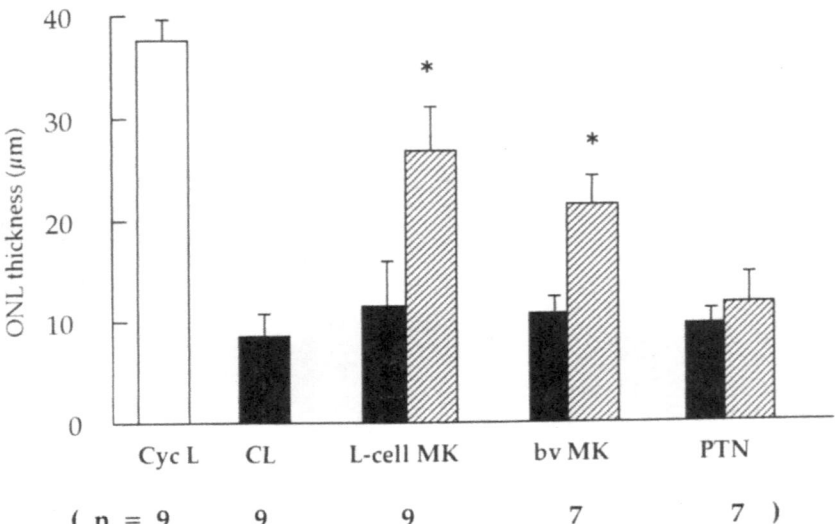

Figure 2. Measurements (mean + standard deviation) of the ONL thickness in eyes exposed to cyclic light (Cyc L), exposed to constant light for 7 days without any injection (CL) and with various agents injected 2 days before light exposure. Midkine from L-cell (L-cell MK), MK from baculo virus (bv-MK), pleiotrophin(PTN). Controls for each agent (PBS) injected 2 days before light exposure. The number of eyes (n) is shown in parentheses. In each case, the bar shows the mean value, and the error bar represents the standard deviation. ONL thickness of eyes injected with L-cell MK, bv-MK shows significant differences in numbers of photoreceptor nuclei surviving (shadowed bars) when compared with their control eye (solid black bars); * indicates P<0.0001. PTN does not show any significant difference from the control.

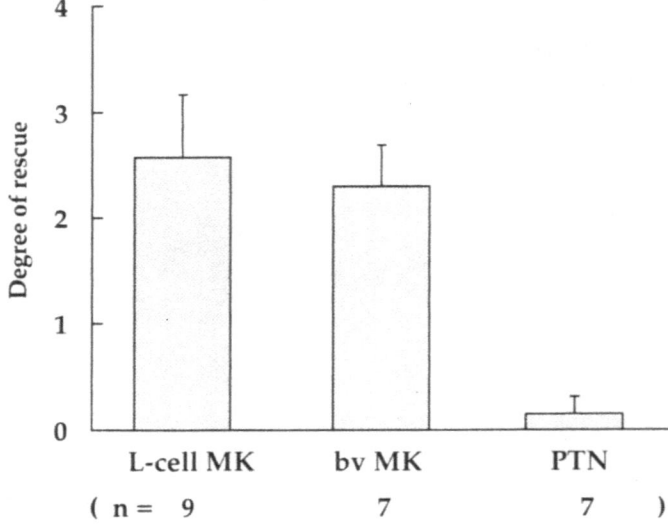

Figure 3. The scores for the degree of photoreceptor rescue (mean + standard deviation) by various agents. The scores of photoreceptor rescued by various agents were similar to those obtained by measuring the ONL thickness. The abbreviations are also the same as those described in the legend to Fig. 2. The number of eyes (n) is shown in parentheses.

COMMENTS

These result demonstrate that MK from two sources, L-cell and baculo virus, showed the neurotrophic activity in vivo. MK is a heparin-binding basic protein of Mr 13,000 and rich in basic amino acids and cysteine residues[1,13]. MK promotes the neurite outgrowth and survival of rat embryonic brain cells in vitro[3,8]. In pressure induced retinal ischemia-reperfusion insult, MK from baculo virus also showed the survival promoting activity in inner retina of the rat(unpublished data). These result suggested that MK has a role in the development and the maintenance of the retina.

Pleiotrophin, known as heparin-binding growth-associated molecule, is also a heparin binding growth factor and 45% of their amino residues are homologous of MK[14,15]. Although both PTN and MK appeared to be important molecules involved in organogenesis[16]. In the light damaged retinal degeneration, PTN has no rescue effect at the concentration examined.

ACKNOWLEDGMENTS

The authors thank Ms. Yoshiko Maeda and Ms. Yuko Oe for technical assistance.

REFERENCES

1. Kadomatsu K., Tomomura M., Muramatsu T. 1988, cDNA cloning and sequencing of a new gene intensely expressed in early differentiation stages of embryonal carcinoma cells and in mid-gestation period of mouse embryogenesis, Biochem. Biophys. Res. Commun. 151: 1312–1318.
2. Kadomatsu K., Huang R.P., Suganuma T., Murata F., Muramatsu T., 1990, A retinoic acid responsive gene MK found in the teratocarcinoma system is expressed in spatially and temporally controlled manner during mouse embryogenesis, J. Cell Biol. 110:607–616.
3. Muramatsu H., Muramatsu T., 1991, Purification of recombinant midkine and examination of its biological activities: functional comparison of new heparin binding factors, Biochem. Biphys. Res. Commun. 177: 652–658.
4. Nurcombe V., Fraser N., Herlaar E., Heath J.K., 1992, MK, a pluripotential embryonic stem cell-derived neuroregulatory factor, Development 116: 1175–1183.
5. Michikawa M., Kikuchi S., Muramatsu H., Muramatsu T., Kim S.U., 1993, Retinoic acid responsive gene product, midkine(MK), has neurotrophic functions for mouse spinal cord and dorsal root ganglion neurons in culture. J. Neurosci. Res. 35:530–539.
6. Unoki K., Ohba N., Arimura H., Muramatsu H., Muramatsu T., 1994, Rescue of photoreceptors from the damaging effects of constant light by midkine, a retinoic acid-responsive gene product, Invest Ophthalmol Vis Sci. 35: 4063–4068.
7. Masuda K., Watanabe I., Unoki K., Ohba N., Muramatsu T., 1995, Functional rescue of photoreceptors from the damaging effects of constant light by survival promoting factors in the rat. Invest. Ophthalmol. Vis. Sci. 36: 2142–2146.
8. Kaneda N., Talukder A.J., Nishiyama H., Koizumi S., Muramatsu T., 1996, Midkine, a heparin-binding growth/differential factor, exhibits nerve cell adhesion and guidance activity for neurite outgrowth in vitro. J.Biochem. 119:1150–1156.
9. LaVail M.M., Unoki K., Yasumura D., Matthes M.T., Yancopoulos G.D., Steinberg R.H., 1992, Multiple growth factors, cytokines and neurotrophins rescue photoreceptors from the damaging effects of constant light. Proc. Natl. Acad. Sci. USA. 89:11249–11253.
10. LaVail M.M., Battelle B.A.. 1975, Influence of eye pigmentation and light deprivation on inherited retinal dystrophy in the rat. Exp. Eye Res. 21: 167–192.
11. LaVail M.M., Gorrin G.M., Repaci M.A., Thomas L.A., Ginsberg H.M., 1987, Genetic regulation of light damage to photoreceptors, Invest Ophthalmol. Vis. Sci. 8:1043–1048.

12. Michon J.J., Li Z.I., Shioura N., Anderson R.J., Tso M.O.M., 1991, A comparative study of methods of photoreceptor morphometry. Invest. Ophthalmol. Vis. Sci.;32: 280- 284.

13. Tomomura M., Kadomatsu K., Matsubara S., Muramatsu T., 1990, A retinoic acid responsive gene, MK, found in the teratocarcinoma system. J. Biol. Chem. 265:10765–10770.

14. Muramatsu T., 1993, Midkine(MK), the product of a retinoic acid responsive gene, and pleiotrophin constitute a new protein family regulating growth and differentiation. Int. J. Dev. Biol. 37: 183–188.

15. Muramatsu T., 1994, The midkine family of growth/differentiation factors, Dev. Growth Differ. 36, 1–8.

16. Mitsiadis T.A., Salmivirta M., Muramatsu T., MuramatsuH., Rauvala H., Lehtonen E., Jalkanen M., Thesleff I., 1995, Expression of the heparin binding cytokines, midkine(MK) and HB-GAM(Pleiotrophin), is associated with epithelial-mesenchymal interactions during fetal development and organogenesis, Development 121: 37–51

40

THE INFLUENCE OF OXYGEN ON THE SURVIVAL AND DEATH OF PHOTORECEPTORS

Evidence from Rat and Mouse

Krisztina Valter, Kyle Mervin, Juliani Maslim, and Jonathan Stone

NSW Retinal Dystrophy Research Centre
Department of Anatomy and Histology
University of Sydney
NSW 2006 Australia

SUMMARY

We have studied the influence of oxygen on the rates of death of photoreceptors in the normogenetic mouse and rat, and in the genetically dystrophic RCS rat. Hyperoxia delays photoreceptor death during development of the normogenetic and RCS rats. Hypoxia accelerated photoreceptor death in the adult normogenetic rat and mouse. Hypoxia also accelerates photoreceptor death during development of the normogenetic rat and mouse, and of the RCS rat. The effects of hypoxia were greater in the juvenile, confirming the concept of a critical period in photoreceptor development, during which photoreceptors are particularly vulnerable to oxygen levels. In the RCS rat, evidence was obtained that the edges of the retina are resistant to hypoxic damage, explaining earlier observations of long-term photoreceptor survival at the edges. It is suggested that nutrient supply is a limiting factor in photoreceptor survival, in both normal and genetically damaged RCS retinas. If these ideas prove relevant also to the human effective therapy for some photoreceptor dystrophies may be achieved by nutrient supplementation.

INTRODUCTION

The genetic basis of retinal dystrophies is now powerfully established. Over 100 mutations in 6–8 genes are known to cause retinal degeneration (1,2). Despite that genetic diversity, dystrophy is largely restricted to one class of retinal cell, the photoreceptor. This paper addresses two consequent questions. One invites speculation: why is the photoreceptor so much more vulnerable to genetic change than other retinal neurones?

The other invites experiment: What steps link the many receptor-lethal mutations to photoreceptor death? From understanding of the vulnerability of the photoreceptor and of the links from mutation to death, opportunities might emerge to intervene, to rescue the photoreceptors.

Many previous authors (e.g. 3–5) have commented on the specialised anatomy and physiology of vertebrate photoreceptors. These cells are both high users of energy and the only neurones of the central nervous system to lack intrinsic blood vessels, being supplied by diffusion from the choriocapillaris. This situation of high energy demand at the end of a long diffusion path is unique to outer retina and raises the possibility that photoreceptors are vulnerable because their energy supply is precarious.

Evidence of this precariousness comes from work which shows that the performance of photoreceptors is sensitive to energy supply. McFarland and colleagues (6,7), working in the wartime context of visual performance at altitude, showed that hypobaria equivalent to 13,000 ft altitude produces substantial (half log unit) rises in dark-adapted (rod) thresholds, readily reversed by either hyperoxia or hyperglycaemia. More generally their work showed that hypobaria and hypoglycaemia cause rises in visual thresholds which can be reversed by either hyperoxia or hyperglycaemia. They reported the striking observation that the fall in blood sugar which occurs in the normal individual overnight causes a measurable rise in visual thresholds, which is reversed by a 'normal breakfast'. More recently, Steinberg (3) noted that the electro-oculogram of the cat is highly sensitive to respiratory parameters and argued that rods function normally on the verge of hypoxia.

We have tested whether the death and survival of photoreceptors are similarly dependent on energy supply. Late in development a wave of death occurs among photoreceptors (8). This wave of death may be induced by an episode of hypoxia, for it occurs as the photoreceptors commence their function and hence their high consumption of oxygen and is reduced by hyperoxia (9). It seems possible that in development the retina overproduces photoreceptors, which are culled to their adult numbers by competition for nutrients. If so, photoreceptor-related genes have presumably evolved to maximise the efficiency of energy delivery and consumption. Mutations which degrade efficiency would then increase the death of photoreceptors and deplete the adult population.

To test the effect of a major nutrient (oxygen) on the death and survival of photoreceptors, we utilised the fact that the partial pressure of oxygen (pO_2) in the retina can be increased or decreased by varying the level of oxygen in the air inspired. Increasing the concentration of oxygen in the inspired air to 100%, for example, raises the pO_2 at the outer surface of the retina of the cat and rat several-fold (11–14) and increases vitreal pO_2 in both species (15, 16). We therefore reared juvenile and adult mice and rats in nitrogen/oxygen mixtures which ranged from hypoxic (10%) to hyperoxic (70%), and examined the retina for evidence that the level of oxygen inspired modulates photoreceptor death. All experiments were done at normobaria.

METHODS

Material

Retinas were obtained from albino Sprague-Dawley and Royal College of Surgeons (RCS) rats and from C57 black mice euthanised with an overdose of sodium pentobarbitone (>60mg/kg, i.p.). The retinas were fixed by immersion in 4% paraformaldehyde in phosphate buffered saline (PBS) at pH7.4 for 2h and were frozen-sectioned at 15μm.

Oxygen Exposure

Litters were exposed to hyperoxia by raising them in air supplemented with oxygen to 70–75% and to hypoxia by raising them air supplemented with nitrogen until oxygen was reduced to 10%. Gas concentrations were controlled by a feedback device (OXYCY-CLER, Reming Bioinstruments, New York). The periods of exposure were chosen to test specific hypotheses of neuronal death and survival. Pup mortality was not noticeably higher than among litters raised in normoxia and the mother tended and groomed the pups normally. However, the pups gained weight more slowly than in normoxia. The litters were kept in standard litter boxes inside a perspex chamber. Lighting was varied in a 12h day, 12h night cycle between darkness and normal room levels (approximately $50cd/m^2$).

TUNEL and Lectin Labelling

Fragmenting DNA was labelled with the TUNEL technique (17) slightly modified (9). Blood vessels and microglia were labelled with the G. simplicifolia lectin (isolectin B4, Sigma), using a form of the lectin conjugated to FITC (9).

Quantitative Analysis

For each age examined, TUNEL-labelled profiles in INL and/or ONL were counted in consecutive 780μm lengths of the section, along its full length. Between 25 and 30 section were counted from each retina. To compare regions of the retina samples of 16–30 fields were taken around the optic disc, at the edge of the retina and half-way between.

RESULTS

Levels of Oxygen in Inspired Air Influence Retinal Development

Figure 1 shows the influence of oxygen concentration in inspired air on the vasculature of the developing rat retina. The calibre of capillaries the periphery of the superficial vasculature of developing retina is increased by hypoxia (compare Figures 1A and C) and the calibre and density of capillaries are decreased by hyperoxia (compare C and E). The large vessels which converge on the optic disc are relatively unaffected, but the capillaries between them are increased by hypoxia (compare B and D, noting that the capillary-free zones apparent along the arteries in D have not formed in B) and are eliminated by hyperoxia (compare D and F). In other contexts we have shown that inspired oxygen levels modulate the expression in the retina of the hypoxia-induced angiogenic factor VEGF (18,19).

Figure 1 is shown to emphasise that variations in oxygen levels in inspired air do reach the retina and can influence developmental events in the retinas under study. Equivalent effects are known to occur in humans, causing for example retinopathy of prematurity (reviewed in 5). These effects do not require hypo- or hyperbaria.

In Adult Retina

Hypoxia Increases Photoreceptor Death. In the retina of the adult normogenetic rat photoreceptors die at a slow rate, slowly depleting the outer nuclear layer (20, 21). In our

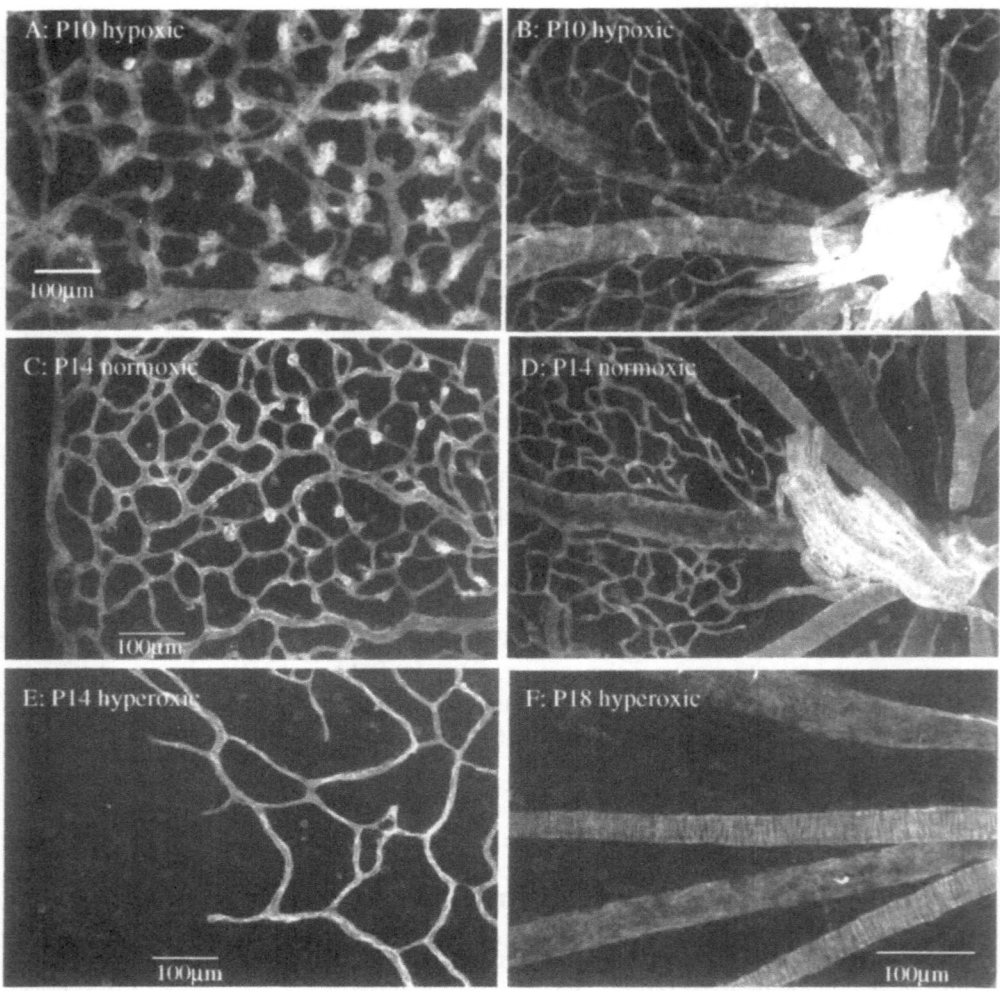

Figure 1. Inspired oxygen affects retinal development. The vessels are labelled in wholemounts with the G. simplicifolia lectin. A,B: Superficial vasculature of a P10 albino rat retina raised since P3 in 10% oxygen, near the edge of the vasculature (A) and near the optic disc (B). C,D: Superficial vasculature of a P14 albino rat retina raised in normoxia, near the edge of the vasculature (C) and near the optic disc (D). E: Superficial vasculature of a P14 albino rat retina raised since P0 in 70% oxygen. F: Superficial vasculature near the optic disc of a P18 rat retina raised 70% oxygen since P10.

material this slow constant death was apparent as scattered TUNEL+ profiles. In both the albino rat and C57 black mouse these TUNEL+ presumably dying cells were overwhelmingly (>99%) located in the ONL.

Figures 2A and B show the effect of 7 days' hypoxia on the frequency of TUNEL+ profiles in sample lengths of a section of two P63 C57 black mice, one raised in normoxia (A) and the other exposed for 8d to 10% oxygen prior to fixation of the retina; the frequency of labelled profiles is increased, specifically in the ONL. The two animals were raised in the same box until the beginning of that period, and were kept in adjacent boxes on the same bench during the 7d period of hypoxia, to ensure identical light exposure.

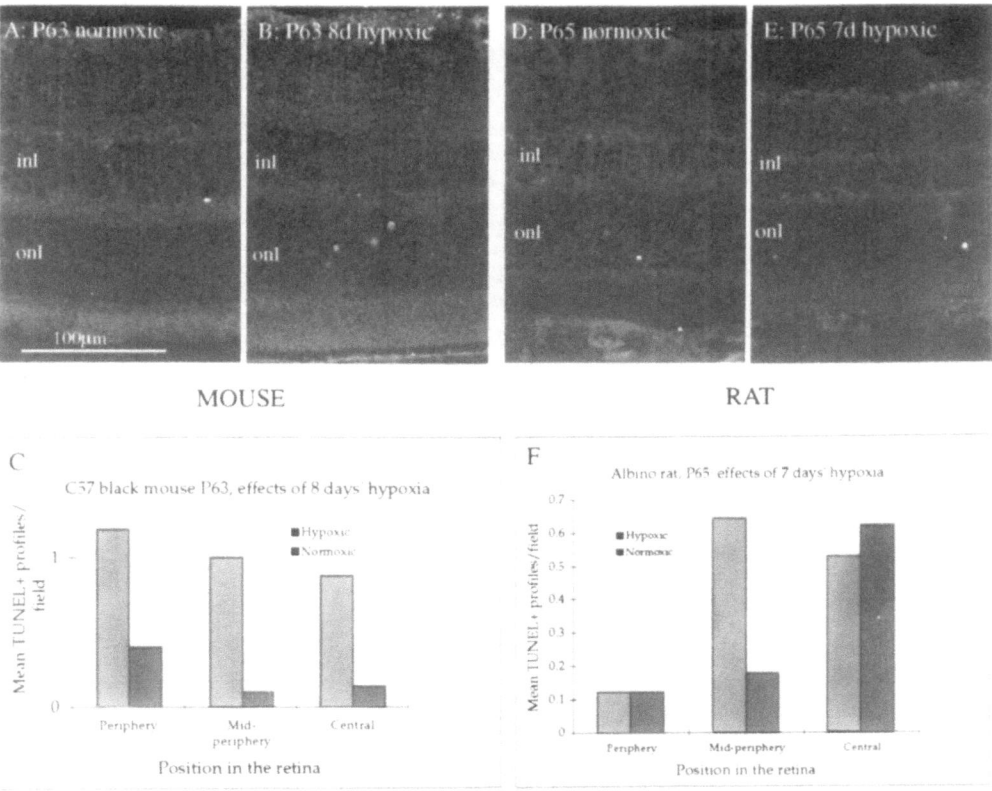

Figure 2. Hypoxia increases photoreceptor death rates in adult retina. A: Section of retina from a P63 C57 black mouse raised in normoxia, and labelled for fragmenting DNA with the TUNEL technique. B: Section of retina from a littermate raised in 10% oxygen for 8d. C: Histograms showing the density of TUNEL+ profiles in the eyes sampled in A and B. D: Section of retina from a P65 albino rat raised in normoxia, and labelled for fragmenting DNA with the TUNEL technique. E: Section of retina from a littermate raised in 10% oxygen for 7d. F: Histograms showing the density of TUNEL+ profiles in the eyes sampled in D and E. TUNEL+ profiles are more common after hypoxia in the mid-peripheral region of retina.

Figure 2C shows the effect in the mouse quantitatively and Figures 2D-F show corresponding data for albino rats aged P65. Considering the retina as a whole, two points can be made for both species. First, the higher frequency of TUNEL+ profiles in the hypoxic material was statistically significant ($p < 0.009$ on a 2-tailed student t-test). Second, the increase was specific to the ONL; no statistically significant increase was detected in inner retina.

Effects of Oxygen Are Pan-Retinal in the Mouse, More Restricted in the Rat. Considering central, mid-peripheral and peripheral regions of retina separately (Figures 2C and F), the higher frequencies of TUNEL+ profiles in hypoxic retina were statistically significant at all regions of the mouse retina ($p < 0.02$). The higher frequencies post-hypoxia were also highly significant in the mid-peripheral region of the rat retina ($p < 0.0002$), but significant differences were not obtained in the peripheral or central samples.

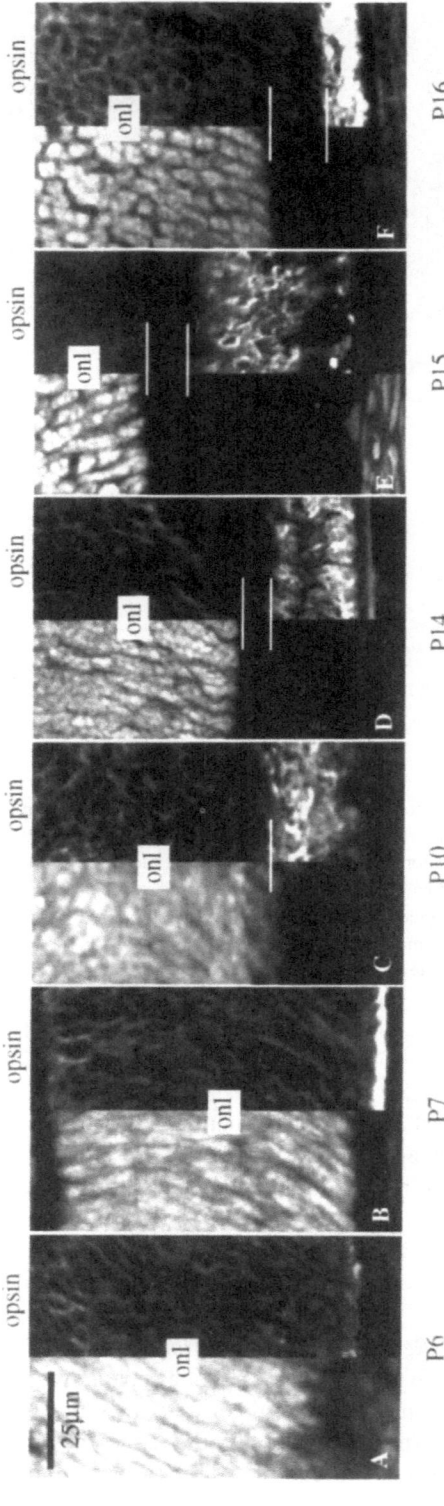

Figure 3. Opsin expression shows receptor development. Each of A-F shows a section of the outer layers of rat retina during development. In each the left panel shows the outer nuclear layer (onl) labelled with propidium iodide, and the right panel shows the same region of retina labelled with an anti-opsin antibody. B: Opsin concentrates in a thin layer at the outer margin of the onl at P7. C: Opsin is distributed along the development inner and outer segments. D-F: Emergence of an opsin-free region (marked by white lines) between the onl and the opsin labelling. This region corresponds to the inner segments.

In Juvenile Retina

Hyperoxia Delays and Hypoxia Accelerates Photoreceptor Death. We have previously noted (9) that photoreceptors are highly sensitive to oxygen supply during a critical period of development which in the rat begins at approximately P15 and lasts approximately one week. Figure 3 shows aspects of photoreceptor development during the onset of this period. In each of A-F, the left panel shows the ONL suing a non-selective nuclear stain (propidium iodide) and the right panel shows labelling for opsin (the rho 4D2 antibody of Hicks and Molday (1986)). Opsin was present in the cytoplasm of photoreceptors at P6, had accumulated at the outer surface of the ONL by P7 (B), and had segregated to a distinct outer layer by P10 (C). By P14, the opsin was sequestered into a layer (the outer segments) which was separate from the ONL by an unlabelled gap, marked by white lines. From P14–16 that gap widened, presumably as a result of growth of the inner segments.

Hyperoxia offered during the critical period (for 4 days from P16–20) reduced the naturally occurring photoreceptor death which occurs during the critical period (9). Figures 4A and B show for the rat that hypoxia during this period (from P15–21) greatly increased the rate of photoreceptor death. Figures 4C and D show a comparable (though lesser) increase in photoreceptor death in the mouse after hypoxia from P8 to P16. In both rat and mouse the dystrophic effect of hypoxia was greater during the critical period than in the adult (cf. Figure 2).

Effects of Oxygen Are Pan-Retinal. Figure 4E and F show a second experiment in the rat in which the pup was exposed to hypoxia for 6 days from P15-P21. As in the experiment shown in Figure 4B, the number of TUNEL+ profiles in the ONL is increased well above normoxic levels; this increase was clear in central (E) and peripheral retina (F).

In the RCS Retina

The death of photoreceptors begins in the RCS rat at the same developmental age as in the non-dystrophic rat, at about P15 (8). One characteristic of this strain, the failure of the pigment epithelium to ingest and dispose of the discarded tips of photoreceptors, is apparent by P15 (23,24), but the photoreceptors themselves are believed to be normal in structure and physiology (25). Because photoreceptor death begins in the RCS (as in the normogenetic) rat as photoreceptors commence their function, and because the accumulation of debris of photoreceptors in the subretinal space might act as a barrier to the diffusion of oxygen, we tested the effect of oxygen levels on the rate of photoreceptor death in the RCS rat.

Hyperoxia Delays and Hypoxia Accelerates Photoreceptor Death. Figures 5A-C show the effect of oxygen on the TUNEL-labelling of photoreceptors in the ONL of the RCS retina at P20. In pups raised in normoxia (B), fragmentation is well above the level seen in a normogenetic rat (compare Figures 4A and 5B); this level of labelling is decreased by 4 days' hyperoxia (Figure 5A (9)) and increased by 4 days' hypoxia (Figure 5C). Figure 5D-F shows the effects of oxygen at P24. By this age the level of TUNEL-labelling in the ONL of the normoxic retina is high (Figure 5E (8, 26)); the level is reduced by 8 days' prior hyperoxia (Figure 5D (9)) and increased by 8 days' prior hypoxia (Figure 5F). Figures 5G and H show the effect of hypoxia at P27. At this age, TUNEL-labelling of the ONL in the normoxic retina is strong (G) and is increased by 11 days' prior hypoxia (H). Moreover the ONL is thinner after hypoxia, indicating an increased rate of photoreceptor disappearance. In addition, hypoxia induced TUNEL-labelled degenerative changes in the INL (Figure 5H); INL degeneration of this form has not been reported in RCS retinas raised in normoxia.

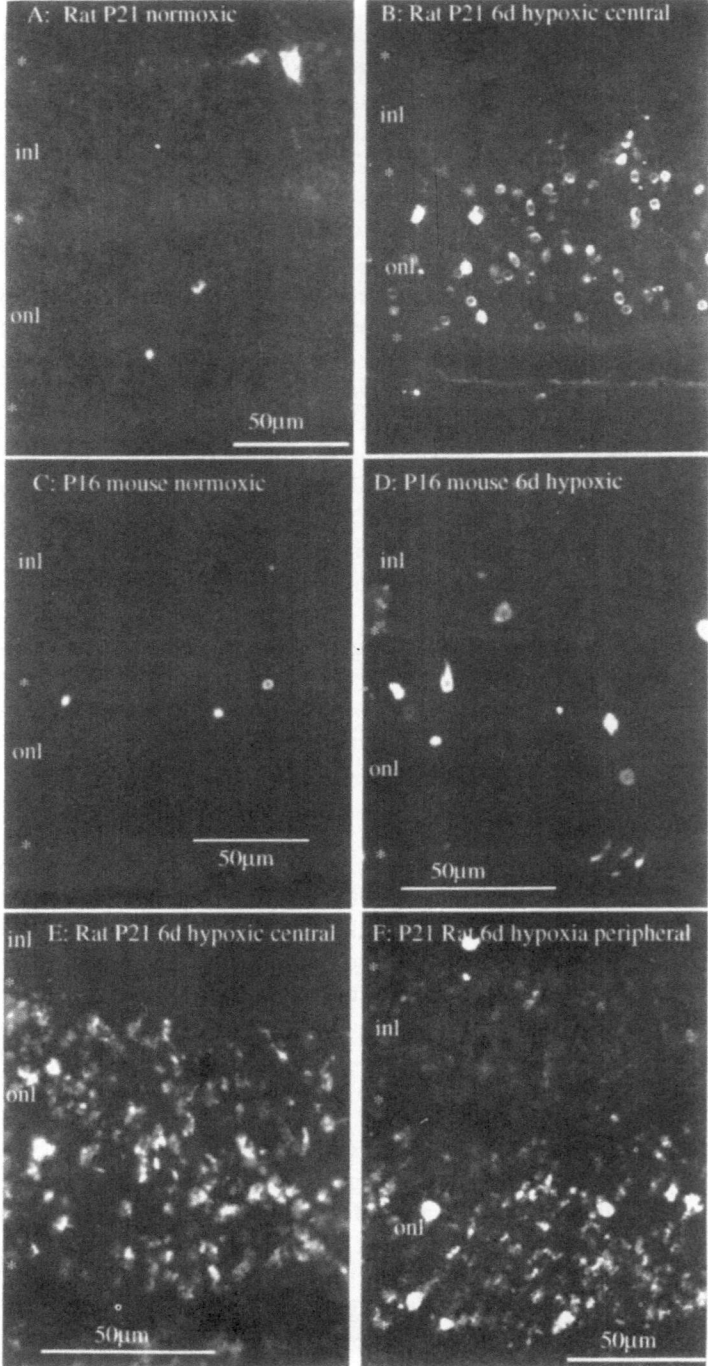

Figure 4. Hypoxia increases photoreceptor death rates in juvenile retina. Sections of retina are labelled for fragmenting DNA with the TUNEL technique. The inner and outer nuclear layers (inl and onl) are labelled, and their borders indicated (*). A,B: Sections of retina from P21 albino rats raised in normoxia (A) and for 6d in hypoxia (B). Both samples are from central retina. C,D: Sections of retina from P16 C57 black mice raised in normoxia (c) and for 6 days in hypoxia (D). E,F: Sections of retina from a P21 albino rat (separate experiment from A,B) raised in hypoxia for 6d. E is a sample from central retina, F from near the edge of the retina.

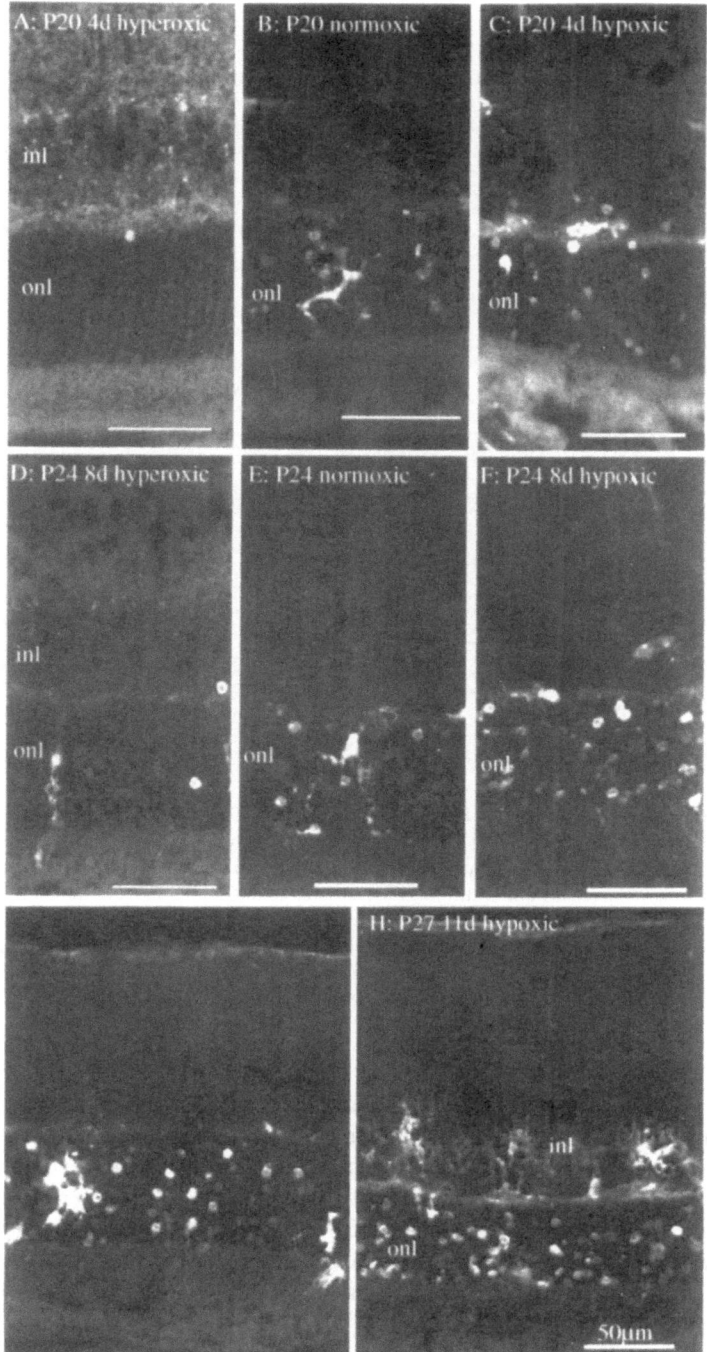

Figure 5. Hyperoxia delays and hypoxia accelerates the RCS dystrophy. Sections of RCS retinas labelled for fragmenting DNA with the TUNEL technique. The scales show 50μm. A-C: Sections of P20 RCS retinas. B shows the TUNEL-labelling of the outer nuclear layer (onl) in a retina raised in normoxia. That labelling is decreased by hyperoxia (A) and increased by 4 days' hypoxia (C). D-F: Sections of P24 RCS retinas. E shows the TUNEL-labelling of the onl in a retina raised in normoxia. That labelling is decreased by hyperoxia (D) and increased by 8 days' hypoxia (F). G,H: Sections of P27 retinas. G shows the TUNEL-labelling in the retina raised in normoxia. After 11 days' hypoxia the onl has thinned and TUNEL-labelling is extensive in the inner nuclear layer (inl) as well as in the onl.

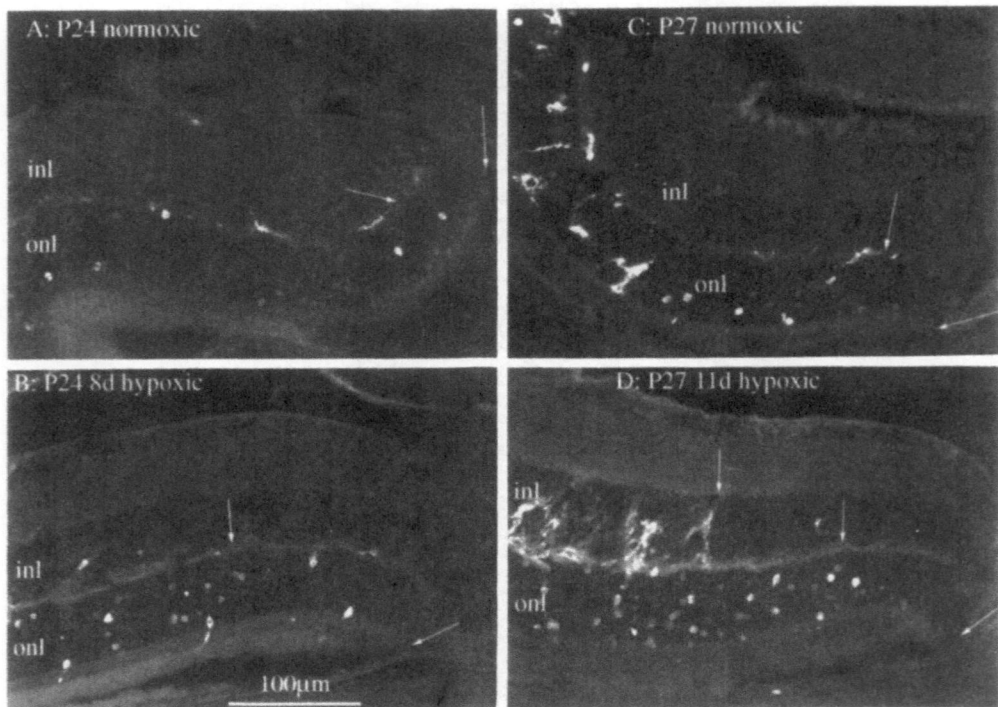

Figure 6. Effects of hypoxia at the edge of the RCS retina. Sections of RCS retinas labelled for fragmenting DNA with the TUNEL technique. In each of A-D the edge of the retina lies to the right. A: In a P24 retina raised in normoxia, labelling is found mainly in the outer nuclear layer (onl). B: In a P24 retina raised for 8 days in 10% oxygen, TUNEL-labelling is increased but is still largely restricted to the onl. C: In a P27 retina raised in normoxia, labelling is higher than at P24 but is still largely restricted to the onl. D: In a P27 retina raised for 11 days in 10% oxygen, TUNEL-labelling of the onl is increased and strong labelling is also seen in the inner nuclear layer (inl). The arrows in A-C and the two lower arrows in D mark TUNEL-sparse regions of the onl at the edge of the retina. The upper two arrows in D mark a TUNEL-sparse region of the inl at the edge of the retina.

Effects of Oxygen Are Pan-Retinal. The rescue of photoreceptors by hyperoxia reported by Maslim et al. (9) was equally apparent in peripheral as in central retina (data not shown). Figure 6 shows evidence that the acceleration of RCS photoreceptor death by hypoxia shown for central retina in Figure 5 also occurs at the retinal periphery. Figures 6A and B show the peripheral margin of P24 RCS retinas raised in normoxia (A) and after 8 days' hypoxia (B). The number of TUNEL+ cells in the ONL is markedly higher in B than in A. Similarly, in a P27 retina raised in normoxia (C) the number of TUNEL+ profiles in the ONL is markedly higher than at P24 (A), showing the progression of the degeneration characteristic of this strain. In a P27 RCS retina after 11 days' hypoxia (D), the number of TUNEL+ profiles in the ONL is further increased, and degenerative structures are prominent in the INL.

Oxygen-around-the-Corner: Effect of Hypoxia Are Limited Near Retinal Edge. Two exceptions were noted to the pan-retinal effects of oxygen on the RCS retina, both anticipated by an earlier study. LaVail and Battelle (27) noted that in the RCS rat photoreceptors survive long periods at the edges of the retina, both at the peripheral margin of the retina and at the margin around the optic disc. Correspondingly, the arrows in Figures 6A-C and the two lower arrows in D delineate TUNEL-sparse regions of the ONL. In the retina shown in D the prolonged (11 day) exposure to hypoxia also produced degeneration in the INL, obvious at left. The upper two arrows in D indicate a TUNEL-sparse region at the edge of the INL.

Figure 7A shows the optic disc region of a P30 RCS retina raised in normoxia, labelled with the TUNEL technique (top panel) and with the G. simplicifolia lectin (lower panel). TUNEL-labelling is prominent in the ONL (o), but not the INL (i). The arrows in the lower panel and the inner arrows in the upper panel indicate the edges of the optic

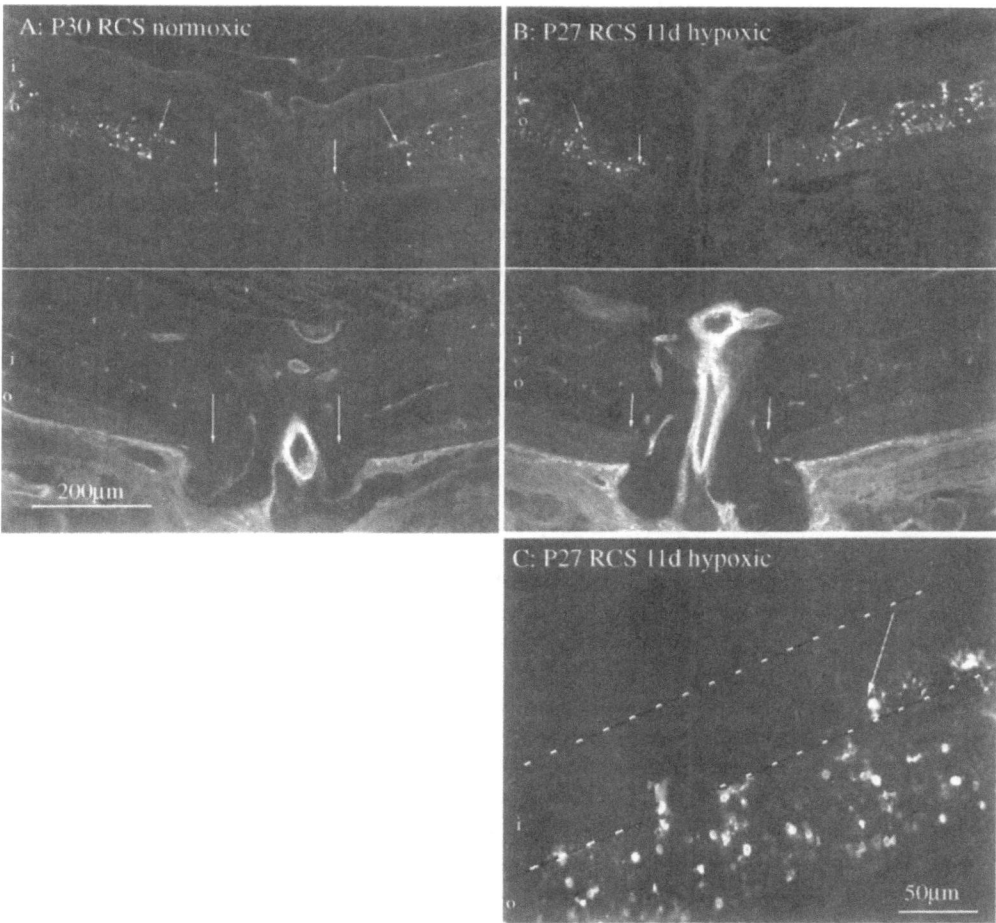

Figure 7. Effects of hypoxia at the optic disc of the RCS retina. A: Upper panel shows TUNEL-labelling of the outer nuclear layer (o) in a P30 retina raised in normoxia. The lower panel shows the same region labelled with the G. simplicifolia lectin, the arrows marking the edges of the head of the optic nerve. Note the large vessel passing through the nerve head. The outer arrows in the upper panel mark TUNEL-sparse regions flanking the nerve head. B: Upper panel shows TUNEL-labelling of the outer nuclear layer (o) in a P27 retina after 11 days in 10% oxygen. The lower panel shows the same region labelled with the G. simplicifolia lectin, the arrows marking the edges of the head of the optic nerve. Note the large vessel passing through the nerve head. The outer arrows in the upper panel mark TUNEL-sparse regions of the ONL flanking the nerve head. C: Region of retina near (just to the right of) the optic disc of the eye shown in B, from an adjacent section. TUNEL-labelling in the inner nuclear layer (i) is suppressed as far as the arrow.

disc. Flanking the disc region, and delineated by the outer arrows in the top panel, are TUNEL-sparse regions. Using the same conventions Figure 7B shows TUNEL-sparse regions of the ONL flanking the optic disc of a P27 retina after 11 days' hypoxia. In this animal, the prolonged hypoxia caused TUNEL-labelling of the INL, illustrated in Figures 5H and 6D, and visible also in Figure 7B. Figure 7C shows at higher power a region just to the right of the optic disc, in a section adjacent to that shown in B. The dotted lines indicate the width of the INL (i); the arrow marks the border of a region near the optic disc in which the INL is largely free of degenerative changes. Apparent in both A and B (lower panels) are some of the large vessels which traverse the optic disc. We suggest in Discussion that oxygen flowing from arteries at the optic nerve head protects peri-disc regions from hypoxic damage, and leads to the preservation of photoreceptors around the optic disc described by LaVail and Battelle (27).

DISCUSSION

Work of long standing (6,7) has provided evidence of the nutrient- (oxygen and glucose) dependence of the performance of human photoreceptors. More recently, Steinberg (3) reviewed evidence of the close oxygen-dependence of the performance of rods in the cat retina, arguing that this dependence stems from the anatomy and physiology of photoreceptors, in particular their separation from the blood vessels which supply them and their high consumption of oxygen. The present paper provides evidence of the importance of a major nutrient, oxygen, in the control of the death and survival of photoreceptors in the normogenetic rat and mouse, and in the dystrophic RCS rat.

Genes vs. Microenvironment: What "Causes" Photoreceptor Dystrophy?

Studies of the genetic basis of inherited retinal dystrophies (reviewed in 1,2) demonstrate why some individuals in a population suffer photoreceptor dystrophy and others do not. They do not address the question why the photoreceptor, rather than some other neurone of the retina, is the casualty of so many genetic mutations. The latter question seems answered by the fact that many receptor-lethal mutations occur in receptor-specific genes: in rhodopsin, in the retinal form of cGMP phosphodiesterase or in the retinal form of peripherin (28). Nevertheless, it has recently emerged that photoreceptors are specifically vulnerable to mutations in ubiquitous genes, in X-linked forms of retinitis pigmentosa (29) and in an AChE transgenic mouse reported by Broide et al. (30). More generally, there is no reason to expect that photoreceptor-related genes are more susceptible to mutation than genes important to all neurones. Mutations in photoreceptor-related genes are more frequently lethal, we suggest, not because they are more frequent or radical, but because the photoreceptor is more vulnerable to mutations.

What makes the photoreceptor vulnerable? Present results suggest that the photoreceptor may be vulnerable to so many mutations because of the precariousness of its nutrient supply. This suggestion complements Steinberg's (3) insight that the vulnerability of the photoreceptors' performance lies in the precariousness of their nutrient supply.

If confirmed, the concept of the nutrient-based vulnerability of photoreceptors will be important for therapy for retinal dystrophy. If the cause of photoreceptor dystrophy is considered to be strictly genetic, it is natural to seek a genetic therapy. If the cause is considered to include aspects of the microenvironment of the cell, such as hypoxia or hypoglycaemia, then a rationale is established for nutrient-based therapy.

What Makes Photoreceptors Die in the RCS Rat?

The failure of the RPE to ingest the outer segment membrane discarded by photoreceptors is part of the chain of events lethal to photoreceptors in this strain (25, 31, 32), but several lines of evidence suggest that the failure of ingestion per se does not cause photoreceptor death, that hypoxia is a more immediate cause.

LaVail and Battelle (27) reported that photoreceptors survive at the margins of the RCS retina, although the RPE abnormality presumably persists at those margins. We have previously reported the effectiveness of hyperoxia in delaying photoreceptor death in the RCS rat (9), and here report the effectiveness of hypoxia in accelerating that death. Figures 6 and 7 show evidence moreover that the edges of the retina are relatively resistant to hypoxic damage. An economical explanation of these observations is that RCS photoreceptors die because the accumulation of debris in the subretinal space makes them hypoxic, perhaps by reducing the diffusion of oxygen from the choriocapillaris. If they receive supplementary oxygen, whether by supplementation of the air they inspire (9, Figure 4), or by diffusion from large arteries funnelling through the optic disc or from the choroid around the peripheral edge of the retina, the photoreceptors survive for long periods. The long-term survival of RCS photoreceptors around a laser lesion shown by Humphrey et al. (33) may also be mediated by the flow of oxygen known to occur through such lesion sites (34). We suggest that RCS photoreceptors die when they become abnormally hypoxic during a critical period (9) of their development; given supplementary oxygen they survive for long periods, despite the malfunction of the RPE.

Photoreceptor Rescue: A Matter of Rates

For an individual photoreceptor rescue seems an all-or-none phenomenon; the cell dies or survives. In a large population of photoreceptors however rescue can be viewed as a matter of rates. Photoreceptor populations of the normogenetic retina suffer slow attrition throughout adulthood, in rodents (20,21) and in humans (34). The rates of death are not trivial; Gao and Hollyfield's data (34) indicate that each human retina loses several hundred thousand rods per year throughout adult life, more in the earlier years of their sample (the teenage years). This loss presumably underlies the steady rise in visual thresholds apparent in later decades of life (35, 36).

Loss of photoreceptors and decline of rod vision are thus normal features of longevity. Since RP sufferers typically retain useful though declining vision for decades, the core difference between normal and dystrophic retinas lies not in the occurrence but in the rate of photoreceptor and visual loss. Effective therapy for RP can then be aimed at reducing the rate of photoreceptor death. Present data suggest that nutrient supply is a factor in the rate of photoreceptor loss in the normogenetic retina, that in some genetic dystrophies compromise to nutrient supply is the factor which increases the rate of loss, and that increasing the supply of nutrient is effective in countering that increase.

ACKNOWLEDGMENTS

The anti-opsin antibody was generously made available by Dr. R. Molday. The skilled technical assistance of Ms Tania Novikova and of the staff of the confocal facility of the Electron Microscope Unit of the University of Sydney is gratefully acknowledged. Supported by the Australian Retinitis Pigmentosa Association, the National Health and Medical Research Council and the Sir Zelman Cowen Universities Fund.

REFERENCES

1. Farrar. G., et al., 1993, Extensive genetic heterogeneity in autosomal dominant retinitis pigmentosa, in: Retinal Degeneration, (eds. J.G. Hollyfield, R.E. Anderson, M.M. LaVail, eds.), pp. 63–77, Plenum Press, New York.

2. Dryja, T., and Berson, E. L., 1995, Retinitis pigmentosa and allied diseases. Implications of genetic heterogeneity, Invest. Ophthal. Vis. Sci. 36: 1197–1200.

3. Steinberg, R.H., 1987, Monitoring communications between photoreceptors and pigment epithelial cells: Effects of "Mild" systemic hypoxia, Invest. Ophthal. Vis. Sci. 28: 1888–1903.

4. Ames, A., Li, Y-Y, Heher, E. C., and Kimble, C. R., 1992, Energy metabolism of rabbit retina as related to function: high cost of Na+ transport, J. Neurosci. 12: 840–853.

5. Chan-Ling, T., and Stone, J., 1993, Retinopathy of prematurity: Origins in the architecture of the retina, Progress in Retinal Research 12: 155–176.

6. McFarland, R., and Forbes, W., 1940, The effects of variations in the concentration of oxygen and of glucose on dark adaptation, J. Gen. Physiol. 24:69–98.

7. McFarland, R., Halperin, M., Niven, J., 1945, Visual thresholds as an index of the modification of the effects of anoxia by glucose, Am. J. Physiol. 44:378–388.

8. Maslim, J., Egensperger, R., Hollander, H., Humphrey, M., and Stone, J., 1995, Receptor degeneration is a normal part of retinal development, in: Degenerative Diseases of the Retina, (G. Anderson, J.G. Hollyfield, M.M. LaVail, eds.), pp 187–196, Plenum Press, New York.

9. Maslim, J., Valter, K., Egensperger, R., Hollšnder, H., and Stone, J. Tissue oxygen during a critical developmental period controls the death and survival of photoreceptors (submitted).

10. Alder, V. A., Cringle, S. J., and Constable, I. J., 1983, The retinal oxygen profile in cats, Invest. Ophthal. Vis. Sci. 24: 30–36.

11. Alder, V. A., Ben-Nun, J., and Cringle, S. J., 1990, PO2 profiles and oxygen consumption in cat retina with an occluded retinal circulation, Invest. Ophthal. Vis. Sci. 31: 1029–1034.

12. Wolbarsht, M. L., Stefansson, E., and Landers, M. B., 1987, Retinal oxygenation from the choroid in hyperoxia, Exp. Biol. 47: 49–52.

13. Cringle, S. J., Yu, D.-Y., and Alder, V.A., 1991, Intraretinal oxygen tension in the rat eye, Graefe's Arch. Clin. Exp. Ophthal. 229: 1–4.

14. Alder, V. A., and Cringle, S. A., 1985, The effect of the retinal circulation on vitreal oxygen tension, Curr. Eye Res. 4: 121–129.

15. Yu, D-Y, Cringle, S. J., and Alder, V. A., 1990, The response of rat vitreal oxygen tension to stepwise increases in inspired percentage oxygen, Invest. Ophthal. Vis. Sci. 31: 2493–2499.

16. Gavrieli, Y., Sherman, Y., and Ben-Sasson, S. A., 1992, Identification of programmed cell death in situ via specific labeling of nuclear DNA fragmentation, J. Cell Biol. 119: 493–501.

17. Stone, J., Itin, A., Alon, T., Pe'er, J., Gnessin, H., Chan-Ling, T., and Keshet, E., 1995, Development of retinal vasculature is mediated by hypoxia-induced vascular endothelial growth factor (VEGF) expression by neuroglia, J. Neurosci. 15: 4738–4747.

18. Stone, J., Chan-Ling, T., Pe'er, J., Itin, A., Gnessin, H., and Keshet, E., 1996, Roles of vascular endothelial growth factor and astrocyte degeneration in the genesis of retinopathy of prematurity, Invest. Ophthal. Vis. Sci. 37: 290–299.

19. Lai, Y-L, Jacoby, R. O., and Jonas, A. M., 1978, Age-related and light-associated retinal changes in Fischer rats, Invest. Ophthal. Vis. Sci 17: 634–8.

20. Shinowara, N. L., London, E. D., and Rapoport, S. I., 1982, Changes in retinal morphology and glucose utilization in aging albino rats, Exp. Eye Res. 34: 517–530.

21. Hicks D., and Molday, R.S., 1986 Differential immunogold-dextran labeling of bovine and frog rod and cone cells using monoclonal antibodies against bovine rhodopsin. Exp. Eye Res. 42:55–71.

22. Dowling, J.E., and Sidman, R.L., 1962, Inherited retinal dystrophy in the rat, J. Cell Biol. 14: 73–109.

23. Bok, D., and Hall, M. O., 1971, The role of the pigment epithelium in the etiology of inherited retinal dystrophy in the rat, J. Cell Biol. 49: 664–682.

24. LaVail, M. M., Li, L., Turner, J. E., and Yasumura, D., 1992, Retinal pigment epithelial cell transplantation in rcs rats: normal metabolism in rescued photoreceptors, Exp. Eye Res. 55: 555–562.

25. Tso, M. O. M., Zhang, C., Abler, A. S., Chang, C-J, Wong, F., Chang, G-Q, and Lam, T. T., 1994, Apoptosis leads to photoreceptor degeneration in inherited retinal dystrophy of RCS rats, Invest. Ophthal. Vis. Sci ..35: 2693–2699.

26. LaVail, M.M., and Battelle, B.A., 1975, Influence of eye pigmentation and light deprivation on inherited retinal dystrophy on the rat, Exp. Eye Res. 21: 167–192.

27. Viczian, A., Sanyal, S., Toffenetti, J., Chader, G.J., and Farber, D.B., 1992, Photoreceptor-specific mRNA in mice carrying different allelic combinations at the rd and rds loci, Exp. Eye Res. 54: 853–860.
28. Meindl, A., Dry, K., Herrmann, K., and Manson, F. et al, 1996, A gene (RPGR) with homology to the RCC1 guanine nucleotide exchange factor is mutated in X-linked retinitis pigmentosa (RP3), Nature Genetics 13: 35–42.
29. Broide, R.S., Stone, J., Patrick, J.W., Shani, M., and Soreq, H. 1996 Selective degeneration of photoreceptors in developing retina of AChE-transgenic mice reveals a non-catalytic function for acetylcholinesterase. Isr. J. Med. Sci., 32, S11.
30. Mullen, R. J., and LaVail, M. M., 1976, Inherited retinal dystrophy: primary defect in pigment epithelium determined with experimental rat chimeras, Science 192: 799–801.
31. Li, L., and Turner, J. E., 1988, Inherited retinal dystrophy in the RCS rat: prevention of photoreceptor degeneration by pigment epithelial cell transplantation, Exp Eye Res. 47: 911–917.
32. Humphrey, M. F., Parker, C., Chu, Y., and Constable, I. J., 1993, Transient preservation of photoreceptors on the flanks of argon laser lesions in the RCS rat, Curr. Eye Res. 12: 367.
33. Gao, H., and Hollyfield, J. G., 1992, Aging of the human retina: differential loss of neurons and retinal pigment epithelial cells, Invest. Ophthal. Vis. Sci 33: 1–17.)
34. Stefansson, E., Landers, M. B., and Wolbarsht, M. L., 1981, Increased retinal oxygen supply following pan-retinal photocoagulation and vitrectomy and lensectomy, Tr. Am. Ophthal. Soc. 74: 307–334.
35. Knoblauch, K., Saunders, F., Kusuda, M., Hynes, R., Podgor, M., Higgins, K.E., and Monasterio, FM, 1987, Age and illuminance effects in the Farnsworth-Munsell 100-hue test, Applied Optics 26: 1441–1448.
36. Sloane, ME, Owsley, C., and Alvarez, SA, 1988, Aging, senile miosis and spatial contrast sensitivity at low luminance, Vision Research 28: 1235–1246.

41

HISTOLOGICAL METHOD TO ASSESS PHOTORECEPTOR LIGHT DAMAGE AND PROTECTION BY SURVIVAL FACTORS

Matthew M. LaVail,[1,3,4] Michael T. Matthes,[3,4] Douglas Yasumura,[1,4] Ella G. Faktorovich,[3,4] and Roy H. Steinberg[2,3,4]

[1]Department of Anatomy
[2]Department of Physiology
[3]Department of Ophthalmology
[4]Beckman Vision Center
University of California at San Francisco
San Francisco, California 94131-0730

ABSTRACT

Constant light damage in albino rats has become a useful model for testing the therapeutic potential of survival factors to protect photoreceptors from cell damage and death. Considerable variation occurs among animals in light damage experiments. An additional complication with the light damage model is that less damaged photoreceptors show greater protection by a given survival factor than those that are more severely damaged by light. A method is described that allows for comparison of relatively small numbers of animals and considers variability, relative degree of protection by survival factors and other qualitative measures of damage.

INTRODUCTION

Quantification of photoreceptor cell loss and survival is important in studies of retinal degenerations. It has been suggested that the most efficient and accurate method is to measure the thickness of the outer nuclear layer (ONL) (1), and this has been used effectively in studies where a degeneration rate is uniform and consistent among animals of a given age (2, 3). Measurement of the ONL thickness, however, does not consider other important factors, such as photoreceptor inner and outer segment length and integrity. Moreover, in some types of degeneration, considerable variation within groups or individual litters of animals is seen, such as in light damage as noted previously (4–6).

Degenerative Retinal Diseases, edited by LaVail *et al.*
Plenum Press, New York, 1997

In experiments studying potential therapeutic measures, such as the use of survival factors, constant light damage has become the most widely used model for assessing photoreceptor protection in the rat (6–9). The considerable variability among animals to the damaging effects of light has necessitated the use of relatively large numbers of animals to achieve statistical significance when using ONL thickness as the measure of protection by various agents (6). Sometimes this is not possible due to limited availability of reagents or animals. One way to counter variability is to compare an experimental eye with the opposite eye of the same animal, since the two eyes of a rat show statistically indistinguishable degrees of light damage to the retina (9). An additional complication with the light damage model is that less damaged photoreceptors show an apparently greater protection by a given survival factor than those that are more severely damaged by light (9), at least when the ONL thickness and photoreceptor inner and outer segment length and integrity are all considered. Thus, a scoring method was needed that took into account this relative degree of protection, as well as the variability in degree of light damage among animals and other qualitative measures, such as inner and outer segment integrity. In this report, we describe such a scoring method, which we have described only incompletely before (9), as well as document the wide range of variability among animals in light damage experiments.

METHODS

In most of the experiments described here, Sprague-Dawley (SD) non-inbred albino rats at 2–5 months of age were obtained from Simonsen Laboratories, Inc. (Gilroy, CA). The rats were maintained in cyclic light for 9 or more days in our laboratory at an illuminance level of less than 25 ft-c, at which time they were exposed for 7 days to constant fluorescent light at an illuminance level of 115–200 ft-c. The rats were then killed by overdose of carbon dioxide followed immediately by vascular perfusion with mixed aldehydes. The eyes were embedded in epoxy resin and sectioned at 1 μm thickness along the vertical meridian, so that superior-inferior hemispheric differences could be studied. These procedures are described elsewhere in detail (2, 6). The superior cornea was trimmed close to the ora serrata to distinguish the superior from the inferior hemisphere in the tissue sections, and the hemispheric identification was confirmed by visualizing the branches of the long posterior ciliary artery located in the choroid on the inferior side of the optic nerve head in the sections (10).

In experiments describing survival factor protection, examples are shown without regard to the specific factor injected. In all cases, however, 1 μl of the agent was injected into one eye of the rat 2 days before it was placed into constant light. At the same time, 1 μl of the phosphate-buffered saline (PBS) carrier was injected into the opposite eye as a control.

For measurement of the ONL, 54 measurements were taken at defined points around the eye, 3 each in 9 sets of 440-μm lengths of retina in each of the superior and inferior hemispheres, with the aid of a Bioquant morphometry system (R and M Biometrics, Nashville, TN) as described elsewhere (6). From these measurements a mean ONL thickness was obtained, as well as a plot of ONL thickness across the eye.

The description of the scoring method for degree of photoreceptor preservation following light damage and the degree of protection provided by survival factors are the subject of this report; thus, they are presented in the Results.

RESULTS

Description of Different Degrees of Light Damage

After 7 days of constant light, retinas of SD rats show significant differences in degrees of light damage. We have arbitrarily chosen the descriptors for the *degree of photoreceptor preservation* as "low," "medium" and "high." The low SD retinas have the least photoreceptor preservation (i.e., greatest damage). The superior hemisphere is more severely damaged by light (11), and it typically has 0–1 row of nuclei in the posterior retina (Figs. 1 and 2B), with 1–2 rows of nuclei present in the ONL of the inferior posterior retina (Figs. 1 and 2B). The far peripheral retina may have a few more rows of nuclei persisting (Fig. 1 and Table 1). Note that in the graph shown in Fig. 1, the low SD does not drop to zero, because each point is the mean of 3 measurements, one or more of which is usually made where at least 1 nucleus is present; each nucleus has a diameter of 5–6 μm. (See also Legend to Fig. 1.) Thus, the ONL measurements usually do not give a precise depiction of the minimal number of nuclei present as seen by light microscopy (Fig. 2B). No rod inner segments are visible in the low SD retinas, except for occasional discrete nubs (defined as <25% of normal length) in the inferior peripheral retina (Table 1), and the outer segments consist of amorphous membranous debris and whorls (large, rounded, membranous profiles about the size of a photoreceptor nucleus), as described elsewhere (12–14) (Figs. 2C and 6A and Table 1).

The medium SD retinas typically have 0–2 rows of nuclei in the superior posterior ONL (Fig. 2D and Table 1), and significantly more in the inferior posterior ONL (Figs. 1

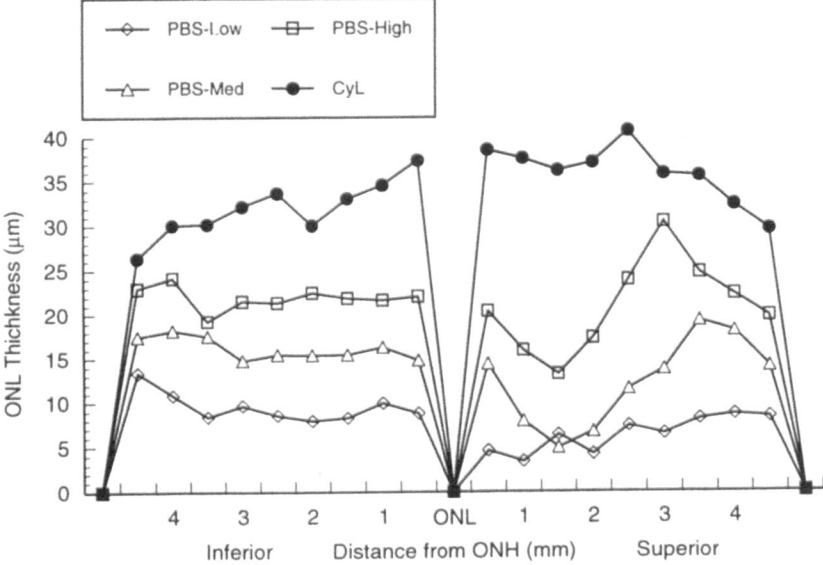

Figure 1. Measurements of the outer nuclear layer (ONL) thickness in SD rats either kept in cyclic light (CyL) or exposed to constant light for 7 days. The graphs are examples of low, medium and high SD retinas from eyes injected with PBS. Each graph shows measurements from a single section from a representative eye taken along the vertical meridian of the eye. Although some of the most degenerated regions of the retina are almost devoid of photoreceptor nuclei in the low and medium SD retinas, the data points on these graphs rarely reach zero, because each data point is the mean of 3 measurements, and one or more nuclei are usually included in at least one of the measurements. Since the diameter of a photoreceptor nucleus is about 5–6 μm, any data point less than 10 μm represents a region of retina where the ONL may consist of less than one photoreceptor nucleus at a given point.

Table 1. Criteria for scoring degree photoreceptor preservation in constant light and photoreceptor protection by survival factors

	PBS	1+	2+	3+	4+
Low SD					
Sup-Post					
ONL	0-2	0-2	1-3	2-4	3-4
RIS	None	None	None	Nubs	Nubs+Short
ROS	Debris	Debris	W	W	W+F
Sup-Periph					
ONL	0-2	3-4	4-5	4-6	4-6
RIS	None	Short	Short	Short	Short+Long
ROS	W	W+F	W+F	W+F	W+F+Short
Inf-Post					
ONL	1-2	2-3	2-3	3-4	5-6
RIS	None	None	None	Nubs	Short
ROS	Debris+W	W	W	W	W+F
Inf-Periph					
ONL	2-3	3-4	3-5	4-6	5-6
RIS	None-Nubs	Short	Short	Short	Short
ROS	Debris+W	W	W	W	W+F
Medium SD					
Sup-Post					
ONL	0-2	1-2	2-3	2-4	3-4
RIS	None	Nubs	Nubs	Nubs	Nubs+Short
ROS	Debris+W	W	W	W	W+F
Sup-Periph					
ONL	3-4	4-5	5-6	5-7	5-7
RIS	Nubs	Short	Short	Short	Short+Long
ROS	W	W+F	W+F	W+F	W+F+Short
Inf-Post					
ONL	2-4	3-5	3-5	5-6	5-7
RIS	Nubs	Nubs+Short	Nubs+Short	Nubs+Short	Short+Long
ROS	W	W	W	W+F	W+F
Inf-Periph					
ONL	3-5	4-5	5-6	5-7	6-8

(continued)

and 2E and Table 1). These retinas almost always exhibit a small area in the superior posterior retina where there are no nuclei (Fig. 2D and Table 1). The far peripheral retina of both hemispheres exhibits a greater number of nuclei than the posterior retina (Fig. 1 and Table 1). The rod inner segments consist of discrete nubs, and rod outer segment membranes consist of whorls in most regions (Table 1).

In the high SD retinas there is significant preservation of the ONL, with 2–4 rows of photoreceptor nuclei in the region of the superior hemisphere that is most severely dam-

RIS	Nubs	Short	Short	Short	Long
ROS	W	W	W+F	W+F+Short	W+F+Short
High SD*					
Sup-Post					
ONL	2-4		3-5	4-6	5-7
RIS	Nubs		Nubs	Nubs+Short	Short
ROS	W+F		W+F	W+F+Short	W+F+Long
Sup-Periph					
ONL	5-7		6-7	6-8	7-9
RIS	Short		Short-Long	Short-Long	Long
ROS	W+F		W+F+Short	W+F+Short	F+Short+Long
Inf-Post					
ONL	5-6		5-7	6-7	6-7
RIS	Nubs		Nubs	Nubs+Short	Short
ROS	W+F		W+F+Short	W+F+Short	W+F+Short
Inf-Periph					
ONL	5-7		5-7	5-7	6-8
RIS	Short		Short	Short	Long
ROS	W+F		W+F+Short	W+F+Short	F+Short+Long

Regions examined are the superior-posterior (Sup-Post) at the point of most severe damage; superior-peripheral (Sup-Periph) approximately 100-400 µm from the ora serrata in the superior hemisphere; inferior-posterior (Inf-Post), at a point approximately 200-600 µm from the optic nerve head; inferior-peripheral (Inf-Periph) approximately 100-400 µm from the ora serrata in the inferior hemisphere.

Scoring: Outer nuclear layer (ONL): Number of rows of nuclei. (Note: The number of nuclei in normal, 2- to 5-month-old rat retinas reared in cyclic light is 8-9 in the posterior retina, 7-9 in the peripheral retina.) Rod inner segments (RIS) : None (most severe degeneration); nubs = <25% of normal length; short = 25-60% of normal length; long = >60% of normal length. Rod outer segments (ROS): Debris = amorphous membranous profiles (most severe degeneration); Whorls (W) = rounded membranous profiles; Fragments (F) = <20% of outer segment length; short = 20-40% of normal length; long = >40% normal length.

*No score of 1+ protection is given with high SD retinas, since the range of protection is significantly less than that in low and medium SD retinas, and it is difficult to distinguish 4 levels of protection between the most degenerated PBS-injected retinas and the most protected (4+) retinas.

aged by light (Fig. 2F), and 5–6 rows in the posterior inferior retina (Fig. 2G and Table 1). In the far peripheral retina in both hemispheres, light damage is usually somewhat less severe, and the number of rows of nuclei in the ONL may approach the number normally found in rats maintained in cyclic light (Fig. 1). The rod inner segments range from nubs to short lengths (defined as 25–60% of normal length) and the rod outer segment mem-

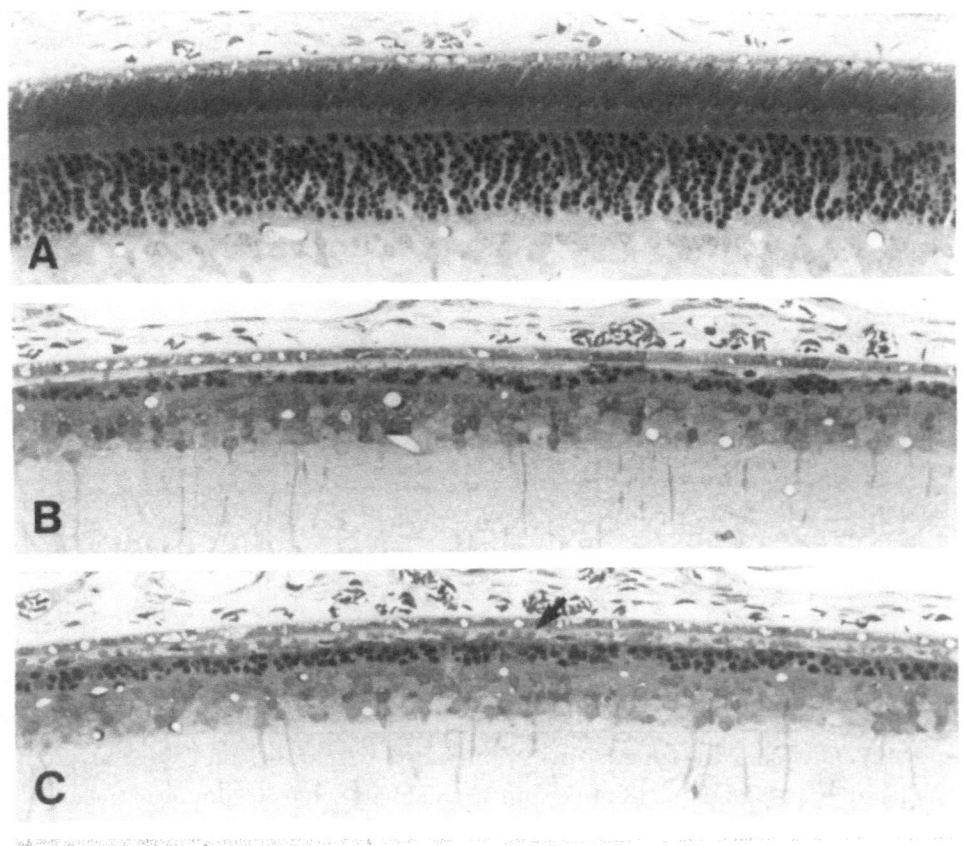

Figure 2. Light micrographs of the posterior retinas of albino SD rats in which the eyes were injected with PBS and the rats were either maintained in cyclic light (A) or exposed to 7 days of constant light (B-G). B and C. Low SD retinas; superior hemisphere (B); inferior hemisphere (C). D-G. On next page. D and E. Medium SD retinas; superior hemisphere (D); inferior hemisphere (E). F and G. High SD retinas; superior hemisphere (F); inferior hemisphere (G). Arrow, outer segment whorl. Toluidine blue stain. X300.

branes range from whorls to fragments, which we define as short lengths of approximately normal diameter (1.5 μm), but <20% the length of normal rod outer segments.

While the superior hemisphere is more severely affected than the inferior with all degrees of degeneration, there often is a small region in the equatorial to peripheral retina that shows a greater number of nuclei (Fig. 1). This is due to protection from photoreceptor injury and death that is induced locally by injury from the needle at the injection site (6).

Not surprisingly there are often retinas where the degree of preservation does not fit into low, medium or high descriptors (i.e., full preservation types), but is either intermediate between two levels, or at one level in one hemisphere and a different level in the other hemisphere. For this reason we needed to expand our scale to include "low-medium" and "medium-high" levels. For instance, a low-medium level would be assigned to a retina where the degree of preservation falls between low and medium, or where one hemisphere is low and the other medium. Even this scale sometimes needed to be extended to describe all retinas (i.e., the scale was still not fine enough). In this case, we underlined one level, for instance a "<u>low</u>-medium" retina was less well preserved than a low-medium retina, and

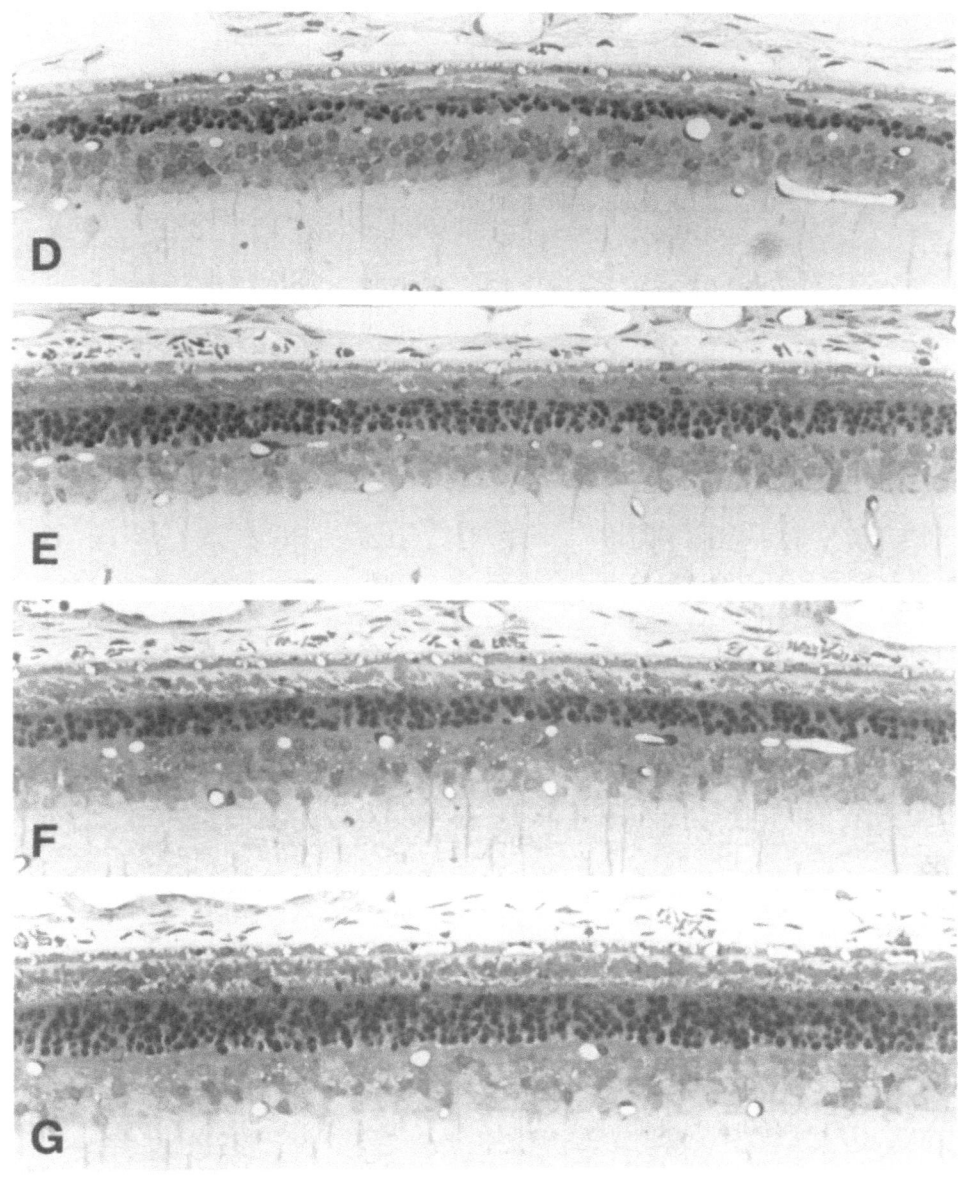

Figure 2. (*Continued*).

so forth. In this latter instance, the reason for the underline was usually a greater or lesser length (representing areal distribution) of one or the other degree of preservation.

Thus, the result was a 5-point scale with a number assigned for each level of preservation, with the full intervals defined by 1.0. For instance, high = 5; medium = 3; and low = 1. Thus, a low-medium SD would have a score of 2. The underlined scores use an interval of 0.5, so for example a low-medium SD would have a score of 1.5. We require that 2–5 observers, usually 3–4, either together or individually, agree on the score of the reti-

Table 2. Variation in degree of light damage*

Scores**	Mean	S.D.	# cages
1.00, 1.00, 1.00	1.00	0	2
2.00, 2.00, 2.00	2.00	0	5
2.50, 2.50, 2.50	2.50	0	1
3.00, 3.00, 3.00	3.00	0	2
5.00, 5.00, 5.00	5.00	0	1
1.00, 1.00, 1.50	1.17	0.29	2
1.00, 1.50, 1.50	1.33	0.29	1
1.50, 1.50, 2.00	1.67	0.29	1
1.50, 2.00, 2.00	1.83	0.29	2
2.00, 2.00, 2.50	2.17	0.29	2
2.00, 2.50, 2.50	2.33	0.29	2
2.50, 2.50, 3.00	2.67	0.29	2
2.50, 3.00, 3.00	2.83	0.29	1
3.00, 3.00, 3.50	3.17	0.29	1
1.00, 1.50, 2.00	1.50	0.50	1
1.50, 2.00, 2.50	2.00	0.50	1
2.00, 2.50, 3.00	2.50	0.50	2
1.00, 1.00, 2.00	1.33	0.58	5
1.00, 2.00, 2.00	1.67	0.58	4
2.00, 2.00, 3.00	2.33	0.58	4
2.00, 3.00, 3.00	2.67	0.58	8
3.00, 3.00, 4.00	3.33	0.58	5
3.00, 4.00, 4.00	3.67	0.58	1
1.50, 2.00, 3.00	2.17	0.76	1
2.00, 3.00, 3.50	2.83	0.76	1
2.50, 3.00, 4.00	3.17	0.76	3
3.00, 3.50, 4.50	3.67	0.76	1
1.50, 3.00, 3.00	2.50	0.87	1
1.00, 2.00, 3.00	2.00	1.00	3

(continued)

nas. With practice, the scoring of the retinas can usually be done within 1–3 minutes, so the scoring procedure is relatively rapid.

Variation in Degree of Light Damage

To illustrate the variation observed in light damage in rats exposed while in the same cage, where the animals presumably would receive very close to the same retinal ir-

Table 2. (*Continued*)

2.00, 3.00, 4.00	3.00	1.00	1
1.00, 2.50, 3.00	2.17	1.04	2
1.50, 2.00, 3.50	2.33	1.04	1
1.50, 3.00, 3.50	2.67	1.04	1
3.00, 3.50, 5.00	3.83	1.04	1
3.00, 4.50, 5.00	4.17	1.04	1
1.00, 1.00, 3.00	1.67	1.15	7
1.00, 3.00, 3.00	2.33	1.15	3
3.00, 3.00, 5.00	3.67	1.15	1
1.50, 3.00, 4.00	2.83	1.26	1
2.00, 3.00, 4.50	3.17	1.26	1
1.00, 2.00, 4.00	2.33	1.53	1
1.00, 3.00, 4.00	2.67	1.53	1
2.00, 4.00, 5.00	3.67	1.53	1
2.00, 2.00, 5.00	3.00	1.73	2
1.00, 3.00, 4.50	2.83	1.76	1
1.50, 3.00, 5.00	3.17	1.76	1
1.50, 2.50, 5.00	3.00	1.80	1
1.00, 3.00, 5.00	3.00	2.00	3
1.50, 1.50, 5.00	2.67	2.02	1
1.00, 2.00, 5.00	2.67	2.08	1
1.00, 5.00, 5.00	3.67	2.31	1

*Sorted in ascending order by variance (S.D, standard deviation).
**Each row represents a single set of degrees of photoreceptor preservation. Each cage contained 3 rats that had received a PBS injection 2 days before a 7-day constant light exposure. These are 100 cages chosen randomly from experiments carried out over a 3-year period, approximately 33% from each year.

radiance, scores were analyzed from 100 cages randomly chosen from experiments carried out in our laboratory from 1994–1996 (approximately 33 from each year). In each instance, 3 rats with the same birthdate were exposed in a single cage (they may or may not have been siblings). Also, the retinas examined always were from PBS-injected control eyes from rats in which the other eye had been injected with a survival factor.

As shown in Table 2, there was significant variation in the degree of light damage in these control retinas. Rats in only 11% of the cages had identical scores in the retinas of all 3 of the rats. Relatively low variation, in which a range of less than one full type was present in a single cage (i.e., a range of less than 2 full score units, such as between low and medium, or between low-medium and medium-high), was seen in 52% of the cages. Moderate variation, where the range was one full degeneration type (i.e., 2 full score units), was seen in 21% of the cages. High variation, where the range was greater than one full degeneration type (i.e., 3 score units), was present in 16% of the cages. Of the latter group, 6 cages had rats with scores that ranged from low to high SD in the same cage, and 3 of these had a low, medium and high SD in the same cage (Table 2).

We have examined light damage in fewer uninjected rat eyes than PBS-injected rat eyes, but the uninjected rat eyes also showed a high degree of variability and all degrees of preservation (data not shown).

Scoring the Degree of Protection by Survival Factors

When we assessed the amount of protection, i.e., the improvement in photoreceptor preservation, produced by the injection of survival factors, we needed to quantify, in some way, the difference in light damage between the control eye and the survival factor-injected eye of a given animal. Ideally, one would separately score each eye in terms of degree of light damage, as described above, and then subtract the two scores. We found, in practice, that this was difficult to do, and instead, adopted a 4-point "pathologist's" scale that described the degree of improvement in the factor-injected eye. Thus, the highest score of 4+ represents the maximal protection we observed for a given level of preservation in the control eye (i.e., baseline levels such as low, medium or high SD). A low score of 1+ represents just perceptible protection, and scores of 2+ and 3+ represent intermediate degrees of protection. When there was no difference between the experimental and control eye (no protection) it was given a zero score. Three features were used to score the degree of protection, the number of rows of nuclei in the ONL; the presence, integrity and length of rod inner segments; and the presence, integrity and length of rod outer segments.

Representative plots of ONL thicknesses of eyes at each of the 4 protection scores are shown for control retinas with low (Fig. 3), medium (Fig. 4) and high (Fig. 5) degrees of photoreceptor preservation. These plots were selected to demonstrate examples of ONL

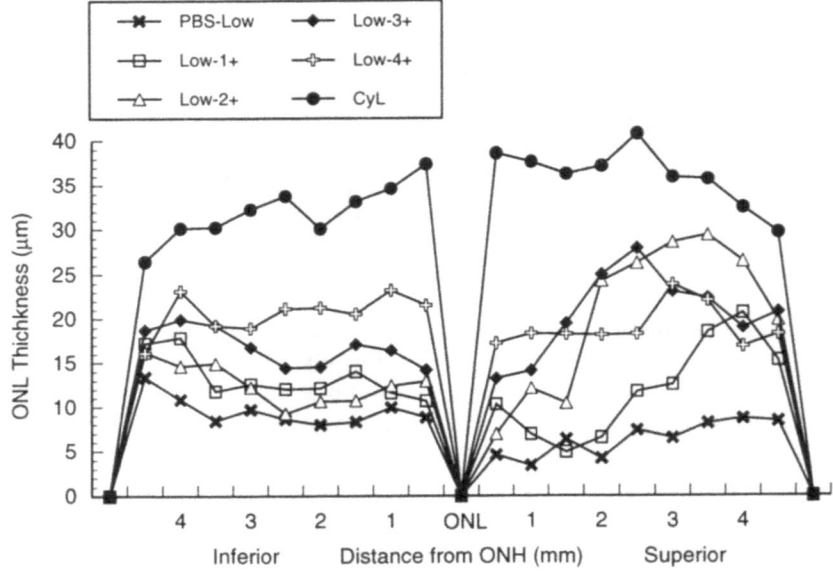

Figure 3. Measurements of ONL thickness in representative sections of eyes showing different degrees of protection from constant light in rats in which the control eyes were scored as low SD (shown as PBS-low) following 7 days of constant light. Retinas with degrees of protection from constant light by one scored 1+, 2+, 3+ and 4+ are illustrated, without regard to the survival factor injected. A graph of a normal retina maintained in cyclic light (CyL) is also shown.

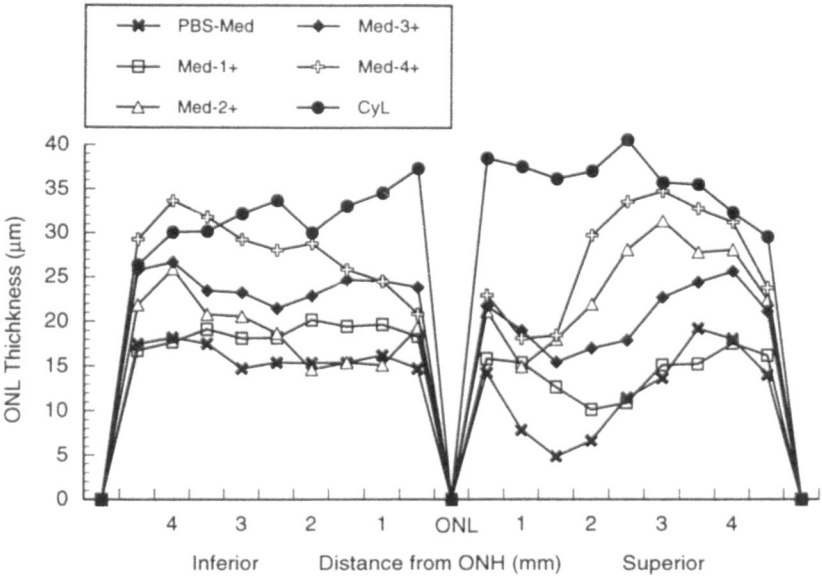

Figure 4. Measurements of ONL thickness in representative sections of eyes showing different degrees of protection from constant light in rats in which the control eyes were scored as medium SD (shown as PBS-medium) following 7 days of constant light. See details in Fig. 3.

thickness for each of the 4 scores. Several general points can be made from these plots (Figs. 3–5). First, ONL plots of many retinas may overlap one another (and have very close mean ONL thicknesses when measured), such as some shown in Figs. 3–5, yet have different protection scores based on other criteria described below.

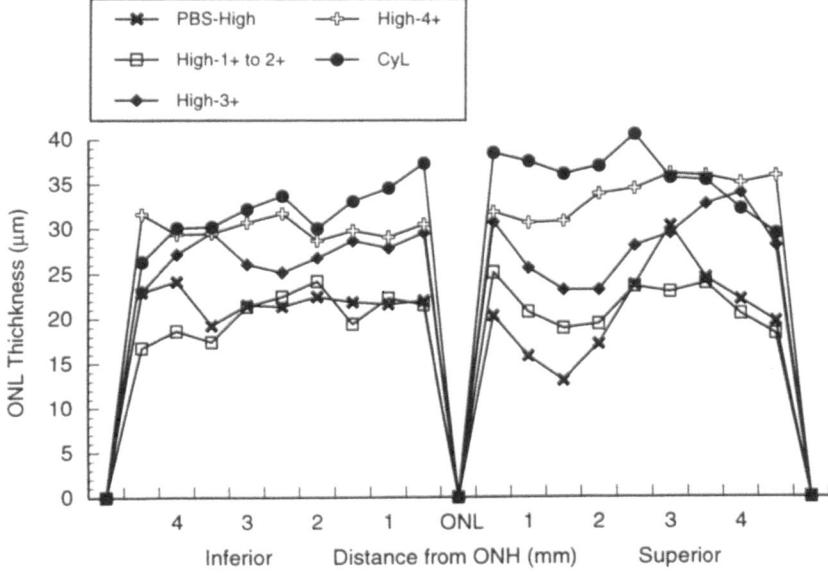

Figure 5. Measurements of ONL thickness in representative sections of eyes showing different degrees of protection from constant light in rats in which the control eyes were scored as high SD (shown as PBS-high) following 7 days of constant light. See details in Fig. 3.

Second, paradoxically, the more severely affected superior hemisphere of the retina appears to respond more dramatically than the inferior hemisphere to the protective effect of a survival factor in many cases (Figs. 3 and 4). This may be due to the interaction of the injection-induced injury response with the exogenous survival factor; or it may be due to a higher concentration of the survival factor, since the eyes were injected in the superior hemisphere; or due to other as yet undefined causes.

Third, the high SD retinas have so many photoreceptors surviving in the control eyes that the range of potential protection between most severely damaged control retinas and normal retinas is less than that for the low and medium SD retinas, and it is therefore significantly more difficult to assign a precise protective score to those retinas. For this reason, we do not use a 1+ score for high SD retinas, and we typically exclude animals with high SD retinas from our quantitative studies on survival factor protection, except to note whether or not protection occurred. Fortunately, only a relatively small percentage of retinas in the experimental population must be discarded, for as shown in Table 2, only about 12% of the control retinas were medium-high SD or higher.

After the PBS-injected control retina is categorized as either low, medium or high, the degree of protection is scored by the criteria given in Table 2. We typically examine the most severely damaged superior-posterior region of the retina and then the rest of the superior hemisphere, followed by the inferior hemisphere, which is the order of the data in Table 2. As noted above, the three features used to score the protection are the number of rows of nuclei in the ONL; the presence, integrity and length of rod inner segments; and the presence, integrity and length of rod outer segments. As described in Table 2, the rod inner segments range from completely missing (most severe damage), to nubs (<25% of normal length), to short lengths (25–60% of normal length) to long lengths (>60% of normal length). The rod outer segments range from debris (most severe damage), to whorls, to fragments, to short lengths (20–40% of normal length) to long lengths (>40% of normal length). Some of these features are shown in Figs. 6A-C, in which examples are given of a rat with a low SD level of photoreceptor preservation in the control eye showing 4+ protection in the factor-injected eye (Fig. 6A), a medium SD showing 3–4+ protection (Fig. 6B) and a high SD showing 1–2+ protection (Fig. 6C).

In many instances, a given retina does not precisely match one of the scoring levels given in Table 2. In these cases, intermediate scores are used (e.g., 2–3+ protection, in which the value 2.5 is used for quantification). Also, if confusion exists on a score, the greater length (representing areal distribution) of a particular degree of protection is used to determine the score

DISCUSSION

Variability in Constant Light Damage

The range of variability in light damage in SD rats is large in a significant fraction of the animals studied. There are clear differences in susceptibility to light damage among strains of mice (15) and rats (14, 16, 17), although the range of variability found in the present study on rats is greater than previously described for either mice or rats, as far as we are aware. There could be several reasons. First, most strains of rats readily available in the United States are non-inbred animals, thus having a greater genetic heterogeneity than inbred strains. At least one genetic factor that regulates the susceptibility of the retina to the damaging effects of light has been found in inbred strains of mice, and to be segre-

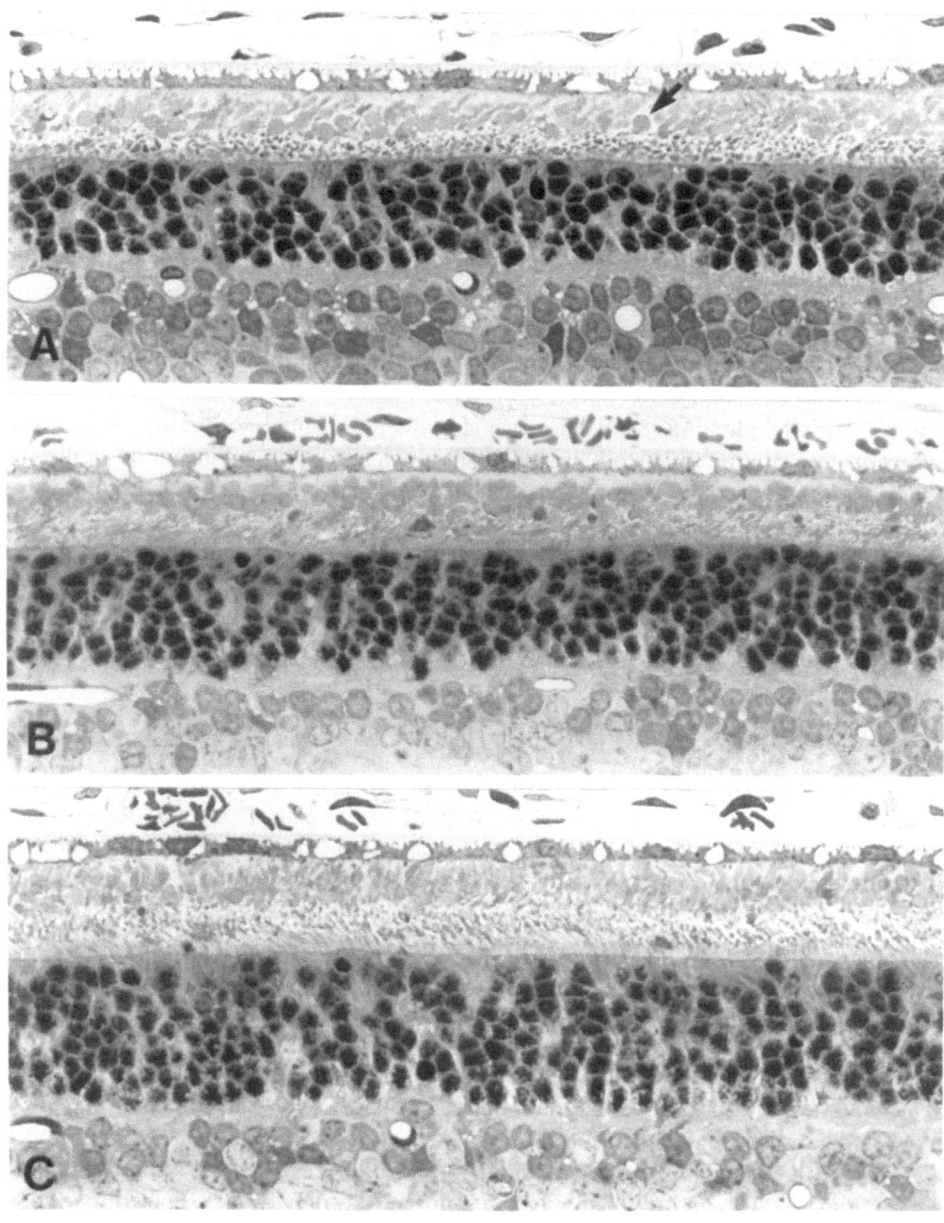

Figure 6. Light micrographs of the posterior retinas of albino SD rats. A. Superior hemisphere of retina showing 4+ protection; the control retina was a low SD. An outer segment whorl (arrow) among fragments of outer segments. B. Inferior hemisphere of a retina showing 3–4+ protection; the control retina was a medium SD. C. Inferior hemisphere of a retina showing 1–2+ protection; the control retina was a high SD. Toluidine blue stain. X600.

gating in at least one inbred mouse strain (15). Thus, a similar genetic factor(s) may exist in the rat. Second, we examined more animals than the number studied in other publications, so it may be that high variability would be found in other strains and stocks of rats if larger populations were examined. Third, it is possible that SD rats are particularly variable. We examined this possibility in one experiment where light damage was compared in uninjected SD, Lewis and F344 rats that were exposed at the same time, and the variance was similar in each of the strains (although the 6 SD rats did not show any high SD retinas in that experiment; unpublished observations). Fourth, it is possible that the SD rats have more variable behavioral traits that would affect light damage (e.g., burrowing beneath one another in constant light) than other strains, but this has not been explored. Fifth, it is possible that the SD rats in different cages were exposed to different levels of cyclic-light illuminance which were great enough to influence the degree of light damage (18, 19), but this would not explain the differences among rats maintained in the same cage, as we have studied. Sixth, it is possible that the injection of PBS resulted in an injury induced protective response that was so large that it, alone, produced the medium and high SD phenotypes. This seems unlikely, however, since in other studies in our laboratory the injection results in a localized protected area and sometimes a slight increase in ONL thickness across the retina, albeit statistically insignificant (6). Furthermore, the survival factors in the opposite eyes gave greater photoreceptor preservation than the control eyes with middle-high and high SD scores in the present experiments, suggesting that the scores of the control eyes were correct. Moreover, we have seen all degrees of photoreceptor preservation, low, medium and high, in uninjected eyes in light damage experiments, as well. Therefore, the causes of variability in light damage are probably complex and remain to be determined.

Utility of Scoring Method for Protection by Survival Factors

The 0–4+ scoring method is a relative scale that takes into consideration the variability of degree of light damage among animals and the greater degree of protection by survival factors in retinas that are less severely damaged when not only the ONL thickness, but also the integrity and length of photoreceptor inner and outer segments, and their abundance and areal distribution are considered.

The measurement of ONL thickness (1, 6) is still a precise way to obtain biologically meaningful data on the number of surviving photoreceptor nuclei. However, the variability noted above demands that large numbers of animals be used in an experiment in order to reach statistical significance. For instance, if a single animal scoring 2+ protection were measured from each of the three major levels of preservation in the control eye (i.e., low, medium and high), then the mean ONL thickness of each of the eyes would be quite different from one another in both the control group (e.g., low = 9.0 μm; medium = 13.0 μm; high = 17 μm) and the factor-injected group (e.g., low [in the control eye] = 11 μm; medium = 16 μm; high = 24 μm). In each case the variance would be extremely high using just 3 animals (i.e., 13.0 ± 4.0 μm [mean ± S.D.] for the control group; 17.0 ± 6.6 for the experimental group), whereas the variance would be zero using our scoring method. The use of large numbers of animals to reduce the variance when an absolute scale is used (e.g., ONL thickness) is not only costly and requires more histological effort, but also is sometimes impractical given the scarcity of some reagents. Moreover, the use of larger numbers of animals should be avoided whenever possible for obvious animal welfare reasons.

In addition to the use of smaller numbers of animals, the relative scoring method also has the advantage of being less time consuming than measuring the ONL, and it takes into consideration other significant measures of photoreceptor protection, such as inner and outer segment presence and integrity, as described above. In addition, the relative scoring method also effectively deals with some potential artifacts that affect ONL measurement. For example, the injection-induced injury response that leads to a localized protection around the needle can be avoided in visually scoring the retinas. Several histological artifacts can also be easily avoided while scoring the retinas, including tissue section folding; breaks in the sections; oblique regions (or entire sections) that give artifactually thick layers; and regions (or entire eyes) with poor or inadequate fixation in which abnormally thick ONLs are present due to vacuolation in the layer. Each of these represents a vexing problem for measuring ONL thickness at regular intervals around the retina.

In our experience, the scoring of light-damaged eyes, first the control eye of a given rat for degree of photoreceptor preservation (i.e., low, medium, etc.) and then the experimental eye for the degree of protection (0–4+), is a rapid and reasonably precise method, especially when the scoring is performed by several individuals. We have found that second and third scoring sessions of the same eyes, with one or more years intervening, results in the same or very close scores. Indeed, this method has been used effectively in studies on various growth factors, cytokines and neurotrophins (9).

It is possible that even greater degrees of protection will be obtained with other survival factors in the future. If so, the scoring will have to be adjusted upward, somewhat, to account for this. The present scoring scale, however, can serve as a basis for such adjustment.

The scoring method of photoreceptor preservation (i.e., low, medium or high converted to a numerical 1–5 scale) can also be used in lieu of ONL measurements, as it was in determining the effect of α2-adrenergic agonists on photoreceptor protection (20). In this case, however, the agent was administered systemically, so that retinas from experimental animals had to be compared to those of control animals. Thus, larger numbers of animals had to be used to avoid the problem of inter-animal variability, as it would had ONL measurements been used.

The relative scoring method is also useful for other species. In our experience, the exposure of BALB/c albino mice to 2 weeks of constant light of the same intensity used in the present study results in about the same degree of damage as the rat eyes seen in the present study, and almost the same degree of variability (21). We have used the method effectively to study the protection of albino mouse retinas from constant light damage by various survival factors (22).

ACKNOWLEDGMENTS

We thank Nancy Lawson and Gloria Riggs for technical and administrative assistance. This work was supported in part by National Institutes of Health Research Grants EY01919, EY06842 and EY01429; Core Grant EY02162; and funds from the Foundation Fighting Blindness, Research to Prevent Blindness, and That Man May See, Inc.

REFERENCES

1. Michon, J.J., Li, Z.L., Shioura, N., Anderson, R.J., and Tso, M.O.M., 1991, A comparative study of methods of photoreceptor morphometry, *Invest. Ophthalmol. Vis. Sci.* **32**: 280–284.

2. LaVail, M.M., and Battelle, B.A., 1975, Influence of eye pigmentation and light deprivation on inherited retinal dystrophy in the rat, *Exp. Eye Res.* **21:** 167–192.

3. LaVail, M.M., White, M.P., Gorrin, G.M., Porrello, K.V., and Mullen, R.J., 1993, Retinal degeneration in the nervous mutant mouse. I. Light microscopic cytopathology and changes in the interphotoreceptor matrix, *J. Comp. Neur.* **333:** 168–181.

4. Birch, D., and Jacobs, G.H., 1977, Effects of constant illumination on vision in the albino rat, *Physiol. Behav.* **19:** 255–259.

5. LaVail, M.M., 1980, Eye pigmentation and constant light damage in the rat retina, in: *The Effects of Constant Light on Visual Processes*, (T.P. Williams, and B. B. Baker, eds.), pp. 357–387, Plenum Press, New York.

6. Faktorovich, E.G., Steinberg, R.H., Yasumura, D., Matthes, M.T., and LaVail, M.M., 1992, Basic fibroblast growth factor and local injury protect photoreceptors from light damage in the rat, *J. Neurosci.* **12:** 3554–3567.

7. Organisciak, D.T., Wang, H.M., Li, Z.Y., and Tso, M.O.M., 1985, The protective effect of ascorbate in retinal light damage of rats, *Invest. Ophthalmol. Vis. Sci.* **26:** 1580–1588.

8. Edward, D.P., Lam, T.T., Shahinfar, S., Li, J., and Tso, M.O.M., 1991, Amelioration of light-induced retinal degeneration by a calcium overload blocker. Flunarizine, *Arch. Ophthalmol.* **109:** 554–562.

9. LaVail, M.M., Unoki, K., Yasumura, D., Matthes, M.T., Yancopoulos, G.D., and Steinberg, R.H., 1992, Multiple growth factors, cytokines and neurotrophins rescue photoreceptors from the damaging effects of constant light, *Proc. Natl. Acad. Sci. USA* **89:** 11249–11253.

10. LaVail, M.M., Matthes, M.T., Yasumura, D., and Steinberg, R.H., 1997, Variability in rate of cone degeneration in the retinal degeneration (*rd/rd*) mouse, *Exp. Eye Res.* (In press).

11. Rapp, L.M., and Williams, T.P., 1980, A parametric study of retinal light damage in albino and pigmented rats, in: *The Effects of Constant Light on Visual Processes*, (T.P. Williams, and B.N. Baker, eds.), pp. 135–159, Plenum Press, New York.

12. Kuwabara, T., and Gorn, R.A., 1968, Retinal damage by visible light, *Arch. Ophthalmol.* **79:** 69–78.

13. Grignolo, A., Orzalesi, N., Castellazzo, R., and Vittone, P., 1969, Retinal damage by visible light in albino rats, *Ophthalmologica* **157:** 43–59.

14. LaVail, M.M., Gorrin, G.M., Repaci, M.A., and Yasumura, D., 1987, Light-induced retinal degeneration in albino mice and rats: strain and species differences., in: *Degenerative Retinal Disorders: Clinical and Laboratory Investigations*, (J.G. Hollyfield, R.E. Anderson, and M.M. LaVail, eds.), pp. 439–454, Alan R. Liss, Inc., New York.

15. LaVail, M.M., Gorrin, G.M., and Repaci, M.A., 1987, Strain differences in sensitivity to light-induced photoreceptor degeneration in albino mice, *Curr. Eye Res.* **6:** 826–834.

16. O'Steen, W.K., and Donnelly, J.E., 1982, Chronologic analysis of variations in retinal damage in two strains of rats after short-term illumination, *Invest. Ophthalmol. Vis. Sci.* **22:** 252–255.

17. Borges, J.M., Edward, D.P., and Tso, M.O., 1990, A comparative study of photic injury in four inbred strains of albino rats, *Curr. Eye Res.* **9:** 799–803.

18. Penn, J.S., Naash, M.I., and Anderson, R.E., 1987, Effect of light history on retinal antioxidants and light damage susceptibility in the rat, *Exp. Eye Res.* **44:** 779–788.

19. Penn, J.S., and Anderson, R.E., 1992, Effects of light history on the rat retina, in: *Progress in Retinal Research*, (N. Osborne, and G. Chader, eds.), pp. 76–98, Vol. 11, Pergamon Press, Oxford.

20. Wen, R., Cheng, T., Li, Y., Cao, W., and Steinberg, R.H., 1996, Alpha 2-adrenergic agonists induce basic fibroblast growth factor expression in photoreceptors in vivo and ameliorate light damage, *J. Neurosci.* **16:** 5986–5992.

21. LaVail, M.M., Gorrin, G.M., Repaci, M.A., Thomas, L.A., and Ginsberg, H.M., 1987, Genetic regulation of light damage to photoreceptors, *Invest. Ophthalmol. Vis. Sci.* **28:** 1043–1048.

22. Yasumura, D., Matthes, M.T., Lau, C., Unoki, K., Steinberg, R.H., and LaVail, M.M., 1995, Attempts to rescue photoreceptors with survival factors in mice with inherited retinal degenerations or constant light damage, *Invest. Ophthalmol. Vis. Sci.* **36:** S252.

42

FURTHER STUDIES ON THE PHAGOCYTOSIS OF PHOTORECEPTOR OUTER SEGMENTS BY RAT RETINAL PIGMENT EPITHELIAL CELLS

Michael O. Hall, Toshka A. Abrams, Barry L. Burgess, and Alexey V. Ershov

Jules Stein Eye Institute
UCLA Medical Center
100 Stein Plaza, Los Angeles, California 90095-7008

INTRODUCTION

The shedding of photoreceptor outer segments (OS) and their phagocytosis by the adjacent retinal pigment epithelium (RPE) is a fundamental process which occurs in all vertebrate species studied to date. Numerous studies have showed that OS shedding and phagocytosis is light entrained and follows a circadian rhythm (1, 2). When this process goes awry, such as in the RCS rat (3–5) and possibly in the vitiligo mouse (6, 7) degeneration of the OS soon follows. There is an accumulating body of evidence (8) that the phagocytosis of OS is a receptor mediated process, whereby a specific receptor on the RPE (the phagocytosis receptor) recognizes a ligand on the surface of the shed OS. After binding to the receptor, the shed OS is internalized by the RPE cell, where it undergoes digestion by lysosomal enzymes.

This relatively simple hypothesis has recently been complicated by reports of a diverse number of molecules on the surface of the RPE which appear to function in the phagocytosis of OS.

Boyle et al. (9) have reported that the mannosylfucosyl (mannose) receptor, which is present on the surface of the RPE, plays a role in this process. Incubation of cultured rat RPE cells with an antibody to the mannose receptor, or with soluble high-mannose containing macromolecules such as horseradish peroxidase, significantly inhibits OS phagocytosis, as does preincubation of OS with purified mannose receptor protein.

Ryeom et al. (10–13) have reported that CD 36, an 88 kD multifunctional cell surface glycoprotein receptor, is present on the surface of normal rat RPE cells, but not on RCS rat RPE cells (10, 11). When this molecule is transfected into a non-phagocytic rat RPE cell line (RPE-J), these cells gained the ability to bind OS; this binding is reduced by 61% when these cells are incubated with a monoclonal antibody to CD 36 (11, 12). Additionally, a fusion protein which spans the amino terminus of CD 36 from amino acid 5 to amino acid 143 competes with the uptake of OS by 60%, while a fusion protein which

Degenerative Retinal Diseases, edited by LaVail *et al.*
Plenum Press, New York, 1997

spans 141 amino acids at the carboxyl terminus of CD 36 has no effect. Soluble CD 36 and anti CD 36 antibody also inhibit OS phagocytosis by a similar amount (13).

Most recently, Miceli et al (14) have shown that removal of vitronectin (VN) from serum reduces the phagocytosis of OS by cultured human RPE cells, and that addition of VN to serum-free medium stimulates OS phagocytosis in a dose-dependent manner. The vitronectin cell surface receptor (VNR) is an integrin (αVβ5 integrin), which is present on the plasma membranes of both OS and RPE cells (15). Miceli et al. (14) have thus examined the effect of treating cultured RPE cells with a monoclonal antibody to VNR; such treatment inhibits serum-stimulated OS uptake by 100%.

Lin and Clegg (16), have further examined the role of the VNR in OS phagocytosis by cultured human RPE cells. Since an RGD sequence in VN is involved in its binding to the VNR, these investigators examined the effect of a peptide containing such a sequence on the ability of RPE cells to phagocytize bovine OS. When cells are incubated with 1 mM GRGDSP, OS binding is reduced by 71%, while the inactive peptide, GRGESP, is a less effective inhibitor of binding.

Finally, we have showed that an antiserum generated to purified rat RPE cell membranes completely inhibits the binding and ingestion of OS to cultured rat RPE cells (17,18). The inhibitory activity of this antiserum can be adsorbed out by a partially purified extract of rat RPE cells. This extract contains neither the mannose receptor, nor CD 36. Thus, if either of these molecules are involved in OS phagocytosis, they cannot be the sole mediators of this process, but may function in combination with one or more as yet undefined cell surface molecules.

Most of the above studies have used passaged human RPE cells and bovine OS to measure phagocytosis. In light of our interest in the process of OS phagocytosis and in the identity of the phagocytosis receptor, we have examined the role of VN and of RGD peptides in the phagocytosis of OS by cultured rat RPE cells. We suggest that the reported stimulation of phagocytosis by VN (14), or its inhibition by RGD containing peptides (16), may be a non-specific process due to dedifferentiation of RPE cells after multiple passage. Additionally, we have measured the ability of rat RPE cells to phagocytize OS after being passaged a number of times. We show that passage of rat RPE cells greatly diminishes their phagocytic ability. We have also examined the phagocytosis of OS during extended incubation with normal or dystrophic RPE cells. We show that RPE cells from the mutant RCS rat, which normally show a much reduced phagocytic ability, acquire the ability to avidly ingest OS if they are left in contact with these organelles for a number of hours. We suggest that this may be a non-specific process.

MATERIALS AND METHODS

Growth of RPE Cells

Retinal pigment epithelial cells are isolated from 9–13 day old normal Long Evans (LE) and dystrophic (RCS; rdy⁻/rdy⁻, p⁺/p⁺) rats as previously described (19), and are grown in a growth medium (GM) consisting of Earle's MEM containing 10% FBS, 40 μg/ml gentamycin, 40 μg/ml kanamycin and 2 mM glutamine (GM). The cells are suspended in GM and seeded onto 18 mm glass discs, at a concentration of 30,000 cells/100μl. The medium is changed every 4 days, and the cells are used when they reach confluence in 6–8 days. The medium is again changed the day before a feeding experiment. Prior to incubation with OS, disks are transferred into individual wells of a 12 well cluster plate.

RPE cells are passaged by incubating confluent primary cultures with calcium and magnesium free balanced salt solution (CMF-BSS) for 20 min at 37°C, followed by incubation with 0.05% trypsin/0.02% EDTA for 20 min at 37°C. The cells are split 1:2 and repassaged when they reach confluence after 6–7 days. Prior to measuring the phagocytic ability of the passaged cells, they are seeded onto glass disks as above.

Quantitation of OS Phagocytosis

Outer segments are isolated from adult Long Evans rat retinas using a sucrose density gradient method (5). The OS are suspended in GM at a concentration of 1×10^7 ROS/ml, and a 1 ml aliquot is added to each well. Incubation is continued for varying periods of time, depending on the experimental protocol. At the end of each incubation period the disks are vigorously rinsed in ice-cold PBS to remove unbound OS and fixed in 3.5% formaldehyde. Bound OS are stained using an antiserum to bovine OS, followed by a biotin conjugated second antibody and finally with fluorescein-streptavidin. The cells are then permeabilized with 47.5% ethanol (5), and ingested OS are stained as described above, except that Texas Red-streptavidin is used. The cells are examined under epifluorescent illumination using a Zeiss Photomicroscope III. When viewed with a fluorescein filter set, bound OS are stained green, while total OS (bound plus ingested) are stained red when viewed with a rhodamine filter set. By subtracting the number of bound OS from the total number of OS, the number of OS which are ingested can be calculated. In order to eliminate bias, confluent areas of cells are randomly selected under transmitted light. Bound and ingested OS are then counted under epifluorescent illumination using a 1 cm^2 ocular grid at a final magnification of 375 X. At this magnification, the grid corresponds to an area of cells of 0.083 mm^2. At least 5 grid areas are quantitated on each disk. Each treatment is carried out on duplicate disks. Thus n=10 for each data point within an experiment. Disks are photographed at final magnifications of 40 X and 100 X using Kodak TMAX 400 black and white film, and epifluorescent illumination with either fluorescein or rhodamine filter sets.

Treatment with Vitronectin

The effect of vitronectin (VN, Sigma Chemical Co., St. Louis, MO) on OS phagocytosis is studied using RPE cells grown as above. Prior to feeding OS, the disks are washed with either GM containing 5% FBS, or with Earle's MEM which contains no FBS. The cells are preincubated for 30 minutes in either GM plus 5% FBS, MEM alone, or MEM containing 1, 5, 10, or 25 µg/ml of bovine VN. A 0.1 ml aliquot of a 10 X concentrated suspension of OS (1×10^8 OS/ml) in GM plus 5% FBS or in MEM is then fed to each disk. Incubation with OS is continued for 1 hour after which the cells are washed, fixed and stained as described above, and bound and ingested OS are quantitated.

Treatment with RGD Containing Peptide

The peptide, GRGDSP (Calbiochem/Novachem, San Diego, CA), is dissolved in complete GM to a concentration of 5 mg/ml. OS are prepared as above and suspended in complete GM containing either 1000, 500, 100, 10 or 1 µg/ml GRGDSP, or in GM alone. RPE cells seeded on disks are washed with GM and incubated for 30 minutes at 37°C in 1.0 ml of GM containing one of the above GRGSDP concentrations, or in GM alone. The medium in each well is replaced by 1 ml of medium containing the same concentration of peptide, and 1×10^7 OS. Incubation with OS is continued for 1 hr at 37°C, after which the disks are washed, fixed, stained and the bound and ingested OS are quantitated.

24-Hour OS Phagocytosis

In order to study the phagocytosis of OS by LE and RCS rats over a 24 hr period, cells are continuously incubated with 1×10^7 OS in 1 ml of GM for 1, 3, 5, 7, 9, 12, 15, 18, and 24 hr. At the end of each incubation period, disks are washed, fixed and stained as above. Due to the difficulty in counting the large numbers of ingested OS on the 18 and 24 hour disks, one-fifth of each grid was counted, and the number of OS thus obtained is multiplied by 5 to obtain an estimate of the total number of ingested OS.

RESULTS

Effect of Passaging Cells on OS Phagocytosis

Primary and first passage RPE cell cultures show no difference in their ability to ingest OS (Fig 1a insert). However the number of OS ingested by second and third passage cells after 1 hour of incubation is reduced by 64% and 92% respectively, and by 69% and 99% after 2 hours of incubation. OS binding is not affected by passage of the cells (Fig 1b insert). If third passage cells are incubated with OS for 12 and 24 hours (Fig 1a), they regain the ability to ingest OS, and do so with an avidity which approximates the phagocytic ability of unpassaged cells. This phenomenon is comparable to the behavior of unpassaged cells after incubation with OS for extended periods of time (see below). These third passage cells also bind large numbers of OS upon extended incubation (Fig 1b).

Effect of Vitronectin on OS Phagocytosis

The ingestion of OS is reduced by 97% when cells are fed OS in the absence of FBS (Fig 2). However the addition of 1 to 25 µg/ml of VN to serum-free MEM does not increase the phagocytosis of OS above that measured in the absence of serum, in either primary cultures or in passaged cells. OS binding is also not affected by VN (data not shown). The addition of 5% FBS to cells which have been preincubated with 5 µg/ml of VN in MEM immediately restores the phagocytic ability of the cells to control levels.

Effect of RGD Containing Peptide on OS Phagocytosis

As can be seen from Fig 3, GRGDSP peptide has no effect on the phagocytosis of OS by primary cultures of rat RPE cells at any concentration from 1 µg/ml to 1000 µg/ml. We also tested the effect of the tripeptide RGD at the same concentrations as the GRGDSP peptide, with similar negative results (data not shown).

Effect of Incubation Time on OS Phagocytosis by Normal and Dystrophic RPE Cells

We have previously showed that normal RPE cells will actively ingest OS for 3–4 hours, after which they enter a refractory period of 4–5 hours during which OS are bound, but ingestion is greatly reduced (20). If OS are removed during this refractory period, the

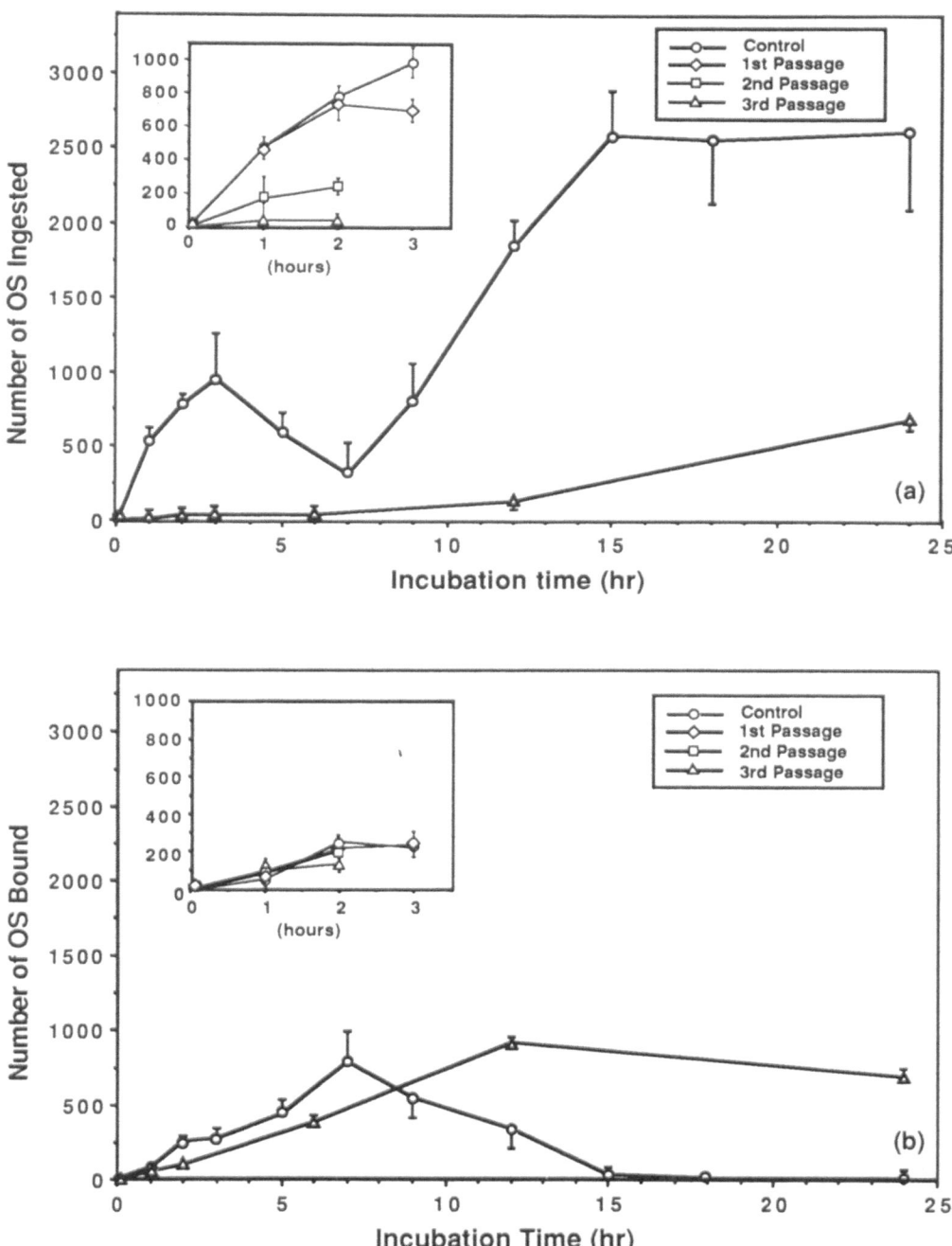

Figure 1. The effect of cell passage on (a) OS ingestion and (b) OS binding. Cells were incubated with OS for 1 to 24 hours. Inserts show the effect of passage during the first three hours of incubation.

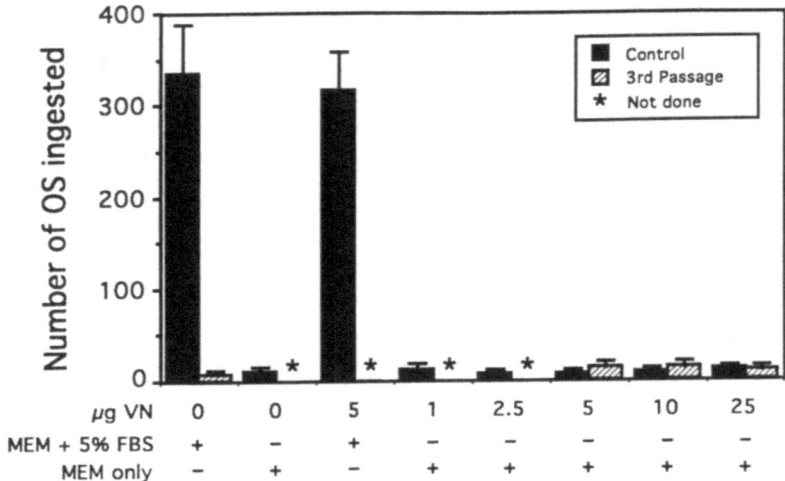

Figure 2. The effect of vitronectin on OS phagocytosis in the presence or absence of FBS by primary RPE cultures, and by third passage cells. Cells were preincubated with VN for 30 minutes and then incubated with OS plus VN for 1 hour.

cells recover their normal phagocytic ability after two hours in the absence of OS. In the experiment reported here, we have incubated normal and dystrophic RPE cells for up to 24 hours in the continuous presence of OS.

As expected from our previous results, normal RPE cells show a rapid ingestion of OS for 3 hours, after which the number of ingested OS decreases for the next 4 hours (Fig 4 a), while the number of bound OS increases (Fig 4 b). Between 7 and 9 hours of incubation, there is a sudden and dramatic increase in the ingestion of OS, until by 15 hr, almost all of the OS associated with the cell monolayer are ingested, with very few bound to the cells. This picture continues unchanged through 24 hours of incubation.

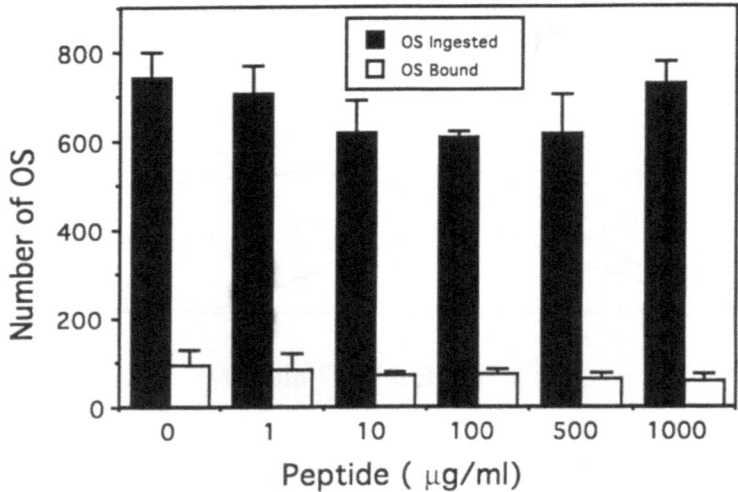

Figure 3. The effect of GRGDSP peptide on the phagocytosis of OS by primary cultures of rat RPE cells. Cells were preincubated with peptide for 30 minutes and then incubated with OS plus peptide for 1 hour.

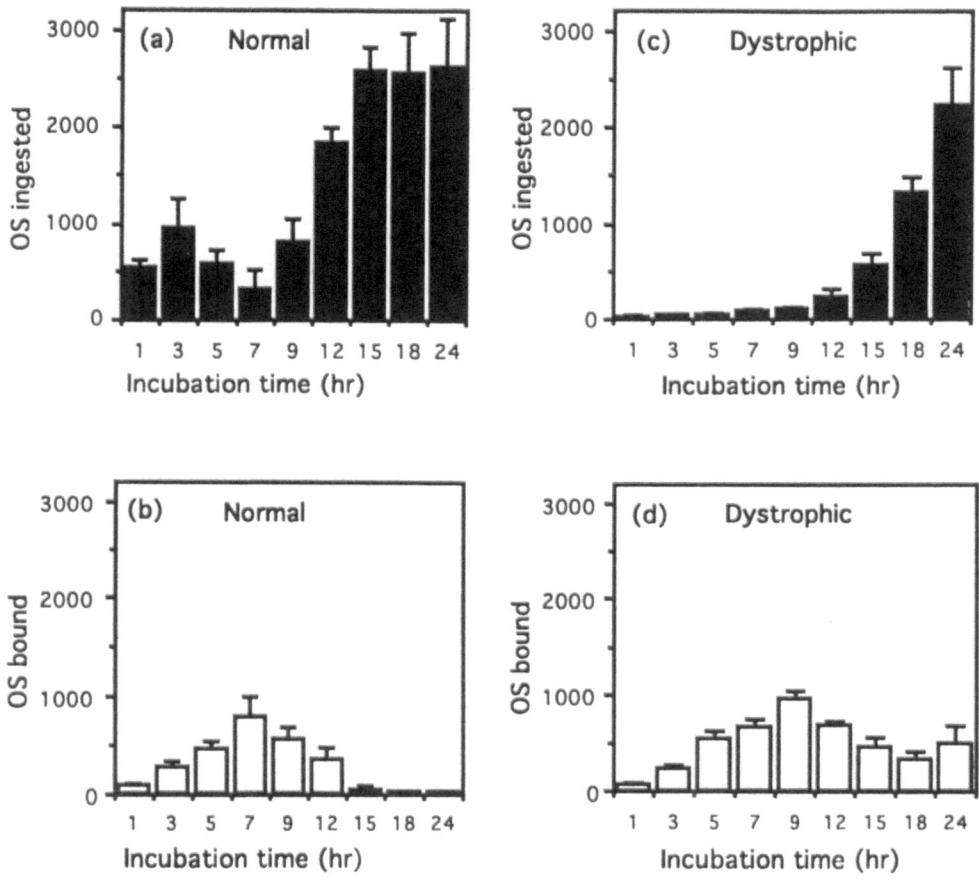

Figure 4. OS ingestion and binding by normal (a, b), and dystrophic (c, d) RPE cells over a 24 hour incubation period.

When dystrophic RPE cells are incubated with OS, the number of OS bound increases for 9 hours (Fig 4 d), with only a slight increase in the number of ingested OS (Fig 4 c), in concurrence with our previous results (20). Between 12–24 hours there is a sharp increase in the number of OS ingested by the dystrophic cells, with a concomitant decrease in the number of OS bound. After 24 hours of incubation with OS, the number of OS ingested by the dystrophic RPE cells is not significantly different from the number ingested by the normal RPE cells; however there are significantly more OS still bound to the surface of the dystrophic RPE cells. It thus seems that prolonged incubation of RPE cells with OS triggers the same mechanism of ingestion in both normal and RCS rat RPE cells (and in passaged cells), the only difference being that the process is delayed by a few hours in the dystrophic cells.

Figures 5 and 6 are fluorescence micrographs which illustrate the temporal pattern of OS binding and ingestion by normal and dystrophic RPE cells.

Figure 5 demonstrates the binding and ingestion of OS by normal RPE cells over the 24 hour incubation period. For the first 7 hours of incubation, binding of OS occurs evenly and increasingly over the entire monolayer (Fig 5 a-f). At 7 hours, when the number of bound OS reaches a maximum, clumping of OS starts to occur, with dense clumps of

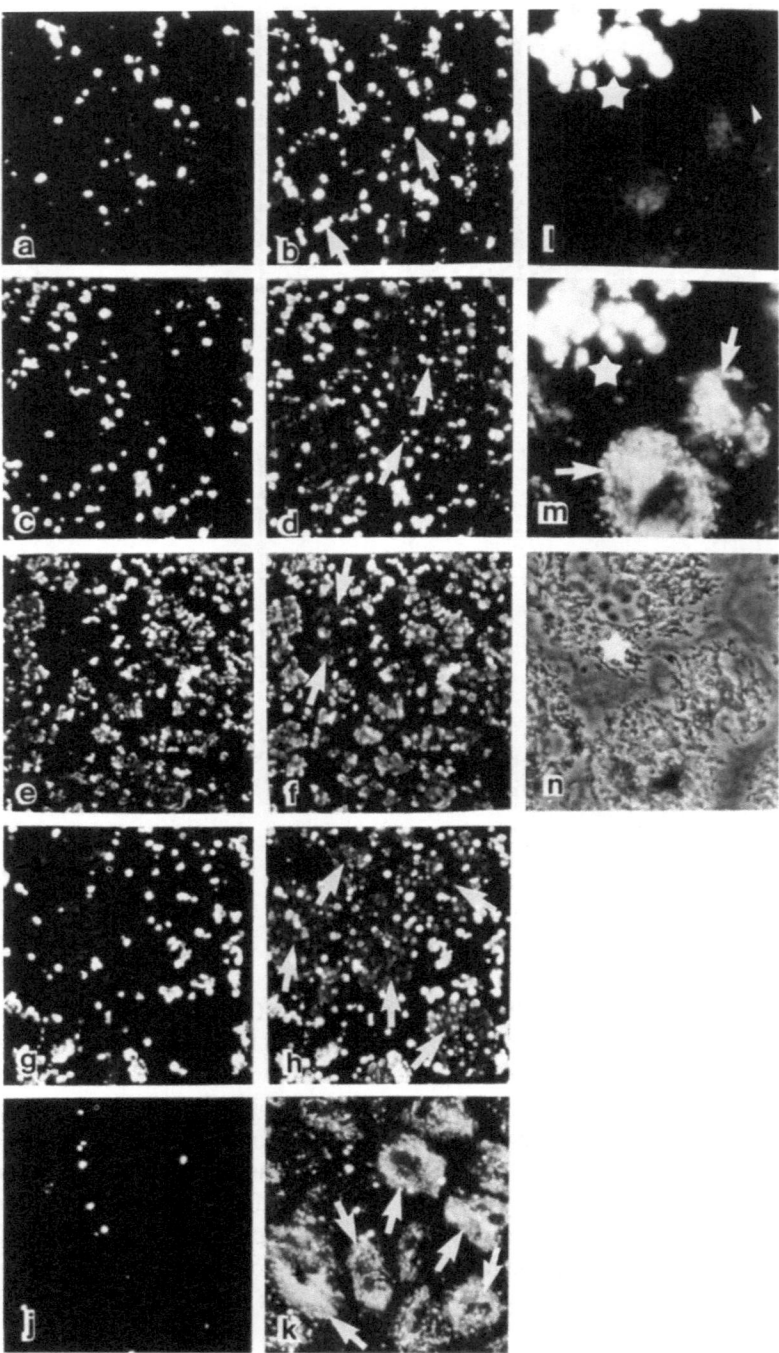

Figure 5. Fluorescence micrographs showing the binding (a,c,e,g,j,l) and ingestion (b,d,f,h,k,m) of OS by normal rat RPE cells after 1 (a, b), 3 (c, d), 7 (e, f), 12 (g, h) and 24 (j-n) hours of incubation. Final magnification is 160 X (a-k) and 400 X (l-n). (n) is a phase contrast micrograph corresponding to the area shown in (l) and (m). Arrows point to either single or groups of ingested OS. Stars indicate the same cell in l, m and n.

bound OS interspersed with areas to which no OS are bound (Fig 5 e, f). This pattern continues up to 24 hours, with a continual increase in the area of the monolayer which is free of bound OS (Fig 5 g-k). At higher magnification it can be seen that these areas which are free of bound OS contain large numbers of ingested OS (Fig 5 l, m). After 24 hours, almost all cells are engorged with OS, while only a few cells have scattered OS bound to them (Fig 5 j-m). It is noteworthy that cells which have ingested large numbers of OS do not bind any more OS.

Figure 6 examines the pattern of OS binding and ingestion by dystrophic RPE cells over the 24 hour incubation period. Initially OS binding appears to be random, with most, but not all cells showing a few bound OS (Fig 6 a-d). After 7 hours in contact with OS, the RPE monolayer is completely covered by a carpet of OS (Fig 6 e, f). By 12 hours, the OS have begun to show a patchy distribution, due in part to the fact that a few cells have begun to avidly ingest the OS, thus creating patches where all of the previously bound OS have been internalised (Fig 6 g-k). Additionally, as reported previously (20), clumping of OS into patches also occurs prior to ingestion occurring. Fig 6 (l, m) shows a higher magnification of cells which have ingested OS after 24 hours of incubation. After 24 hours (Fig 5 j-m) many cells have ingested all of the OS which are bound to their surface, however numerous cells are still covered with patches of OS. Presumably, if the incubation had been continued for longer, these bound OS would have been ingested, as is seen with normal RPE cells.

Figures 5 n and 6 n are phase contrast pictures of cells after 24 hours of incubation with OS. In each of the corresponding fluorescence micrographs (Fig 5 l, m and Fig 6 l, m) a cell is shown which still has OS bound to it, surrounded by cells which have avidly ingested OS. No obvious morphological differences are seen in the phase contrast micrographs, between the cells which have ingested OS, and those to which OS are still bound.

DISCUSSION

The intimate and correct interaction of the photoreceptor cell outer segments with the RPE is structurally and functionally important for the maintenance of photoreceptor viability. When this interaction breaks down, as in the RCS rat (3–5) or the vitiligo mouse (6, 7), death of the photoreceptors occurs. Amongst the many important functions of the RPE is the phagocytosis of the tips of OS, which are shed on a diurnal rhythm. This process is widely considered to be mediated through the interaction of a ligand on the plasma membrane of the OS with a receptor on the apical surface of the RPE (8). Neither the ligand nor the receptor have been identified, although at least three specific cell surface molecules have been proposed to serve as the receptor, either singly or in combination with some other unidentified molecule(s) (9, 10–13, 14, 16).

Since it is difficult to study the kinetics of OS phagocytosis in vivo, or to pharmacologically manipulate this process in the whole eye, sophisticated in vitro methods have been developed in a number of laboratories to measure the uptake of OS by monolayer cultures of RPE cells. Such studies have largely used RPE cells from rat or human eyes; however cells from pig, cow and monkey eyes have been used to a lesser extent. RPE cells are generally obtained from 10 day old rats, and are used as primary cultures, while cells obtained from human eyes (of varying ages) have been used as primary cultures, or more usually, as passaged cell lines.

Results which have been obtained using primary cultures of rat RPE cells are quite different from those obtained using passaged human cells. Thus primary cultures of rat RPE

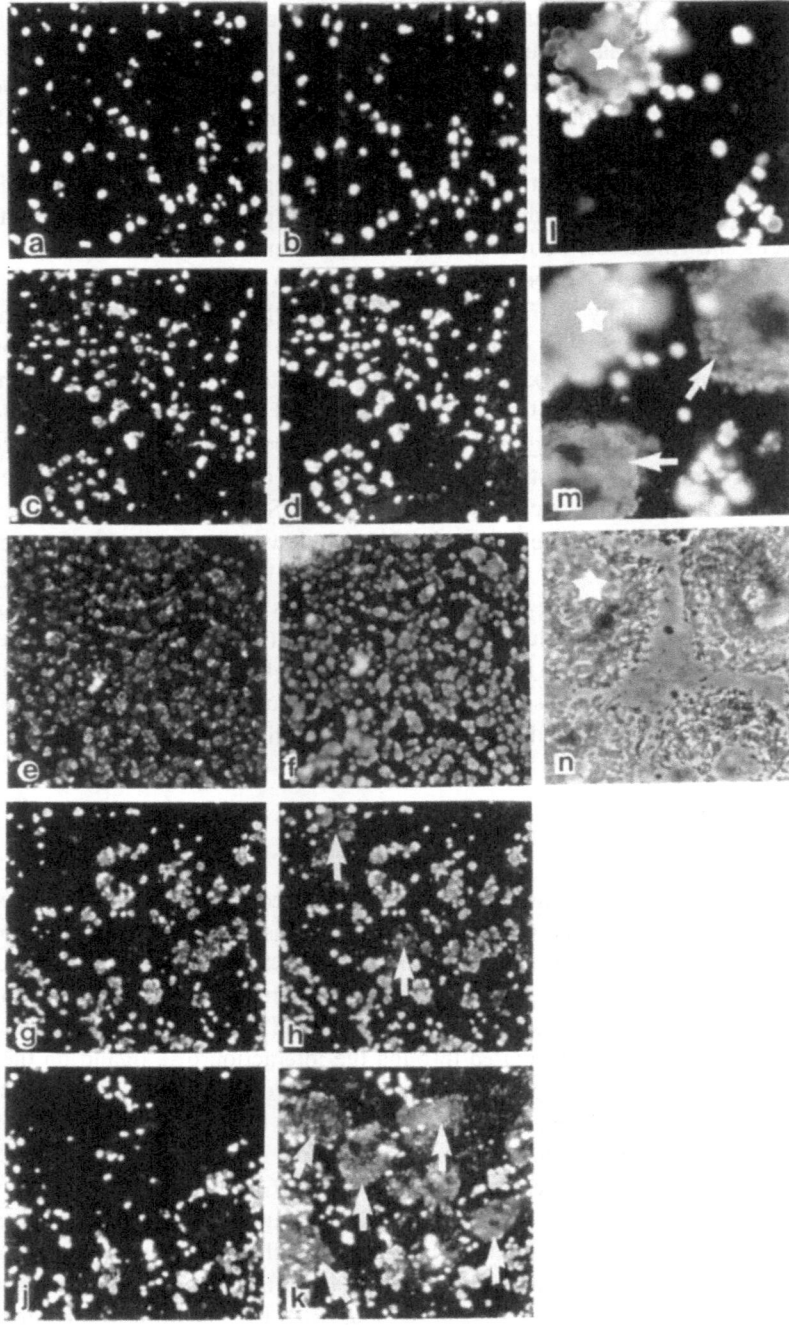

Figure 6. Fluorescence micrographs showing the binding (a,c,e,g,j,l) and ingestion (b,d,f,h,k,m) of OS by dystrophic rat RPE cells after 1 (a, b), 3 (c, d), 7 (e, f), 12 (g, h) and 24 (j-n) hours of incubation. Final magnification is 160 X (a-k) and 400 X (l-n). (n) is a phase contrast micrograph corresponding to the area shown in (l) and (m). Arrows point to groups of ingested OS. Stars indicate the same cell in l, m and n.

cells will ingest significant numbers of RPE cells during the first hour of incubation, and uptake is saturated after 3–4 hours (Fig. 4a, and 5, 20), while passaged human RPE cells show limited ingestion during the first few hours of incubation with OS but continue to ingest OS for 24 to 36 hours (21, 22) and may or may not show saturation of OS uptake with time (23). Primary and first passage cultures of rat RPE cells ingest OS in equivalent amounts and with similar kinetics; however ingestion is markedly decreased as the cells undergo further passage. After more than one passage, rat RPE cells certainly do not show the rapid uptake of OS during the first 2–3 hours of incubation, as is shown by primary cultures; however they will resume phagocytosis after incubation with OS for 12 or more hours.

Normal rat RPE cells show a biphasic OS ingestion curve (Fig 4a), while dystrophic RPE cells, and passaged rat RPE cells show only a single, delayed phase of ingestion, which is acquired after prolonged incubation with OS (Figs 1 and 4c). McLaren (24) has also recently reported that dystrophic RPE cells acquire the ability to ingest OS after prolonged incubation. Our long term incubations with primary cultures of normal and dystrophic RPE cells, and with passaged cells, suggest that the initial uptake of OS by normal RPE cells is due to a specific, receptor-mediated mechanism, (since it does not operate in dystrophic or passaged RPE cells), while the uptake observed after more than 9 hours of incubation is probably non-specific (since it operates in normal, dystrophic and passaged RPE cells), or at least is not mediated by the same receptor system responsible for the initial uptake of OS. The delayed phase of OS phagocytosis may be due to a signaling mechanism which is only activated after a number of hours, and is thus probably different from the specific mechanism which is operative immediately upon the presentation of OS to the cells. Recently it has been reported that OS phagocytosis may be transcriptionally controlled, as it is inhibited by incubation with actinomycin D (25). However this only affects the delayed phase of phagocytosis. Additionally the zinc finger-containing transcription factor, zif 268, is upregulated during OS phagocytosis, and may be responsible for modifying the expression of other genes which contain zif 268 binding sites in their regulatory sequences (26). The activation of such genes, and the synthesis of a number of gene products may be responsible for the delayed phase of OS phagocytosis.

In contrast to two recent reports, we find no evidence that VN plays a role in the phagocytosis of OS by rat RPE cells. Addition of VN to serum-free medium did not enhance the ingestion of OS as reported by Miceli et al (14), while an RGD containing peptide did not inhibit OS uptake as observed by Lin and Clegg (16). It is clear, as reported by numerous investigators, that serum greatly stimulates the uptake of OS; in our experiments, serum increased the ingestion of OS almost 30-fold. However the active component does not seem to be VN, at least in our system.

What might be the reason for the different results reported by Miceli and Newsome (14) and Lin and Clegg (16), and the results which we report here? One obvious difference is the species used—human RPE cells as opposed to rat RPE cells. It is unlikely however, that the process of OS phagocytosis, which is so highly conserved in all animal species studied, would use a completely different receptor and ligand in different species to achieve the same end. Another major difference is the use of passaged human cells as opposed to primary cultures of rat cells. Our results clearly show that passage greatly decreases the ingestion of OS. It is quite likely that this is due to dedifferentiation of the cells, such that the relatively small amount of ingestion which is observed with passaged cells is due to nonspecific mechanisms—which are known to occur during OS phagocytosis (27).

In addition to the integrin receptor, both a mannose receptor (9) and the CD 36 molecule (10–13) have been implicated to play a role in the phagocytosis of OS. It is obvious that all of these molecules cannot be the *primary* receptor for this process, and yet all have been conclusively shown to be present on the surface of the RPE, and when each is blocked with its specific antibody, OS ingestion is markedly decreased. It is possible that each of these receptors interacts with its ligand through a common mechanism involving (perhaps) a specific sugar residue(s) of an oligosaccharide side-chain. Such a residue(s) may be the component of the ligand on the OS which is specifically recognized by the receptor, as well as non-specifically by all of the above molecules. This might explain the small amount of OS phagocytosis which is normally seen in the RCS rat.

It is obvious that the mechanism which controls OS phagocytosis by the RPE is far from resolved. That this is a receptor mediated process is fairly well established, since it shows all of the requirements for such a process (8, 20, 22). However, it is obviously not as simple as a single receptor interacting with a specific ligand, as is the case in so many receptor mediated processes. Perhaps the fact that the particle which is being ingested is so large, necessitates the utilization of numerous different cell surface molecules to bind the OS to the RPE, only one of which is responsible for initiating a signal which is responsible for ingestion. This might explain why the RCS rat, which suffers from a single recessive mutation, is able to bind OS, but not to ingest them. In this animal, the molecule which is affected by the mutation may be the specific receptor that is responsible for the signaling step of OS phagocytosis. Alternatively the mutation may affect an intracellular molecule which is linked to the signaling receptor.

ACKNOWLEDGMENTS

This work was supported by grants EY 00046 and EY 00331 from the USPHS and by a grant from the Foundation Fighting Blindness, Inc., Baltimore MD, to MOH.

REFERENCES

1. Young, R. W., and Bok, D., 1969, Participation of the retinal pigment epithelium in the rod outer segment renewal process, J. Cell Biol. **42**: 392–403.
2. LaVail, M. M., 1976, Rod outer segment disk shedding in rat retina: relationship to cyclic lighting. Science **194**: 1071–1074.
3. Dowling, J. E., and Sidman, R.L., 1962, Inherited retinal dystrophy in the rat, J. Cell Biol. **14**: 73–109.
4. Bok D., and Hall, M. O., 1971, The role of the pigment epithelium in the etiology of inherited retinal dystrophy in the rat, J. Cell Biol. **49**: 664–682.
5. Chaitin, M. H., and Hall, M. O., 1983, Defective ingestion of rod outer segments by cultured dystrophic rat pigment epithelial cells. Invest. Ophthalmol. Vis. Sci. **24**: 812–820.)
6. Nir, I., Ransom, N., and Smith, S.B., 1995, Ultrastructural features of retinal dystrophy in mutant vitiligo mice. Exp. Eye Res. **60**: 364–377.
7. Kosaras B., and Sidman, R.L., 1996, Phagosome number and distribution in retinal pigment epithelial cells of vitiligo mutant mice, Exp. Eye Res.**63**: 151–158.
8. McLaughlin, B. J., Cooper, N.G.F., and Shepherd, V.L., 1994, How good is the evidence to suggest that phagocytosis of ROS by RPE is receptor mediated? Prog. Retinal Eye Res. **13**: 147–164.
9. Boyle, D., Tien, L., Cooper, N.G.F., Shepherd, V., and McLaughlin, B.J., 1995, A mannose receptor is involved in retinal phagocytosis. Invest. Ophthalmol. Vis. Sci. **32**: 1464–1470.
10. Ryeom, S. W., Sparrow, J. R., and Silverstein, R. L., 1994, CD 36 is expressed on retinal pigment epithelium (RPE) and mediates phagocytosis of photoreceptor outer segments. Invest. Ophthalmol. Vis. Sci.**35**: 2140.
11. Ryeom, S. W., Silverstein, R. L., and Sparrow, J. R., 1995, The absence of CD 36, a molecule mediating phagocytosis of photoreceptor outer segments (ROS), correlates with a lack of ROS uptake by retinal pigment epithelial cells (RPE). Invest. Ophthalmol. Vis. Sci.**36**: S815.

12. Ryeom, S. W., Sparrow, J. R., and Silverstein, R. L., 1996a, Photoreceptor outer segment (ROS) interactions with CD 36: structure/function relationships, Invest. Ophthalmol. Vis. Sci.**37**: S920.

13. Ryeom, S. W., Sparrow, J. R., and Silverstein, R. L., 1996b, CD 36 participates in the phagocytosis of rod outer segments by retinal pigment epithelium, J. Cell Sci. **109**: 387–395.

14. Miceli, M. V., Newsome, D. A., and Tate, D. J., 1996, Vitronectin is responsible for the serum induced stimulation of ROS uptake by RPE cells. Invest. Ophthalmol. Vis. Sci.**37**: S920.

15. Anderson D. H., Johnson, L. V., and Hageman G. S., 1995, Vitronectin receptor (VnR) is expressed on the apical surfaces of photoreceptor and RPE cells. Invest. Ophthalmol. Vis. Sci.**36**: S512.

16. Lin, H., and Clegg, D. O., 1996, Arg-Gly-Asp containing peptide inhibits ROS phagocytosis by cultured human RPE. Invest. Ophthalmol. Vis. Sci.**37**: S378.

17. Gregory, C. Y., and Hall, M. O., 1992, The phagocytosis of ROS by RPE cells is inhibited by an antiserum to rat RPE cell plasma membranes. Exp. Eye Res. **54**: 843–851.

18. Hall, M. O., Burgess, B. L., Abrams, T. A., Ershov, A. E., and Gregory, C. Y. 1996, Further studies on the identification of the phagocytosis receptor of rat retinal pigment epithelial cells. Exp. Eye Res. **63**: 265–275.

19. Mayerson, P. L., Hall, M. O., Clark, V. M., and Abrams, T. A., 1985, An improved method for the isolation and culture of rat retinal pigment epithelial cells. Invest. Ophthalmol. Vis. Sci. **26**: 1599–1609.

20. Hall, M. O., and Abrams, T. A., 1987, Kinetic studies of ROS binding and ingestion by cultured rat RPE cells. Exp. Eye. Res. **45**: 907–922.

21. Boulton, M., Marshall, J., and Mellerio, J.. 1984, Retinitis Pigmentosa: a quantitative study of the apical membrane of normal and dystrophic human retinal pigment epithelial cells in tissue culture in relation to phagocytosis, Graefe's Arch. Clin. Exp. Ophthalmol. **221**: 214–229.

22. Kennedy, C. J., Rakoczy, P. E., Robertson, T. A., Papadimitriou, J. M., and Constable, I. J., 1994, Kinetic studies on phagocytosis and lysosomal digestion of rod outer segments by human retinal pigment epithelial cells in vitro. Exp. Cell Res. **210**: 209–214.

23. Miceli, M. V., and Newsome, D. A., 1994. Insulin stimulation of retinal outer segment uptake by cultured human retinal pigment epithelial cells. determined by a flow cytometric method, Exp. Eye Res. **59**: 271–280.

24. McLaren, M. J., 1996. Kinetics of rod outer segment phagocytosis by cultured retinal pigment epithelial cells. Invest. Ophthalmol. Vis. Sci. **37**: 1213–1224.

25. McLaren, M. J., and Inana, G., 1995. ROS phagocytosis by cultured normal rat RPE and bFGF-treated dystrophic RPE is transcriptionally controlled. Invest. Ophthalmol. Vis. Sci.**36**: S815.

26. Ershov, A.V., Lukiw, W. J., and Bazan, N. G., 1995. Phagocytosis of rod outer segments (ROS) triggers the expression of zinc finger-containing transcription factor zif 268 in cultured rat retinal pigment epithelium (RPE) cells. Invest. Ophthalmol. Vis. Sci.**36**: S815.

27. Mayerson, P. L., and Hall, M. O., 1986. Rat retinal pigment epithelial cells show specificity of phagocytosis in vitro, J. Cell Biol. **103**: 299–308.

43

PURIFICATION AND CHARACTERIZATION OF MATRIX METALLOPROTEINASE-3 (STROMELYSIN-1) FROM BOVINE INTERPHOTORECEPTOR MATRIX

Abdelkrim Smine and James J. Plantner

Center for Vision Research
Department of Ophthalmology
Case Western Reserve University and University Hospitals of Cleveland
10900 Euclid Ave., Cleveland, Ohio 44106-5068

SUMMARY

Recently, we reported the presence of stromelysin-1 (matrix metalloproteinase 3 (MMP-3)) in human and bovine vitreous and interphotoreceptor matrix (IPM). In the present study, we have purified and partially characterized a 46 kDa metalloproteinase from bovine retinal pigment epithelial (RPE)-IPM using gelatin-agarose affinity chromatography and FPLC Mono-QHR ion exchange chromatography. Zymogram analysis of purified MMP-3 showed gelatinolytic and caseinolytic activities which were completely inhibited by EDTA and o-phenanthroline. Western immunoblot analysis confirmed that the enzyme was stromelysin-1.

Treatment of the purified stromelysin with trypsin or APMA did not lead to the production of lower molecular weight activated forms, suggesting that most of the enzyme in the IPM was already present in its final, activated form in vivo.

INTRODUCTION

Stromelysin-1 (MMP-3, EC3.4.24.17) is a member of the matrixin family. These metalloproteinases play a pivotal role in the degradation and remodeling of the extracellular matrix. MMP-3 degrades a number of extracellular matrix constituents such as aggrecan core protein; fibronectin; laminin and collagen types IV (1, 2) IX, X, and XI (1, 3, 4).

Like most other matrixins, MMP-3 is secreted from cells as an inactive zymogen (1, 5). ProMMP-3 is comprised of an NH_2-terminal propeptide of 82 amino acids, a catalytic domain of 165 amino acids, and a COOH-terminal hemopexin/vitronectin-like domain of 213 amino acids (6). All of the MMPs need proteolytic activation to function. These enzymes can be activated in vitro by serine proteinases such as plasmin, neutrophil elastase and trypsin or by detergents, organomercurial compounds and chaotropic agents. The '"cysteine-switch" model has been proposed to explain non-proteolytic activation of proMMPs by SH-reacting reagents (7, 8). In this model, a cysteine in the propeptide region complexes with the Zn^{+2} at the active site, preventing its interaction with substrate. When this complex is broken by one of the activators, the enzyme assumes an active form.

MMP-3 has been implicated in connective tissue diseases such as rheumatoid arthritis and osteoarthritis (9, 10, 11, 12) and in tumor cell invasion and metastasis (13).

Several ocular tissues secrete MMPs and TIMPs in culture including cornea (14), trabecular meshwork (15) and retinal pigment epithelium (RPE) (16, 17). Studies indicate that an imbalance of MMPs and their inhibitors may be involved in the pathogenesis of glaucoma (15), corneal ulceration and keratoconus(14). Our laboratory has recently shown that metalloproteinases and metalloproteinase inhibitors are present in bovine and human IPM (18, 19, 20). This has been confirmed by others (21). MMPs are likely to play an important role in the turnover of the protein-containing components of the IPM as they do in other systems. An imbalance in their level may be associated with the pathological processes of a number of retinal diseases such as diabetic retinopathy, retinopathy of prematurity, proliferative vitreoretinopathy and age-related macular degeneration.

In the present study, we report the purification and the partial characterization of MMP-3 from bovine interphotoreceptor matrix.

MATERIALS AND METHODS

Preparation of Interphotoreceptor Matrix (IPM)

Bovine eyes were obtained from a local slaughterhouse within one hour of death and stored on ice until processing. IPM was prepared by the method of Adler and Severin (22) as reported previously (23). Briefly, after removal of the cornea, lens, aqueous and vitreous humor and retina, cold buffer (50 mM Tris-HCl, pH 7.4, containing 0.1 M NaCl, 3 mM PMSF, 5mg/ml benzamidine, 3mg/ml pepstatin A and 2mg/ml leupeptin) was added to the eyecup and gently pipetted against the apical surface of the RPE a number of times to produce the RPE-IPM. This was centrifuged at 48,000g for 30 min to remove debris, and the supernatant fluid was concentrated to the equivalent of about 4 to 5 eyes/ml by ultrafiltration using Amicon (Beverly, MA) YM-10 membranes and stored at -20° C.

Purification Procedure

Gelatin-Agarose Affinity Chromatography. RPE-IPM was dialyzed against 50 mM Tris-HCl, pH 7.4, containing 5 mM $CaCl_2$, 0.005% Brij-35, 0.5 M NaCl and 0.002% NaN_3. Four milliliters were applied to a 1.5x15 cm gelatin-agarose column (Sigma, St. Louis, MO), equilibrated with the same buffer. At a flow rate of 40 ml/h, the column was washed with the starting buffer until the absorbance at 280 nm reached base-line. The bound material containing the gelatinolytic activity was then eluted with 50 mM Tris-HCl, pH 7.4; containing 1M NaCl; 5 mM $CaCl_2$; 0.005% Brij-35; 0.002% NaN_3 and 5% dimethyl sulfoxide (v/v). The elution was monitored by measurement of absorbance at 280 nm and proteolytic activity by zymography (see below). The bound active fractions were pooled, concentrated to one ml by ultrafiltration using a 3 kDa cutoff filter and dialyzed against 20 mM Tris-HCl buffer, pH 7.4.

FPLC Ion Exchange Chromatography. Aliquots of the dialyzed and concentrated fractions (200 µl) were injected onto a Mono-QHR column (Pharmacia, Uppsala) equilibrated with 20 mM Tris-HCl buffer, pH 7.4. Cationic proteins were removed by rinsing the column with the same buffer. Enzyme was then eluted by applying a NaCl gradient (0–1M) in this buffer. Absorbance was monitored at 226 nm, and the flow rate was 0.8 ml/min. The eluted material was pooled as described in the Results, and each pool was concentrated to 0.5 ml by ultrafiltration as above. Enzyme activity was analyzed using gelatin and casein zymography.

SDS-Page

Electrophoresis was performed as described by Laemmeli (24) using 10% polyacrylamide gels. Samples (0.5 to 2 µg of protein) were denatured in the presence of 2.5% β-mercaptoethanol at room temperature for 30 min. Following electrophoresis, the gels were silver stained.

Zymography

Zymography was performed as described by Hibbs et al. (25). Samples were mixed with SDS sample buffer in the absence of reducing agents and incubated for 30 min at room temperature. Electrophoresis was performed on 10% polyacrylamide gels containing 0.1% gelatin or 0.1% casein as described by Laemmeli (24). Following electrophoresis, gels were washed in 2.5% Triton X-100 for 1h and then gelatinolysis and caseinolysis was carried out by incubating the gel in 50 mM Tris-HCl, pH 7.4, containing 5 mM $CaCl_2$; 1mM $ZnCl_2$ and 0.02% Na azide at 37° C for 48h. The gels were stained with coomassie brilliant blue R250, and the location of activity was detected as clear bands in the background of uniform staining. Molecular weight standards (Sigma) were included to calibrate the gels.

Western Immunoblotting

Electrophoresis was performed on samples denatured with SDS sample buffer containing β-mercaptoethanol as described by Laemmeli (24). Prestained molecular weight standard mixtures (Sigma) were included for calibration.

Following electrophoresis, proteins were electrophoretically transferred to PVDF sheets (MSI, Westboro, MA). Transfer conditions were 100 V for 90 min, at 4°C, using a Tris-glycine transfer buffer (25 mM Tris and 250 mM glycine in 17% methanol). The blots were blocked at room temperature with 5% nonfat dry milk in TBS-T buffer (50 mM Tris-HCl, pH 7.2; 0.15M NaCl; 0.05% Tween 20) for 2h, incubated in 1:2000 diluted primary antibody for 2h and in 1:2000 diluted horseradish peroxidase-labeled donkey anti-sheep IgG (Sigma) for 2h. Bound antibody was visualized by the use of ECL reagent (Amersham, Arlington Hts., IL). After each incubation, the blot was extensively washed (6X, 10min) with TBS-T buffer, pH 7.0. The polyclonal sheep anti-human MMP-3 used as the primary antibody was a generous gift from C. Brinckerhoff, Dartmouth Medical School, New Hampshire.

Standards for the MMP-3 used in the immunoblots were obtained from cultured HT-1080 human fibrosarcoma cell medium. The cells were obtained from the American Tissue Culture Center (ATCC, Rockville, MD) and grown in RPMI-1640 medium supplemented with 10% fetal bovine serum and penicillin/streptomycin (500 units/ml, 500 µg/ml, respectively). At confluence, the cells were rinsed and cultured overnight with serum-free medium before being stimulated with phorbol 12-myristate 13-acetate (60 ng/ml) (26). The cell suspension was centrifuged at 3000g for 15 min and the supernatant fluid was concentrated 20 fold by ultrafiltration using a 3 kDa cutoff filter. The MMP-3 was then partially purified from the concentrated medium by preparative SDS-PAGE on 10% gels. The portion of the gel corresponding to the migration of a 25 kDa to 66 kDa protein was excised, extracted with 50 mM Tris-HCl, pH 7.4, concentrated, and used as standard for the western blots.

In Vitro Activation

Aliquots of purified stromelysin-1 (pool 3 from the Mono Q column) were incubated with 2 mM aminophenylmercuric acetate (APMA) or 10 µg/ml TPCK-trypsin at 37° C for 2h or 4h.

RESULTS AND DISCUSSION

Stromelysin-1 was purified from RPE-IPM prepared from fresh bovine eyes. The concentrated and dialyzed RPE-IPM was first subjected to gelatin-agarose affinity chromatography. After washing the column, the material which bound to the gel was then eluted with high salt and dimethyl sulfoxide and subjected to FPLC mono-QHR ion exchange chromatography as indicated in materials and methods. As can be seen in figure 1, one unbound and three bound fractions were separated by FPLC.

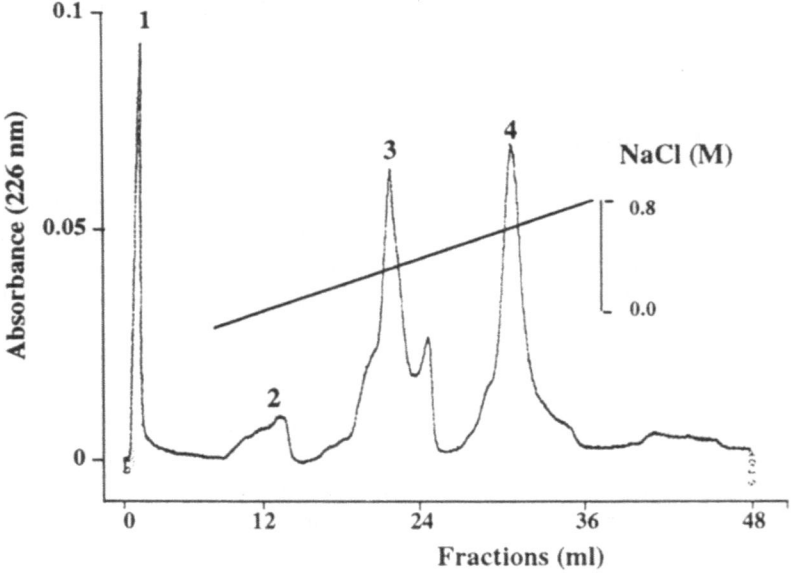

Figure 1. FPLC Mono-QHR ion exchange chromatography. Following gelatin affinity chromatography of dialyzed RPE-IPM, the retained fraction was dialyzed against 20 mM Tris-HCl buffer, pH 7.4, and concentrated to 1 ml by ultrafiltration using a 3 kDa cutoff filter, as described in materials and methods. A 200 µl aliquot was injected onto a Mono-QHR column equilibrated with the same buffer. Cationic proteins were removed by rinsing the column with this buffer, and the column was then eluted at a flow rate of 0.8 ml/min with a NaCl gradient (0–1M) in this buffer. Absorbance was monitored at 226 nm, and enzyme activity was analyzed using gelatin and casein zymography (see figure 3).

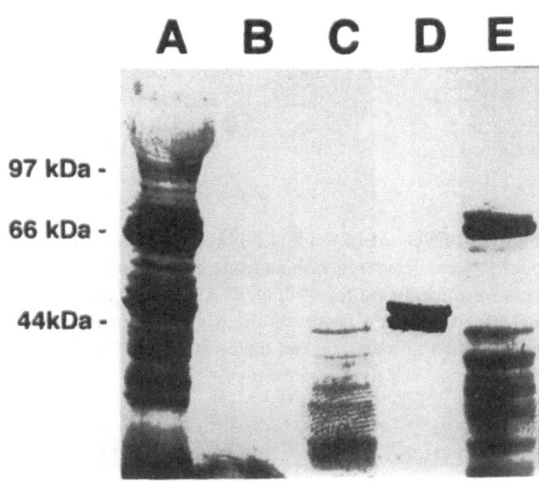

Figure 2. SDS-PAGE of pools from Mono-QHR ion exchange chromatography. Aliquots of the starting material (RPE-IPM) and the 4 concentrated pools obtained from ion exchange chromatography (see Figure 1) were denatured in the presence of 2.5% ß-mercaptoethanol, electrophoresed on a 10% gel and silver stained as described in materials and methods. The molecular weights refer to the migration of standard proteins. Lane: A] RPE-IPM, B] Pool 1, C] Pool 2, D] Pool 3, E] Pool 4.

Fractions were pooled on the basis of absorbance at 226 nm as follows: pool 1] 0.3 to 0.95 ml, pool 2] 4.8 to 7.5 ml, pool 3] 10.8 to 12.5 ml, pool 4] 16 to 18 ml. SDS-PAGE of the material in each pool (figure 2) was performed on a 10% silver stained gel. The purified material in pool three migrated as a doublet at a molecular weight of about 44 and 46 kDa, corresponding to the activated unglycosylated and glycosylated forms typical for stromelysin-1, respectively (27). This molecular weight is similar to that of other stromelysins from other tissues (12, 28, 29). No higher or lower molecular weight forms were seen.

Zymography indicated that pool three had the stronger gelatinolytic and caseinolytic activities, as shown in figure 3. These activities were completely inhibited by 20 mM EDTA or 5 mM o-phenanthroline (figure 4) confirming the metalloproteinase nature of the purified enzyme. The unretained material (pool 1) contained no gelatinolytic or caseinolytic activities, while pool 2 had very weak activity (data not shown).

Western immunoblotting performed with the mono-QHR pools (figure 5) showed strong reactivity of pool three with sheep anti-human MMP-3 polyclonal antibody as a doublet centered at about 45 kDa, as seen on SDS-PAGE (see above). Note that even though pool 4 contained gelatinase activity at about 46 kDa (see figure 3), it showed little cross-reactivity with the anti-MMP-3 antibody. Semi-purified MMP-3 from HT-1080 fibrosarcoma cell culture media was used as a control.

Stromelysins are secreted like most other matrixins as latent proenzymes and must be activated prior to displaying their action on components of the extracellular matrices.

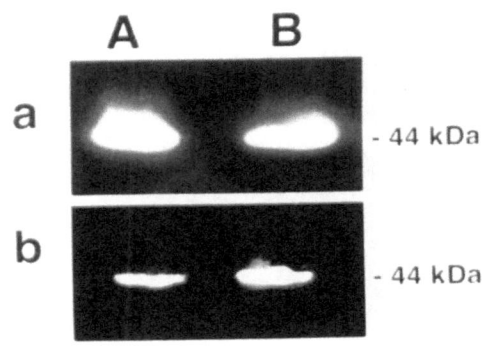

Figure 3. Zymography of material from Mono-QHR ion exchange chromatography. Aliquots of concentrated pools obtained from ion exchange chromatography were denatured without ß-mercaptoethanol and electrophoresed on 10% gels containing either 0.1% gelatin (a) or 0.1% casein (b). The gels were incubated for 48 h at 37° C and visualized with coomassie blue as described in materials and methods. The molecular weight refers to the migration of an ovalbumin standard. Lane: A] Pool 3, B] Pool 4.

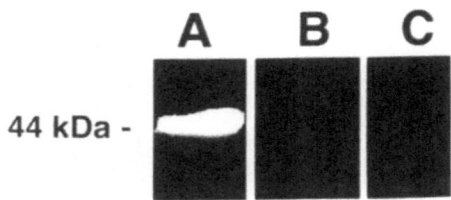

Figure 4. Effect of EDTA and o-phenanthroline on purified stromelysin. Aliquots of purified stromelysin (from pool 3) were denatured without ß-mercaptoethanol and electrophoresed on 10% gels containing 0.1% gelatin. The gels were incubated for 48 h at 37°C and visualized with coomassie blue as described in figure 3, except that EDTA (20 mM) or o-phenanthroline (5mM) were added to the denaturing solution, washing and incubation buffers as described in materials and methods. Lane: A] gelatin activity without metal chelate, B] EDTA C] o-phenanthroline.

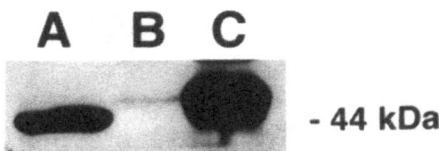

Figure 5. Western immunoblot analysis of pools 3 and 4 from Mono-QHR ion exchange chromatography. Aliquots of concentrated pools 3 and 4 were denatured in the presence of 2.5% ß-mercaptoethanol, electrophoresed on a 10% gel and electrotransferred to PVDF membranes. After blocking, the blots were probed with 1:2000 sheep anti-human MMP-3 as described in materials and methods. Lane: A] Positive control (semi-purified MMP-3 fraction from HT-1080 culture media), B] pool 4, C] pool 3.

Activation of the 56 kDa pro-form produces an active intermediate of 46 kDa which may then be converted to a 28 kDa form by splitting off a portion of the C-terminal domain (30). An alternative activation pathway leading to an active 21-kDa form of stromelysin has also been reported by Kolkenbrock et al. (31). In our case no other lower molecular weight active forms were found either in IPM itself or after treatment of the active 46 kDa form of purified stromelysin with trypsin or APMA, as it can be seen in figure 6. The same result was obtained after 4h activation (data not shown).

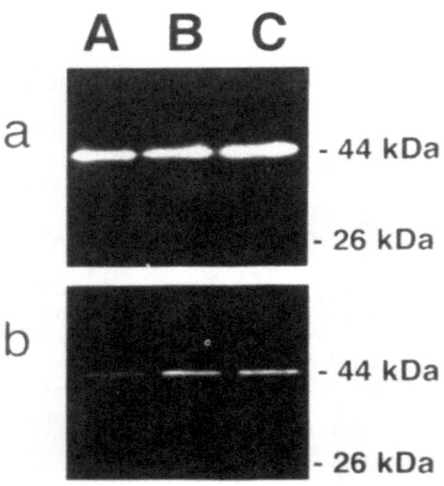

Figure 6. Treatment of purified stromelysin with trypsin and APMA. Potential conversion of stromelysin was assessed by zymography following treatment with 2 mM APMA or 10 μg/ml TPCK-trypsin at 37°C for 2h. Aliquots of treated material were denatured without ß-mercaptoethanol and electrophoresed on 10% gels containing (a) 0.1% gelatin or (b) 0.1% casein. The gels were incubated for 48 h at 37° C and visualized with coomassie blue as described in materials and methods. Lane: A] without activation, B] APMA, C] trypsin.

CONCLUDING REMARKS

The rapid growth of information on new members of the MMP family has resulted in an explosion of studies on the role of specific MMPs in different diseases. These studies, performed both in vivo and in vitro, consistently demonstrated a direct correlation between high-level expression of MMPs and/or low level expression of TIMPs and pathological conditions (11, 13, 32, 33). The presence of both the MMPs and TIMPs in ocular tissues suggests the need for a closely regulated balance in the level of expression of the active form of these enzymes in the eye. In a manner similar to that shown in other systems, an imbalance of MMPs and TIMPs might be associated with pathological conditions as manifest in a number of ocular diseases. Indeed, Sorsby's fundus dystrophy, an inherited disease having similarities to exudative age-related macular degeneration (34, 35), has been mapped to a defect in the TIMP-3 gene locus (36), and changes in the level of expression of TIMP-3 have been found in simplex retinitis pigmentosa (37).

In the present study, we report the purification and the partial characterization of stromelysin-1 from bovine IPM, which we first reported to be present in ocular tissue (20). Other procedures have been used for the purification of MMP-3 from other sources. Gunja-Smith et al. (38) used three chromatography steps, including, DEAE-Sephacel, gel filtration and anti-MMP-3 immunoadsorbent to purify human articular cartilage stromelysin; Stack et al. (39) used N-carboxyalkyl peptides for the affinity purification of porcine Stromelysin; and Obata et al. (40) used a one-step immunoaffinity procedure to isolate stromelysin from the culture medium of synovial fibroblasts. The relatively simple two-step procedure we have employed here seems to produce pure stromelysin-1 without resorting to immunoaffinity techniques.

Only the 46 kDa activated form was found to be present. This is reasonable, since IPM is at a "steady state." It is not an acutely stimulated material like tissue culture cell medium. It thus reflects the normal, in vivo situation much more accurately. Studies are currently in progress to purify other MMPs present in human IPM and to determine their modes of activation in vivo as well as their endogenous substrates.

ACKNOWLEDGMENTS

This work was supported in part by NIH grant EY 10184 (JJP), the Ohio Lions Eye Research Foundation and Research to Prevent Blindness, Inc.

REFERENCES

1. Okada, Y., Nagase, H. and Harris, Jr., E.D., 1986, A metalloproteinase from human rheumatoid synovial fibroblasts that digests connective tissue matrix components. Purification and characterization, *J. Biol. Chem.* **261**:14245–14255.
2. Murphy, G.J.P., Murphy, G. and Reynolds J.J., 1991, The origin of matrix metalloproteinases and their familial relationships, *FEBS Lett.* **289**:4–7.
3. Okada, Y., Konomi, H., Toshikazu, Y., Kimata, K. and Nagase, H., 1986, Degradation of type IX collagen by matrix metalloproteinase 3 (stromelysin) from human rheumatoid synovial cells, *FEBS Lett.* **244**:473–476.
4. Wu, J.J., Lark, M.W., Chun, L.E. and Eyre, D.R., 1991, Molecular sites of stromelysin cleavage in collagen types II, IX, X and XI of cartilage, *J. Biol. Chem.* **266**:5625–5628.
5. Okada, Y., Harris, Jr., E.D. and Nagase, H., 1988, The precursor of a metalloendopeptidase from human rheumatoid synovial fibroblasts. Purification and mechanism of activation by endopeptidases and 4-aminophenylmercuric acetate, *Biochem.* **29**:731–741

6. Woessner, Jr., J.F., 1991, Matrix metalloproteinases and their inhibitors in connective tissue remodeling, *FASEB J.* **5**:2145–2154.
7. Springman, E.B., Angelton, E.L., Birkedal-Hansen, H. and VanWart, H.E., 1990, Multiple modes of activation of latent human fibroblast collagenase: evidence for the role of Cys 73 active-site zinc complex in latency and a "cysteine switch" mechanism for activation, *Proc. Natl. Acad. Sci. U. S. A.* **87**:364–368.
8. VanWart, H.E. and Birkedal-Hansen, H., 1990, The cysteine switch: A principle of regulation of metalloproteinase activity with potential applicability to the entire matrix metalloproteinase gene family, *Proc. Natl. Acad. Sci. U. S. A.* **87**:5578–5582.
9. Okada Y., Takeuchi N., Tomita K., Nakanishi, I. and Nagase, H., 1989, Immunolocalization of matrix metalloproteinase 3 (Stromelysin) in rheumatoid synovioblasts (Bcell): Correlation with arthritis, *Ann. Rheum. Dis.* **48**:645–653.
10. Hasty, K.A., Reife, R.A., Kang, A.H. and Stuart, J.M., 1990, The role of stromelysin in the cartilage destruction that accompanies inflammatory arthritis, *Arthritis Rheum.* **33**:388–397.
11. Case, J.P., Lafyatis, R., Remmers, E.F., Kumkumian, G.K. and Wilder, R.L., 1989, Transin/stromelysin expression in rheumatoid synovium. A transformation-associated metalloproteinase secreted by phenotypically invasive synoviocytes, *Am. J. Pathol.* **135**:1055–1064.
12. Walakovits, L., Moore, V., Bhardwaj, N., Gallick, G. and Lark, M., 1992, Detection of stromelysin and collagenase in synovial fluid from patients with rheumatoid arthritis and posttraumatic knee injury, *Arthritis and Rheum.* **35**:35–42.
13. Stetler Stevenson, W.G., Liotta, L.A. and Kleiner, Jr., D.E., 1993, Extracellular matrix 6: Role of matrix metalloproteinases in tumor invasion and metastasis, *FASEB J.* **7**:1434–1441.
14. Fini, M.E., Yue, B.Y.J.T. and Sugar, J., 1992, Collagenolytic/gelatinolytic metalloproteinases in normal and keratoconus corneas, *Curr. Eye Res.* **11**:849–862.
15. Alexander, J.P., Samples, J.R., Van Buskirk, E.M. and Acott, T.S., 1991, Expression of matrix metalloproteinases and inhibitor by human trabecular meshwork, *Invest. Ophthalmol. Vis. Sci.* **32**:172–180.
16. Alexander, J.P., Bradley, J.M., Gabourel, J.D. and Acott, T.S., 1990, Expression of matrix metalloproteinases and inhibitor by human retinal pigment epithelium, *Invest. Ophthalmol. Vis. Sci.* **31**:2520–2528.
17. Hunt, R.C., Fox, A., Al Pakalnis, V., Sigel, M.M., Kosnosky, W., Choudhury, P. and Black, E.P., 1993, Cytokines cause cultured retinal pigment epithelial cells to secrete metalloproteinases and to contract collagen gels, *Invest. Ophthalmol. Vis. Sci.* **34**:3179–3786.
18. Plantner, J.J., 1992, The presence of neutral metalloproteolytic activity and metalloproteinase inhibitors in the interphotoreceptor matrix, *Curr. Eye Res.* **11**:91–101.
19. Plantner, J.J. and Drew, T.A., 1994, Polarized distribution of metalloproteinases in the bovine interphotoreceptor matrix. *Exp. Eye Res.* **59**:577–585.
20. Plantner, J.J., Quinn, T.A., Drew, T.A., Dwyer, G.J. and Rambasek, T.E., 1995, Metalloproteinases (MPs) and MP inhibitors and implications for retinal degeneration, *Invest. Ophthalmol. Vis. Sci.* **36**:S658.
21. Jones, B.E., Moshyedi, P., Gallo, S., Tombran-Tink, J., Arand, G., Reid, D.A., Thompson, E.W., Chader, G.J. and Waldbillig, R.J., 1994, Characterization and novel activation of 72-kDa metalloproteinase in retinal interphotoreceptor matrix and Y-79 culture medium, *Exp. Eye Res.* **59**:257–269.
22. Adler, A.J. and Severin, K.M., 1981, Proteins of the bovine interphotoreceptor matrix: tissues of origin, *Exp. Eye Res.* **32**:755–69
23. Plantner, J.J., 1992, High molecular weight mucin-like glycoproteins of the bovine interphotoreceptor matrix, *Exp. Eye Res.* **54**:113–125
24. Laemmeli, U.K., 1970, Cleavage of structural proteins during the assembly of the head of bacteriophage T4, *Nature* **227**:680–685.
25. Hibbs, M.S., Hoidal, J. and Kang, A.H., 1987, Expression of a metalloproteinase that degrades native type V collagen and denatured collagens by cultured alveolar macrophages, *J. Clin. Invest.* **80**:1644–1650
26. Huhtala, P., Tuuttila, A., Chow, L.T., Lohi, J., Keski-Oja, J. and Tryggvason, K., 1991, Complete structure of the human gene for 92-kDa type IV collagenase, *J. Biol. Chem.* **266**:16485–16490.
27. Bayne, E., Hutchinson, N., Walakovits, S., Macnaul, K., Harper, C., Cameron, P., Moore, V. and Lark, M., 1992, Production, purification and characterization of canine prostromelysin, *Matrix* **12**:173–184.
28. Murphy, G., Segain, J.P., O'Shea, M., Cockett, M., Ioannou, C., Lefebvre, O., Chambon, P. and Basset, P., 1993, The 28-kDa N-terminal domain of mouse Stromelysin-3 has the general properties of a weak metalloproteinase, *J. Biol. Chem.* **268**:15435–15441.
29. Okada, Y., Harris, Jr., E.D. and Nagase, H., 1988, The precursor of a metalloendopeptidase from human rheumatoid synovial fibroblasts, *Biochem. J.* **254**:731–741.
30. Nagase, H., Suzuki, K., Morodomi, T., Enghild, J.J. and Salvesen, G., 1992, Activation mechanisms of the precursors of matrix metalloproteinases 1,2 and 3, *Matrix suppl.* **1**:237–244.

31. Kolkenbrock, H., Hecker-Kia, A., Orgel, D., Huser, H., Schroder, W. and Ulbrich, N., 1994, Isolation of latent 31-kDa-truncated stromelysin and 21-kDa stromelysin from rabbit synovial fibroblasts: an alternative action pathway for stromelysin, *Biol. Chem.* **375**:241–247.

32. Hepper, K. J., Matrisian, L. M., Jensen, R.A. and Rodgers, W.H., 1996, Expression of most matrix metalloproteinase family members in breast cancer represents a tumor-induced host response, *Am. J. Pathol.* **149**:273–281.

33. Stetler-Stevenson, W.G., Hewitt, R. and Corcoran, M., 1996, Matrix metalloproteinases and tumor invasion: from correlation and causality to the clinic, *Cancer Biol.* **7**:147–154.

34. Hamilton, W.K., Ewing, C.C., Ives, E.I. and Carruthers, J.D., 1989, Sorsby's fundus dystrophy, *Ophthalmol.* **96**:1755–1762.

35. Polkinghorne, P.J., Caon, M.R.C., Berninger, T., Lyness A.L., Sehmi, K. and Bird, A.C., 1995, Sorsby's dystrophy. A clinical study, *Ophthalmol.* **96**:1763–1768.

36. Weber, B.H.F., Vogt, G., Pruett, R.C., Stohr, H. and Felbor, U., 1994, Mutations in the tissue inhibitor of metalloproteinases-3 (TIMP-3) in patients with sorsby's fundus dystrophy, *Nature Genetics* **8**:352–358.

37. Jones, S.E., Jomary, C. and Neal, M.J., 1994, Expression of TIMP3 mRNA is elevated in retinas affected by simplex retinitis pigmentosa, *FEBS Lett.* **352**:171–174.

38. Gunja-Smith, Z., Nagase, H. and Woessner, Jr., J.F., 1989, Purification of the neutral proteoglycan-degrading metalloproteinase from human articular cartilage tissue and its identification as stromelysin matrix metalloproteinase-3, *Biochem. J.* **258**:115–119.

39. Stack, S., Emberts, C., and Gray, R., 1991, Application of N-carboxyalkyl peptides to the inhibition and affinity purification of the porcine matrix metalloproteinases collagenase, gelatinase, and stromelysin, *Arch. Biochem. Biophys.* **287**:240–249.

40. Obata, K., Iwata, K., Okada, Y., Kohrin, Y., Ohuchi, E., Yoshida S., Shinmei M. and Hayakawa T., 1992, A one-step sandwich enzyme immunoassay for human matrix metalloproteinase 3 (stromelysin-1) using monoclonal antibody, *Clin. Chim. Acta* **211**:59–72.

44

SENILE RETINAL DEFICIENCIES IN ASTROCYTES AND BLOOD VESSELS

Elisabeth Rungger-Brändle and Peter M. Leuenberger

Electron Microscopy Laboratory
University Eye Clinic
22 rue Alcide-Jentzer, CH-1211 Geneva 14, Switzerland

ABSTRACT

Vessels in the mammalian retina are surrounded by astrocytes and Müller cells forming the glial limitans. The spatial correlation between astrocytes and blood vessel in the inner retina has been largely recognized and the functional implication of astrocytes in vessel permeability has been suggested.

In the senile retinas examined by electron microscopy, arteries and veins exhibited loss of smooth muscle cells, while the endothelium was still intact. By confocal laser scan fluorescence microscopy, the corresponding deficiencies in smooth muscle α-actin staining were spotty or extended over longer segments of the vascular wall. Vessel damage was associated with an atrophy of the glia limitans, visualized by confocal microscopy as weak staining for GFAP and, by electron microscopy, as exfoliation and degeneration of the glial sheath. Chronic hypoxia might represent one possible aggravating factor in the development of the disease.

INTRODUCTION

In the vascularized mammalian retina, blood vessels are surrounded by the glia limitans, formed by astrocytes and Müller cells (for review, see 1). The particular spatial correlation between vessels and astrocytes in the inner retina has been largely documented (e.g. 2,3) and interpreted to be the consequence of correlated spreading of mesenchymal precursor cells and astrocytes immigrating from the optic nerve during retinal development (4–7).

The fact that astrocytes are associated with blood vessels has suggested the functional implication of the former in vessel permeability. Indeed, endothelial cells in coculture with astrocytes differentiate tight junctions (8). Similarly in vivo, non-neural endothelial cells form tight junctions in conjunction with astrocytes but not with meningeal cells grafted in close vicinity (9). By contrast, neovessels lacking a glial sheath are leaky (10).

Degenerative Retinal Diseases, edited by LaVail *et al.*
Plenum Press, New York, 1997

We have tried to establish a correlation between vessel pathology and deficiencies of the surrounding glia limitans in aged retinas. Electron microscopy gives evidence for a deficient glia limitans in those vessels with severe damage of the medial cell layer. Confocal laser scan fluorescence microscopy provides the possibility to unambiguously analyze the spatial relationship between the vascular wall and the surrounding glia over extended vessel segments. Here we illustrate the principle of this approach, while results are dealt with in a more detailed investigation on the pathology of chorio-retinal blood vessels in the elderly (E. Rungger-Brändle and P.M. Leuenberger, in preparation).

MATERIALS AND METHODS

Tissue

The age of the donors was over 80 years. All patients had suffered intellectual dementia, some of them of the Alzheimer type but none had experienced diabetes or hypertension during life history. Enucleation was performed within 2 hours postmortem.

Electron Microscopy

Posterior segments were fixed in 2.5% glutaraldehyde in the presence of 0.2% tannic acid, osmicated, dehydrated, and the retina thin sectionned transversally as previously described (11).

Confocal Laser Scan Fluorescence Microscopy

Isolated retinas were flat-mounted and processed for indirect immunofluorescence microscopy as decribed (3) using a rabbit serum to GFAP (glial fibrillary acidic protein) and a monoclonal antibody to smooth muscle α-actin. GFAP was indirectly visualized with fluorescein, smooth muscle α-actin with Texas Red. Flat-mounted tissue was viewed, vitreous side towards the objective, in a Zeiss LSM 410 inverted microscope, equipped with 2 different lasers that were used simultaneously to observe fluorescein and Texas Red, respectively. Extended focus images of the partial (Fig. 2) or complete (Fig. 3) data set were performed. The axis of single sections in the z dimension was set along the vessel, i.e. running obliquely to the x/y coordinates of the raw data set (Fig. 3).

RESULTS

Electron microscopic examination of a large artery close to the inner limiting membrane discloses a near to complete local absence of smooth muscle cells from the medial vascular layer and an increased thickness of the extracellular matrix (Fig. 1). The presence of vesicular deposits and multiple basement membranes indicates that disappearance of smooth muscle cells is due to their degeneration. Müller cells of the glia limitans are vacuolized and portions of their cytoplasm separated from the neural parenchyma by intercalated sheets of extracellular matrix. Such "exfoliation" of the glia limitans eventually leads to glial cell degeneration, which occurs before alterations in neurons become detectable (not shown). In the material examined, we never observed hypertrophy, such as an increase of intermediate filaments, in the glia limitans lining the vessels. Astrocytes participating in the formation of the

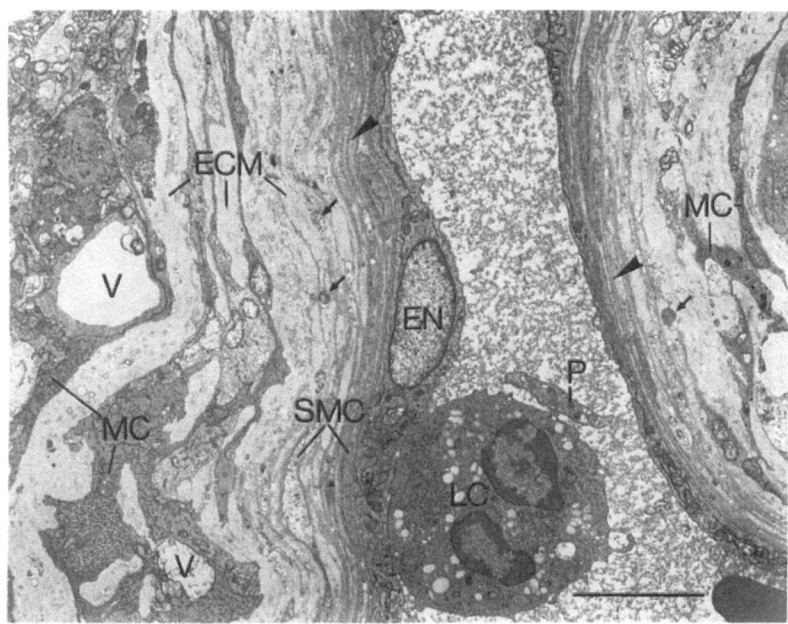

Figure 1. Electron micrograph of an artery close to the inner limiting membrane of a senile retina. The medial layer shows heavy loss of smooth muscle cells (SMC), whereas the endothelium (EN) looks intact. Müller cells (MC) forming the glia limitans are vacuolized (V) and spaced by sheets of extracellular matrix (ECM). Note multiple basement membranes (arrowheads) and vesicular deposits (small arrows) in the thickened ECM. LC, leukocyte firmly adhering to the endothelial surface. P, platelet. Bar, 5 μm.

glia limitans show signs of degeneration similar to those observed in Müller cells (not shown) suggesting that both glial types suffer in a similar manner.

The spatial distribution of vascular lesions and glial elements was assessed by confocal double fluorescence microscopy. Figs. 2 and 3 are based on a data set of 75 optical sections through a branching artery close to the inner limiting membrane in the mid-peripheral retina. The gallery of extended focus images starting at the vitreo-retinal interface (Fig. 2, top) and going into the inner plexiform layer (Fig. 2, bottom) reveals massive gliosis confined to the vitreo-retinal interface (Fig. 2a'). Deficiencies in smooth muscle α-actin staining in several segments of the vascular wall may be small and focal or extend over a length of more than 50μm. Large deficiencies may involve the entire circumference of the vessel (Fig. 2b,c). The corresponding GFAP staining patterns (Fig. 2b',c') appear normal with a preferential orientation of the glial filaments in the optic nerve fiber layer along the ganglion cell axons (Fig. 2b') and, as shown on deeper optical sections (Fig. 2c), they also appose to the vessel surface (Fig. 2c'). The deepest images into the retina (Fig. 2d,d') show a number of glial filaments crossing over the retinal aspect of the vessel wall.

Optical sectioning of the entire data set (Fig. 3a,a') through the z axis along the 2 arterial branches (axes b and c in Fig. 3a) confirms that deficiencies in smooth muscle α-actin staining most often extend over both the vitreal and retinal side of the vascular wall (Fig. 3b,c). The corresponding GFAP images (Fig. 3b',c') reveal intense staining of the gliotic tissue at the vitreo-retinal interface, whereas the one of the glia limitans lining the vessel is particularly weak. Strongly stained dots at the vitreal side of the vessel represent cross-sectioned filament bundles along optic nerve fibers (cf. Fig. 2b').

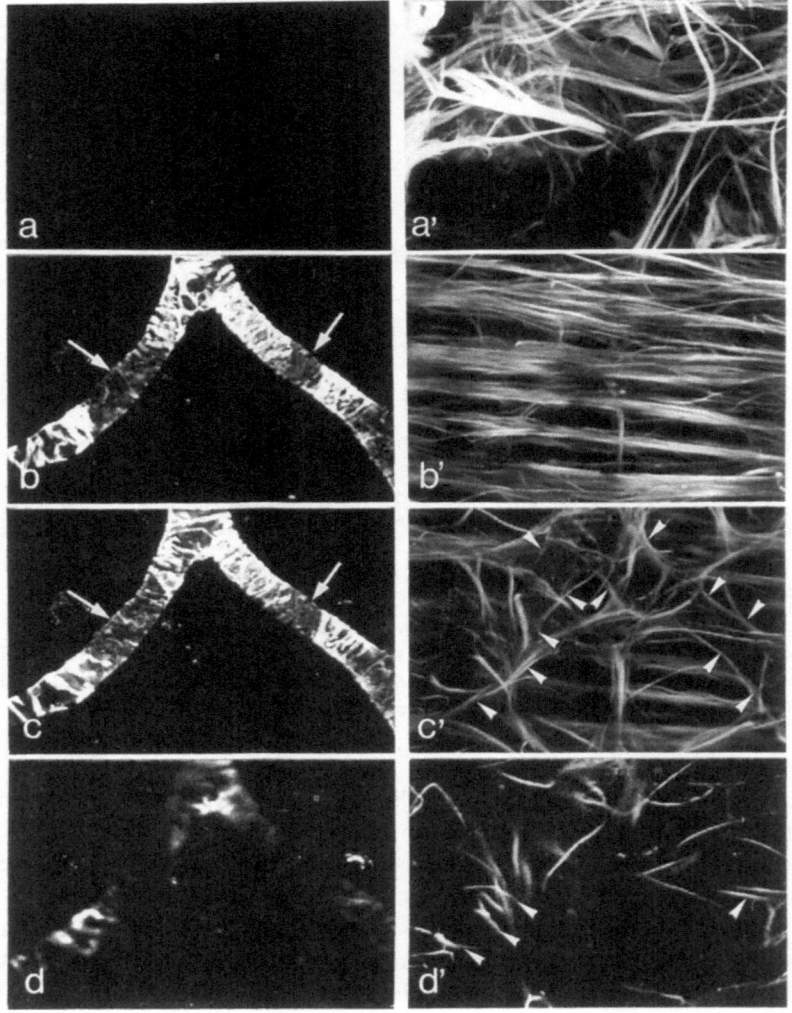

Figure 2. Gallery of 4 extended focus images of a bifurcating artery and astrocytes in a midperipheral senile retina, going from vitreous (a,a') into the retina (d,d'). Confocal double fluorescence microscopy. The media of the arterial wall was stained with monoclonal antibody to smooth muscle α-actin (Texas Red, left panels a-d), astrocytes with polyclonal antibody to GFAP (FITC, right panels a'-d'). Laser scanning of the raw data set was performed simultaneously for both fluorochromes. The reconstructions are based on a data set of 75 optical sections (stepsize 0.4μm) from the vitreoretinal interface (top; a,a') into the retina (bottom; d,d'). The images are irregularly spaced and the staining intensities are corrected to show the typical patterns of glial filaments (a'-d'). a,a', Extended focus image of sections 3–12 showing pronounced gliosis (a') vitreal to the vessel (a). b,b', Extended focus image of sections 36–56 including the vitreal aspect of the vessel (b) with glial filaments arranged essentially along the nerve fibers (b'). c,c', Extended focus image of sections 50–75 including the entire thickness of the vessel (c). Glial filaments accompany the vessel wall (c', arrowheads). d,d', Extended focus image of sections 70–75 showing the most retinal aspect of the vessel wall (d) and glial filaments apposed to it (d', arrowheads). The diameter of the 2 vessel branches in the x and y axes measures 48 μm.

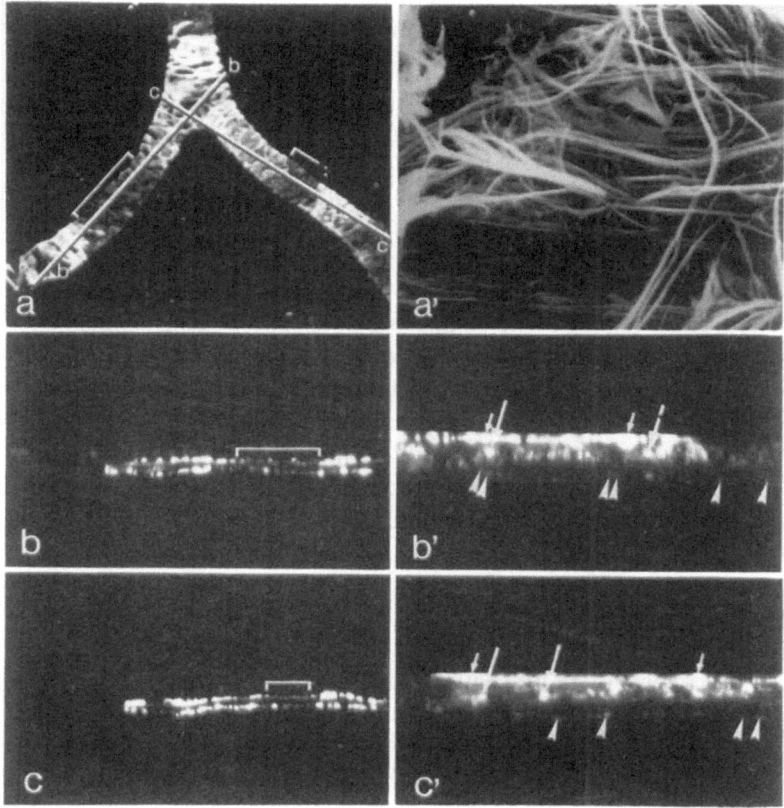

Figure 3. Extended focus reconstructions (a,a′) and single oblique (axes x and y) optical sections (b-c′) using the same confocal data set as in Fig. 2. The left panels (a-c) show staining of smooth muscle α-actin, the right panels (a′-c′) the corresponding GFAP staining. Unlike in Fig. 2, staining intensities were not corrected here. a,a′, Projections along the axes x and y are reconstructed from the total data set of 75 optical sections. b,c, Oblique optical sections along the z axis through the left (b) and the right arterial branch (c) show heavy deficiencies (brackets) in smooth muscle α-actin staining. b′,c′, The corresponding staining pattern for glial filaments (GFAP) at the retinal aspect of the vessel wall (arrowheads) is weak. Note strong staining (small arrows) reflecting important gliosis at the vitreo-retinal interface. The dotted pattern (large arrows) intercalated between vitreo-retinal interface and vessel represents cross-sectioned glial filament bundles along the optic nerve fibers (cf. Fig. 2b′). Due to immersion-fixation, the vessel is collapsed in the z axis (vertical in b,c).

DISCUSSION

Immunofluorescent staining of flat-mounted retina with monoclonal antibody to smooth muscle α-actin reveals smooth muscle cells in first and second order arteries and veins, whereas the cells of the microvasculature such as small arterioles, capillaries, and venules, are inconsistenly stained depending on the species examined (own unpublished observations). In humans, the microvasculature is not revealed by anti-smooth muscle α-actin staining but larger vessels are strongly and reproducibly stained. We thus have used this antibody to investigate the state of the medial vascular layer in larger vessels of the inner retina under various pathological conditions.

Staining for GFAP, as it was used in this study in the flat-mounted human retina, detects astrocytes but not Müller cells. However, since the electron microscopic observations suggest that both glial cell types involved in the formation of the glia limitans are susceptible to degeneration, we think that the immunofluorescent data on astrocytes hold true also for Müller cells.

The analysis of the senescent retina by confocal fluorescence microscopy has revealed pronounced, focally or segmentally extended deficiencies in smooth muscle α-actin staining within an artery close to the inner limiting membrane. Based on accompanying electron microscopic observations, such staining deficiencies are not due to the expression of a different actin isoform, but indeed represent degeneration and disappearance of smooth muscle cells. Similar lesions were found in postmortem tissue from patients having suffered from longstanding diabetes (12). In both pathological situations, these vascular lesions go along with an increase in thickness of the extracellular matrix.

Interestingly, the confocal data set presented here has revealed only weak GFAP staining associated with the vessel wall and electron microscopy of a heavily damaged vessel has shown degeneration of the glia ensheathing the vessel. Taken together, these observations indicate that an atrophy of the glia limitans is closely associated with vessel damage. At this final stage of the disease, atrophy rather than hypertrophy is characteristic of glia surrounding the vessels, whereas typical gliosis is restricted to the vitreo-retinal interface (cf. Fig. 2a').

Glia ensheathing blood vessels appear remarkably susceptible to hypoxia in the genesis of retinopathy of prematurity in the kitten (10) and their damage precedes overt alterations in neurons. It is likely that, similar to the situation in the prematurity model, longstanding tissue hypoxia in the senile retina plays an important role in the pathogenesis of alterations in both, smooth muscle cells and glia limitans ultimately leading to an increase in vessel permeability.

ACKNOWLEDGMENTS

We thank Ms. C. Alliod for her help with confocal microscopy and Mr. A. Conti for skillful assistance in electron microscopy and photographic work. This study was supported by the Swiss National Science Foundation, grants 31-32047.91 and 31-041909.94.

REFERENCES

1. Schnitzer, J., 1988, Astrocytes in mammalian retina, in: *Progress in Retinal Research*, Volume 7 (N.N. Osborne, and G.J. Chader, eds.), pp. 209–231, Pergamon Press, Oxford.
2. Holländer, H., Makarov, F., Dreher, Z., van Driel, D., Chan-Ling, T., and Stone J., 1991, Structure of the macroglia of the retina: Sharing and division of labour between astrocytes and Müller cells, *J. Comp. Neurol.* **313**: 587–603.
3. Rungger-Brändle, E., Messerli, J.M., Niemeyer, G., and Eppenberger, H.M., 1993, Confocal microscopy and computer-assisted image reconstruction of astrocytes in the mammalian retina. *Eur. J. Neurosci.* **5**: 1093–1106.
4. Ling, T. and Stone, J., 1988, The development of astrocytes in the cat retina: Evidence of migration from the optic nerve. *Dev. Brain Res.* **44**: 73–85.
5. Watanabe, T. and Raff, M.C., 1988, Retinal astrocytes are immigrants from the optic nerve. *Nature* **332**: 834–837.
6. Ling, T., Mitrofanis, J., and Stone, J., 1989, Origin of retinal astrocytes in the rat: Evidence of migration from the optic nerve. *J. Comp. Neurol.* **286**: 345–352.
7. Chan-Ling, T. and Stone, J., 1991, Factors determining the migration of astrocytes into the developing retina: Migration does not depend on intact axons and patent vessels. *J. Comp. Neurol.* **303**: 375–386.

8. Tao-Chen, J.-H., Nagy, Z., and Brightman, M.W., 1987, Tight junctions of brain endothelium in vitro are enhanced by astroglia. *J. Neurosci.* **7**: 3293–3299.

9. Janzer, R.C. and Raff, M. C., 1987, Astrocytes induce blood-brain barrier properties in endothelial cells. *Nature* **325**: 253–257.

10. Chan-Ling, T. and Stone, J., 1992, Degeneration of astrocytes in feline retinopathy of prematurity causes failure of the blood-retinal barrier. *Invest. Ophthalmol. Vis. Sci.* **33**: 2148–2159.

11. Rungger-Brändle, E., Englert, U., and Leuenberger, P.M., 1987, Exocytic clearing of degraded membrane material from pigment epithelial cells in frog retina. *Invest. Ophthalmol. Vis. Sci.* **28**: 2026–2037.

12. Rungger-Brändle, E., Dosso, A.A., and Leuenberger, P.M., 1997, Smooth muscle lesions in large-caliber arteries and veins of the human diabetic retina. *Invest. Ophthalmol. Vis. Sci.* (abstract ARVO, in press).

CORRESPONDING AUTHORS BY CHAPTER NUMBER

1. **Dr. Gregory Hageman**
Anheuser-Busch Eye Institute
St. Louis University
1755 S. Grand Blvd.
St. Louis, MO 63104 USA
TEL 314-865-8329
FAX 314-771-0596
E-MAIL hagemang@sluvca.slu.edu

2. **Dr. Joe Hollyfield**
Research Institute, FFb-29
Cleveland Clinic Foundation
9500 Euclid Avenue
Cleveland, OH 44195 USA
TEL (216) 445-3252
FAX (216) 445-3670
E-MAIL hollyfj@cesmtp.ccf.org

3. **Dr. Joe Hollyfield**
Research Institute, FFb-29
Cleveland Clinic Foundation
9500 Euclid Avenue
Cleveland, OH 44195 USA
TEL (216) 445-3252
FAX (216) 445-3670
E-MAIL hollyfj@cesmtp.ccf.org

4. **Dr. Enzo Vingolo**
Institute of Ophthalmology
University of Rome "La Sapienza"
Via G. Dandini 5
Roma 00154 Italy
TEL 39-6-578 2744
FAX 39-6-578 2744

5. **Professor Yozo Miyake**
Department of Ophthalmology
Nagoya University
65 Tsuruma-cho, Aichi
Nagoya-shi 466 Japan
TEL81-52-744-2279
FAX 81-52-744-2279

6. **Dr. Mutsuko Hayakawa**
Department of Ophthalmology
Juntendo University
3-1-3 Hongo Bunkyo-ku
Tokyo 113 Japan
TEL 81-3-3813-3111
FAX 81-33817-0260

7. **Dr. Paul Hargrave**
Department of Ophthalmology
University of Florida
Box J-100284/JHMHC
Gainesville, FL 32610-0284 USA
TEL 352-392-9098
FAX 352-392-0573
E-MAIL hargrave@eye1.eye.ufl.edu

8. **Dr. Norio Ohba**
Department of Ophthalmology
Kagoshima University
8-35-1 Sakuragaoka
Kagoshima
Kagoshima-shi 890 Japan
TEL 81-992-75-5402
FAX 81-99-264-1387

9. **Dr. Elizabeth Rakoczy**
Department of Molecular Biology
Lions Eye Institute
2 Verdun Street
Nedlands 6009 Western Australia
TEL 61-9-346-2811
FAX 61-9-382-1171
E-MAIL rakoczy@uniwa.uwa.edu.au

10. **Dr. Marco Zarbin**
Department of Ophthalmology
New Jersey Medical School
90 Bergen Street, Suite 6155
Newark, NJ 07103-2499
TEL 201-982-2036; FAX 201-468-2068
E-MAIL zarbin@umdnj.edu

11. **Dr. Andreas Gal**
Institut für Humangenetik
Universität Hamburg
Butenfeld 32
Hamburg 33529 Germany
TEL 49 40 47 17 21 20
FAX 49 40 47 17 5138

12. **Dr. Muna Naash**
Depts. of Ophthalmology and Visual
   Science/UIC Eye Center
1855 West Taylor St.
Chicago, IL 60612 USA
TEL 312-413-1164; FAX 312-996-7773
E-MAIL munanaas@uicvm.uic.edu

13. **Dr. Olaf Strauss**
Neurobiologie Battelle-Institut e.V.
Am Roemerhof 35
d-600 Frankfurt Germany
TEL 49 30 8445 2536
FAX 49 30 8445 4239
E-MAIL wiederho@zedat.fu-berlin.de

14. **Dr. Edward Kean**
Department of Ophthalmology
Case Western Reserve University
2074 Abington Rd., Rm 653 Wearn
Cleveland, OH 44106 USA
TEL 216-844-3613
FAX 216-844-7899
E-MAIL elk2@po.cwru.edu

15. **Dr. Katsuhiro Yamaguchi**
Department of Ophthalmology
Tohoku University School of Medicine
1-1 Seiryocho, Aobaku
Sendai Miyagi 980 Japan
TEL 81-22-717-7294
FAX 81-236-28-5376

16. **Dr. Margaret McLaren**
6500 S.W. 133rd Drive
Miami, FL 33156
TEL 305-667-9929
FAX 305-607-9929

17. **Dr. William Stark**
Department of Biology
St. Louis University
3507 Laclede Avenue
St. Louis, MO 63103-2010 USA
TEL 314-658-7151
FAX 314-658-7151
E-MAIL starkws@sluvca.slu.edu

18. **Dr. David Hyde**
Department of Biological Science
University of Notre Dame
Galvin Life Sciences Building
Notre Dame, IN 46556 USA
TEL 219-631-8054
FAX 219-631-7413
E-MAIL david.r.hyde.1@nd.edu

19. **Dr. Joachim Bentrop**
Department of Zoologie I
Universität Karlsruhe (T.H.)
Kornblumenstr. 13, Postfach 69 80
Karlsruhe 1 FRG 776128
TEL 49-721-608-2218
FAX 49-721-608-4848
E-MAIL dc02@rz.uni-karlsruhe.de

20. **Dr. Ian Morgan**
Visual Sciences Group
Centre for Visual Science & Research
School of Biological Sci. GPO Box 475
Canberra City ACT 2601 Australia
TEL 616-249-4671; FAX 616-249-3808
E-MAIL
   morgan@rsbs-central.anu.edu.au

21. **Dr. Fumiyuki Uehara**
Department of Ophthalmology
Kagoshima University
8-35-1 Sakuragaoka
Kagoshima
Kagoshima-shi 890 Japan
TEL 992-75-5402
FAX 992-65-4894

22. **Dr. Farhad Hafezi**
Department of Ophthalmology
University Hospital
Ramistrasse 100
Zurich CH-8091 Switzerland
TEL 411 255 3276/3719
FAX 411 255 4385
E-MAIL
   100577.1207@compuserve.com

23. **Dr. Satoru Kato**
Department of Neurobiology/NIRI
University of Kanazawa
13-1 Takara machi
Ishikawa Kanazawa-shi 920 Japan
TEL 81-762-34-4235
FAX 81-762-34-4235
E-MAIL
   satoru@med.kanazawa-u.ac.jp

24. **Dr. George Inana**
Laboratory of Molecular Genetics
Bascom Palmer Eye Institute
1638 NW 10th Ave.
Miami, FL 33136 USA
TEL 305-326-6509
FAX 305-326-6306
E-MAIL
   ginana@mednet.med.miami.ed

25. **Dr. Debora Farber**
Jules Stein Eye Institute
JSEI, UCLA School of Medicine
100 Stein Plaza
Los Angeles, CA 90024-7008 USA
TEL 310-206-6800
FAX 310-206-3652
E-MAIL farber@jsei.ucla.edu

26. **Dr. Andreas Gal**
Institut für Humangenetik
Universität Hamburg
Butenfeld 32
Hamburg 33529 Germany
TEL 49 40 47 17 21 20
FAX 49 40 47 17 5138

27. **Dr. Joyce Tombran-Tink**
Dept. of Psychiatry & Neuroscience
Children's National Medical Center
111 Michigan Avenue NW
Washington, DC 20010
TEL 202-884-2584; FAX 301-884-2588

28. **Dr. Yoshihiro Hotta**
Department of Opthalmology
Juntendo University
3-1-3 Hongo
Toyko Bunkyo-ku 113 Japan
TEL 81-3-3813-3111
FAX 81-3-3817-0260

29. **Dr. Roser González-Duarte**
Departmento of Genética
Universidad de Barcelona
Av. Diagonal 645
Barcelona 645 Spain
TEL 34-3-4021498
FAX 34-3-4110969
E-MAIL roser@porthos.bio.ub.es

30. **Dr. Stephen Daiger**
Human Genetics Center
The University of Texas HSC
P.O. Box 20334
Houston, TX 77225 USA
TEL 713-500-0900; FAX 713-500-9829
E-MAIL
   sdaiger@kiwi.imgen.bcm.tmc.edu

31. **Dr. Jacquie Greenberg**
Department of Human Genetics
University of Cape Town
Medical School Observatory
Cape Town 7925 South Africa
TEL 021-406-6299
FAX 021 47 7703
E-MAIL jg@anat.uct.ac.za

32. **Dr. Cheryl Gregory**
Department of Molecular Genetics
Institute of Ophthalmology
Bath Street
London EC1V 9EL England
TEL 44 171-608-6823
FAX 44 171-608-6863
E-MAIL cgregory@hgmp.mrc.ac.uk

33. **Dr. Radha Ayyagari**
Kellogg Eye Center
1000 Wall Street, Rm. 325
Ann Arbor, MI 48105
TEL 313-647-6345; FAX 313-936-7231
E-MAIL AYYAGARI@umich.edu

34. **Dr. Marion Maw**
Department of Biochemistry
University of Otago
P.O. Box 56, Dunedin, New Zealand
TEL 64 3 479 7863
FAX 64 3 479 7866
E-MAIL
    marion.maw@stonebow.otago.ac.nz

35. **Dr. Yuko Wada**
Department of Ophthalmology
Tohoku University School of Medicine
1-1 Seiryo-Machi
Sendai Miyagi 980 Japan
FAX 81-22-717-7298

36. **Professor Makoto Tamai**
Department of Ophthalmology
Tohoku University School of Medicine
1-1 Seiryo-Machi
Sendai Miyagi 980 Japan
TEL 81-22-717-7294
FAX 81-22-717-7298
E-MAIL j23463@cctu.cc.tohoku.ac.jp

37. **Dr. Kristina Narfström**
Department of Medicine and Surgery
Swedish Univ. of Agricultural Sciences
Box 7018, Uppsala S-750 Sweden
TEL 46-18-671471
FAX 46-18-672919
E-MAIL
    kristina.narfstrom@kirmed.slu.se

38. **Dr. Devjani Lahiri-Munir**
Department of Ophthalmology
University of Texas HSC
6411 Fannin
Houston, TX 77030 USA
TEL 713-757-7453
FAX 713-756-5399
E-MAIL
    dlahiri@mail-opht.med.uth.tmc.edu

39. **Dr. Kazuhiko Unoki**
Department of Ophthalmology
Kagoshima University Faculty of
    Medicine
Sakuragaoka 8-35-1
Kagoshima-shi 890 Japan
TEL 81 992 75 5402
FAX 81 992 65 4894
E-MAIL
    kazuhiko@med5.kufm.kagoshima-u.
    ac.jp

40. **Dr. Jonathan Stone**
Department of Anatomy and Histology
University of Sydney, NSW 2006
Sydney 2006 Australia
TEL 61 2 351 2496
FAX 61 2 9351-6556
E-MAIL jonstone@anatomy.su.oz.au

41. **Dr. Matthew M. LaVail**
Beckman Vision Center
10 Kirkham St., Room K-120
UCSF School of Medicine
San Francisco, CA 94143-0730
TEL 415-476-4233
FAX 415-476-0709
E-MAIL mmlv@itsa.ucsf.edu

42. **Dr. Michael Hall**
Jules Stein Eye Institute
UCLA School of Medicine
100 Stein Plaza
Los Angeles, CA 90024-7008 USA
TEL 310-825-6669
FAX 310-206-8583
E-MAIL hall@jsei.ucla.edu

43. Dr. James Plantner
Department of Ophthalmology
Case Western Reserve University,
    Wearn Bldg 647
2074 Abington Rd.
Cleveland, OH 44106 USA
TEL 216-844-3612; FAX 216-844-5297
E-MAIL jjp3@po.cwru.edu

44. Dr. Elisabeth Rungger-Brändle
Electron Microscopy Laboratory
University Eye Clinic
22, rue Alcide-Jentzer
Geneve 14 CH 1211 Switzerland
FAX 41 22 38 28 382

INDEX

adRP (autosomal dominant retinitis pigmentosa)
   positional cloning of RP1 (8q11-q21), RP10
      (7q31.3), and VMD1 (8q24), 277–290
   VPP mouse model, 89–98
Age-related macular degeneration: *see* AMD
AMD (age-related macular degeneration)
   histochemistry of drusen in monkeys and humans,
      1–10, 61–70
   model for drusen, lipofuscin, basal laminar deposits
      in Bruch's membrane, 61–70
   photoreceptor rosettes, 17–22
   TIMP-3 accumulation in drusen and Bruch's mem-
      brane, 11–16
Apoptotic cells, 181–192, 193–198, 353–368
Arrestin gene
   chromosome 2, CSNB (congenital stationary night
      blindness), Oguchi disease, 319–322, 313–
      318
   X-chromosome, 205–226
arRP (autosomal recessive retinitis pigmentosa)
   candidate gene and positional cloning approaches in
      Spanish families, 263–276
   PDE6A (rod PDE-α, chromosome 5), 237–244
   RCV1 (recoverin, chromosome 2), 263–276
   SAG (S-antigen, chromosome 17), 263–276
Astrocytes, 409–416
Autoimmune retinopathy, cystoid macular degenera-
      tion and anti-retinal protein antibody, 51–56

bFGF
   neonatal RCS rats, RPE and choroidal angiogenesis,
      121–134
   transfected in iris pigment epithelial cells, 323–329
BLD (basal laminar deposits), 1–10, 61–70
Blood vessels, 11–16, 115–120, 121–134, 409–416
Bovine
   MMP-3 in RPE-IPM, 399–408
   RPE and interphotoreceptor matrix, 399–408
Bruch's membrane, in age-related macular degenera-
      tion, 11, 17, 61

c-fos –/– mouse, light-induced photoreceptor degen-
      eration in, 193–198

Candidate genes for retinal degeneration
   Arrestin gene (chromosome 2), 313–318
   PDE-α'(cone PDE α' gene, chromosome 10),
      227–236
   COD1 (cone dystrophy, X-chromosome), 43–50
   CORD2 (cone rod dystrophy, chromosome 19),
      295–302
   HRG-4 (human retinal gene-4), 205–226
   ND (X-linked Norrie Disease, X-chromosome),
      57–60
   PDE6A (phosphodiesterase α, chromosome 5),
      237–244
   PEDF (pigment epithelium-derived factor, chromo-
      some 17), 245–254, 291–294
   peripherin/rds, 255–263, 277–290
   recoverin (chromosome 17), 205–226, 263–
      276
   rhodopsin, 255–263, 277–290
   ROM-1 (peripherin/rds, chromosome 11), 205
   SAG (S-antigen, chromosome 2), 263–276
   Ush 1C (Ushers, chromosome 11), 303–312
   X-Arrestin (X-chromosome), 205–226
Cat
   Iodoacetate-induced retinopathy, 71–80
   RPE transplantation in Abyssinian, 329–338
CD-36
   RPE cells and glycoprotein, 385–398
cGMP
   nitric oxide and NAT (serotonin N-Acetyltrans-
      ferase), 171–180
   cone PDE-α' gene, 227–236
   rod PDE- α gene, PDE6A gene, 237–244
Choroid
   angiogenesis and neonatal RCS rat, 121–134
   angiogenesis/neovascularization and TIMP-3 in
      AMD, 11–16
   ICGV (indocyanine green videoangiography) of
      choriocapillaris in RCS rat, 115–120
Chromosome 2
   arrestin gene, 313–318
   CSNB, Oguchi disease, arrestin gene, 313–318,
      319–322
   SAG (S-antigen), 263–276

Chromosome 5, PDE6A, 238
Chromosome 8
    RP1, 277–290
    VMD1, 277–290
Chromosome 10
    cone PDE α' Gene, 227–236
    RP10, 277–290
Chromosome 11
    ROM-1, 205
    Usher syndrome Type 1C, 303–312
Chromosome 17
    PEDF, RP13 locus, 291–294
    recoverin, 205–226, 263–276
Chromosome 19, CORD2, YAC Contig, 295–301
Chromosome-X
    X-arrestin, 205–226
    X-linked cone dystrophy
    X-linked ND (Norrie Disease) gene, 57–60
CME (cystoid macular edema), in retinitis pigmentosa
    and autoimmunity, 51–56
Cone dystrophy
    cone PDE-α' gene, 227–236
    X-linked cone dystrophy with tapetal like reflex,
        43–50
CORD2 (cone-rod dystrophy), chromosome 19, YAC
    contig, 295–301
CSNB (congenital stationary night blindness)
    foveal cone densitometry and ERG, 31–42
    glutamate analogs (APB and KYN) in bipolar cells
        of monkeys, 31–42
    model of "Off-retina" and "On-retina," 31–42
    Oguchi disease, arrestin gene, chromosome 2 in,
        313–318, 319–322

DCF (2,7 dichlorofluorescein), fish retina, 199–204
Dogs, Swedish Briard, 81–88
DRBP (drosophila retinoid binding protein), in ret-
    inoid deprivation and replacement, 135–
    144
Drosophila
    mutations in cytoplasmic domains of rhodopsin,
        159–170
    characterization of rdgB rhodopsin mutations,
        144–158.
    rhodopsin, DRBP and PLC in retinoid deprivation
        and replacement, 135–144
Drusen
    age-related macular degeneration, 1–10, 11–22,
        61
    histochemistry of drusen in monkeys and humans,
        1–10, 61–70
    model for drusen, lipofuscin, basal laminar deposits
        in Bruch's membrane, 61–70
    TIMP-3 accumulation in drusen and Bruch's mem-
        brane, 11–16

Fish retina, ischemia, GSH (glutathione), DCF (2,7 di-
    chlorofluorescein), 199–204

Glutamate analogs, APB and KYN in bipolar cells of
    monkeys, CSNB, 31–42
GSH (glutathione), fish retina, 199–204

HR-4 (human retinal gene-4), 205–226
Human donor
    age-related macular degeneration, 1–10, 11–16,
        17–22
    RPE cell transplantation, 339–346
    Blood vessels and astrocytes, 409–416
Hyperoxia, 353
Hypoxia, 353, 409

ICGV (indocyanine green videoangiography), chorio-
    capillaris in RCS rat, 115–120
Iodoacetate, induced retinal degeneration in the cat,
    71–80
Iris pigment epithelial cells (bFGF transfected), in
    transplantation in RCS rat, 323–329
Ischemia, in fish retina, 199–204

LCA (Lebers congenital amaurosis), PEDF gene ex-
    cluded, 252
Light-induced photoreceptor degeneration in the rat,
    histological assessment, 369–384
    Midkine, 347–352
    scoring survival factor rescue, 369–384
    PTN, 347–352
    sialoglycoconjugates, 181–193
Light-induced photoreceptor degeneration in the
    mouse VPP (P23H) mouse, 89–98
Light-induced photoreceptor rescue by the c-fos –/–
    mouse, 193–198

MAOA, polymorphism of X-chromosome linked cone
    gene, 43–50
Midkine, in light-induced photoreceptor rescue in the
    rat, 347–352
MMP (Matrix Metalloproteinase), in AMD donor tis-
    sue and drusen formation, 11–16
MMP-3 (Matrix Metalloproteinase 3), in bovine RPE-
    IPM, 399–408
Monkey
    as model for CSNB and bipolar cell function, 31–
        42
    as model for drusen formation in AMD, 1–10
Mouse
    c-fos –/–, 193–198
    C57BL/6, 193–198, 353–368
    generic, 135–144
    VPP, 89–98

NAT (serotonin N-acetyltransferase), cGMP and nitric
    oxide, 171–180
Nitric oxide, in retinal cGMP and NAT (serotonin N-
    acetyltransferase), 171–180
Norrie disease, X-linked pattern congenital blindness,
    57–60

Oguchi disease, congenital stationary night blindness and arrestin gene, 319–322, 313–318
Oligosaccharides, in rhodopsin of normal and RCS rat, 107–114

PDE6A, arRP-human screening, 237–244
PEDF (pigment epithelium-derived factor)
    RP13 locus, 245–254, 291–294
    RPE and IPM, 246
    serpin, 246
Peripherin/rds mutation, 205–226, 255–262, 277–290
Phagocytosis, in RPE cells, 385–398
Photoreceptor rosettes, histology in age-related macular degeneration, 17–22
PKC (protein kinase C), in RCS rat RPE cells, 385–398
PLC (phospholipase C), in Drosophila retinoid deprivation and replacement, 135–144
PTN (rat pleiotrophin), in light-induced photoreceptor degeneration, 347–352

Rat
    albino Sprague–Dawley, 339–346, 347–352, 353–368, 369–384
    albino Wistar, 181–192
    generic, 135–144
    Long Evans, 385–398
    RCS, 99–108, 109–114, 121–134, 323–329, 353–368, 385–398
RCV1 (recoverin), chromosome 2, 205–226, 263–276
rdgB (recessive and dominant), characterization of Drosophila rhodopsin mutations, 144–158
Recoverin
    candidate genes for retinal degenerations, chromosome 17, 205–226
    RCV1 (recoverin), chromosome 2, 205–226, 263–276
Retinal pigment epithelial cells
    deposit formation using ROS and cultured RPE cells in AMD, 61–70
    TIMP-3 in AMD, 11–16
    in vitro phagocytosis of ROS; vitronectin, RGD peptide, CD-36, RCS rat, 385–398
    neonatal RCS rats, bFGF and choroidal angiogenesis, 121–134
    polymer substrate and transplantation, 339–346
    regulation of ion channels, PKC (Protein Kinase C), RCS rat, 385–398
retinal degeneration
    Drosophila retinoid deprivation and replacement in, 135–144
    Drosophila rhodopsin with cytoplasmic domains mutations in, 159–170
    Drosophila with rdgB rhodopsin mutations, rhodopsin mediated pathway, 144–158.
    Iodoacetate-induced degeneration in the cat in, 71–80

Retinal pigment epithelial cells (cont.)
    light-induced photoreceptor degeneration in the rat, histological assessment, 369–384
        Midkine, 347–352
        scoring survival factors, 369–384
        PTN, 347–352
        sialoglycoconugates, 181–193
    light-induced photoreceptor degeneration in the mouse VPP (P23H) mouse, 89–98
    light-induced photoreceptor rescue by the c-fos –/– mouse, 193–198
    Swedish Briard dogs in, 81–88
    VPP mouse biochemical, histochemical, and functional studies, 89–98
    Nitric oxide in, 176
    rescue by iris pigment epithelial cells in, 323–329
retinal dystrophy
    Japanese families, 255–262
    Spanish families, 263–276
Retinitis Pigmentosa
    autoimmune and cystoid macular degeneration, 51–56
    Oguchi disease and phototransduction, 313–318
    Usher Syndrome Type 1C and Yag contig, 303–312
    visual fields as clinical measure, 23–30
Retinoid, deprivation and replacement in Drosophila, 135–144
RGD peptides, in RPE cells, 385–398
Rhodopsin
    mutation in (P23H), 277–290
    mutation in Japanese families, 255–262
    mutations in cytoplasmic domains of Drosophila rhodopsin, 159–170
    retinoid deprivation and replacement in Drosophila, 135–144
    rhodopsin-mediated pathway degeneration in Drosophila, 145–158
    RCS rat, 107–114
Rodents, comparison to retinoid deprivation and replacement in Drosophila, 135–144
ROM-1, 205–226
RP1, 281
RP10, 281
RP13, 291–294

SAG (S-antigen), 263–276
Sialoglycoconugates, in light-induced photoreceptor degeneration, 181–193
Sorsby's fundus dystrophy, TIMP-3, 12
Swedish Briard dog, exclusion of six candidate genes, 81–88

Tapetal-like reflex, in X-linked cone dystrophy, 43–50
TIMP-3
    in age-related macular degeneration and Bruch's membrane, 11–16

Transplantation
    neonatal neural retina in Abyssinian cats, 329–338
    polymer substrates and RPE cells, 339–346

Usher Syndrome Type 1C, chromosome 11, YAC
    contig, 303–312

Visual fields, clinical measure of retinal sensitivity in
    retinitis pigmentosa 23–30

Vitronectin, in RPE cells, 385–398
VMD1 (villiform macular dystrophy), 286

YAC Contig
    CORD2 Cone-Rod Dystrophy, chromosome 19,
        295–301
    Usher Syndrome Type 1C, chromosome 11, 303–
        312